MATHEMATICS RESEARCH DEVELOPMENTS

ADVANCED FRACTIONAL DIFFERENTIAL AND INTEGRAL EQUATIONS

MATHEMATICS RESEARCH DEVELOPMENTS

MATHEMATICS RESEARCH DEVELOPMENTS

ADVANCED FRACTIONAL DIFFERENTIAL AND INTEGRAL EQUATIONS

SAID ABBAS,
MOUFFAK BENCHOHRA
AND
GASTON MANDATA N'GUEREKATA

nova publishers
New York

Library of Congress Cataloging-in-Publication Data

N'Guerekata, Gaston M., 1953- author.
Advanced fractional differential and integral equations / Said Abbas (Morgan State University, MD, US), Mouffak Benchohra and Gaston Mandata N'Guerekata.
pages cm. -- (Mathematics research developments)
Includes bibliographical references and index.
ISBN 978-1-63463-109-9 (hardcover)
1. Fractional differential equations. 2. Fractional integrals. I. Abbas, Said, author. II. Benchohra, M. (Mouffak), author. III. Gaston Mandata N'Guerekata Title.
QA314.N48 2014
515'.83--dc23
2014036539

Published by Nova Science Publishers, Inc. † New York

Dedication

We dedicate this book to our family members. In particular, Saïd Abbas dedicates to the memory of his father Abdelkader Abbas; Mouffak Benchohra makes his dedication to the memory of his father Yahia Benchohra; and Gaston N'Guérékata to the memory of his father Jean N'Guérékata.

Contents

Preface

The fractional calculus deals with extensions of derivatives and integrals to non-integer orders. It represents a powerful tool in applied mathematics to study a myriad of problems from different fields of science and engineering, with many break-through results found in mathematical physics, finance, hydrology, biophysics, thermodynamics, control theory, statistical mechanics, astrophysics, cosmology and bioengineering. This book is devoted to the existence and uniqueness of solutions and some Ulam's type stability concepts for various classes of functional differential and integral equations of fractional order.

Some equations present delay which may be finite, infinite or state-dependent. Others are subject to multiple time delay effect. The tools used include classical fixed points theorems. Other tools are based of the measure of non-compactness together with appropriates fixed point theorems. Each chapter concludes with a section devoted to notes and bibliographical remarks and all abstract results are illustrated by examples.

The content of the book is new and complements the existing literature in Fractional Calculus. It is useful for researchers and graduate students for research, seminars and advanced graduate courses, in pure and applied mathematics, engineering, biology and all other applied sciences.

We are grateful to D. Baleanu, A. Cabada, T. Diagana, J.R. Graef, J. Henderson, K. Ezzinbi, S. Litimein, F-Z. Mostefai, J.J. Nieto, M. Rivero, S. Sivasundaram, J.J. Trujillo and M.P. Velasco for their collaboration in research related to the problems considered in this book.

S. Abbas,
Saida, Algeria
M. Benchohra,
Sidi Bel Abbes, Algeria,
and Jeddah, Saudi Arabia
G.M. N'Guérékata,
Morgan, USA

Introduction

Fractional calculus is a generalization of the ordinary differentiation and integration to arbitrary non-integer order, it is a field of mathematic study that grows out of the traditional definitions of the calculus integral and derivative operators in much the same way fractional exponents is an outgrowth of exponents with integer value. Consider the physical meaning of the exponent. According to our primary school teachers exponents provide a short notation for what is essentially a repeated multiplication of a numerical value. This concept in itself is easy to grasp and straight forward. However, this physical definition can clearly become confused when considering exponents of non integer value. While almost anyone can verify that $x^3 = x.x.x$, how might one describe the physical meaning of $x^{3.4}$, or moreover the transcendental exponent x^{π}. One cannot conceive what it might be like to multiply a number or quantity by itself 3.4 times, or π times, and yet these expressions have a definite value for any value x, verifiable by infinite series expansion, or more practically, by calculator. Now, in the same way consider the integral and derivative. Although they are indeed concepts of a higher complexity by nature, it is still fairly easy to physically represent their meaning. Once mastered, the idea of completing numerous of these operations, integrations or differentiations follows naturally. Given the satisfaction of a very few restrictions (e.g. function continuity), completing n integrations can become as methodical as multiplication. But the curious mind can not be restrained from asking the question what if n were not restricted to an integer value? Again, at first glance, the physical meaning can become convoluted (pun intended), but as this report will show, fractional calculus flows quite naturally from our traditional definitions. And just as fractional exponents such as the square root may find their way into innumerable equations and applications, it will become apparent that integrations of order $\frac{1}{2}$ and beyond can find practical use in many modern problems.

The subject of Fractional Calculus is as old as the differential calculus since, starting from some speculations of G.W. Leibniz and L'Hôspital (1695) [260], J. Wallis and J. Bernoulli (1697), L. Euler (1730), P. S. Laplace (1812), S.F. Lacroix (1819) [251], Fourier (1822) [185] and N.H. Abel (1823) [54], it has been developed up to nowadays. The first great theory of fractional derivation is due to J. Liouville (in 1832) [265]. But the formula most often used today, said integral of Riemann-Liouville, was given by G.F.B. Riemann (1847). His definition of Fractional Integral is

$$D^{-\nu} f(x) = \frac{1}{\Gamma(\nu)} \int_c^x (x-t)^{\nu-1} f(t)dt + \psi(t),$$

where $\Gamma(.)$ is the (Euler's) Gamma function defined by $\Gamma(\varsigma) = \int_0^{\infty} t^{\varsigma-1} e^{-t} dt;\ \varsigma > 0$. On these accepts historical, one will be able to refer to work of S. Dugowson [171]. According to Riemann-Liouville the notion of fractional integral of order α, $\alpha > 0$ for a function $f(t)$, is a natural consequence of the well known formula (Cauchy-Dirichlet), that reduces the calculation of the $n-$fold primitive of a function $f(t)$ to a single integral of convolution type.

$$I^n_{a^+}f(t) = \frac{1}{(n-1)!}\int_a^t (t-\tau)^{n-1}f(\tau)d\tau,\ n \in \mathbb{N},$$

vanishes at $t = a$ with its derivatives of order $1, 2, ..., n-1$. Require $f(t)$ and $I^n_{a^+}f(t)$ to be causal functions, that is, vanishing for $t < 0$. Extend to any positive real value by using the Gamma function, $(n-1)! = \Gamma(n)$. So, the left-sided mixed Riemann-Liouville integral of order α of f is defined by

$$I^\alpha_{a^+}f(t) = \frac{1}{\Gamma(\alpha)}\int_a^t (t-\tau)^{\alpha-1}f(\tau)d\tau.$$

The derivative of order $\alpha < 0$ a causal function $f(t)$ is given by the Riemann-Liouville integral:

$$D^\alpha f(t) = \int_0^t \frac{\xi^{-\alpha-1}}{\Gamma(-\alpha)}f(t-\xi)d\xi. \tag{1}$$

If $\alpha > 0$, we can pose

$$D^\alpha f(t) = PF\int_0^t \frac{\xi^{-\alpha-1}}{\Gamma(-\alpha)}f(t-\xi)d\xi,\ \alpha > 0,\ \alpha \notin \mathbb{N},$$

where PF represents the finite part of the integral (L. Schwartz).
The Laplace transform is a function transformation commonly used in the solution of complicated differential equations. With the Laplace transform it is frequently possible to avoid working with equations of different differential order directly by translating the problem into a domain where the solution presents itself algebraically. The formal definition of the laplace transform is given by

$$\mathcal{L}[f](t) := \int_0^\infty e^{-st}f(t)dt = \tilde{f}(s).$$

The Laplace transform of the function $f(t)$ is said to exist if the above integral is convergent. The requirement for this is that $f(t)$ does not grow at a rate higher than the rate at which the exponential term r^{-st} decreases. The operator of fractional derivative $D^\alpha f(t)$, can be defined by the transform of Laplace integrals as follows:

$$\mathcal{L}[D^\alpha f](z) = z^\alpha \mathcal{L}[f](z) - \sum_{k=0}^{n-1} f^{(k)}(0)z^{\alpha-k-1},$$

where $\alpha > 0$, and n be the integer such that $n-1 < \alpha < n$.
In 1867, Grünwald and Letnikov joined this definition which is sometimes useful

$${}_aD^\alpha_t f(t) = \lim_{h\to 0} h^{-\alpha}\sum_{k=0}^{[\frac{t-a}{h}]}(-1)^k\binom{\alpha}{k}f(t-kh) = \sum_{k=0}^{\infty}(-1)^k\binom{\alpha}{k}f(t). \tag{2}$$

This definition of fractional derivative of a function $f(t)$ based on finite differences is obtained from the classical definition of integer order derivative (A. K. Grünwald [205]). We can get an idea of equivalence definitions (1) and (2), using the defined in the factorial function $\Gamma(\alpha)$ by Gauss: $\Gamma(\alpha) = \lim_{k\to\infty} \frac{k!k^{\alpha}}{\alpha(\alpha+1)...(\alpha+k)}$.

A list of mathematicians, who have provided important contributions up to the middle of the last century, includes N.Ya. Sonin (1869), A.V. Letnikov (1872), H. Laurent (1884), P.A. Nekrassov (1888), A. Krug (1890), J. Hadamard (1892) , O. Heaviside (1892-1912), S. Pincherle (1902), G.H. Hardy and J.E. Littlewood (1917-1928), H. Weyl (1917), P. Lévy (1923), A. Marchaud (1927), H.T. Davis (1924-1936), E. L. Post (1930), A. Zygmund (1935- 1945), E.R. Love (1938-1996), A. Erdélyi (1939-1965), H. Kober (1940), D.V. Widder (1941), M. Riesz (1949), W. Feller (1952) and K. Nishimoto (1987-). They considered to the Cauchy Integral formula

$$f^{(n)}(z) = \frac{h!}{2\pi i}\int_c \frac{f(t)}{(t-z)^{n+1}}dt,$$

and substituted n by ν to obtain

$$D^{\nu} f(x) = \frac{\Gamma(\nu+1)}{2\pi i}\int_c^{x^+} \frac{f(t)}{(t-z)^{\nu+1}}dt.$$

The Riemann-Liouville definition of fractional calculus given by

$${}_aD_t^{\alpha} f(t) = \frac{1}{\Gamma(n-\alpha)}\left(\frac{d}{dt}\right)^n \int_a^t \frac{f(\tau)d\tau}{(t-\tau)^{\alpha-n+1}}; \quad n-1 \le \alpha < n,$$

is the popular definition, it has certain disadvantages when trying to model real-world phenomena with fractional differential equations. Therefore, we shall introduce a modified fractional differential operator ${}_a^cD_t^{\alpha}$ proposed by M. Caputo (1967) first in his work on the theory of viscoelasticity [131], and two years later in his book [132]. Caputo's definition can be written as

$${}_a^cD_t^{\alpha} = \frac{1}{\Gamma(n-\alpha)}\int_a^t \frac{f^{(n)}(\tau)d\tau}{(t-\tau)^{\alpha-n+1}}; \quad n-1 \le \alpha < n.$$

The notion of Riemann-Liouville derivative is the preferred fractional derivative among mathematicians, while Caputo fractional derivative is the preferred one among engineers. If $f(a) = f'(a) = ... = f^{(n-1)}(a) = 0$, then both Riemann-Liouville and Caputo derivatives coincide. In particular, for $\alpha \in (0,1)$ and $f(a) = 0$ one has ${}_a^cD_t^{\alpha} f(t) = {}_aD_t^{\alpha} f(t)$.

The Mittag-Leffler function is a generalization of the exponential function, it is an important function that finds widespread use in the world of fractional calculus. Just as the exponential naturally arises out of the solution to integer order dfferential equations, the Mittag-Leffler function plays an analogous role in the solution of non-integer order differential equations. In fact, the exponential function itself is a very specific form, one of an infinite set, of this seemingly ubiquitous function. The function was developed by the Scandinavian mathematician G.M. Mittag-Leffler (1846-1927) [278, 279], who was a contemporary of Oliver Heaviside(1850-1925). The standard definition of the Mittag-Leffler is given by

$$E_\alpha(z) = \sum_{k=0}^{\infty} \frac{z^k}{\Gamma(\alpha k+1)}; \; \alpha > 0.$$

It is also common to represent the Mittag-Leffler function in two arguments α and β, such that

$$E_{\alpha,\beta}(z) = \sum_{k=0}^{\infty} \frac{z^k}{\Gamma(\alpha k+\beta)}; \; \alpha, \beta > 0.$$

This is the more generalized form of the function, however not always a necessity when used with fractional differential equations.
In 1993, K. S. Miller and B. Ross used differential operator D as

$$D^{\overline{\alpha}} f(t) = D^{\alpha_1} D^{\alpha_2} ... D^{\alpha_n} f(t); \; \overline{\alpha} = (\alpha_1, \alpha_2, ..., \alpha_n),$$

which D^{α_i} is Riemann-Liouville or Caputo definitions.

It is clear that within the $20th$ century especially numerous applications and physical manifestations of fractional calculus have been found. However, these applications and the mathematical background surrounding fractional calculus are far from paradoxical. While the physical meaning is difficult (arguably impossible) to grasp, the definitions themselves are no more rigorous than those of their integer order counterparts. Applied problems require definitions of fractional derivative allowing the utilization of physically interpretable initial conditions. Caputo's fractional derivative, originally introduced by Caputo [130] and afterwards adopted in the theory of linear viscoelasticity, satisfies this demand. For a consistent bibliography on this topic, historical remarks and examples we refer to [55, 67, 68, 112, 314, 315, 354, 355].

Differential equations are one of the most useful mathematical tools in both pure and applied analysis. This is particularly true of problems in mechanical vibrations and the related fields of engineering and mathematical physics. We can find numerous applications in rheology, porous media, viscoelasticity, electrochemistry, electromagnetism, signal processing, dynamics of earthquakes, optics, geology, viscoelastic materials, bio-sciences, bioengineering, medicine, economics, probability and statistics, astrophysics, chemical engineering, physics, splines, tomography, fluid mechanics, electromagnetic waves, nonlinear control, control of power electronic, converters, chaotic dynamics, polymer science, proteins, polymer physics, electrochemistry, statistical physics, thermodynamics, neural networks, etc. [166, 193, 196, 224, 271, 269, 275, 296, 300, 314, 315, 330, 340, 344, 355, 374]. The problem of the existence of solutions of Cauchy-type problems for ordinary differential equations of fractional order and without delay, in spaces of integrable functions was studied in some works [242, 339]. The similar problem, in spaces of continuous functions was studied in [359]. Recently several papers have been devoted to the study of hyperbolic partial integer order differential equations and inclusions with local and nonlocal conditions; see for instance [125, 126, 127, 128, 255], the nonlocal conditions of this type can be applied in the theory of elasticity with better effect that the initial or Darboux conditions. For similar results with set-valued right hand side we refer to [109, 110, 111, 129, 239, 306]. During the last ten years, hyperbolic ordinary and partial differential equations and inclusions of fractional order have been intensely studied by many mathematicians; see for instance [2, 3, 11, 360, 361, 362, 363]. Implicit differential equations involving the regularized fractional derivative was analyzed by many authors, in the last year; see for instance [12, 13, 50, 362, 363] and the references therein. The method of upper and lower solutions has been successfully applied to study the existence of multiple solutions for initial and boundary value problems of the first and second order partial differential equations. This method plays an important role in the investigation of solutions for differential and partial differential equations and inclusions. We refer to the monographs by Benchohra *et al.* [98], the papers of Abbas and Benchohra [4, 5, 9, 10], Heikkila and Lakshmikantham [213], Ladde *et al.* [252], Lakshmikantham and Pandit [255], Lakshmikantham *et al.* [256], Pandit [306] and the references cited therein.

Integral equations are one of useful mathematical tools in both pure and applied analysis, in particular in the modeling phenomena from vehicular traffic theory, mathematical physics, engineering, economics, biology, queuing theory [87, 121, 124, 133, 136, 141, 154, 227, 373]. There has been a significant development in ordinary and partial fractional integral equations in recent years; see the monographs of Miller and Ross [276], Podlubny [314], Abbas *et al.* [14, 15, 16, 17, 18, 19, 36, 37, 38, 39], Banaś *et al.* [85], Darwish *et al.* [146], Dhage [157, 158, 160, 159, 161] and the references therein. The method of upper and lower solutions has been successfully applied to study the existence of multiple solutions for initial and boundary value problems of the first and second order partial differential equations.

In recent years, there has been a significant development in fractional calculus techniques in ordinary and partial functional differential and integral equations and inclusions, some recent contributions can be seen in the monographs of Anastassiou [71], Abbas *et al.* [40], Baleanu *et al.* [82], Diethelm [164], Kaczorek [237], Kilbas *et al.* [244], Lakshmikantham *et al.* [254], Miller and Ross [276], Podlubny [314], Samko *et al.* [337], Zhou [378], the papers of Abbas *et al.*[2, 3, 6, 7, 32, 33, 34, 35, 42, 47, 50, 53], Agarwal*et al.* [56, 65, 60, 64], Ahmad and Nieto [66], Ait Dads *et al.* [68], Almeida and Torres [69, 70], Araya and Lizama [72], Arena *et al.* [73], Arshad and Lupulescu [76], Balachandran*et al.* [80, 81], Baleanu and Vacaru [83], Bazhlekova [90], Belarbi *et al.* [91], Benchohra *et al.* [94, 93, 96, 99, 108], Burton [119], Chang and Nieto [137], Clarke*et al.* [139], Danca and Diethelm [145], Darwish *et al.* [146], Das [147, 148], Dawidowski and Kubiaczyk [149], Debbouche [150], Debbouche and Baleanu [151], Delbosco and Rodino [155], Denton and Vatsala [156], Diagana *et al.* [162], Diethelm [165, 166], Ding *et al.* [167], Dong *et al.* [170], El-Borai [175, 176], El-Borai *et al.* [177, 178], El-Sayed [179, 180, 181], Furati and Tatar[191, 192], Glóckle *et al.* [196, 197], Gorenflo *et al.* [198, 199, 200], Hawking [211], Henderson and Ouahab [215, 216, 217], Herzallah *et al.* [221, 222], Heymans *et al.* [223], Ibrahim [231], Kadem and Baleanu [238], Kamont [239], Kamont and Kropielnicka [240], Kaufmann and Mboumi [241], Kilbas and Marzan [243], Kirane *et al.* [245], Kiryakova and Luchko [246], Klafter *et al.* [248], Li *et al.* [262], Labidi and Tatar [250], Lakshmikanthan [253], Lakshmikantham and Vatsala [257], Li and N'Guérékata [261], Libertiaux *et al.* [263], Luchko [268], Magin *et al.* [270], Mainardi *et al.* [271, 272, 273], Moaddy *et al.* [281], Mophou *et al.* [282, 283, 284, 285, 286, 287, 288], Mura *et al.* [289], Muslih and Agrawal [292], Muslih *et al.*[293], Nieto [295], Ouahab [301], Petrovic *et al.* [310, 311], Podlubny *et al.* [316], Povstenko [317], Ramirez and Vatsala [324], Razminia *et al.* [325], Rivero *et al.* [326], Sabatier *et al.* [331], Salem [332, 333, 334], Samko *et al.* [338],

Tarasov [345], Tarasov and Edelman [346], Tenreiro Machado [347, 348, 349, 350], Tenreiro Machado *et al.* [351, 352], Trigeassou *et al.* [353], Vazquez [357], Wang *et al.* [365], Vityuk [358], Vityuk and Golushkov [360], Yu and Gao [372], Zhang [375], Zhou *et al.* [377, 379, 380] and the references therein.

There is a close connection between differential and integral equations, and some problems may be formulated either way. See, for example, Maxwells equations. Volterra integral equations were introduced by Vito Volterra and then studied by Traian Lalescu in 1908, they have developed application in demography, the study of viscoelastic materials, and in insurance mathematics through the renewal equation. An integro-differential equation is an equation which involves both integrals and derivatives of an unknown function. It arises in many fields like electronic, fluid dynamics, biological models, and chemical kinetics. A well-known example is the equations of basic electric circuit analysis. In recent years, the theory of various integro-differential equations in Banach spaces has been studied deeply due to their important values in sciences and technologies, and many significant results have been established (see, e.g., [174, 370] and references therein).

Functional differential and integral equations with state-dependent delay appear frequently in applications as model of equations and for this reason the study of this type of equations has received great attention in the last year; see for instance [209, 210] and the references therein. The literature related to partial functional differential equations with state-dependent delay is limited; see for instance [8, 53, 220]. The literature related to ordinary and partial functional differential equations with delay for which $\rho(t,.) = t$ or $(\rho_1(x,y,.),\rho_2(x,y,.)) = (x,y)$ is very extensive; see for instance [2, 3, 208] and the references therein.

The stability of functional equations was originally raised by Ulam in 1940 in a talk given at Wisconsin University. The problem posed by Ulam was the following: Under what conditions does there exist an additive mapping near an approximately additive mapping? (for more details see [356]). The first answer to Ulam's question was given by Hyers in 1941 in the case of Banach spaces in [229]. Thereafter, this type of stability is called the Ulam-Hyers stability. In 1978, Rassias [323] provided a remarkable generalization of the Ulam-Hyers stability of mappings by considering variables. The concept of stability for a functional equation arises when we replace the functional equation by an inequality which acts as a perturbation of the equation. Thus, the stability question of functional equations is how do the solutions of the inequality differ from those of the given functional equation? Considerable attention has been given to the study of the Ulam-Hyers and Ulam-Hyers-Rassias stability of all kinds of functional equations; one can see the monographs of [230, 234]. Bota-Boriceanu and Petrusel [115], Petru *et al.* [308, 309], and Rus [328, 329] discussed the Ulam-Hyers stability for operatorial equations and inclusions. Castro and Ramos [135] considered the Hyers-Ulam-Rassias stability for a class of Volterra integral equations. Ulam stability for fractional differential equations with Caputo derivative are proposed by Wang *et al.* [366, 367]. Some stability results for fractional integral equation are obtanied by Wei *et al.* [368]. More details from historical point of view, and recent developments of such stabilities are reported in the monographs [230, 234, 236] and the papers [235, 328, 48, 366, 367, 368].

In this book we are interesting by initial value problems (IVP for short) for partial functional differential and integral equations and inclusions with Caputo's fractional derivative.

In the proof of our theorems we essentially use several fixed point techniques. This book is arranged and organized as follows:

In Chapter 1, we introduce notations, definitions and some preliminary notions. In *Section 1.1*, we give some notations from the theory of Banach spaces and Banach algebras. *Section 1.2* is concerned to recall some basic definitions and facts on partial fractional calculus theory. In *Section 1.3*, we give some preliminaries from general topology. *Section 1.4* is concerned to recall some properties of set-valued maps. In *Section 1.5*, we give some results about the Stability. *Section 1.6* is devoted to fixed points theory, here we give the main theorems that will be used in the following chapters. In *Section 1.7*, we give some other auxiliary lemmas.

In Chapter 2, we shall be concerned by Nonlinear differential equations of fractional order In *Section 2.2*, we prove some results concerning the existence and the attractivity of solutions for a system of delay differential equations of fractional order, *Section 2.3* is concerned with the existence of solutions for fractional order partial neutral functional differential equations by using Krasnoselskii's fixed point theorem. This section initiate the question of global attractivity of the solution. In *Section 2.4*, we establish sufficient conditions for the existence and the stability of solutions of a system of nonlinear delay differential equations of fractional order. In *Section 2.5*, we establish a sufficient condition for the existence and the stability of solutions of a system of nonlinear delay differential equations of fractional.

In Chapter 3, we shall be concerned by some functional integral equations of Riemann-Liouville fractional order. In *Section 3.2*, we present some results concerning the existence, the uniqueness and the attractivity of solutions for some functional integral equations of Riemann-Liouville fractional order, by using some fixed point theorems. In *Section 3.3*, by means of integral inequalities and the fixed point approach, we investigate a partial integral equation of Riemann-Liouville fractional order. *Section 3.4* deals with the existence of solutions to partial integral equation of Riemann-Liouville fractional order.

In Chapter 4, we shall present existence results for some classes of Riemann-Liouville Volterra-Stieltjes quadratic integral equations of fractional order by using some fixed point theorems. In *Section 4.2*, we establish some sufficient conditions for the existence of solutions of fractional order Riemann-Liouville Volterra-Stieltjes quadratic integral equations. *Section 4.3* deals with the existence and the stability of solutions to nonlinear fractional order Riemann-Liouville Volterra-Stieltjes quadratic integral equations. *Section 4.4* deals with the existence of solutions to partial integral equations of Riemann-Liouville fractional order. In *Section 4.5*, we concerned by the existence of solutions to nonlinear quadratic Volterra functional Riemann-Liouville Volterra-Stieltjes integral equations of fractional order. *Section 4.6* deals with the existence and the attractivity of solutions to the following nonlinear fractional order Riemann-Liouville Volterra-Stieltjes quadratic partial integral equations.

In Chapter 5 , we shall present existence results for some classes of Riemann-Liouville Volterra-Stieltjes quadratic delay integral equations of fractional order. In *Section 5.2*, we establish some sufficient conditions for the existence and the stability of solutions of a system of delay fractional order Riemann-Liouville Volterra-Stieltjes quadratic integral equations of the form. *Section 5.3* deals with the existence and the attractivity of solutions to a system of nonlinear fractional order Riemann-Liouville Volterra-Stieltjes quadratic multi-delay partial integral equations.

In Chapter 6, we shall present existence results for some classes of Riemann-Liouville integro-differential equations of fractional order by using some fixed point theorems. In *Section 6.2*, we investigate the existence and uniqueness of solutions for integro-differential equations of fractional order. *Section 6.3* deals with the existence of solutions for a system of perturbed integro-differential equations of fractional order. In *Section 6.4*, we prove some results for a system of impulsive integro-differential equations of fractional order.

In Chapter 7, we shall present existence results for some classes of Riemann-Liouville integro-differential equations of fractional order with delay. *Section 7.2* deals with the existence of solutions for two systems of neutral integro-differential equations of fractional order with delay. In *Section 7.3*, we investigate the existence and uniqueness of solutions for fractional order integro-differential equations. In *Section 7.4*, we established sufficient conditions for the existence and the attractivity of solutions of a system of delay integro-differential equations of fractional order. In *Section 7.5*, by means of integral inequalities and the fixed point approach, we initiate the question of global attractivity of the solution for a system of partial integro-differential equations of fractional order.

In Chapter 8, we shall present local existence and uniqueness results for integral equations of Volterra type. In *Section 8.2*, we present existence results on a compact interval $J = [0,b]$. *Section 8.3* deals with the global existence of mild solutions, defined on the positive real line.

In Chapter 9, we prove sufficient conditions for the existence and uniqueness of mild solutions for semilinear integro-differential equations of fractional order with infinite delay. In *Section 9.2*, we are concerned with the local existence and uniqueness of solutions for semilinear integro-differential equations of fractional order, defined on a compact real interval $[0,b]$. In *Section 9.3*, we are concerned with the global existence and uniqueness of solutions defined on $[0,b]$. *Section 9.4* is concerned to the existence of solutions defined on a compact real interval for semilinear integro-differential equations of fractional order. *Section 9.5* deals with the existence of mild solutions, defined on the positive semi-infinite real interval $[0,+\infty)$, for semilinear integro-differential equations of fractional order.

In Chapter 10, we prove sufficient conditions for the existence and uniqueness of solutions for Abstract integro-differential inclusions. *Section 10.2* studied the existence of mild solutions for semilinear integro-differential inclusions of fractional order. In *Section 10.3*,a recent nonlinear alternative of Leray-Schauder type due to Frigon for contractive multivalued maps in Fréchet spaces [186] is used to investigate the existence and uniqueness solutions for fractional integro-differential inclusions in Fréchet spaces.

In Chapter 11, we present some results about Fractional calculus in Banach spaces with weak topologies. In *Section 11.2*, we shall be concerned with fractional calculus in Banach spaces with weak toplogies. In *Section 11.3*, we investigate the existence of solutions for boundary value problem with fractional order differential equations and nonlinear integral conditions. *Section 11.4* deals with the existence of weak solutions for boundary value problems with fractional order differential equations on the half-line in a Banach space. In *Section 11.5*, we investigate the existence of weak solutions, for boundary value problems, for fractional differential equations with mixed boundary conditions. In *Section 11.6*, we will investigate the existence of solutions of an initial value problem for hyperbolic fractional differential equations. *Section 11.7* deals with the investigation of the existence of solutions for fractional integral equations.

In Chapter 12, we are concerned with the existence of weak solutions for some classes of Nonlinear fractional differential inclusions. *Section 12.2* is concerned with the existence of solutions of boundary value problems with fractional order differential inclusions and nonlinear integral conditions. In *Section 12.3*, we investigate the existence of solutions for hyperbolic fractional differential inclusions.

In Chapter 13, we investigate some Ulam's type stability concepts for the Darboux problem of partial fractional differential equations and inclusions. In *Section 13.2* we investigate some Ulam's type stability concepts for the Darboux problem of partial differential and integro-differential equations of Caputo's fractional order derivative. *Section 12.3* is concerned by Ulam's type stability concepts for the Darboux problem of partial fractional implicit differential equations. In *Section 13.4* we investigate some Ulam's type stability concepts for the Darboux problem of partial differential equations of fractional order on unbounded domain. We use a nonlinear alternative of Leray-Schauder type for contraction maps in Fréchet spaces and a fractional version of Gronwall's inequality. In *Section 13.5* we investigate some Ulam's type stability concepts for the Darboux problem of partial fractional differential inclusions with a nonconvex valued right hand side. Our results are based upon Covitz-Nadler fixed point theorem. In *Section 13.6* we investigate some Ulam's type stability concepts for the Darboux problem of partial differential equations of Caputo's fractional order derivative. Our results are based upon the Banach contraction principle. An example will be presented for each problem.

Keywords and phrases: Differential and integral equations, integro-differential equations, fractional order, left-sided mixed Riemann-Liouville integral, Riemann-Liouville and Caputo fractional-order derivatives, mixed regularized derivative, convex and nonconvex valued multivalued, solution, mild solution, weak solution, initial value problem, nonlocal conditions, contraction, existence, uniqueness, Fréchet space, phase space, impulses, time delay, state-dependent delay, attractive, stable, fixed point

AMS (MOS) Subject Classifications: 26A33, 34A60, 34A37, 39A12, 34A37, 39A12, 34A60, 34B15, 26A33, 34K05

Chapter 1

Preliminary Background

In this chapter, we introduce notations, definitions, and preliminary facts that will be used in the remainder of this book.

1.1. Notations and Definitions

Let $J := [0,a] \times [0,b]$; $a,b > 0$, and let E be a (real or complex) separable Banach space. Denote $L^1(J)$ the space of Lebesgue-integrable functions $u : J \to E$ with the norm

$$\|u\|_{L^1} = \int_0^a \int_0^b \|u(x,y)\|_E dydx,$$

where $\|.\|_E$ denotes a complete norm on E. Also, $L^1(J,\mathbb{R}^n)$ denotes the space of Lebesgue-integrable functions $u : J \to \mathbb{R}^n$ with the norm

$$\|u\|_{L^1} = \int_0^a \int_0^b \|u(x,y)\|dydx,$$

where $\|.\|$ denotes a suitable complete norm on $\mathbb{R}^n$.

Let $L^\infty(J)$ be the Banach space of measurable functions $u : J \to E$ which are essentially bounded, equipped with the norm

$$\|u\|_{L^\infty} = \inf\{c > 0 : \|u(x,y)\|_E \le c, \ a.e. \ (x,y) \in J\}.$$

Also, $L^\infty(J,\mathbb{R}^n)$ is equipped with the norm

$$\|u\|_{L^\infty} = \inf\{c > 0 : \|u(x,y)\| \le c, \ a.e. \ (x,y) \in J\}.$$

As usual, by $AC(J)$ we denote the space of absolutely continuous functions from J into E, and $C := C(J)$ is the Banach space of all continuous functions from J into E with the norm $\|.\|_\infty$ defined by

$$\|u\|_\infty = \sup_{(x,y)\in J} \|u(x,y)\|_E.$$

Also $C(J,\mathbb{R}^n)$ is endowed with norm $\|u\|_\infty = \sup_{(x,y)\in J} \|u(x,y)\|$, and $C(J,\mathbb{R})$ is endowed with norm $\|u\|_\infty = \sup_{(x,y)\in J} |u(x,y)|$.

By $BC := BC([-\alpha,\infty)\times[-\beta,b])$; $\alpha,\beta,b>0$ we denote the Banach space of all bounded and continuous functions from $[-\alpha,\infty)\times[-\beta,b]$ into $\mathbb{R}$ equipped with the standard norm

$$\|u\|_{BC} = \sup_{(t,x)\in[-\alpha,\infty)\times[-\beta,b]} |u(t,x)|.$$

For $u_0 \in BC$ and $\eta \in (0,\infty)$, we denote by $B(u_0,\eta)$, the closed ball in BC centered at u_0 with radius η.

1.2. Preliminaries from General Topology

In this section, we are given a set X, a collection of topological spaces $(Y_i)_{i\in I}$ and a collection of maps $(f_i)_{i\in I}$ such that each f_i maps X into Y_i.

We wish to define a topology on X that makes all the functions f_i continuous. And we want to do this in the way that: there should be no more open sets in X than required for this purpose.

Obviously, all the $f_i^{-1}(O_i)$, where O_i is an open set in Y_i should be open in X. Then a finite intersections of those should be also be open. And then, any union of finite intersections should be open. By this process, we have created a few open sets as required. Yet it is not clear that the collection obtained is closed under finite intersections. It actually is, as a consequence of the following lemma:

Lemma 1.1. *Let X be a set and let $O \subset \mathcal{P}(X)$ be a collection of subsets of X, such that*

- *$\emptyset$ and X are in O;*
- *O is closed under finite intersections.*

Then $\mathcal{J} = \{\cup_{S\in\mathcal{U}} S : \mathcal{U} \subset O\}$ is a topology on X.

It's easy to check that $\mathcal{J}$ is closed under finite intersections.

Corollary 1. *The collection of all unions of finite intersection of sets of the form $f_i^{-1}(O_i)$ where $i \in I$ and O_i is an open set in Y_i is a topology. it is called the weak topology on X generated by the $(f_i)_{i\in I}$'s and we denote it by $\sigma(X,(f_i)_{i\in i})$. By definition, the functions $(f_i)_{i\in I}$ are continuous for this topology.*

Theorem 1.2. *Let $(x_n)_{n\in\mathbb{N}}$ be a sequence in X. It converges in the topology $\sigma(X,(f_i)_{i\in I})$ to some $x \in X$ if and only if*

$$\forall i \in I, \qquad \lim_{n\to\infty} f_i(x_n) = f_i(x).$$

Notice that if the topology $\mathcal{J}$ is not Hausdorff, there is no unicity for the limit of a sequence. For example, if x and y are distinct points that cannot be separeted by open sets, the constant sequence equal to x will converge both to x and y.

Theorem 1.3. *Let $(Z, \mathcal{J})$ be a topological space, and $\varphi : Z \to X$ be a map. Then φ is continuous for the topologies $\mathcal{J}$ and $\sigma(X, (f_i)_{i\in I})$ if and only if for every $i \in I$, $f_i \circ \varphi$ is continuous.*

As an example of weak topology is the the topology of pointwise convergence. It is defined as follows: Let A be any set and let X be the set of functions $A \to \mathbb{R}$. For every $a \in A$, define the function $\varphi_a : X \mapsto \mathbb{R}$ by

$$\forall f \in X, \qquad \varphi_a(f) = f(a).$$

The topology of pointwise convergence is $\sigma(X, (f_a)_{a\in A})$. In this topology, a sequence of functions converges if and only if it converges pointwise, in view of Theorem 1.2.

1.2.1. The Weak Topology $\sigma(X, X^*)$

In this section, X is a normed space and X^* its dual. Unless stated otherwise, we do not assume that it is complete.

Definition 1.4. *The weak topology on X is the topology $\sigma(X, (f)_{f\in X^*})$. For convenience, it is simply denoted $\sigma(X, X^*)$.*

The first thing we want to check is that the weak topology on X is Hausdorff, which will guarantee us the unicity of limits.

Theorem 1.5. *The topology $\sigma(X, X^*)$ is Hausdorff.*

Proposition 1.6.

1. *The weak topology is weaker than the norm topology: every weakly open (resp. closed) set is strongly open (resp.closed).*
2. *A sequence $(x_n)_{n\in\mathbb{N}}$ converges weakly to x if and only if*

$$\forall f \in X^*, \qquad \lim_{n\to\infty} (f, x_n) = (f, x).$$

3. *A strongly converging sequence converges weakly.*
4. *If $(x_n)_{n\in\mathbb{N}}$ is a sequence in X converging weakly to x, then $(x_n)_{n\in\mathbb{N}}$ is bounded and*

$$\|x\| \le \liminf_{n\to\infty} \|x_n\|.$$

5. *If $(x_n)_{n\in\mathbb{N}}$ is a sequence in X converging weakly to x and $(f_n)_{n\in\mathbb{N}}$ is a sequence in X^* converging strongly to f, then*

$$\lim_{n\to\infty} (f_n, x_n) \text{ exists and equals } (f, x).$$

The next step is to identify a basis of neighborhoods for $\sigma(X, X^*)$.

Theorem 1.7. *Let $x_0 \in X$. A basis of neighborhood of x_0 for the weak topology is given by the collection of the sets of the form*

$$W_{\varepsilon, f_1, \cdot, \cdot, \cdot, f_n} = \{x \in X : \forall i \in \{1, \cdot, \cdot, \cdot, n\}, |(f_i, x) - (f_i, x_0)| < \varepsilon\}$$

$n \in \mathbb{N}$, $\varepsilon > 0$, $f_1, \cdot, \cdot, \cdot, f_n \in X^*$.

Now that we have two topologies on X, we may wonder if by any chance they coincide. The answer is given by the following proposition.

Proposition 1.8. *The weak topology and the strong topology on X coincide if and only if X is finite dimensional.*

The next theorem answers the important question: If the weak topology is strictly weaker than the strong topology in infinite dimension, are there sets for which we can guarantee that strongly closed implies weakly closed?

Theorem 1.9. *Let C be a nonempty convex set in X. Then C is strongly closed if and only if it is weakly closed.*

1.2.2. The Space $C([0,T],E_w)$

Let E the real Banach space with norm $\|\cdot\|$ and dual E^*. Also $E_w = (E, \sigma(E,E^*))$ denotes the space E with its weak topology. We will let $C([0,T],E)$ the space of strong continuous functions $x : [0,T] \to E$ endowed by the norm $\|x\|_\infty = \sup_{t\in[0,T]} \|x(t)\|$. Further, we recall the following

Definition 1.10. *The mapping x is said to be weakly continuous (measurable) at $t_0 \in [0,T]$ if for every $\varphi \in E^*$, $\varphi(x)$ is continuous (measurable) at t_0.*

We denote $C([0,T],E_w)$ the space of weakly continuous functions defined on [0, T] with values in the Banach space E_w.

Definition 1.11. *Let $f : [0,T] \times E \to E$. Then $f(t,x)$ is said to be weakly continuous at (t_0,x_0) if given $\varepsilon > 0$, $\varphi \in E^*$, there exists a $\delta > 0$ and a weakly open set U containing x_0 such that*

$$|\varphi(f(t,x) - f(t_0,x_0))| < \varepsilon, \quad \textit{whenever} \quad |t - t_0| < \delta \quad \textit{and } x \in U.$$

Definition 1.12. *A function $h : E \to E$ is said to be weakly sequentially continuous if h takes each weakly convergent sequence in E to weakly convergent sequence in E (i.e., for any $(x_n)_n$ in E with $x_n \to x$ in E_w then $h(x_n) \to h(x)$ in E_w).*

Remark 1.13. *The notion of sequentially weak continuity seems to be the most convenient in use. It is not always possible to show that a given operator between Banach spaces is weakly continuous, quite often its sequentially weak continuity presents no problem. This follows from fact that Lebesgue Dominated Convergence Theorem is valid for sequences but not for nets. (We recall the following counter example taken from Bourbaki [116]: If A denotes the family of finite subsets of $[0,1]$ ordered by inclusion, and if x_α is the characteristic function of $\alpha \in A$, then $(x_\alpha) \to 1$ but $\int x_\alpha = 0$ for each $\alpha \in A$).*

Remark 1.14. *One of the advantages of the weak topology of a Banach space E is the fact that if a set D is weakly compact, then every sequentially weakly continuous map $F : D \to E$ is weakly continuous. This is an immediately consequence of Eberlein-Smulian's Theorem.*

Proposition 1.15. *If the function $\varphi(x(\cdot))$ is differentiable for every $\varphi \in E^*$ and if there is a function $y : [0,T] \to E$ such that $\frac{d}{dt}(\varphi(x(t))) = \varphi(y(t))$ for every $\varphi \in E^*$ and every $t \in [0,T]$, then x is weakly differentiable and we write $x'(t) = y(t)$, where $x'(t)$ denotes the weak derivative of the function x.*

1.3. Fréchet Spaces

Let X be a Fréchet space with a family of semi-norms $\{\|\cdot\|_n\}_{n\in\mathbb{N}^*}$. We assume that the family of semi-norms $\{\|\cdot\|_n\}$ verifies:

$$\|x\|_1 \leq \|x\|_2 \leq \|x\|_3 \leq \ldots \quad for\ every\ x \in X.$$

Let $Y \subset X$, we say that Y is bounded if for every $n \in \mathbb{N}$, there exists $\overline{M}_n > 0$ such that

$$\|y\|_n \leq \overline{M}_n \quad for\ all\ y \in Y.$$

To X we associate a sequence of Banach spaces $\{(X^n, \|\cdot\|_n)\}$ as follows: For every $n \in \mathbb{N}$, we consider the equivalence relation $\sim_n$ defined by: $x \sim_n y$ if and only if $\|x-y\|_n = 0$ for $x, y \in X$. We denote $X^n = (X|_{\sim_n}, \|\cdot\|_n)$ the quotient space, the completion of X^n with respect to $\|\cdot\|_n$. To every $Y \subset X$, we associate a sequence $\{Y^n\}$ of subsets $Y^n \subset X^n$ as follows : For every $x \in X$, we denote $[x]_n$ the equivalence class of x of subset X^n and we defined $Y^n = \{[x]_n : x \in Y\}$. We denote $\overline{Y^n}$, $int_n(Y^n)$ and $\partial_n Y^n$, respectively, the closure, the interior and the boundary of Y^n with respect to $\|\cdot\|_n$ in X^n. For more information about this subject see [187].

Definition 1.16. *Let X be a Fréchet space. A function $N : X \longrightarrow X$ is said to be a contraction if for each $n \in \mathbb{N}^*$ there exists $k_n \in [0,1)$ such that*

$$\|N(u) - N(v)\|_n \leq k_n\|u - v\|_n \ for\ all\ u, v \in X.$$

1.4. Properties of Fractional Calculus

In this section, we introduce notations, definitions and preliminary Lemmas concerning to partial fractional calculus theory.

Definition 1.17. *[316, 337] The Riemann-Liouville fractional integral of order $\alpha \in (0,\infty)$ of a function $h \in L^1([0,b],\mathbb{R}^n)$; $b > 0$ is defined by*

$$I_0^\alpha h(t) = \frac{1}{\Gamma(\alpha)}\int_0^t (t-s)^{\alpha-1}h(s)ds.$$

Definition 1.18. *[316, 337] The Riemann-Liouville fractional derivative of order $\alpha \in (0,1]$ of a function $h \in L^1([0,b],\mathbb{R}^n)$ is defined by*

$$\begin{aligned} D_0^\alpha h(t) &= \frac{d}{dt} I_0^{1-\alpha} h(t) \\ &= \frac{1}{\Gamma(1-\alpha)}\frac{d}{dt}\int_0^t (t-s)^{-\alpha}h(s)ds; \ for\ almost\ all\ t \in [0,b]. \end{aligned}$$

Definition 1.19. *[316, 337] The Caputo fractional derivative of order $\alpha \in (0,1]$ of a function $h \in L^1([0,b],\mathbb{R}^n)$ is defined by*

$$\begin{aligned} {}^cD_0^\alpha h(t) &= I_0^{1-\alpha}\frac{d}{dt}h(t) \\ &= \frac{1}{\Gamma(1-\alpha)}\int_0^t (t-s)^{-\alpha}\frac{d}{ds}h(s)ds; \ for\ almost\ all\ t \in [0,b]. \end{aligned}$$

Definition 1.20. *[244, 337] Let $\alpha \in (0,\infty)$ and $u \in L^1(J,\mathbb{R}^n)$. The partial Riemann-Liouville integral of order α of $u(x,y)$ with respect to x is defined by the expression*

$$I^{\alpha}_{0,x}u(x,y) = \frac{1}{\Gamma(\alpha)}\int_0^x (x-s)^{\alpha-1}u(s,y)ds; \textit{ for almost all } (x,y) \in J.$$

Analogously, we define the integral

$$I^{\alpha}_{0,y}u(x,y) = \frac{1}{\Gamma(\alpha)}\int_0^y (y-s)^{\alpha-1}u(x,s)ds; \text{ for almost all } (x,y) \in J.$$

Definition 1.21. *[244, 337] Let $\alpha \in (0,1]$ and $u \in L^1(J,\mathbb{R}^n)$. The Riemann-Liouville fractional derivative of order α of $u(x,y)$ with respect to x is defined by*

$$(D^{\alpha}_{0,x}u)(x,y) = \frac{\partial}{\partial x}I^{1-\alpha}_{0,x}u(x,y); \textit{ for almost all } (x,y) \in J.$$

Analogously, we define the derivative

$$(D^{\alpha}_{0,y}u)(x,y) = \frac{\partial}{\partial y}I^{1-\alpha}_{0,y}u(x,y); \text{ for almost all } (x,y) \in J.$$

Definition 1.22. *[244, 337] Let $\alpha \in (0,1]$ and $u \in L^1(J,\mathbb{R}^n)$. The Caputo fractional derivative of order α of $u(x,y)$ with respect to x is defined by the expression*

$${}^cD^{\alpha}_{0,x}u(x,y) = I^{1-\alpha}_{0,x}\frac{\partial}{\partial x}u(x,y); \textit{ for almost all } (x,y) \in J.$$

Analogously, we define the derivative

$${}^cD^{\alpha}_{0,y}u(x,y) = I^{1-\alpha}_{0,y}\frac{\partial}{\partial y}u(x,y); \text{ for almost all } (x,y) \in J.$$

Definition 1.23. *[360] Let $r = (r_1,r_2) \in (0,\infty)\times(0,\infty)$, $\theta = (0,0)$ and $u \in L^1(J,\mathbb{R}^n)$. The left-sided mixed Riemann-Liouville integral of order r of u is defined by*

$$(I^r_{\theta}u)(x,y) = \frac{1}{\Gamma(r_1)\Gamma(r_2)}\int_0^x\int_0^y (x-s)^{r_1-1}(y-t)^{r_2-1}u(s,t)dtds.$$

In particular,

$$(I^{\theta}_{\theta}u)(x,y) = u(x,y),\ (I^{\sigma}_{\theta}u)(x,y) = \int_0^x\int_0^y u(s,t)dtds; \text{ for almost all } (x,y) \in J,$$

where $\sigma = (1,1)$.
For instance, $I^r_{\theta}u$ exists for all $r_1, r_2 > 0$, when $u \in L^1(J,\mathbb{R}^n)$. Note also that when $u \in C(J,\mathbb{R}^n)$, then $(I^r_0 u) \in C(J,\mathbb{R}^n)$, moreover

$$(I^r_{\theta}u)(x,0) = (I^r_{\theta}u)(0,y) = 0;\ x \in [0,a],\ y \in [0,b].$$

Example 1.24. *Let $\lambda, \omega \in (-1,\infty)$ and $r = (r_1,r_2) \in (0,\infty)\times(0,\infty)$, then*

$$I^r_{\theta}x^{\lambda}y^{\omega} = \frac{\Gamma(1+\lambda)\Gamma(1+\omega)}{\Gamma(1+\lambda+r_1)\Gamma(1+\omega+r_2)}x^{\lambda+r_1}y^{\omega+r_2}; \textit{ for almost all } (x,y) \in J.$$

By $1-r$ we mean $(1-r_1, 1-r_2) \in [0,1) \times [0,1)$. Denote by $D^2_{xy} := \frac{\partial^2}{\partial x \partial y}$, the mixed second order partial derivative.

Definition 1.25. *[360] Let $r \in (0,1] \times (0,1]$ and $u \in L^1(J, \mathbb{R}^n)$. The Caputo fractional-order derivative of order r of u is defined by the expression $({}^cD^r_\theta u)(x,y) = (I^{1-r}_\theta D^2_{xy} u)(x,y)$ and the mixed fractional Riemann-Liouville derivative of order r of u is defined by the expression $(D^r_\theta u)(x,y) = (D^2_{xy} I^{1-r}_\theta u)(x,y)$.*

The case $\sigma = (1,1)$ is included and we have

$$({}^cD^\sigma_\theta u)(x,y) = (D^\sigma_\theta u)(x,y) = (D^2_{xy} u)(x,y); \text{ for almost all } (x,y) \in J.$$

Example 1.26. *Let $\lambda, \omega \in (-1, \infty)$ and $r = (r_1, r_2) \in (0,1] \times (0,1]$, then*

$$D^r_\theta x^\lambda y^\omega = \frac{\Gamma(1+\lambda)\Gamma(1+\omega)}{\Gamma(1+\lambda-r_1)\Gamma(1+\omega-r_2)} x^{\lambda-r_1} y^{\omega-r_2}; \textit{ for almost all } (x,y) \in J.$$

Definition 1.27. *[363] For a function $u : J \to \mathbb{R}^n$, we set*

$$q(x,y) = u(x,y) - u(x,0) - u(0,y) + u(0,0).$$

By the mixed regularized derivative of order $r = (r_1, r_2) \in (0,1] \times (0,1]$ of a function $u(x,y)$, we name the function

$$\overline{D}^r_\theta u(x,y) = D^r_\theta q(x,y).$$

The function

$$\overline{D}^{r_1}_{0,x} u(x,y) = D^{r_1}_{0,x}[u(x,y) - u(0,y)],$$

is called the partial r_1-order regularized derivative of the function $u(x,y) : J \to \mathbb{R}^n$ with respect to the variable x. Analogously, we define the derivative

$$\overline{D}^{r_2}_{0,y} u(x,y) = D^{r_2}_{0,y}[u(x,y) - u(x,0)].$$

Let $a_1 \in [0,a]$, $z^+ = (a_1, 0) \in J$, $J_z = [a_1, a] \times [0,b]$. For $w \in L^1(J_z, \mathbb{R}^n)$, the expression

$$(I^r_{z^+} w)(x,y) = \frac{1}{\Gamma(r_1)\Gamma(r_2)} \int_{a_1}^x \int_0^y (x-s)^{r_1-1} (y-t)^{r_2-1} w(s,t) dt ds,$$

is called the left-sided mixed Riemann-Liouville integral of order r of w. The Caputo fractional-order derivative of order r of w is defined by $({}^cD^r_{z^+} w)(x,y) = (I^{1-r}_{z^+} D^2_{xy} w)(x,y)$ and the mixed fractional Riemann-Liouville derivative of order r of w is defined by $(D^r_{z^+} w)(x,y) = (D^2_{xy} I^{1-r}_{z^+} w)(x,y)$.

1.4.1. Preliminary Lemmas

Let $f, g \in L^1(J, \mathbb{R}^n)$.

Lemma 1.28. *[2, 3] A function $u \in AC(J,\mathbb{R}^n)$ such that its mixed derivative D^2_{xy} exists and is integrable on J is a solution of problem*

$$\begin{cases} ({}^cD^r_\theta u)(x,y) = f(x,y);\ (x,y)\in J,\\ u(x,0)=\varphi(x);\ x\in[0,a],\ u(0,y)=\psi(y);\ y\in[0,b],\\ \varphi(0)=\psi(0),\end{cases}$$

if and only if $u(x,y)$ satisfies

$$u(x,y) = \mu(x,y) + (I^r_\theta f)(x,y);\ (x,y)\in J,$$

where

$$\mu(x,y) = \varphi(x)+\psi(y)-\varphi(0).$$

Lemma 1.29. *[51] A function $u \in AC(J,\mathbb{R}^n)$ such that the mixed derivative $D^2_{xy}(u-g)$ exists and is integrable on J is a solution of problem*

$$\begin{cases} {}^cD^r_\theta[u(x,y)-g(x,y)] = f(x,y);\ (x,y)\in J,\\ u(x,0)=\varphi(x);\ x\in[0,a],\ u(0,y)=\psi(y);\ y\in[0,b],\\ \varphi(0)=\psi(0),\end{cases}$$

if and only if $u(x,y)$ satisfies

$$u(x,y) = \mu(x,y)+g(x,y)-g(x,0)-g(0,y)+g(0,0)+I^r_\theta(f)(x,y);\ (x,y)\in J.$$

1.4.2. Riemann-Liouville and Caputo Partial Fractional Derivatives

For a function $h\in L^1([0,b])$; $b>0$ and $\alpha\in(0,1]$. The connection between D^α_0 and ${}^cD^\alpha_0$ is given by

$${}^cD^\alpha_0 h(t) = D^\alpha_0 h(t) - \frac{t^{-\alpha}}{\Gamma(1-\alpha)}h(0^+),\ \text{for almost all } t\in[0,b]. \tag{1.1}$$

For more detail see [244].

Corollary 2. *For a function $u\in L^1(J)$ and $r=(r_1,r_2)\in(0,1]\times(0,1]$. The connection between $D^{r_1}_{0,x}u(x,y)$ and ${}^cD^{r_1}_{0,x}u(x,y)$ with respect to x is given by*

$$({}^cD^{r_1}_{0,x}u)(x,y) = (D^{r_1}_{0,x}u)(x,y) - \frac{x^{-r_1}}{\Gamma(1-r_1)}u(0^+,y). \tag{1.2}$$

Analogously, the connection between $D^{r_2}_{0,y}u(x,y)$ and ${}^cD^{r_2}_{0,y}u(x,y)$ with respect to y is given by

$$({}^cD^{r_2}_{0,y}u)(x,y) = (D^{r_2}_{0,y}u)(x,y) - \frac{y^{-r_2}}{\Gamma(1-r_2)}u(x,0^+). \tag{1.3}$$

Theorem 1.30. *[50]. For $u(x,y)\in AC(J)$ and $r=(r_1,r_2)\in(0,1]\times(0,1]$ we have*

$$\begin{aligned}({}^cD^r_\theta u)(x,y) &= \overline{D}^r_\theta u(x,y) = (D^r_\theta u)(x,y) - \frac{x^{-r_1}}{\Gamma(1-r_1)}(D^{r_2}_{0,y}u)(0,y)\\ &\quad - \frac{y^{-r_2}}{\Gamma(1-r_2)}(D^{r_1}_{0,x}u)(x,0) + \frac{x^{-r_1}y^{-r_2}}{\Gamma(1-r_1)\Gamma(1-r_2)}u(0,0).\end{aligned}$$

1.4.3. Fractional Pettis Integral

In this section we give the definitions of the Pettis fractional integrals on a finite interval of the real line and present some of their properties. Let $L^1(I)$ be the class of Lebesgue-integrable functions on the interval I. E is a Banach space and dual E^*. We will denote by E_w the space E endowed with the weak topology generated by the continuous functionals on E.

Definition 1.31. *Let $x : I \to E$ be a weakly continuous function and let $\alpha > 0$. Then the fractional weak Riemann-integral $I^\alpha x(t)$is defined by*

$$I^\alpha x(t) = \int_0^t \frac{(t-s)^{\alpha-1}}{\Gamma(\alpha)} x(s)ds.$$

In the above definition the sign "$\int$" denotes the weak Riemann-integral.

Theorem 1.32. *Let $x : I \to E$ be a weakly continuous function and let $\alpha > 0$. The fractional weak Riemann-integral $I^\alpha x(t)$is defined by*

$$I^\alpha x(t) = \int_0^t \frac{(t-s)^{\alpha-1}}{\Gamma(\alpha)} x(s)ds.$$

exists for every $t \in I$ and it is also a weakly Riemann-integrable function.

Proof. Since $x : I \to E$ is weakly continuous, x is weakly Riemann-integrable in $[0,T]$ for every $t \in I$. It follows from the definition and the properties of the fractional order integrals in the Banach space $L^1(I)$ (see e.g. [276, 314, 336, 337]) that

$$I^\alpha \varphi(x(t)) = \int_0^t \frac{(t-s)^{\alpha-1}}{\Gamma(\alpha)} \varphi(x(s))ds = \int_0^t \varphi\left(\frac{(t-s)^{\alpha-1}}{\Gamma(\alpha)} x(s)\right) ds.$$

exists for every $t \in I$ and is an element in $C(I)$. That is, for $t > s$ the continues function $\varphi\left(\frac{(t-s)^{\alpha-1}}{\Gamma(\alpha)} x(s)\right) = \frac{(t-s)^{\alpha-1}}{\Gamma(\alpha)} \varphi(x(s))$ is Riemann-integrable hence $\frac{(t-s)^{\alpha-1}}{\Gamma(\alpha)} x(s)$ is weakly Riemann-integrable. Thus from the definition of weak Riemann-integral there exists a function, denoted by $I^\alpha x(\cdot)$, from I into E that satisfies

$$\varphi(I^\alpha x(t)) = \int_0^t \frac{(t-s)^{\alpha-1}}{\Gamma(\alpha)} \varphi(x(s))ds = I^\alpha \varphi(x(t))$$

for all $\varphi \in E^*$ and each $t \in I$. By definition

$$I^\alpha x(t) = \int_0^t \frac{(t-s)^{\alpha-1}}{\Gamma(\alpha)} x(s)ds.$$

Moreover, it's clear that $I^\alpha x(\cdot)$ is weakly continuous (because $\varphi(I^\alpha(x(\cdot)))$ is continuous), hence it is again a weakly Riemann-integrable function. □

Definition 1.33. *[335] Let $x : I \to E$ be a weakly measurable function, such that $\varphi(x(\cdot)) \in L^1(I)$ for each $\varphi \in E^*$ and let $\alpha > 0$. Then the fractional Pettis-integral $I^\alpha x(t)$ is defined by*

$$I^\alpha x(t) = \int_0^t \frac{(t-s)^{\alpha-1}}{\Gamma(\alpha)} x(s)ds$$

where the sign "$\int$" denotes the Pettis-integral and Γ is the Gamma function.

Such an integral is well defined:

Theorem 1.34. *Let $x : I \to E$ be a weakly measurable function, such that $\varphi(x(\cdot)) \in L^1(I)$ for each $\varphi \in E^*$ and let $\alpha > 0$. The fractional Pettis-integral*

$$I^\alpha x(t) = \int_0^t \frac{(t-s)^{\alpha-1}}{\Gamma(\alpha)} x(s)ds$$

exists for almost every $t \in I$ as a function from I into E and $\varphi(I^\alpha x(t)) = I^\alpha \varphi(x(t))$.

Proof. Since $x : I \to E$ is weakly measurable and $\varphi(x(\cdot)) \in L^1(I)$, x is Pettis-integrable function (see Lemma 11.34). From the definition of integrals of fractional order it follows that

$$I^\alpha \varphi(x(t)) = \int_0^t \frac{(t-s)^{\alpha-1}}{\Gamma(\alpha)} \varphi(x(s))ds = \int_0^t \varphi\left(\frac{(t-s)^{\alpha-1}}{\Gamma(\alpha)} x(s)\right) ds.$$

exists for almost every $t \in I$ and is an element in $L^1(I)$. That is, for almost every $t \in I, s \in (0,t)$ the measurable function $\varphi\left(\frac{(t-s)^{\alpha-1}}{\Gamma(\alpha)} x(s)\right) = \frac{(t-s)^{\alpha-1}}{\Gamma(\alpha)} \varphi(x(s))$ is Lebesgue-integrable, hence $\frac{(t-s)^{\alpha-1}}{\Gamma(\alpha)} x(s)$ is Pettis-integrable on I. Thus from the definition of the Pettis-integral there exists a function, denoted by $I^\alpha x(\cdot)$, from I into E that satisfies

$$\varphi(I^\alpha x(t)) = \int_0^t \frac{(t-s)^{\alpha-1}}{\Gamma(\alpha)} \varphi(x(s))ds = I^\alpha \varphi(x(t))$$

for all $\varphi \in E^*$ and each $t \in I$. By definition

$$I^\alpha x(t) = \int_0^t \frac{(t-s)^{\alpha-1}}{\Gamma(\alpha)} x(s)ds.$$

for a.e. $t \in I$. □

Theorem 1.35. *Let E be a Banach space, and let $x : I \to B$ be a Pettis-integrable function such that $\varphi(x(\cdot)) \in L^\infty(I)$ for each $\varphi \in E^*$. If $\alpha > 0$ then the fractional Pettis integral*

$$I^\alpha x(t) = \int_0^t \frac{(t-s)^{\alpha-1}}{\Gamma(\alpha)} x(s)ds$$

exists as a function from I into E and $\varphi(I^\alpha x(t)) = I^\alpha \varphi(x(t))$.

Proof. Set $h(s) = \frac{(t-s)^{\alpha-1}}{\Gamma(\alpha)}, t \in I$ and $s < t$. Since x is Pettis-integrable function and $\varphi(x(\cdot)) \in L^\infty(I)$, then according to Proposition 11.35, the function $x(\cdot)h(\cdot)$ is Pettis integrable, hence from the definition of Pettis-integral there exists a function, denoted by $I^\alpha x(\cdot)$, from I into E so that

$$\varphi(I^\alpha x(t)) = \int_0^t \frac{(t-s)^{\alpha-1}}{\Gamma(\alpha)} \varphi(x(s))ds = I^\alpha \varphi(x(t)) \qquad \text{for all } \varphi \in E^*.$$

By definition

$$I^\alpha x(t) = \int_0^t \frac{(t-s)^{\alpha-1}}{\Gamma(\alpha)} x(s)ds, \qquad t \in I.$$

□

Corollary 3. *Let E be a reflexive banach space, and let $x : I \to E$ be a weakly measurable function such that $\varphi(x(\cdot)) \in L^{\infty}(I)$ for each $\varphi \in E^*$. If $\alpha > 0$, then the fractional Pettis integral $I^{\alpha}x(\cdot)$ exists as a function from I into E and$\varphi(I^{\alpha}x(t)) = I^{\alpha}\varphi(x(t))$.*

Proof. E is reflexive, thus according to Lemma 11.34 x is Pettis integrable and the result immediate follows from Theorem 1.35. □

Remark 1.36.

1. *It is clear that if x is weakly continuous (hence weakly measurable) and $\varphi(x(\cdot)) \in L^1(I)$, then $I^{\alpha}x(t)$ exits for every $t \in I$ as a weakly continuous function.*
2. *We cannot define $I^{\alpha}x(\cdot)$ for arbitrary bounded weakly measurable function x since from the Phillip example (see [313]) it follows that a bounded weakly measurable function need not be Pettis-integrable.*

Now we have the following lemma for the properties of fractional order integrals in reflexive spaces.

Lemma 1.37. *Let $x : I \to E$ be weakly continuous. If $\alpha, \beta > 0$ then for almost every $t \in I$ we have*

1. $I^{\alpha}I^{\beta}x(t) = I^{\alpha+\beta}x(t)$,
2. $\lim_{\alpha \to n} I^{\alpha}x(t) = I^n x(t), n = 1, 2, \cdots$ *weakly uniformly,*
3. $\lim_{\alpha \to 0} I^{\alpha}x(t) = x(t)$, *weakly,*
4. *If $\varphi(x(t))$ is bounded for each $\varphi \in E^*$ then* $\lim_{\alpha \to 0} I^{\alpha}x(t) = 0$.

Proof. for any $\varphi \in E^*$ we have (from the properties of the fractional calculus)

1.
$$\begin{aligned}\varphi(I^{\alpha}I^{\beta}x(t)) &= I^{\alpha}\varphi(I^{\beta}x(t)) = I^{\alpha}I^{\beta}\varphi(x(t)) \\ &= I^{\alpha+\beta}\varphi(x(t)) = \varphi(I^{\alpha+\beta}x(t)),\end{aligned}$$
which implies $\varphi(I^{\alpha}I^{\beta}x(t) - I^{\alpha+\beta}x(t)) = 0$, hence we obtain $I^{\alpha}I^{\beta}x(t) = I^{\alpha+\beta}x(t)$

2.
$$|\varphi(I^{\alpha}x(t)) - \varphi(I^n x(t))| = |I^{\alpha}\varphi(x(t)) - I^n\varphi(x(t))|,$$
since $\varphi x(\cdot) \in L^1(I)$, we have $\varphi(I^{\alpha}x(t)) \to \varphi(I^n x(t))$ uniformly on I, hence $I^{\alpha}x(t) \to I^n x(t)$ weakly uniformly on I.

3.
$$|\varphi(I^{\alpha}x(t)) - \varphi(x(t))| = |I^{\alpha}\varphi(x(t)) - \varphi(x(t))|,$$
since $\varphi(x(\cdot))$is continuous, we have $\varphi(I^{\alpha}x(t)) \to \varphi(x(t))$ on I (cf.[314] page 66),hence $I^{\alpha}x(t) \to x(t)$ weakly on I.

4. The proof of this part follows immediately from the fact that if $x_0 \in E$ such that $\varphi(x_0) = 0$ for every $\varphi \in E^*$ then $x_0 = 0$ and from the inequality

$$|\varphi(I^\alpha x(t))| = |I^\alpha \varphi(x(t))| \leq K \frac{t^\alpha}{\Gamma(1-\alpha)},$$

where $K = \sup|\varphi(x(t))|$. □

□

Now, we will prove a result which is an immediate consequence of the Lebesgue Dominated Convergence theorem for the Pettis-integral [[194], Corollary 4].

Theorem 1.38. *Let $x : I \to E$. Suppose there is a sequence (x_n) of weakly continuous functions from I into E such that $\lim_{n\to\infty} \varphi(x_n) = \varphi(x)$ a.e. for $\varphi \in E^*$. If there is a scalar function $\psi \in L^1(I)$ with $\|x_n(\cdot)\| < \psi(\cdot)$ a.e for all n, then x is Pettis integrable and $\varphi(I^\alpha x_n(t)) \to \varphi(I^\alpha x(t))$ for each $\varphi \in E^*$ and $t \in I$.*

Proof. Since (x_n) is weakly continuous, $I^\alpha x(t)$ exists for almost every $t \in I$. Moreover, for $t \in I$ and a.e. $s \in (0,t)$ we have

$$\left\| \frac{(t-s)^{\alpha-1}}{\Gamma(\alpha)} x_n(s) \right\| \leq \frac{(t-s)^{\alpha-1}}{\Gamma(\alpha)} \psi(s) \quad \text{for all } n$$

and

$$\begin{aligned} \lim_{n\to\infty} \varphi\left(\frac{(t-s)^{\alpha-1}}{\Gamma(\alpha)} x_n(s)\right) &= \lim_{n\to\infty} \frac{(t-s)^{\alpha-1}}{\Gamma(\alpha)} \varphi(x_n(s)) \\ &= \frac{(t-s)^{\alpha-1}}{\Gamma(\alpha)} \varphi(x(s)) \\ &= \varphi\left(\frac{(t-s)^{\alpha-1}}{\Gamma(\alpha)} x(s)\right) \text{ a.e. for all } \varphi \in E^*. \end{aligned}$$

By applying the Lebesgue dominated convergence theorem for the Pettis integral we complete the proof. □

Example. Let $x : I \to L^\infty(I)$ be defined by $x(t) = \chi_{[0,t]}$. This function is weakly measurable and for each $\varphi \in L^{\infty*}$, we have $\varphi x \in L^1$ (each φx is a function of bounded variation)(cf. [172, 173, 195]). Thus, according to Theorem 1.34, $I^\alpha x$ exists. To calculate the fractional Pettis integral of x, let $\rho \in L^1$ and let φ be the element in $L^{\infty*}$ corresponding to ρ. For any $t > 0$, we have

$$\begin{aligned} &\int_0^t \varphi\left(\frac{(t-s)^{\alpha-1}}{\Gamma(\alpha)} x(s)\right) ds \\ = &\int_0^t \frac{(t-s)^{\alpha-1}}{\Gamma(\alpha)} \varphi(x(s)) ds = \int_0^t \frac{(t-s)^{\alpha-1}}{\Gamma(\alpha)} \int_0^1 \rho(\theta)\chi_{[0,s]} d\theta ds \end{aligned}$$

$$
\begin{aligned}
&= \int_0^t \frac{(t-s)^{\alpha-1}}{\Gamma(\alpha)} \int_0^s \rho(\theta)d\theta ds = \int_0^t \int_0^s \frac{(t-s)^{\alpha-1}}{\Gamma(\alpha)} \rho(\theta)d\theta ds \\
&= \int_0^t \int_0^t \frac{(t-s)^{\alpha-1}}{\Gamma(\alpha)} \rho(\theta)dsd\theta = \int_0^t \frac{(t-\theta)^{\alpha}}{\Gamma(1+\alpha)} \rho(\theta)d\theta \\
&= \int_0^1 \frac{(t-\theta)^{\alpha}}{\Gamma(1+\alpha)} \rho(\theta)\chi_{[0,t]}(\theta)d\theta = \varphi\left(\frac{(t-s)^{\alpha}}{\Gamma(1+\alpha)}\chi_{[0,t]}(s)\right).
\end{aligned}
$$

Hence

$$I^{\alpha}x(t) = \int_0^t \frac{(t-s)^{\alpha-1}}{\Gamma(\alpha)} x(s)ds = \frac{(t-\cdot)^{\alpha}}{\Gamma(1+\alpha)}\chi_{[0,t]}(\cdot) \in L^{\infty}.$$

Remark 1.39. *As $\alpha \to 1$, we get the same result due to R. Geitz (see [195]).*

1.4.4. The Weak Derivative of Fractional-Order

Here we give the definition of weak derivative of fractional order.

Definition 1.40. *Let $x : I \to E$ be a weakly differentiable function and x' be weakly continuous. We define the weak derivative of x of order $\beta \in (0,1]$ by*

$$D^{\beta}x(t) = I^{1-\beta}Dx(t),$$

D denotes the weak differential operator.

For the properties of the weak fractional order derivative we have the following lemmas.

Lemma 1.41. *Let $x : I \to E$ be weakly differentiable and x' weakly continuous. If $\beta \in (0,1]$ then we have*

1. $\lim_{\beta \to 0} D^{\beta}x(t) = x(t) - x(0)$,
2. $I^{\beta}D^{\beta}x(t) = x(t) - x(0)$,
3. $D^{\beta}I^{\beta}x(t) = x(t)$.

For the proof of this Lemma, see [335].

Lemma 1.42. *Let $x : I \to E$ be weakly differentiable and let x' be weakly continuous. If $\alpha, \beta \in (0,1]$, then we have*

1. $\beta \leq \alpha \Rightarrow I^{\beta}D^{\alpha}x(t) = D^{\alpha-\beta}x(t)$,
2. *if* $x(0) = 0$ *then*
   (a) $\beta \leq \alpha \Rightarrow D^{\alpha}I^{\beta}x(t) = D^{\alpha-\beta}x(t)$,
   (b) $I^{\beta}D^{\alpha}x(t) = I^{\beta-\alpha}x(t)$.

Proof. 1. $I^{\beta}D^{\alpha}x(t) = I^{\beta}I^{1-\alpha}Dx(t) = I^{1-(\alpha-\beta)}Dx(t) = D^{\alpha-\beta}x(t)$.

2. Also, we have

(a)

$$D^{\alpha}I^{\beta}x(t) = I^{1-\alpha}DI^{\beta}x(t) = I^{1-\alpha}I^{\beta}Dx(t)$$
$$= I^{1-(\alpha-\beta)}Dx(t) = D^{\alpha-\beta}x(t).$$

(b)

$$I^{\beta}D^{\alpha}x(t) = I^{\beta}I^{1-\alpha}Dx(t) = I^{1+\beta-\alpha}$$
$$= I^{\beta-\alpha}I^{1}Dx(t) = I^{\beta-\alpha}x(t).$$

□

□

Lemma 1.43. *Let $x : I \to E$ be weakly differentiable and let x' be weakly continuous. If $\beta \in (0,1]$, then we have*

$$\lim_{\beta \to 1} D^{\beta}x(t) = x'(t).$$

Proof. Since $D^{\beta}x(t) = I^{1-\alpha}Dx(t)$, by taking the limit of both sides, we get by Lemma 1.37

$$\lim_{\beta \to 1} I^{1-\beta}Dx(t) = \lim_{\alpha \to 0} I^{\alpha}Dx(t) = Dx(t).$$

□

Definition 1.44. *[244] For a function $h : I \to E$, the Caputo fractional-order derivative of h, is defined by*

$$^{c}D^{\alpha}x(t) = \frac{1}{\Gamma(n-\alpha)} \int_0^t \frac{x^{(n)}(s)ds}{(t-s)^{1-n+\alpha}}$$

here $n = [\alpha] + 1$ and $[\alpha]$ denotes the integer part of α.

1.4.5. Riemann-Liouville Stieltjes Fractional Integral

If u is a real function defined on the interval $[a,b]$, then the symbol $\bigvee_a^b u$ denotes the variation of u on $[a,b]$. We say that u is of bounded variation on the interval $[a,b]$ whenever $\bigvee_a^b u$ is finite. If $w : [a,b] \times [c,b] \to \mathbb{R}$, then the symbol $\bigvee_{t=p}^q w(t,s)$ indicates the variation of the function $t \to w(t,s)$ on the interval $[p,q] \subset [a,b]$, where s is arbitrarily fixed in $[c,d]$. In the same way we define $\bigvee_{s=p}^q w(t,s)$. For the properties of functions of bounded variation we refer to [294].

If u and φ are two real functions defined on the interval $[a,b]$, then under some conditions (see [294]) we can define the Stieltjes integral (in the Riemann-Stieltjes sense)

$$\int_a^b u(t)d\varphi(t)$$

of the function u with respect to φ. In this case we say that u is Stieltjes integrable on $[a,b]$ with respect to φ. Several conditions are known guaranteeing Stieltjes integrability [294]. One of the most frequently used requires that u is continuous and φ is of bounded variation on $[a,b]$.

In what follows we use the following properties of the Stieltjes integral ([327], section 8.13)
If u is Stieltjes integrable on the interval $[a,b]$ with respect to a function φ of bounded variation, then

$$\left|\int_a^b u(t)d\varphi(t)\right| \leq \int_a^b |u(t)|d\left(\bigvee_a^t \varphi\right).$$

If u and v are Stieltjes integrable functions on the interval $[a,b]$ with respect to a nondecreasing function φ such that $u(t) \leq v(t)$ for $t \in [a,b]$. Then

$$\int_a^b u(t)d\varphi(t) \leq \int_a^b v(t)d\varphi(t).$$

We also consider Stieltjes integrals of the form

$$\int_a^b u(t)d_s g(t,s),$$

and Riemann-Liouville Stieltjes integrals of fractional order of the form

$$\frac{1}{\Gamma(r)}\int_0^t (t-s)^{r-1}u(s)d_s g(t,s),$$

where $g : \mathbb{R}_+ \times \mathbb{R}_+ \to \mathbb{R}$, $r \in (0,\infty)$ and the symbol d_s indicates the integration with respect to s.

1.5. Multi-valued Analysis

Let $(X, \|\cdot\|)$ be a Banach space. Denote $\mathcal{P}(X) = \{Y \in X : Y \neq \emptyset\}$, $\mathcal{P}_{cl}(X) = \{Y \in \mathcal{P}(X) : Y \text{ closed}\}$, $\mathcal{P}_b(X) = \{Y \in \mathcal{P}(X) : Y \text{ bounded}\}$, $\mathcal{P}_{cp}(X) = \{Y \in \mathcal{P}(X) : Y \text{ compact}\}$ and $\mathcal{P}_{cp,cv}(X) = \{Y \in \mathcal{P}(X) : Y \text{ compact and convex}\}$.

Definition 1.45. *A multivalued map $T : X \to \mathcal{P}(X)$ is convex (closed) valued if $T(x)$ is convex (closed) for all $x \in X$. T is bounded on bounded sets if $T(B) = \cup_{x\in B} T(x)$ is bounded in X for all $B \in \mathcal{P}_b(X)$ (i.e. $\sup_{x\in B}\sup_{y\in T(x)} \|y\| < \infty$). T is called upper semi-continuous (u.s.c.) on X if for each $x_0 \in X$, the set $T(x_0)$ is a nonempty closed subset of X, and if for each open set N of X containing $T(x_0)$, there exists an open neighborhood N_0 of x_0 such that $T(N_0) \subseteq N$. T is lower semi-continuous (l.s.c.) if the set $\{x \in X : T(x) \cap A \neq \emptyset\}$ is open for any open subset $A \subseteq X$. T is said to be completely continuous if $T(\mathcal{B})$ is relatively compact for every $\mathcal{B} \in P_b(X)$. T has a fixed point if there is $x \in X$ such that $x \in T(x)$. The fixed point set of the multivalued operator T will be denoted by $FixT$. A multivalued map $G : X \to \mathcal{P}_{cl}(\mathbb{R}^n)$ is said to be measurable if for every $v \in \mathbb{R}^n$, the function $x \longmapsto d(v, G(x)) = \inf\{\|v-z\| : z \in G(x)\}$ is measurable.*

For more details on multivalued maps see the books of Aubin and Cellina [78], Aubin and Frankowska [79], Deimling [153], Gorniewicz [202], Hu and Papageorgiou [228] and Kisielewicz [247].

Lemma 1.46. *([228]) Let G be a completely continuous multivalued map with nonempty compact values, then G is u.s.c. if and only if G has a closed graph (i.e. $u_n \to u$, $w_n \to w$, $w_n \in G(u_n)$ imply $w \in G(u)$).*

Definition 1.47. *A multivalued map $F : J \times \mathbb{R}^n \to \mathcal{P}(\mathbb{R}^n)$ is said to be Carathéodory if*

(i) $(x,y) \longmapsto F(x,y,u)$ *is measurable for each* $u \in \mathbb{R}^n$;

(ii) $u \longmapsto F(x,y,u)$ *is upper semicontinuous for almost all* $(x,y) \in J$.

F is said to be L^1-Carathéodory if $(i), (ii)$ and the following condition holds;

(iii) for each $c > 0$, there exists $\sigma_c \in L^1(J,\mathbb{R}_+)$ such that

$$\begin{aligned} \|F(x,y,u)\|_{\mathcal{P}} &= \sup\{\|f\| : f \in F(x,y,u)\} \\ &\leq \sigma_c(x,y) \ \text{for all } \|u\| \leq c \text{ and for a.e. } (x,y) \in J. \end{aligned}$$

For each $u \in C(J,\mathbb{R}^n)$, define the set of selections of F by

$$S_{F,u} = \{w \in L^1(J,\mathbb{R}^n) : w(x,y) \in F(x,y,u(x,y)) \ \text{a.e.} \ (x,y) \in J\}.$$

Definition 1.48. *Let $\mathcal{A}$ be a subset of $J \times \mathbb{R}^n$. $\mathcal{A}$ is $\mathcal{L} \otimes \mathcal{B}$ measurable if $\mathcal{A}$ belongs to the $\sigma-$algebra generated by all sets of the form $\mathcal{N} \otimes \mathcal{D}$ where $\mathcal{N}$ is Lebesgue measurable in J and $\mathcal{D}$ is Borel measurable in $\mathbb{R}^n$. A subset I of $L^1(J,\mathbb{R}^n)$ is decomposable if for all $u,v \in I$ and $N \subset J$ measurable, $u\chi_I + v\chi_{J-\mathcal{N}} \in I$, where χ_J stands for the characteristic function of J.*

Definition 1.49. *Let $F : J \times \mathbb{R}^n \to \mathcal{P}(\mathbb{R}^n)$ be a multi-valued map with nonempty compact values. Assign to F the multi-valued operator $\mathcal{F} : C(J,\mathbb{R}^n) \to \mathcal{P}(L^1(J,\mathbb{R}^n))$ by letting $\mathcal{F}(u) = S_{F,u}$. We say F is l.s.c. if $\mathcal{F}$ is l.s.c. and has nonempty closed and decomposable values.*

The above operator $\mathcal{F}$ is called the Niemytzki operator associated to F.

Definition 1.50. *Let X be a separable metric space and $N : X \to \mathcal{P}(L^1(J,\mathbb{R}^n))$ be a multi-valued operator. We say N has property (BC) if*
1) N is lower semi-continuous (l.s.c.) ;
2) N has nonempty closed and decomposable values.

Next we state a selection theorem due to Bressan and Colombo.

Lemma 1.51. *(Bressan-Colombo) ([117]) Let X be a separable metric space and $N : X \to \mathcal{P}(L^1(J,\mathbb{R}^n))$ be a multi-valued operator which has property (BC). Then N has a continuous selection , i.e. there exists a continuous function (single-valued) $g : X \to L^1(J,\mathbb{R}^n)$ such that $g(u) \in N(u)$ for every $u \in X$.*

Let (X,d) be a metric space induced from the normed space $(X,\|\cdot\|)$. Consider H_d : $\mathcal{P}(X)\times\mathcal{P}(X)\longrightarrow\mathbb{R}_+\cup\{\infty\}$ given by

$$H_d(A,B)=\max\left\{\sup_{a\in A}d(a,B),\sup_{b\in B}d(A,b)\right\},$$

where $d(A,b)=\inf_{a\in A}d(a,b)$, $d(a,B)=\inf_{b\in B}d(a,b)$. Then $(\mathcal{P}_{b,cl}(X),H_d)$ is a metric space and $(\mathcal{P}_{cl}(X),H_d)$ is a generalized metric space (see [247]).

Definition 1.52. *A multivalued operator $N:X\to\mathcal{P}_{cl}(X)$ is called*

a) γ-Lipschitz if and only if there exists $\gamma>0$ such that

$$H_d(N(u),N(v))\leq\gamma d(u,v)\ \textit{for each}\ \ u,\ v\in X,$$

b) a contraction if and only if it is γ-Lipschitz with $\gamma<1$.

1.6. Concept of the Stability

Let $\emptyset\neq\Omega\subset BC$, and let $G:\Omega\to\Omega$, and consider the solutions of equation

$$(Gu)(t,x)=u(t,x). \tag{1.4}$$

Inspired by the definition of the attractivity of solutions of integral equations (for instance [85]), we introduce the following concept of attractivity of solutions for equation (1.4).

Definition 1.53. *Solutions of equation (1.4) are locally attractive if there exists a ball $B(u_0,\eta)$ in the space BC such that, for arbitrary solutions $v=v(t,x)$ and $w=w(t,x)$ of equations (1.4) belonging to $B(u_0,\eta)\cap\Omega$, we have that, for each $x\in[0,b]$,*

$$\lim_{t\longrightarrow\infty}(v(t,x)-w(t,x))=0. \tag{1.5}$$

When the limit (1.5) is uniform with respect to $B(u_0,\eta)\cap\Omega$, solutions of equation (1.4) are said to be uniformly locally attractive (or equivalently that solutions of (1.4) are locally asymptotically stable).

Definition 1.54. *[85] The solution $v=v(t,x)$ of equation (1.4) is said to be globally attractive if (1.5) hold for each solution $w=w(t,x)$ of (1.4). If condition (1.5) is satisfied uniformly with respect to the set Ω, solutions of equation (1.4) are said to be globally asymptotically stable (or uniformly globally attractive).*

Definition 1.55. *[138] The zero solution $u(t,x)$ of equation (1.4) is globally attractive if every solution of (1.4) tends to zero as $t\to\infty$.*

1.7. Fixed Point Theorems

By $\overline{U}$ and ∂U we denote the closure of U and the boundary of U respectively. Let us start by stating a well known result, the Nonlinear Alternative.

Theorem 1.56. *[204] (Nonlinear alternative of Leray-Schauder type) Let X be a Banach space and C a nonempty convex subset of X. Let U a nonempty open subset of C with $0 \in U$ and $T : \overline{U} \to C$ continuous and compact operator.*
Then either

(a) T has fixed points. Or

(b) There exist $u \in \partial U$ and $\lambda \in [0,1]$ with $u = \lambda T(u)$.

The multi-valued version of Nonlinear Alternative:

Lemma 1.57. *[204] Let X be a Banach space and C a nonempty convex subset of X. Let U a nonempty open subset of C with $0 \in U$ and $T : \overline{U} \to \mathcal{P}(C)$ an upper semicontinuous and compact multivalued operator.*
Then either

(a) T has fixed points. Or

(b) There exist $u \in \partial U$ and $\lambda \in [0,1]$ with $u \in \lambda T(u)$.

Theorem 1.58. *(Schaefer) [218] Let X be a Banach space and $N : X \to X$ completely continuous operator. If the set*

$$E(N) = \{u \in X : \ u = \lambda N(u) \text{ for some } \lambda \in [0,1]\}$$

is bounded, then N has fixed points.

Other single-valued fixed point theorems:

Theorem 1.59. *(Burton-Kirk) [123] Let X be a Banach space, and $A, B : X \to X$ two operators satisfying:*

(i) A is a contraction, and

(ii) B is completely continuous.

Then either

(a) the operator equation $u = A(u) + B(u)$ has a solution, or

(b) the set $\mathcal{E} = \{u \in X : u = \lambda A(\frac{u}{\lambda}) + \lambda B(u)\}$ is unbounded for $\lambda \in (0,1)$.

Theorem 1.60. *(Krasnoselskii) [63] Let X be a Banach space, let D be a bounded closed convex subset of X and let A, B be maps of D into X such that $Au + Bv \in D$ for every $(u,v) \in D$. If A is a contraction and B is completely continuous, then the equation $Aw + Bw = w$ has a solution w on D.*

We need the following fixed point theorem, including the improvement of a fixed point theorem of Krasnoselskii [120] due to Burton [122].

Theorem 1.61. *(Krasnoselskii's Fixed Point Theorem) [120, 122] Let S be a nonempty, closed, convex and bounded subset of the Banach space X and let* $A : X \to X$ *and* $B : S \to X$ *be two operators such that*

(a) *A is a contraction with constant* l,

(b) *B is continuous,* $B(S)$ *resides in a compact subset of* X,

(c) $[u = A(u) + B(v);\ v \in S] \Rightarrow u \in S$.

Then the operator equation $A(u) + B(u) = u$ *has a solution in S.*

Next we state two multi-valued fixed point theorems

Lemma 1.62. *(Bohnenblust-Karlin)([114]) Let X be a Banach space and* $K \in \mathcal{P}_{cl,cv}(X)$ *and suppose that the operator* $G : K \to \mathcal{P}_{cl,cv}(K)$ *is upper semicontinuous and the set* $G(K)$ *is relatively compact in* X*. Then G has a fixed point in* K.

Lemma 1.63. *(Covitz-Nadler) [142] Let* (X, d) *be a complete metric space. If* $N : X \to \mathcal{P}_{cl}(X)$ *is a contraction, then N has fixed points.*

Now, we present fixed point results involving the weak topology of a Banach space. The following fixed point theorem is a variant of the Darbo-Mönch fixed point theorem given by O'Regan [298] in 1998, which was motived by some ideas used in [74].

Theorem 1.64. *[298] Let Q be a closed convex and equicontinuous subset of a metrizable locally convex vector space* $C(J, E)$ *such that* $0 \in Q$*. Assume that* $T : Q \to Q$ *is wk-sequentially continuous. If the implication (the so-called Mönch condition)*

$$\overline{V} = \overline{conv}(\{0\} \cup T(V)) \Rightarrow V \text{ is relatively weakly compact,} \tag{1.6}$$

holds for every subset $V \subset Q$*, then T has a fixed point.*

The multivalued version of the previous theorem is the following

Theorem 1.65. *Let E be a Banach space with Q a nonempty, bounded, closed, convex, equicontinuous subset of* $C([0, T], E)$*. Suppose* $F : Q \to \mathcal{P}_{cl,cv}(Q)$ *has a weakly sequentially closed graph. If the implication,*

$$\overline{V} = \overline{conv}(\{0\} \cup F(V)) \Longrightarrow V \text{ is relatively weakly compact,} \tag{1.7}$$

holds for every subset $V \subset Q$*, then the operator inclusion* $x \in F(x)$ *has a solution in Q.*

Let us state the nonlinear alternative of Leray-Schauder type for contraction maps in Fréchet spaces.

Theorem 1.66. *[187] Let X be a Fréchet space and $Y \subset X$ a closed subset in X. Let $N : Y \longrightarrow X$ be a contraction such that $N(Y)$ is bounded. Then one of the following statements holds:*

(S1) The operator N has a unique fixed point;

S2) There exists $\lambda \in [0,1)$, $n \in \mathbb{N}^*$ and $u \in \partial_n Y^n$ such that

$$\|u - \lambda N(u)\|_n = 0.$$

The multi-valued version of Theorem 1.66

Theorem 1.67. *[186] Let X be a Fréchet space and U an open neighborhood of the origin in X and let $N : \overline{U} \to \mathcal{P}(X)$ be an admissible multivalued contraction. Assume that N is bounded. Then one of the following statements holds :*

$(S1)$ N has a fixed point ;

$(S2)$ There exists $\lambda \in [0,1)$ and $x \in \partial U$ such that $x \in \lambda\, N(x)$.

1.8. Auxiliary Lemmas

In the sequel we will make use of the following generalization of Gronwall's lemma for two independent variables and singular kernel.

Lemma 1.68. *(Gronwall lemma) [214] Let $\upsilon : J \to [0,\infty)$ be a real function and $\omega(.,.)$ be a nonnegative, locally integrable function on J. If there are constants $c > 0$ and $0 < r_1, r_2 < 1$ such that*

$$\upsilon(x,y) \leq \omega(x,y) + c\int_0^x \int_0^y \frac{\upsilon(s,t)}{(x-s)^{r_1}(y-t)^{r_2}}dtds,$$

then there exists a constant $\delta = \delta(r_1, r_2)$ such that

$$\upsilon(x,y) \leq \omega(x,y) + \delta c\int_0^x \int_0^y \frac{\omega(s,t)}{(x-s)^{r_1}(y-t)^{r_2}}dtds,$$

for every $(x,y) \in J$.

Also, we will use the following lemma:

Lemma 1.69. *[140] Let $D \subset BC$. Then D is relatively compact in BC if the following conditions hold:*
(a) D is uniformly bounded in BC.
(b) The functions belonging to D are almost equicontinuous on $\mathbb{R}_+ \times [0,b]$,
i.e., *equicontinuous on every compact subset of $\mathbb{R}_+ \times [0,b]$.*
(c) The functions from D are equiconvergent, that is, given $\varepsilon > 0$, $x \in [0,b]$ there corresponds $T(\varepsilon,x) > 0$ such that $|u(t,x) - \lim_{t\to\infty} u(t,x)| < \varepsilon$ for any $t \geq T(\varepsilon,x)$ and $u \in D$.

Chapter 2

Nonlinear Differential Equations of Fractional Order

2.1. Introduction

In this chapter, we shall present existence results for some classes of nonlinear differential equations of fractional order by using some fixed point theorems.

2.2. Local Attractivity for Delay Partial Differential Equations

2.2.1. Introduction

In this section, we prove some results concerning the existence and the attractivity of solutions for the following system of delay differential equations of fractional order of the form

$$^{c}D^{r}_{\theta}u(t,x) = f(t,x,u(t-\tau_1,x-\xi_1),\dots,u(t-\tau_m,x-\xi_m)); \quad \text{for } (t,x)\in J, \tag{2.1}$$

$$u(t,x) = \Phi(t,x); \quad \text{for } (t,x)\in \tilde{J} := [-T,\infty)\times[-\xi,b]\backslash(0,\infty)\times(0,b], \tag{2.2}$$

$$\begin{cases} u(t,0) = \mu(t); & t\in[0,\infty), \\ u(0,x) = \mu(0); & x\in[0,b], \end{cases} \tag{2.3}$$

where $J = \mathbb{R}_+ \times [0,b]$, $b > 0$, $\mathbb{R}_+ = [0,\infty)$, $\theta = (0,0), \tau_i, \xi_i \geq 0$; $i = 1\dots,m$, $T = \max\limits_{i=1\dots,m}\{\tau_i\}$, $\xi = \max\limits_{i=1\dots,m}\{\xi_i\}$, $^{c}D^{r}_{\theta}$ is the Caputo fractional derivative of order $r = (r_1,r_2) \in (0,\infty)\times(0,\infty)$, $f : J\times\mathbb{R}^m \to \mathbb{R}$ is a given continuous function, $\mu : \mathbb{R}_+ \to \mathbb{R}$ is absolutely continuous function, and $\Phi : \tilde{J} \to \mathbb{R}^n$ is continuous with $\mu(t) = \Phi(t,0)$ for each $t\in\mathbb{R}_+$ and $\mu(0) = \Phi(0,x)$; for each $x\in[0,b]$.

2.2.2. Main Results

Let us start by defining what we mean by a solution of the problem (2.1)-(2.3).

Definition 2.1. *A function $u \in BC$ is said to be a solution of (2.1)-(2.3) if u satisfies equation (2.1) on J, equation (2.2) on $\tilde{J}$ and condition (2.3) is satisfied.*

Lemma 2.2. *[2] Let $f \in L^1([0,a] \times [0,b])$; $a,b > 0$. A function $u \in AC([0,a] \times [0,b])$ is a solution of problem*

$$\begin{cases} ({}^cD_\theta^r u)(t,x) = f(t,x); \ (t,x) \in [0,a] \times [0,b], \\ u(t,0) = \varphi(t); \ t \in [0,a], \ u(0,x) = \psi(x); \ x \in [0,b], \\ \varphi(0) = \psi(0), \end{cases}$$

if and only if $u(t,x)$ satisfies

$$u(t,x) = \lambda(t,x) + (I_\theta^r f)(t,x); \ (t,x) \in [0,a] \times [0,b],$$

where

$$\lambda(t,x) = \varphi(t) + \psi(x) - \varphi(0).$$

Now, we shall prove the following theorem concerning the existence and the attractivity of a solution of problem (2.1)-(2.3).

Theorem 2.3. *Assume that the following hypotheses hold*

(2.3.1) $\lim_{t\to\infty} \mu(t) = 0$,

(2.3.2) *The function f is continuous and there exists a continuous function $p : \mathbb{R}_+ \times [0,b] \to \mathbb{R}_+$ such that*

$$\left(1 + \sum_{i=1}^m |u_i|\right) |f(t,x,u_1,u_2,\ldots,u_m)| \leq p(t,x);$$

for $(t,x) \in \mathbb{R}_+ \times [0,b]$ and for $u_i \in \mathbb{R}$; $i = 1\ldots m$. Moreover, assume that

$$\lim_{t\to\infty} \int_0^t \int_0^x (t-\tau)^{r_1-1}(x-s)^{r_2-1} p(\tau,s)\,ds\,d\tau = 0; \ x \in [0,b].$$

Then the problem (2.1)-(2.3) has at least one solution in the space BC. Moreover, solutions of problem (2.1)-(2.3) are locally asymptotically stable.

Proof. Set

$$\Phi^* := \sup_{(t,x)\in\tilde{J}} \Phi(t,x), \ \mu^* := \sup_{t\in\mathbb{R}_+} \mu(t)$$

and $d^* := \sup_{(t,x)\in\mathbb{R}_+\times[0,b]} d(t,x)$, where

$$d(t,x) = \frac{1}{\Gamma(r_1)\Gamma(r_2)} \int_0^t \int_0^x (t-\tau)^{r_1-1}(x-s)^{r_2-1} p(\tau,s)\,ds\,d\tau.$$

From the hypotheses, we infer that μ^* and d^* are finite. Let us define the operator N such that, for any $u \in BC$,

$$(Nu)(t,x) = \begin{cases} \Phi(t,x); & (t,x) \in \tilde{J}, \\ \mu(t) + I_\theta^r f(t,x,u(t-\tau_1,x-\xi_1),..., \\ u(t-\tau_m,x-\xi_m)); & (t,x) \in J. \end{cases} \tag{2.4}$$

From the assumptions, we infer that $N(u)$ is continuous on $[-T,\infty)\times[-\xi,b]$. Now we prove that $N(u) \in BC$ for any $u \in BC$. For arbitrarily fixed $(t,x) \in J$ we have

$$\begin{aligned} & |(Nu)(t,x)| \leq |\mu(t)| \\ + & |I_\theta^r f(t,x,u(t-\tau_1,x-\xi_1),...,u(t-\tau_m,x-\xi_m))| \\ \leq & |\mu(t)| + \frac{1}{\Gamma(r_1)\Gamma(r_2)} \int_0^t \int_0^x \frac{(t-\tau)^{r_1-1}(x-s)^{r_1} p(\tau,s)}{1+\sum_{i=1}^m |u(\tau-\tau_i,s-\xi_i)|} ds d\tau. \\ \leq & \mu^* + d^*, \end{aligned}$$

and for $(t,x) \in \tilde{J}$, we have

$$|(Nu)(t,x)| = |\Phi(t,x)| \leq \Phi^*.$$

Thus,

$$\|Nu\|_{BC} \leq \max\{\Phi^*, \mu^* + d^*\} := \eta. \tag{2.5}$$

Hence, $N(u) \in BC$. This proves that the operator N maps BC into itself.
By Lemma 2.2, the problem of finding the solutions of the problem (2.1)-(2.3) is reduced to finding the solutions of the operator equation $N(u) = u$. Equation (2.5) yields that N transforms the ball $B_\eta := B(0,\eta)$ into itself. We shall show that $N : B_\eta \to B_\eta$ satisfies the assumptions of Schauder's fixed point theorem [204]. The proof will be given in several steps.

Step 1: *N is continuous.*
Let $\{u_n\}_{n\in\mathbb{N}}$ be a sequence such that $u_n \to u$ in B_η. Then, for each $(t,x) \in [-T,\infty)\times[-\xi,b]$, we have

$$\begin{aligned} & |(Nu_n)(t,x) - (Nu)(t,x)| \leq \tfrac{1}{\Gamma(r_1)\Gamma(r_2)} \int_0^t \int_0^x (t-\tau)^{r_1-1}(x-s)^{r_2-1} \\ \times & |f(\tau,s,u_n(\tau-\tau_1,s-\xi_1),...,u_n(\tau-\tau_m,s-\xi_m)) \\ - & f(\tau,s,u(\tau-\tau_1,s-\xi_1),...,u(\tau-\tau_m,s-\xi_m)))| ds d\tau. \end{aligned} \tag{2.6}$$

Case 1. If $(t,x) \in \tilde{J} \cup ([0,T_0]\times[0,b])$, $T_0 > 0$, then, since $u_n \to u$ as $n \to \infty$ and f is continuous, (2.6) gives

$$\|N(u_n) - N(u)\|_{BC} \to 0 \quad \text{as } n \to \infty.$$

Case 2. If $(t,x) \in (T_0,\infty)\times[0,b]$, $T_0 > 0$, then from (2.3.2) and (2.6), we get

$$\begin{aligned} & |(Nu_n)(t,x) - (Nu)(t,x)| \leq \tfrac{2}{\Gamma(r_1)\Gamma(r_2)} \int_0^t \int_0^x \frac{(t-\tau)^{r_1-1}(x-s)^{r_2-1} p(\tau,s)}{1+\sum_{i=1}^m |u(\tau-\tau_i,s-\xi_i)|} ds d\tau \\ \leq & \tfrac{2}{\Gamma(r_1)\Gamma(r_2)} \int_0^t \int_0^x (t-\tau)^{r_1-1}(x-s)^{r_2-1} p(\tau,s) ds d\tau. \end{aligned} \tag{2.7}$$

Since $u_n \longrightarrow u$ as $n \longrightarrow \infty$ and $t \longrightarrow \infty$, then (2.7) gives

$$\|N(u_n) - N(u)\|_{BC} \rightarrow 0 \quad \text{as } n \longrightarrow \infty.$$

Step 2: $N(B_\eta)$ *is uniformly bounded.*
This is clear since $N(B_\eta) \subset B_\eta$ and B_η is bounded.

Step 3: $N(B_\eta)$ *is equicontinuous on every compact subset* $[-T,a] \times [-\xi,b]$ *of* $[-T,\infty) \times [-\xi,b]$, $a > 0$.
Let $(t_1,x_1),(t_2,x_2) \in [0,a] \times [0,b]$, $t_1 < t_2$, $x_1 < x_2$ and let $u \in B_\eta$. Thus we have

$$\begin{aligned}
&|(Nu)(t_2,x_2) - (Nu)(t_1,x_1)| \leq |\mu(t_2) - \mu(t_1)| \\
&+ \frac{1}{\Gamma(r_1)\Gamma(r_2)} \int_0^{t_1} \int_0^{x_1} [(t_2-\tau)^{r_1-1}(x_2-s)^{r_2-1} - (t_1-\tau)^{r_1-1}(x_1-s)^{r_2-1}] \\
&\times |f(\tau,s,u(\tau-\tau_1,s-\xi_1),\dots,u(\tau-\tau_m,s-\xi_m))|dsd\tau \\
&+ \frac{1}{\Gamma(r_1)\Gamma(r_2)} \int_{t_1}^{t_2} \int_{x_1}^{x_2} (t_2-\tau)^{r_1-1}(x_2-s)^{r_2-1} \\
&\times |f(\tau,s,u(\tau-\tau_1,s-\xi_1),\dots,u(\tau-\tau_m,s-\xi_m))|dsd\tau \\
&+ \frac{1}{\Gamma(r_1)\Gamma(r_2)} \int_0^{t_1} \int_{x_1}^{x_2} (t_2-\tau)^{r_1-1}(x_2-s)^{r_2-1} \\
&\times |f(\tau,s,u(\tau-\tau_1,s-\xi_1),\dots,u(\tau-\tau_m,s-\xi_m))|dsd\tau \\
&+ \frac{1}{\Gamma(r_1)\Gamma(r_2)} \int_{t_1}^{t_2} \int_0^{x_1} (t_2-\tau)^{r_1-1}(x_2-s)^{r_2-1} \\
&\times |f(\tau,s,u(\tau-\tau_1,s-\xi_1),\dots,u(\tau-\tau_m,s-\xi_m))|dsd\tau.
\end{aligned}$$

Thus

$$\begin{aligned}
&|(Nu)(t_2,x_2) - (Nu)(t_1,x_1)| \leq |\mu(t_2) - \mu(t_1)| \\
&+ \frac{1}{\Gamma(r_1)\Gamma(r_2)} \int_0^{t_1} \int_0^{x_1} [(t_2-\tau)^{r_1-1}(x_2-s)^{r_2-1} - (t_1-\tau)^{r_1-1}(x_1-s)^{r_2-1}] \\
&\times p(\tau,s)dsd\tau + \frac{1}{\Gamma(r_1)\Gamma(r_2)} \int_{t_1}^{t_2} \int_{x_1}^{x_2} (t_2-\tau)^{r_1-1}(x_2-s)^{r_2-1} p(\tau,s)dsd\tau \\
&+ \frac{1}{\Gamma(r_1)\Gamma(r_2)} \int_0^{t_1} \int_{x_1}^{x_2} (t_2-\tau)^{r_1-1}(x_2-s)^{r_2-1} p(\tau,s)dsd\tau \\
&+ \frac{1}{\Gamma(r_1)\Gamma(r_2)} \int_{t_1}^{t_2} \int_0^{x_1} (t_2-\tau)^{r_1-1}(x_2-s)^{r_2-1} p(\tau,s)dsd\tau,
\end{aligned}$$

From continuity of μ, p and as $t_1 \longrightarrow t_2$ and $x_1 \longrightarrow x_2$, the right-hand side of the above inequality tends to zero. The equicontinuity for the cases $t_1 < t_2 < 0$, $x_1 < x_2 < 0$ and $t_1 \leq 0 \leq t_2$, $x_1 \leq 0 \leq x_2$ is obvious.

Step 4: $N(B_\eta)$ *is equiconvergent.*

Let $(t,x) \in \mathbb{R}_+ \times [0,b]$ and $u \in B_\eta$, then we have

$$\begin{aligned}|(Nu)(t,x)| &\leq |\mu(t)| + |\frac{1}{\Gamma(r_1)\Gamma(r_2)}\int_0^t\int_0^x (t-\tau)^{r_1-1}(x-s)^{r_2-1} \\ &\quad \times f(\tau,s,u(\tau-\tau_1,s-\xi_1),\ldots,u(\tau-\tau_m,s-\xi_m))dsd\tau| \\ &\leq |\mu(t)| + \frac{1}{\Gamma(r_1)\Gamma(r_2)}\int_0^t\int_0^x (t-\tau)^{r_1-1}(x-s)^{r_2-1}p(\tau,s)dsd\tau \\ &\leq |\mu(t)| + d(t,x).\end{aligned}$$

Thus, for each $x \in [0,b]$, we get

$$|(Nu)(t,x)| \longrightarrow 0, \ as\ t \longrightarrow +\infty.$$

Hence,

$$|(Nu)(t,x) - (Nu)(+\infty,x)| \longrightarrow 0, \ as\ t \longrightarrow +\infty.$$

As a consequence of Steps 1 to 4 together with the Lemma 1.69, we can conclude that $N : B_\eta \to B_\eta$ is continuous and compact. From an application of Schauder's theorem [204], we deduce that N has a fixed point u which is a solution of the problem (2.1)-(2.3).

Now we investigate the stability for solutions of problem (2.1)-(2.3). Let us assume that u_0 is a solution of problem (2.1)-(2.3) with the conditions of this theorem. Consider the ball $B(u_0,\eta^*)$, where

$$\begin{aligned}\eta^* : &= \frac{1}{\Gamma(r_1)\Gamma(r_2)}\sup_{(t,x)\in J}\left\{\int_0^t\int_0^x (t-\tau)^{r_1-1}(x-s)^{r_2-1}\right. \\ &\quad \times|f(\tau,s,u(\tau-\tau_1,s-\xi_1),\ldots,u(\tau-\tau_m,s-\xi_m)) \\ &\quad \left. - f(\tau,s,u_0(\tau-\tau_1,s-\xi_1),\ldots,u_0(\tau-\tau_m,s-\xi_m))|dsd\tau;\ u \in BC\right\}.\end{aligned}$$

Taking $u \in B(u_0,\eta^*)$, we have

$$\begin{aligned}|(Nu)(t,x) - u_0(t,x)| &= |(Nu)(t,x) - (Nu_0)(t,x)| \\ &\leq \frac{1}{\Gamma(r_1)\Gamma(r_2)}\int_0^t\int_0^x (t-\tau)^{r_1-1}(x-s)^{r_2-1} \\ &\quad \times|f(\tau,s,u(\tau-\tau_1,s-\xi_1),\ldots,u(\tau-\tau_m,s-\xi_m)) \\ &- f(\tau,s,u_0(\tau-\tau_1,s-\xi_1),\ldots,u_0(\tau-\tau_m,s-\xi_m))|dsd\tau \\ &\leq \eta^*.\end{aligned}$$

Thus we observe that N is a continuous function such that $N(B(u_0,\eta^*)) \subset B(u_0,\eta^*)$. Moreover, if u is a solution of problem (2.1)-(2.3), then

$$\begin{aligned}|u(t,x) - u_0(t,x)| &= |(Nu)(t,x) - (Nu_0)(t,x)| \\ &\leq \tfrac{1}{\Gamma(r_1)\Gamma(r_2)}\int_0^t\int_0^x (t-\tau)^{r_1-1}(x-s)^{r_2-1} \\ &\quad \times|f(\tau,s,u(\tau-\tau_1,s-\xi_1),\ldots,u(\tau-\tau_m,s-\xi_m)) \\ &- f(\tau,s,u_0(\tau-\tau_1,s-\xi_1),\ldots,u_0(\tau-\tau_m,s-\xi_m))|dsd\tau \\ &\leq \tfrac{2}{\Gamma(r_1)\Gamma(r_2)}\int_0^t\int_0^x (t-\tau)^{r_1-1}(x-s)^{r_2-1}p(\tau,s)dsd\tau \\ &\leq 2d(t,x).\end{aligned} \tag{2.8}$$

By using (2.8) and the fact that $d(t,x) \longrightarrow 0$ as $t \longrightarrow \infty$, we deduce that

$$\lim_{t\longrightarrow\infty} |u(t,x) - u_0(t,x)| = 0.$$

Consequently, all solutions of problem (2.1)-(2.3) are locally asymptotically stable. □

2.2.3. An Example

As an application and to illustrate our results, we consider the following system of delay differential equations of fractional order

$$^cD_\theta^r u(t,x) = f\left(t,x,u\left(t-1,x-\frac{1}{4}\right),u\left(t-\frac{1}{2},x-\frac{1}{8}\right),u\left(t-\frac{2}{3},x-\frac{1}{5}\right)\right);$$

$$\text{for } (t,x) \in J := [0,\infty)\times[0,1], \tag{2.9}$$

$$u(t,x) = \frac{1}{1+t}; \quad \text{for } (t,x) \in \tilde{J} := [-1,\infty)\times[-\frac{1}{4},1]\backslash(0,\infty)\times(0,1], \tag{2.10}$$

$$\begin{cases} u(t,0) = \dfrac{1}{1+t}; & t\in[0,\infty), \\ u(0,x) = 1; & x\in[0,1], \end{cases} \tag{2.11}$$

where $r = (r_1, r_2) = (\frac{1}{4}, \frac{1}{2})$ and

$$\begin{cases} f(t,x,u,v,w) = \dfrac{xt^{\frac{-3}{4}}\sin t}{2+|u|+|v|+|w|}; & (t,x)\in(0,\infty)\times[0,1] \text{ and } u,v,w\in\mathbb{R}, \\ f(t,x,u,v,w) = 0; & (t,x)\in\{0\}\times[0,1] \text{ and } u,v,w\in\mathbb{R}. \end{cases}$$

We can easily check that the assumptions of Theorem 2.3 are satisfied. In fact, the function μ satisfies assumption (2.3.1).
Next, let us notice that the function f satisfies assumption (2.3.2), where

$$\begin{cases} p(t,x) = xt^{\frac{-3}{4}}|\sin t|; & (t,x)\in(0,\infty)\times[0,1], \\ p(0,x) = 0; & x\in[0,1]. \end{cases}$$

Also, for each $x \in [0,1]$, we get

$$\begin{aligned}
&\int_0^t\int_0^x (t-\tau)^{r_1-1}(x-s)^{r_2-1}p(\tau,s)\,ds\,d\tau \\
&\quad = \int_0^t\int_0^x (t-\tau)^{\frac{-3}{4}}(x-s)^{\frac{-1}{2}}s\tau^{\frac{-3}{4}}|\sin\tau|\,ds\,d\tau \\
&\quad \le \int_0^t\int_0^x (t-\tau)^{\frac{-3}{4}}(x-s)^{\frac{-1}{2}}s\tau^{\frac{-3}{4}}\,ds\,d\tau \\
&\quad = \frac{\Gamma^2(\frac{1}{4})\Gamma(\frac{1}{2})\Gamma(2)}{\Gamma(\frac{5}{4})\Gamma(\frac{5}{2})}t^{\frac{-1}{2}}x^{\frac{3}{2}}.
\end{aligned}$$

Thus

$$\lim_{t\longrightarrow\infty}\int_0^t\int_0^x (t-\tau)^{r_1-1}(x-s)^{r_2-1}p(\tau,s)\,ds\,d\tau = 0.$$

Hence by Theorem 2.3, the problem (2.9)-(2.11) has a solution defined on $[-1,\infty)\times[-\frac{1}{4},1]$ and all solutions are uniformly locally asymptotically stable on $[-1,\infty)\times[-\frac{1}{4},1]$.

2.3. Existence and Attractivity for the Darboux Problem of Neutral Differential Equations

2.3.1. Introduction

This section concerned with the existence and the attractivity of solutions to the following IVP for partial neutral functional differential equations

$$^{c}D_{\theta}^{r}\left(u(t,x)-g(t,x,u_{(t,x)})\right)=f(t,x,u_{(t,x)});\text{ if }(t,x)\in J:=\mathbb{R}_{+}\times[0,b], \tag{2.12}$$

$$u(t,x)=\phi(t,x);\text{ if }(t,x)\in\tilde{J}:=[-\alpha,\infty)\times[-\beta,b]\backslash(0,\infty)\times(0,b], \tag{2.13}$$

$$\begin{cases}u(t,0)=\varphi(t);\ t\in[0,\infty),\\ u(0,x)=\psi(x);\ x\in[0,b],\\ \varphi(0)=\psi(0),\end{cases} \tag{2.14}$$

where $\alpha,\beta,b>0$, $\theta=(0,0)$, $^{c}D_{\theta}^{r}$ is the standard Caputo's fractional derivative of order $r=(r_1,r_2)\in(0,1]\times(0,1]$, $f,g:J\times C\to\mathbb{R}$ are given functions, $\phi:\tilde{J}\to\mathbb{R}^n$ is continuous and $\varphi:[0,\infty)\to\mathbb{R}$, $\psi:[0,b]\to\mathbb{R}$ are given absolutely continuous functions with $\varphi(t)=\phi(t,0)$ for each $t\in\mathbb{R}_+$, $\psi(x)=\phi(0,x)$ for each $x\in[0,b]$, and $C:=C([-\alpha,0]\times[-\beta,0])$ is the space of continuous functions on $[-\alpha,0]\times[-\beta,0]$ with the standard norm

$$\|u\|_C=\sup_{(t,x)\in[-\alpha,0]\times[-\beta,0]}|u(t,x)|.$$

If $u\in C:=C([-\alpha,\infty)\times[-\beta,b])$, then for any $(t,x)\in J$ define $u_{(t,x)}$ by

$$u_{(t,x)}(\tau,\xi)=u(t+\tau,x+\xi),\ \text{for }(\tau,\xi)\in[-\alpha,0]\times[-\beta,0].$$

This section initiates the question of global attractivity of the solution of problem (2.12)-(2.14). We present some existing results for the problem (2.12)-(2.14) using Krasnoselskii's fixed point theorem.

2.3.2. Main Results

Let us start by defining what we mean by a solution of the problem (2.12)-(2.14).

Definition 2.4. *A function $u\in BC$ with its mixed derivative D_{tx}^{2} exists and is integrable is said to be a solution of (2.12)-(2.14) if u satisfies equations (2.12) and (2.14) on J and the condition (2.13) on $\tilde{J}$.*

For each $\rho>0$ we consider the following set

$$A_\rho=\left\{u\in BC: u|_{\tilde{J}}=\phi\ \text{ and }\|u-\mu\|_{BC}\leq\frac{t^\rho}{1+t^\rho},\ \text{ for }t\in\mathbb{R}_+\right\}.$$

Theorem 2.5. *Assume that the following hypotheses are satisfied.*

(2.6.1) *The functions Φ and μ are in BC,*

(2.6.2) *The function* $(t,x) \mapsto f(t,x,u)$ *is measurable on* J *for each* $u \in BC$, *and the function* $u \mapsto f(x,y,u)$ *is continuous on* BC *for a.e.* $(t,x) \in J$,

(2.6.3) *There exists a continuous function* $p : J \to \mathbb{R}_+$ *such that*

$$|f(t,x,u)| \leq \frac{t^\rho p(t,x)\|u\|_{BC}}{1+t^\rho+\|u\|_{BC}};\ \text{for any } u \in A_\rho \text{ and for } (t,x) \in J.$$

Moreover, assume that

$$\lim_{t\to\infty} I_\theta^r p(t,x) = 0;\ x \in [0,b],$$

(2.6.4) *The function* g *is continuous and there exists a constant* $L > 0$ *such that, for any* $u,v \in A_\rho$ *and for* $(t,x) \in J$, *we have*

$$|g(t,x,u) - g(t,x,v)| \leq \frac{Lt^\rho\|u-v\|_{BC}}{(1+t^\rho+t^{2\rho})(1+\|u-v\|_{BC})}.$$

Moreover, assume that the function $(t,x) \to g(t,x,0)$ *is bounded on* J *with*

$$|g(t,x,0)| \leq \frac{Lt^\rho}{1+t^\rho}.$$

If

$$6L + p^* < 1, \tag{2.15}$$

where

$$p^* := \sup_{(t,x)\in\mathbb{R}_+\times[0,b]} I_\theta^r p(t),$$

then the IVP (2.12)-(2.14) has at least one solution on $[-\alpha,\infty)\times[-\beta,b]$. *Moreover, solutions of equation (2.12)-(2.14) are uniformly globally attractive.*

Proof. First, according to conditions (2.6.1) and (2.6.2), it is easy to obtain that f is Lebesgue measurable on J.

Let $\widetilde{\phi} \in A_\rho$ be defined as $\widetilde{\phi}|_{\tilde{J}} = \phi$, $\widetilde{\phi}(t,x) = \mu(t,x)$ for all $(t,x) \in J$. If u is a solution of the IVP (2.12)-(2.14), let $u(t,x) = \widetilde{\phi}(t,x) + v(t,x)$, $(t,x) \in [-\alpha,\infty)\times[-\beta,b]$, then we have $u_{(t,x)} = \widetilde{\phi}_{(t,x)} + v_{(t,x)}$, $(t,x) \in [-\alpha,\infty)\times[-\beta,b]$. Thus according to Lemma 2.2, the function v satisfies the equation

$$\begin{aligned} v(t,x) &= \frac{1}{\Gamma(r_1)\Gamma(r_2)}\int_0^t\int_0^x (t-s)^{r_1-1}(x-y)^{r_2-1} \\ &\times\ f(s,y,\widetilde{\phi}_{(s,y)}+v_{(s,y)})dyds \\ &+\ g(t,x,\widetilde{\phi}_{(t,x)}+v_{(t,x)}) - g(t,0,\widetilde{\phi}_{(t,0)}+v_{(t,0)}) \\ &-g(0,x,\widetilde{\phi}_{(0,x)}+v_{(0,x)}) + g(0,0,\widetilde{\phi}_{(0,0)}+v_{(0,0)});\ (t,x)\in J. \end{aligned} \tag{2.16}$$

Define E_ρ by, $E_\rho = \Big\{v \in BC : v(t,x) = 0,\ \text{for } (t,x) \in \tilde{J}$

$$\text{and } \|v\|_{BC} \leq \frac{t^\rho}{1+t^\rho},\ \text{for each } t \in \mathbb{R}_+\Big\}.$$

Then E_ρ is a closed bounded and convex subset of BC.

Remark 2.6. *From* (2.6.4), *for each* $(t,x) \in \mathbb{R}_+ \times [0,b]$ *and* $v \in E_\rho$, *we get*

$$|g(0,x,\widetilde{\phi}_{(0,x)}+v_{(0,x)}) - g(0,0,\widetilde{\phi}_{(0,0)}+v_{(0,0)})| \leq \frac{2Lt^\rho}{1+t^\rho}.$$

Consider the operators $A : BC \to BC,\ B : E_\rho \to BC$ defined by,

$$A(v)(t,x) = \begin{cases} 0, & (t,x) \in \tilde{J}, \\ g(t,x,\widetilde{\phi}_{(t,x)}+v_{(t,x)}) - g(t,0,\widetilde{\phi}_{(t,0)}+v_{(t,0)}) & \\ -g(0,x,\widetilde{\phi}_{(0,x)}+v_{(0,x)}) + g(0,0,\widetilde{\phi}_{(0,0)}+v_{(0,0)}) & (t,x) \in J, \end{cases}$$

and

$$B(v)(t,x) = \begin{cases} 0, & (t,x) \in \tilde{J}, \\ \frac{1}{\Gamma(r_1)\Gamma(r_2)} \int_0^t \int_0^x (t-s)^{r_1-1}(x-y)^{r_2-1} & \\ \times f(s,y,\widetilde{\phi}_{(s,y)}+v_{(s,y)})dyds & (t,x) \in J. \end{cases}$$

It is easy to see that,the operator equation

$$v = A(v) + B(v), \tag{2.17}$$

has a solution $v \in E_\rho$ if and only if v is a solution of equation (2.16). Thus $u(t,x) = \widetilde{\phi}(t,x) + v(t,x)$, is a solution of (2.12)-(2.14) on J. Therefore, the existence of a solution of the IVP (2.12)-(2.14) is equivalent that equation (2.17) has a fixed point in E_ρ. Now we show that $A+B$ has a fixed point in E_ρ. The proof is divided into three steps.

Step 1: *A is a contraction on BC.*
For any $v, \overline{v} \in BC$ and each $(t,x) \in \mathbb{R}_+ \times [0,b]$, by (2.6.4), we get

$$\begin{aligned} |(Av)(t,x) - (A\overline{v})(t,x)| &\leq |g(t,x,\widetilde{\phi}_{(t,x)}+v_{(t,x)}) - g(t,x,\widetilde{\phi}_{(t,x)}+\overline{v}_{(t,x)})| \\ &+ |g(t,0,\widetilde{\phi}_{(t,0)}+v_{(t,0)}) - g(t,0,\widetilde{\phi}_{(t,0)}+\overline{v}_{(t,0)})| \\ &+ |g(0,x,\widetilde{\phi}_{(0,x)}+v_{(0,x)}) - g(0,x,\widetilde{\phi}_{(0,x)}+\overline{v}_{(0,x)})| \\ &+ |g(0,0,\widetilde{\phi}_{(0,0)}+v_{(0,0)}) - g(0,0,\widetilde{\phi}_{(0,0)}+\overline{v}_{(0,0)})| \\ &\leq \frac{4Lt^\rho}{1+t^\rho}\|v-\overline{v}\|_{BC}, \end{aligned}$$

which implies that

$$\|A(v) - A(\overline{v})\|_{BC} \leq 4L\|v-\overline{v}\|_{BC}.$$

In view of (2.15), A is a contraction on BC.

Step 2: *$B(E_\rho)$ resides in a compact subset of E_ρ.*
The proof is divided into several Claims.

Claim 1: *B maps E_ρ in E_ρ.*

For any $(t,x)\in\mathbb{R}_+\times[0,b]$, and $v\in E_\rho$ we have

$$\begin{aligned}|(Bv)(t,x)| &\leq \frac{1}{\Gamma(r_1)\Gamma(r_2)}\int_0^t\int_0^x (t-s)^{r_1-1}(x-y)^{r_2-1}\\ &\times |f(s,y,\widetilde{\phi}_{(s,y)}+v_{(s,y)})|dyds\\ &\leq \frac{1}{\Gamma(r_1)\Gamma(r_2)}\int_0^t\int_0^x (t-s)^{r_1-1}(x-y)^{r_2-1}\frac{s^\rho p(s,y)}{1+s^\rho}dyds\\ &\leq \frac{t^\rho}{1+t^\rho}I_\theta^r p(t,x)\\ &\leq \frac{t^\rho p^*}{1+t^\rho}.\end{aligned}$$

By (2.15), we get

$$|(Bv)(t,x)|\leq\frac{t^\rho}{1+t^\rho},$$

which implies that $B(E_\rho)\subset E_\rho$.

Claim 2: *B is continuous on E_ρ.*

Let $\{v_n\}_{n\in\mathbb{N}}$ be a sequence such that $v_n\to u$ in E_ρ. Then, for each $(t,x)\in\mathbb{R}_+\times[0,b]$, we have

$$\begin{aligned}|(Bv_n)(t,x)-(Bv)(t,x)\| &\leq \frac{1}{\Gamma(r_1)\Gamma(r_2)}\int_0^t\int_0^x (t-s)^{r_1-1}(x-y)^{r_2-1}\\ &\times|f(s,y,\widetilde{\phi}_{(s,y)}+v_{n(s,y)})-f(s,y,\widetilde{\phi}_{(s,y)}+v_{(s,y)})|dyds. \qquad (2.18)\end{aligned}$$

Case 1. If $(t,x)\in[0,a]\times[0,b]$; $a>0$, then, since $v_n\to v$ as $n\to\infty$ and f is continuous, (2.18) gives

$$\|B(v_n)-B(v)\|_{BC}\to 0 \quad \text{as } n\to\infty.$$

Case 2. If $(t,x)\in(a,\infty)\times[0,b]$; $a>0$, then from (2.6.3) and (2.18), for each $(x,y)\in\mathbb{R}_+\times[0,b]$ we get

$$|(Bv_n)(x,y)-(Bv)(x,y)|\leq\frac{2t^\rho}{1+t^\rho}I_\theta^r p(t,x). \qquad (2.19)$$

Since $u_n\to u$ as $n\to\infty$ and $t\to\infty$, then (2.19) gives

$$\|B(v_n)-B(v)\|_{BC}\to 0 \quad \text{as } n\to\infty.$$

Claim 3: *$B(E_\rho)$ is uniformly bounded.*

This is clear since $B(E_\rho)\subset E_\rho$ and E_ρ is bounded.

Claim 4: *$B(E_\rho)$ is equicontinuous on every compact $[0,a]\times[0,b]$; $a>0$ of $[-\alpha,\infty)\times[-\beta,b]$.*

Let $(t_1,x_1),(t_2,x_2)\in[0,a]\times[0,b]$, $t_1<t_2$, $x_1<x_2$ and let $v\in B_\eta$, thus we have

$$\begin{aligned}&|(Bv)(t_2,x_2)-(Bv)(t_1,x_1)|\leq\\ &+\frac{1}{\Gamma(r_1)\Gamma(r_2)}\int_0^{t_1}\int_0^{x_1}[(t_2-s)^{r_1-1}(x_2-y)^{r_2-1}-(t_1-s)^{r_1-1}(x_1-y)^{r_2-1}]\\ &\times p(s,y)dyds+\frac{1}{\Gamma(r_1)\Gamma(r_2)}\int_{t_1}^{t_2}\int_{x_1}^{x_2}(t_2-s)^{r_1-1}(x_2-y)^{r_2-1}p(s,y)dyds\end{aligned}$$

$$
\begin{aligned}
&+ \quad \frac{1}{\Gamma(r_1)\Gamma(r_2)}\int_0^{t_1}\int_{x_1}^{x_2}(t_2-s)^{r_1-1}(x_2-y)^{r_2-1}p(s,y)dyds \\
&+ \quad \frac{1}{\Gamma(r_1)\Gamma(r_2)}\int_{t_1}^{t_2}\int_0^{x_1}(t_2-\tau)^{r_1-1}(x_2-s)^{r_2-1}p(s,y)dyds.
\end{aligned}
$$

From continuity of p and as $t_1 \longrightarrow t_2$ and $x_1 \longrightarrow x_2$, the right-hand side of the above inequality tends to zero. The equicontinuity for the cases $t_1 < t_2 < 0$, $x_1 < x_2 < 0$ and $t_1 \leq 0 \leq t_2$, $x_1 \leq 0 \leq x_2$ is obvious.

Claim 5: *$B(E_\rho)$ is equiconvergent.*
Let $(t,x) \in \mathbb{R}_+ \times [0,b]$ and $v \in E_\rho$, then we have

$$|(Bv)(t,x)| \leq \frac{t^\rho}{1+t^\rho} I_\theta^r p(t,x).$$

Thus, for each $x \in [0,b]$, we get

$$|(Bv)(t,x)| \longrightarrow 0, \text{ as } t \longrightarrow +\infty.$$

As a consequence of Claims 1 to 5 together with the Lemma 1.69, we can conclude that $B(E_\rho)$ resides in a compact subset of E_ρ.

Step 3: *For $v \in E_\rho$ and for all $u \in BC$, $u = Au + Bv \Rightarrow u \in E_\rho$ holds.*
If $u = Au + Bv$, apply our conditions and Remark 13.36, we obtain that

$$
\begin{aligned}
|u(t,x)| &\leq |(Au)(t,x)| + |(Bv)(t,x)| \\
&\leq |g(t,x,\widetilde{\phi}_{(t,x)} + v_{(t,x)}) - g(t,x,0)| \\
&+ |g(t,0,0) - g(t,0,\widetilde{\phi}_{(t,0)} + v_{(t,0)})| \\
&+ |g(0,x,\widetilde{\phi}_{(0,x)} + v_{(0,x)}) - g(0,0,\widetilde{\phi}_{(0,0)} + v_{(0,0)})| \\
&+ |g(t,x,0)| + |g(t,0,0)| \\
&+ \frac{1}{\Gamma(r_1)\Gamma(r_2)}\int_0^t\int_0^x (t-s)^{r_1-1}(x-y)^{r_2-1} \\
&\times |f(s,y,\widetilde{\phi}_{(s,y)} + u_{(s,y)})|dyds \\
&\leq \frac{6Lt^\rho}{1+t^\rho} + \frac{t^\rho}{1+t^\rho} I_\theta^r p(t,x) \\
&\leq \frac{(6L+p^*)t^\rho}{1+t^\rho}.
\end{aligned}
$$

By (2.15), we get

$$|u(t,x)| \leq \frac{t^\rho}{1+t^\rho},$$

which implies that $u \in E_\rho$. From an application Theorem 1.61, we deduce that $A+B$ has a fixed point u which is a solution of the problem (2.12)-(2.14).

Finally, we show the uniform global attractivity of the solutions of the problem (2.12)-(2.14) on $\mathbb{R}_+ \times [0,b]$. Let u and v be any two solutions of problem (2.12)-(2.14) in E_ρ. Then for each $(x,y) \in \mathbb{R}_+ \times [0,b]$ we have

$$\begin{aligned} |u(t,x) - v(t,x)| &\leq |(Au)(t,x) - (Av)(t,x)| + |(Bu)(t,x) - (Bv)(t,x)| \\ &\leq \frac{4L}{1+t^\rho} + \frac{2t^\rho}{1+t^\rho} I_\theta^r p(t,x). \end{aligned} \tag{2.20}$$

By using (2.20), we deduce that, for each $x \in [0,b]$, we get

$$\lim_{t\to\infty}(u(t,x) - v(t,x)) = 0.$$

Consequently, problem (2.12)-(2.14) has at least one solution and all these solutions are uniformly globally attractive. □

2.3.3. An Example

As an application of our results we consider a problem for a fractional order neutral differential equation of the form,

$${}^cD_\theta^r(u(t,x) - g(t,x,u_{(t,x)})) = f(t,x,u_{(t,x)}); \text{ if } (t,x) \in J := [0,1] \times [0,\infty), \tag{2.21}$$

$$u(t,0) = \frac{t}{1+t^2},\ t \in [0,\infty),\ u(0,x) = x^2;\ x \in [0,1], \tag{2.22}$$

$$u(t,x) = x^2 + \frac{t}{1+t^2},\ (t,x) \in [-1,\infty) \times [-2,1] \backslash (0,\infty) \times (0,1], \tag{2.23}$$

where $r = (r_1, r_2) = (\frac{1}{4}, \frac{1}{2})$,

$$\begin{cases} f(t,x,u_{(t,x)}) = \dfrac{xt^{\frac{5}{4}} \sin t}{c(1+t^{\frac{-1}{2}})(1+t^2+|u(t-1,x-2)|)}; \\ \qquad (t,x) \in (0,\infty) \times [0,1], \\ f(t,x,u_{(t,x)}) = 0; \qquad (t,x) \in \{0\} \times [0,1], \end{cases}$$

$c = \frac{32}{3\sqrt{\pi}}$, and

$$\begin{cases} g(t,x,u_{(t,x)}) = \dfrac{t^2}{3e^{x+2}(1+t^2+t^4)(1+|u(t-1,x-2)|)}; \\ \qquad (t,x) \in (0,\infty) \times [0,1], \\ g(t,x,u_{(t,x)}) = 0; \qquad (t,x) \in \{0\} \times [0,1], \end{cases}$$

First, we can see that condition (2.6.1) is satisfied with

$$\Phi(t,x) = \mu(t,x) = x^2 + \frac{t}{1+t^2}.$$

It is clear that conditions (2.6.2) and (2.6.3) hold with

$$\begin{cases} p(t,x) = \dfrac{xt^{\frac{-3}{4}}}{c(1+t^{\frac{-1}{2}})}|\sin t|; \ (t,x) \in (0,\infty)\times[0,1], \\ p(0,x) = 0; \qquad\qquad\qquad\qquad x\in[0,1]. \end{cases}$$

For each $(t,x)\in[0,\infty)\times[0,1]$ we have

$$|g(t,x,u_{(t,x)}) - g(t,x,\overline{u}_{(t,x)})| \le \frac{1}{3e^2}\frac{t^2\|u-\overline{u}\|_{BC}}{(1+t^2+t^4)(1+\|u-\overline{u}\|_{BC})}.$$

Hence condition (2.6.4) is satisfied with $L = \dfrac{1}{3e^2}$. Next, we notice that the function p satisfies (2.15). Indeed, for each $x\in[0,1]$, we get

$$\begin{aligned} I_\theta^r p(t,x) &= \frac{1}{\Gamma(r_1)\Gamma(r_2)}\int_0^t\int_0^x (t-\tau)^{r_1-1}(x-s)^{r_2-1}p(\tau,s)dsd\tau \\ &= \frac{1}{\Gamma(\frac{1}{4})\Gamma(\frac{1}{2})}\int_0^t\int_0^x \frac{(t-\tau)^{\frac{-3}{4}}(x-s)^{\frac{-1}{2}}s\tau^{\frac{-3}{4}}}{c(1+t^{\frac{-1}{2}})}|\sin\tau|dsd\tau \\ &\le \frac{1}{\Gamma(\frac{1}{4})\Gamma(\frac{1}{2})}\int_0^t\int_0^x \frac{(t-\tau)^{\frac{-3}{4}}(x-s)^{\frac{-1}{2}}s\tau^{\frac{-3}{4}}}{c(1+t^{\frac{-1}{2}})}dsd\tau \\ &= \frac{t^{\frac{-1}{2}}x^{\frac{3}{2}}}{1+t^{\frac{-1}{2}}} \longrightarrow 0 \ as\ t\to\infty. \end{aligned}$$

Thus $p^* \le \frac{1}{2}$. Hence

$$6L+p^* \le \frac{2}{e^2}+\frac{1}{2} < 1.$$

Consequently, Theorem 2.5 implies that problem (2.21)-(2.23) has at least one solution defined on $[-1,\infty)\times[-2,1]$, and all solutions are uniformly globally attractive.

2.4. Asymptotic Stability in Nonlinear Delay Differential Equations

2.4.1. Introduction

In this section we establish sufficient conditions for the existence and the stability of solutions of the following system of nonlinear delay differential equations of fractional order of the form

$$\begin{aligned} {}^cD_\theta^r u(t,x) &= p(t,x)u(t,x)\,[f(t,x,u(t,x)) \\ &- g(t,x,u(t-\tau_1,x-\xi_1),\dots,u(t-\tau_m,x-\xi_m))]; \quad \text{for } (t,x)\in J, \end{aligned} \tag{2.24}$$

$$u(t,x) = \Phi(t,x); \quad \text{for } (t,x)\in\tilde{J} := [-T,\infty)\times[-\xi,b]\backslash(0,\infty)\times(0,b], \tag{2.25}$$

$$\begin{cases} u(t,0) = \varphi(t); & t\in[0,\infty), \\ u(0,x) = \psi(x); & x\in[0,b], \end{cases} \tag{2.26}$$

where $J=\mathbb{R}_+\times[0,b]$, $b>0$, $\theta=(0,0)$, $\mathbb{R}_+=[0,\infty)$, $\tau_i,\xi_i\geq 0$; $i=1\ldots,m$, $T=\max_{i=1\ldots,m}\{\tau_i\}$, $\xi=\max_{i=1\ldots,m}\{\xi_i\}$, ${}^cD^r_\theta$ is the Caputo fractional derivative of order $r=(r_1,r_2)\in(0,1]\times(0,1]$, $p:J\to\mathbb{R}$, $f:J\times\mathbb{R}\to\mathbb{R}$, $g:J\times\mathbb{R}^m\to\mathbb{R}$ are given continuous functions, $\varphi:\mathbb{R}_+\to\mathbb{R}$, $\psi:[0,b]\to\mathbb{R}$ are absolutely continuous functions with $\lim_{t\to\infty}\varphi(t)=0$, and $\psi(x)=\varphi(0)$ for each $x\in[0,b]$, and $\Phi:\tilde{J}\to\mathbb{R}^n$ is continuous with $\varphi(t)=\Phi(t,0)$ for each $t\in\mathbb{R}_+$, and $\psi(x)=\Phi(0,x)$ for each $x\in[0,b]$.

2.4.2. Main Results

Let us start by defining what we mean by a solution of of the problem (2.24)-(2.26).

Definition 2.7. *A function $u\in BC$ is said to be a solution of (2.24)-(2.26) if u satisfies equation (2.24) on J, equation (2.25) on $\tilde{J}$ and condition (2.26) is satisfied.*

By $BC:=BC([-T,\infty)\times[-\xi,b])$ we denote the Banach space of all bounded and continuous functions from $[-T,\infty)\times[-\xi,b]$ into $\mathbb{R}$ equipped with the standard norm

$$\|u\|_{BC}=\sup_{(t,x)\in[-T,\infty)\times[-\xi,b]}|u(t,x)|.$$

The following hypotheses will be used in the sequel:

(2.9.1) The function Φ is in BC and p is continuous and bounded on J.

(2.9.2) There exist a nondecreasing function $\Lambda:\mathbb{R}_+\to\mathbb{R}_+$ and a continuous function $d:\mathbb{R}_+\times[0,b]\to\mathbb{R}_+$ such that

$$|f(t,x,u)|\leq d(t,x)\Lambda(|u|),\ \text{for } (t,x)\in J \text{ and } u\in\mathbb{R}.$$

Moreover, assume that

$$\lim_{t\to\infty}I^r_\theta d(t,x)=0;\ x\in[0,b],$$

(2.9.3) There exist continuous functions $q_i:\mathbb{R}_+\times[0,b]\to\mathbb{R}_+$ such that

$$\left(1+\sum_{i=1}^m|u_i|\right)|g(t,x,u_1,u_2,\ldots,u_m)|\leq\sum_{i=1}^m|u_i|q_i(t,x);$$

for $(t,x)\in\mathbb{R}_+\times[0,b]$ and for $u_i\in\mathbb{R}$; $i=1\ldots m$. Moreover, assume that

$$\lim_{t\to\infty}I^r_\theta q_i(t,x)=0;\ x\in[0,b];\ i=1\ldots m.$$

Remark 2.8. *Set*

$$\Phi^*:=\sup_{(t,x)\in\tilde{J}}\Phi(t,x),\ \varphi^*:=\sup_{t\in\mathbb{R}_+}\varphi(t),\ p^*:=\sup_{(t,x)\in\mathbb{R}_+\times[0,b]}p(t,x),$$

$$d^*:=\sup_{(t,x)\in\mathbb{R}_+\times[0,b]}I^r_\theta d(t,x)\ \text{ and }\ q_i^*:=\sup_{(t,x)\in\mathbb{R}_+\times[0,b]}I^r_\theta q_i(t,x);\ i=1\ldots m.$$

From the hypotheses, we infer that Φ^*, φ^*, p^*, d^* and q_i^*; $i=1\ldots m$ are finite.

Now, we shall prove the following theorem concerning the existence and the stability of a solution of problem (2.24)-(2.26).

Theorem 2.9. *Assume that* $(2.9.1)-(2.9.3)$ *and the following hypothesis hold*

$(2.11.1)$ *There exists a constant* $\eta > 0$, *such that*

$$\max\left\{\Phi^*, \varphi^* + p^*\eta\left(d^*\Lambda(\eta) + \sum_{i=1}^{m} q_i^*\right)\right\} \leq \eta.$$

Then the problem (2.24)-(2.26) has at least one solution in the space BC. Moreover, if there exists a constant $\eta^* > 0$, *such that*

$$p^*d^*\Lambda(\eta+\eta^*) + p^*\sum_{i=1}^{m} q_i^* < 1,$$

and

$$2p^*\eta\left(d^*\Lambda(\eta+\eta^*) + \sum_{i=1}^{m} q_i^*\right) \leq \eta^*\left(1 - p^*d^*\Lambda(\eta+\eta^*) - p^*\sum_{i=1}^{m} q_i^*\right),$$

then solutions of problem (2.24)-(2.26) are locally asymptotically stable.

Proof. Let us define the operator N such that, for any $u \in BC$,

$$(Nu)(t,x) = \begin{cases} \Phi(t,x); & (t,x) \in \tilde{J}, \\ \varphi(t) + I_\theta^r[p(t,x)u(t,x)\,(f(t,x,u(t,x)) & \\ -g(t,x,u(t-\tau_1,x-\xi_1),\ldots, & \\ u(t-\tau_m,x-\xi_m)))]; & (t,x) \in J. \end{cases} \tag{2.27}$$

The operator N maps BC into BC. Indeed the map $N(u)$ is continuous on $[-T,\infty)\times[-\xi,b]$ for any $u \in BC$, and for each $(t,x) \in J$ we have

$$\begin{aligned}
|(Nu)(t,x)| &\leq |\varphi(t)| + I_\theta^r|p(t,x)u(t,x)(f(t,x,u(t,x)) \\
&- g(t,x,u(t-\tau_1,x-\xi_1),\ldots,u(t-\tau_m,x-\xi_m)))| \\
&\leq |\varphi(t)| + \frac{1}{\Gamma(r_1)\Gamma(r_2)}\int_0^t\int_0^x (t-\tau)^{r_1-1}(x-s)^{r_1} \\
&\times |p(t,x)||u(t,x)|\left[|d(t,x)|\Lambda(|u(t,x)|)\right. \\
&+ \left(\sum_{i=1}^{m}|u(\tau-\tau_i,s-\xi_i)|q_i(\tau,s)\right) \\
&\times \left.\left(1+\sum_{i=1}^{m}|u(\tau-\tau_i,s-\xi_i)|\right)^{-1} dsd\tau\right] \\
&\leq \varphi^* + p^*\|u\|_{BC}\left(d^*\Lambda(\|u\|_{BC}) + \sum_{i=1}^{m} q_i^*\right),
\end{aligned}$$

and for $(t,x) \in \tilde{J}$, we have

$$|(Nu)(t,x)| = |\Phi(t,x)| \leq \Phi^*.$$

Thus,

$$\|N(u)\|_{BC} \leq \max\left\{\Phi^*, \varphi^* + p^*\|u\|_{BC}\left(d^*\Lambda(\|u\|_{BC}) + \sum_{i=1}^{m} q_i^*\right)\right\}. \tag{2.28}$$

Hence, $N(u) \in BC$. This proves that the operator N maps BC into itself.
By Lemma 2.2, the problem of finding the solutions of the problem (2.24)-(2.26) is reduced to finding the solutions of the operator equation $N(u) = u$. Equation (2.28) and hypothesis (2.12.1) implies that N transforms the ball $B_\eta := B(0,\eta)$ into itself. We shall show that $N : B_\eta \to B_\eta$ satisfies the assumptions of Schauder's fixed point theorem [204]. The proof will be given in several steps.

Step 1: *N is continuous.*
Let $\{u_n\}_{n\in\mathbb{N}}$ be a sequence such that $u_n \to u$ in B_η. Then, for each $(t,x) \in [-T,\infty) \times [-\xi, b]$, we have

$$\begin{aligned}
& |(Nu_n)(t,x) - (Nu)(t,x)| \\
& \leq I_\theta^r[|p(t,x)|\,|u_n(t,x)\,(f(t,x,u_n(t,x)) \\
- & \quad g(t,x,u_n(t-\tau_1,x-\xi_1),\dots,u_n(t-\tau_m,x-\xi_m)) \\
- & \quad u(t,x)\,(f(t,x,u(t,x)) - g(t,x,u(t-\tau_1,x-\xi_1),\dots,u(t-\tau_m,x-\xi_m)))|] \\
& \leq I_\theta^r[|p(t,x)||u_n(t,x) - u(t,x)||f(t,x,u_n(t,x)) \\
- & \quad g(t,x,u_n(t-\tau_1,x-\xi_1),\dots,u_n(t-\tau_m,x-\xi_m))|] \\
+ & \quad I_\theta^r[|p(t,x)u(t,x)|\,|f(t,x,u_n(t,x)) - f(t,x,u(t,x)) \\
- & \quad g(t,x,u_n(t-\tau_1,x-\xi_1),\dots,u_n(t-\tau_m,x-\xi_m)) \\
+ & \quad g(t,x,u(t-\tau_1,x-\xi_1),\dots,u(t-\tau_m,x-\xi_m))|] \\
& \leq p^* I_\theta^r[|u_n(t,x) - u(t,x)||f(t,x,u_n(t,x)) \\
- & \quad g(t,x,u_n(t-\tau_1,x-\xi_1),\dots,u_n(t-\tau_m,x-\xi_m))|] \\
+ & \quad p^*\eta I_\theta^r|f(t,x,u_n(t,x)) - f(t,x,u(t,x))| \\
+ & \quad p^*\eta I_\theta^r|g(t,x,u_n(t-\tau_1,x-\xi_1),\dots,u_n(t-\tau_m,x-\xi_m)) \\
- & \quad g(t,x,u(t-\tau_1,x-\xi_1),\dots,u(t-\tau_m,x-\xi_m))| \\
& \leq p^*\left(d^*\Lambda(\eta) + \sum_{i=1}^{m} q_i^*\right)\|u_n - u\|_{BC} \\
+ & \quad p^*\eta I_\theta^r|f(t,x,u_n(t,x)) - f(t,x,u(t,x))| \\
+ & \quad p^*\eta I_\theta^r|g(t,x,u_n(t-\tau_1,x-\xi_1),\dots,u_n(t-\tau_m,x-\xi_m)) \\
& \qquad -g(t,x,u(t-\tau_1,x-\xi_1),\dots,u(t-\tau_m,x-\xi_m)).
\end{aligned} \tag{2.29}$$

Case 1. If $(t,x) \in \tilde{J} \cup ([0,a] \times [0,b])$; $a > 0$, then, since $u_n \to u$ as $n \to \infty$ and f, g are continuous, (2.29) gives

$$\|N(u_n) - N(u)\|_{BC} \to 0 \quad \text{as } n \to \infty.$$

Case 2. If $(t,x) \in (a,\infty) \times [0,b]$; $a > 0$, then from our hypotheses and (2.29), we get

$$|(Nu_n)(t,x) - (Nu)(t,x)| \leq p^* \left(d^*\Lambda(\eta) + \sum_{i=1}^{m} q_i^* \right) \|u_n - u\|_{BC}$$

$$+2p^*\eta\Lambda(\eta) I_\theta^r d(t,x) + 2p^*\eta \sum_{i=1}^{m} (I_\theta^r q_i(t,x)). \tag{2.30}$$

Since $u_n \to u$ as $n \to \infty$ and $t \to \infty$, then (2.30) gives

$$\|N(u_n) - N(u)\|_{BC} \to 0 \quad \text{as } n \to \infty.$$

Step 2: *$N(B_\eta)$ is uniformly bounded.*
This is clear since $N(B_\eta) \subset B_\eta$ and B_η is bounded.

Step 3: *$N(B_\eta)$ is equicontinuous on every compact subset $[-T,a] \times [-\xi,b]$ of $[-T,\infty) \times [-\xi,b]$, $a > 0$.*
Let $(t_1,x_1),(t_2,x_2) \in [0,a] \times [0,b]$, $t_1 < t_2$, $x_1 < x_2$ and let $u \in B_\eta$. Thus we have

$$\begin{aligned}
& |(Nu)(t_2,x_2) - (Nu)(t_1,x_1)| \leq |\varphi(t_2) - \varphi(t_1)| \\
+ \; & I_\theta^r [(p(t_2,x_2)|u(t_2,x_2)| - p(t_1,x_1)|u(t_1,x_1)|)\,(f(t_2,x_2,u(t_2,x_2)) \\
- \; & g(t_2,x_2,u(t_2-\tau_1,x_2-\xi_1),\ldots,u(t_2-\tau_m,x_2-\xi_m)))] \\
+ \; & I_\theta^r [p(t_1,x_1)|u(t_1,x_1)||f(t_2,x_2,u(t_2,x_2)) - f(t_1,x_1,u(t_1,x_1))| \\
+ \; & |g(t_2,x_2,u(t_2-\tau_1,x_2-\xi_1),\ldots,u(t_2-\tau_m,x_2-\xi_m)) \\
- \; & g(t_1,x_1,u(t_1-\tau_1,x_1-\xi_1),\ldots,u(t_1-\tau_m,x_1-\xi_m))|)].
\end{aligned}$$

Thus

$$\begin{aligned}
& |(Nu)(t_2,x_2) - (Nu)(t_1,x_1)| \leq |\varphi(t_2) - \varphi(t_1)| \\
+ \; & I_\theta^r \left[\eta |p(t_2,x_2) - p(t_1,x_1)| \left(d^*\Lambda(\eta) + \sum_{i=1}^{m} q_i(t_2,x_2) \right) \right] \\
+ \; & p^*\eta I_\theta^r [|f(t_2,x_2,u(t_2,x_2)) - f(t_1,x_1,u(t_1,x_1))| \\
+ \; & |g(t_2,x_2,u(t_2-\tau_1,x_2-\xi_1),\ldots,u(t_2-\tau_m,x_2-\xi_m)) \\
- \; & g(t_1,x_1,u(t_1-\tau_1,x_1-\xi_1),\ldots,u(t_1-\tau_m,x_1-\xi_m))|)].
\end{aligned}$$

From continuity of φ, p, f, g and as $t_1 \to t_2$ and $x_1 \to x_2$, the right-hand side of the above inequality tends to zero. The equicontinuity for the cases $t_1 < t_2 < 0$, $x_1 < x_2 < 0$ and $t_1 \leq 0 \leq t_2$, $x_1 \leq 0 \leq x_2$ is obvious.

Step 4: *$N(B_\eta)$ is equiconvergent.*
Let $(t,x)\in\mathbb{R}_+\times[0,b]$ and $u\in B_\eta$, then we have

$$\begin{aligned}|(Nu)(t,x)| &\leq |\varphi(t)|+\frac{p^*\eta\Lambda(\eta)}{\Gamma(r_1)\Gamma(r_2)}\int_0^t\int_0^x(t-\tau)^{r_1-1}(x-s)^{r_2-1}d(\tau,s)dsd\tau\\ &+ \frac{p^*\eta}{\Gamma(r_1)\Gamma(r_2)}\sum_{i=0}^m\left(\int_0^t\int_0^x(t-\tau)^{r_1-1}(x-s)^{r_2-1}q_i(\tau,s)dsd\tau\right)\\ &\leq |\varphi(t)|+p^*\eta\Lambda(\eta)I_\theta^r d(t,x)+p^*\eta\sum_{i=0}^m I_\theta^r p_i(t,x).\end{aligned}$$

Thus, for each $x\in[0,b]$, we get

$$|(Nu)(t,x)|\to 0,\ as\, t\to+\infty.$$

Hence,

$$|(Nu)(t,x)-(Nu)(+\infty,x)|\to 0,\ as\, t\to+\infty.$$

As a consequence of Steps 1 to 4 together with the Lemma 1.69, we can conclude that $N:B_\eta\to B_\eta$ is continuous and compact. From an application of Schauder's theorem [204], we deduce that N has a fixed point u which is a solution of the problem (2.24)-(2.26).

Step 5: *The uniform local attractivity for solutions.*
Let us assume that u_0 is a solution of problem (2.24)-(2.26) with the conditions of this theorem. Taking $u\in B(u_0,\eta^*)$, we have

$$\begin{aligned}|(Nu)(t,x)-u_0(t,x)| &= |(Nu)(t,x)-(Nu_0)(t,x)|\\ &\leq p^*I_\theta^r[|u(t,x)-u_0(t,x)||f(t,x,u(t,x))\\ &- g(t,x,u(t-\tau_1,x-\xi_1),\dots,u(t-\tau_m,x-\xi_m))|]\\ &+ p^*\eta I_\theta^r|f(t,x,u(t,x))-f(t,x,u_0(t,x))|\\ &+ p^*\eta I_\theta^r|g(t,x,u(t-\tau_1,x-\xi_1),\dots,u(t-\tau_m,x-\xi_m))\\ &- g(t,x,u_0(t-\tau_1,x-\xi_1),\dots,u_0(t-\tau_m,x-\xi_m))|\\ &\leq p^*(2\eta+\eta^*)\left(d^*\Lambda(\eta+\eta^*)+\sum_{i=1}^m q_i^*\right)\\ &\leq \eta^*.\end{aligned}$$

Thus we observe that N is a continuous function such that

$$N(B(u_0,\eta^*))\subset B(u_0,\eta^*).$$

Moreover, if u is a solution of problem (2.24)-(2.26), then

$$\begin{aligned}|u(t,x)-u_0(t,x)| &= |(Nu)(t,x)-(Nu_0)(t,x)|\\ &\leq p^*\left(\Lambda(\eta+\eta^*)I_\theta^r d(t,x)+\sum_{i=1}^m(I_\theta^r q_i(t,x))\right)\|u-u_0\|_{BC}\\ &+ 2p^*\eta\Lambda(\eta+\eta^*)I_\theta^r d(t,x)+2p^*\eta\sum_{i=1}^m(I_\theta^r q_i(t,x)).\end{aligned}$$

Thus

$$|u(t,x)-u_0(t,x)| \le p^*(2\eta+\eta^*)\left(\Lambda(\eta+\eta^*)I_\theta^r d(t,x)+\sum_{i=1}^{m}(I_\theta^r q_i(t,x))\right). \tag{2.31}$$

By using (2.31) and the fact that $\lim_{n\to\infty} I_\theta^r d(t,x) = \lim_{n\to\infty} I_\theta^r q_i(t,x) = 0;\ i = 0\ldots m$ we deduce that

$$\lim_{t\to\infty}|u(t,x)-u_0(t,x)| = 0.$$

Consequently, all solutions of problem (2.24)-(2.26) are locally asymptotically stable. □

2.4.3. An Example

As an application of our results we consider the following system of Lotka-Volterra type model of nonlinear delay differential equations of fractional order of the form

$$^cD_\theta^r u(t,x) = p(t,x)u(t,x)\left[f(t,x,u(t,x)) - g\left(t,x,u\left(t-1,x-\frac{1}{4}\right),\right.\right.$$

$$\left.\left.u\left(t-\frac{2}{3},x-\frac{1}{5}\right)\right)\right];\quad \text{for } (t,x)\in J := \mathbb{R}_+\times[0,1], \tag{2.32}$$

$$u(t,x) = \frac{1}{1+t^2};\quad \text{for } (t,x)\in \tilde{J} := [-1,\infty)\times[-\frac{1}{4},1]\backslash(0,\infty)\times(0,1], \tag{2.33}$$

$$\begin{cases} u(t,0) = \dfrac{1}{1+t^2}; & t\in[0,\infty),\\ u(0,x) = 1; & x\in[0,1], \end{cases} \tag{2.34}$$

where $(r_1,r_2) = (\frac{1}{4},\frac{1}{2})$, $p(t,x) = \frac{1}{1+t^2+x^2}$,

$$\begin{cases} f(t,x,u) = \dfrac{3cxt^{\frac{-1}{4}}|u|\sin t}{64(1+\sqrt{t})}; & (t,x)\in(0,\infty)\times[0,1],\ u\in\mathbb{R},\\ f(0,x,u) = 0; & x\in[0,1],\ u\in\mathbb{R}, \end{cases}$$

$c = \frac{3\sqrt{\pi}}{16}$, and

$$\begin{cases} g(t,x,u,v) = \dfrac{cxt^{\frac{-1}{4}}\left(|u|\sin t+|v|e^{-\frac{1}{t}}\right)}{4(1+\sqrt{t})(2+|u|+|v|)}; & (t,x)\in(0,\infty)\times[0,1]\ and\ u,v\in\mathbb{R},\\ g(0,x,u,v) = 0; & x\in[0,1]\ and\ u,v\in\mathbb{R}. \end{cases}$$

First, we can see that the hypothesis (2.9.1) is satisfied with $\Phi^* = \varphi^* = p^* = 1$. The function f is continuous and satisfies assumption (2.9.2), witch $\Lambda(w) = w$ and

$$\begin{cases} d(t,x) = \dfrac{3cxt^{\frac{-1}{4}}|\sin t|}{64(1+\sqrt{t})}; & (t,x)\in(0,\infty)\times[0,1],\\ d(0,x) = 0; & x\in[0,1]. \end{cases}$$

The function g satisfies assumption (2.9.3), with $q_1(t,x) = \frac{16}{3}d(t,x)$, and

$$\begin{cases} q_2(t,x) = \dfrac{cxt^{\frac{-1}{4}}e^{-\frac{1}{t}}}{4(1+\sqrt{t})}; \ (t,x) \in (0,\infty)\times[0,1], \\ q_2(0,x) = 0; \qquad\qquad\qquad\qquad x \in [0,1]. \end{cases}$$

Also, for each $x \in [0,1]$, we get

$$\begin{aligned} I_\theta^r q_1(t,x) &= \frac{1}{\Gamma(r_1)\Gamma(r_2)}\int_0^t\int_0^x (t-\tau)^{r_1-1}(x-s)^{r_2-1}q_1(\tau,s)dsd\tau \\ &\leq \frac{c}{4(1+t^{-\frac{1}{2}})\Gamma(\frac{1}{4})\Gamma(\frac{1}{2})}\int_0^t\int_0^x (t-\tau)^{\frac{-3}{4}}(x-s)^{\frac{-1}{2}}s\tau^{\frac{-3}{4}}|\sin\tau|dsd\tau \\ &\leq \frac{c}{4(1+t^{-\frac{1}{2}})\Gamma(\frac{1}{4})\Gamma(\frac{1}{2})}\int_0^t\int_0^x (t-\tau)^{\frac{-3}{4}}(x-s)^{\frac{-1}{2}}s\tau^{\frac{-3}{4}}dsd\tau \\ &= \frac{1}{4}\frac{t^{\frac{-1}{2}}x^{\frac{3}{2}}}{(1+t^{-\frac{1}{2}})} \\ &\leq \frac{1}{4}t^{\frac{-1}{2}}x^{\frac{3}{2}} \longrightarrow 0 \ as \ t \to \infty, \end{aligned}$$

and

$$\begin{aligned} I_\theta^r q_2(t,x) &= \frac{1}{\Gamma(r_1)\Gamma(r_2)}\int_0^t\int_0^x (t-\tau)^{r_1-1}(x-s)^{r_2-1}q_2(\tau,s)dsd\tau \\ &\leq \frac{c}{4(1+t^{-\frac{1}{2}})\Gamma(\frac{1}{4})\Gamma(\frac{1}{2})}\int_0^t\int_0^x (t-\tau)^{\frac{-3}{4}}(x-s)^{\frac{-1}{2}}s\tau^{\frac{-3}{4}}e^{-\frac{1}{\tau}}dsd\tau \\ &\leq \frac{16c}{12\sqrt{\pi}}\frac{t^{\frac{-1}{2}}x^{\frac{3}{2}}}{(1+t^{-\frac{1}{2}})} \\ &\leq \frac{1}{4}t^{\frac{-1}{2}}x^{\frac{3}{2}} \longrightarrow 0 \ as \ t \to \infty. \end{aligned}$$

Thus

$$\lim_{t\to\infty} I_\theta^r d(t,x) = \lim_{t\to\infty} I_\theta^r q_i(t,x) = 0; \ x \in [0,1]; \ i = 0,1,2,$$

and $d^* = \frac{3}{64}$, $q_1^* = q_2^* = \frac{1}{4}$. Finally, we can see that the hypothesis (2.11.1) is satisfied. Indeed,

$$\varphi^* + p^*\eta\left(d^*\Lambda(\eta) + \sum_{i=1}^m q_i^*\right) = 1 + \frac{1}{2}\eta + \frac{3}{64}\eta^2.$$

So,

$$\varphi^* + p^*\eta\left(d^*\Lambda(\eta) + \sum_{i=1}^m q_i^*\right) \leq \eta,$$

implies that

$$1 - \frac{1}{2}\eta + \frac{3}{64}\eta^2 \leq 0,$$

which is satisfies for $\eta = 3$. Thus,

$$\max\left\{\Phi^*, \varphi^* + p^*\eta\left(d^*\Lambda(\eta) + \sum_{i=1}^{m} q_i^*\right)\right\} \leq \eta.$$

Also,

$$p^*d^*\Lambda(\eta+\eta^*) + p^*\sum_{i=1}^{m} q_i^* < 1,$$

implies that $\eta^* < \frac{23}{3}$. We can take $\eta^* = 5$, because the inequality

$$2p^*\eta\left(d^*\Lambda(\eta+\eta^*) + \sum_{i=1}^{m} q_i^*\right) \leq \eta^*\left(1 - p^*d^*\Lambda(\eta+\eta^*) - p^*\sum_{i=1}^{m} q_i^*\right),$$

is satisfies with $\eta = 3$ and $\eta^* = 5$. This inequality gives $\frac{21}{4} \leq \frac{45}{8}$. Hence by Theorem 2.9, the problem (2.32)-(2.34) has a solution defined on $[-1,\infty) \times [-\frac{1}{4}, 1]$ and solutions of this problem are locally asymptotically stable.

2.5. Global Asymptotic Stability for Nonlinear Multi-delay Differential Equations

2.5.1. Introduction

In this section, we establish a sufficient condition for the existence and the stability of solutions of the following system of nonlinear delay differential equations of fractional order of the form

$$^cD_\theta^r u(t,x) = p(t,x) - q(t,x)f(t,x,u(t-\tau_1, x-\xi_1), \ldots, u(t-\tau_m, x-\xi_m));$$

$$\text{for } (t,x) \in J := \mathbb{R}_+ \times [0,b], \tag{2.35}$$

$$u(t,x) = \Phi(t,x); \quad \text{for } (t,x) \in \tilde{J} := [-T,\infty) \times [-\xi, b] \backslash (0,\infty) \times (0,b], \tag{2.36}$$

$$\begin{cases} u(t,0) = \varphi(t); & t \in [0,\infty), \\ u(0,x) = \psi(x); & x \in [0,b], \end{cases} \tag{2.37}$$

where $b > 0$, $\theta = (0,0)$, $\mathbb{R}_+ = [0,\infty)$, $\tau_i, \xi_i \geq 0$; $i = 1\ldots,m$, $T = \max_{i=1\ldots,m}\{\tau_i\}$, $\xi = \max_{i=1\ldots,m}\{\xi_i\}$, $^cD_\theta^r$ is the Caputo fractional derivative of order $r = (r_1, r_2) \in (0,1] \times (0,1]$, $p, q : J \to \mathbb{R}$, $f : J \times \mathbb{R}^m \to \mathbb{R}$ are given continuous functions, $\varphi : \mathbb{R}_+ \to \mathbb{R}$, $\psi : [0,b] \to \mathbb{R}$ are absolutely continuous functions with $\lim_{t\to\infty}\varphi(t) = 0$, and $\psi(x) = \varphi(0)$ for each $x \in [0,b]$, and $\Phi : \tilde{J} \to \mathbb{R}^n$ is continuous with $\varphi(t) = \Phi(t,0)$ for each $t \in \mathbb{R}_+$, and $\psi(x) = \Phi(0,x)$ for each $x \in [0,b]$.

2.5.2. Main Results

Definition 2.10. *A function $u \in BC$ is said to be a solution of (2.35)-(2.37) if u satisfies equation (2.35) on J, equation (2.36) on $\tilde{J}$ and condition (2.37) is satisfied.*

The following hypotheses will be used in the sequel:

(2.13.1) The functions Φ, p and q are in BC. Moreover, assume that

$$\lim_{t\longrightarrow\infty} I_\theta^r p(t,x) = 0; \ x \in [0,b],$$

(2.13.2) There exist continuous functions $d_i : \mathbb{R}_+ \times [0,b] \to \mathbb{R}_+$ such that

$$\left(1+\sum_{i=1}^{m}|u_i|\right)|f(t,x,u_1,u_2,...,u_m)| \leq \sum_{i=1}^{m}|u_i|d_i(t,x);$$

for $(t,x) \in \mathbb{R}_+ \times [0,b]$ and for $u_i \in \mathbb{R}$; $i = 1\dots m$. Moreover, assume that

$$\lim_{t\longrightarrow\infty} I_\theta^r d_i(t,x) = 0; \ x \in [0,b]; \ i = 1\dots m.$$

Remark 2.11. *Set*

$$\Phi^* := \sup_{(t,x)\in\tilde{J}} \Phi(t,x), \ \varphi^* := \sup_{t\in\mathbb{R}_+} \varphi(t), \ p^* := \sup_{(t,x)\in J} I_\theta^r p(t,x),$$

$$q^* := \sup_{(t,x)\in J} q(t,x) \ \textit{and} \ d_i^* := \sup_{(t,x)\in J} I_\theta^r d_i(t,x); \ i = 1\dots m.$$

From hypotheses, we infer that Φ^*, φ^*, p^*, q^* and d_i^*; $i = 1\dots m$ are finite.

Now, we shall prove the following theorem concerning the existence and the stability of a solution of problem (2.35)-(2.37).

Theorem 2.12. *Assume that the hypotheses (2.13.1) and (2.13.2) hold, then the problem (2.35)-(2.37) has at least one solution in the space BC. Moreover, solutions of problem (2.35)-(2.37) are globally asymptotically stable.*

Proof. Let us define the operator N such that, for any $u \in BC$,

$$(Nu)(t,x) = \begin{cases} \Phi(t,x); & (t,x) \in \tilde{J}, \\ \varphi(t) + I_\theta^r[p(t,x) & \\ -q(t,x)f(t,x,u(t-\tau_1,x-\xi_1),..., & \\ u(t-\tau_m,x-\xi_m))]; & (t,x) \in J. \end{cases} \tag{2.38}$$

The operator N maps BC into BC. Indeed the map $N(u)$ is continuous on $[-T,\infty)\times[-\xi,b]$ for any $u\in BC$, and for each $(t,x)\in J$ we have

$$\begin{aligned}
|(Nu)(t,x)| &\leq |\varphi(t)|+I_\theta^r|p(t,x)-q(t,x)\\
&\times\ g(t,x,u(t-\tau_1,x-\xi_1),\dots,u(t-\tau_m,x-\xi_m))|\\
&\leq |\varphi(t)|+\frac{1}{\Gamma(r_1)\Gamma(r_2)}\int_0^t\int_0^x(t-\tau)^{r_1-1}(x-s)^{r_1}\\
&\times\ \left[|p(t,x)|+|q(t,x)|\left(\sum_{i=1}^m|u(\tau-\tau_i,s-\xi_i)|d_i(\tau,s)\right)\right.\\
&\times\ \left.\left(1+\sum_{i=1}^m|u(\tau-\tau_i,s-\xi_i)|\right)^{-1}\right]dsd\tau\\
&\leq \varphi^*+p^*+q^*\sum_{i=1}^m d_i^*,
\end{aligned}$$

and for $(t,x)\in\tilde{J}$, we have

$$|(Nu)(t,x)|=|\Phi(t,x)|\leq\Phi^*.$$

Thus,

$$\|N(u)\|_{BC}\leq\max\left\{\Phi^*,\varphi^*+p^*+q^*\sum_{i=1}^m d_i^*\right\}:=\eta. \tag{2.39}$$

Hence, $N(u)\in BC$. This proves that the operator N maps BC into itself.
By Lemma 2.2, the problem of finding the solutions of the problem (2.35)-(2.37) is reduced to finding the solutions of the operator equation $N(u)=u$. Equation (2.39) implies that N transforms the ball $B_\eta:=B(0,\eta)$ into itself. We shall show that $N:B_\eta\to B_\eta$ satisfies the assumptions of Schauder's fixed point theorem [204]. The proof will be given in several steps.

Step 1: *N is continuous.*
Let $\{u_n\}_{n\in\mathbb{N}}$ be a sequence such that $u_n\to u$ in B_η. Then, for each $(t,x)\in[-T,\infty)\times[-\xi,b]$, we have

$$\begin{aligned}
&|(Nu_n)(t,x)-(Nu)(t,x)|\\
\leq\ &\tfrac{1}{\Gamma(r_1)\Gamma(r_2)}\int_0^t\int_0^x(t-\tau)^{r_1-1}(x-s)^{r_2-1}|q(\tau,s)|\\
\times\ &|f(\tau,s,u_n(\tau-\tau_1,s-\xi_1),\dots,u_n(\tau-\tau_m,s-\xi_m))\\
-\ &f(\tau,s,u(\tau-\tau_1,s-\xi_1),\dots,u(\tau-\tau_m,s-\xi_m)))|dsd\tau.
\end{aligned}\tag{2.40}$$

Case 1. If $(t,x)\in\tilde{J}\cup([0,a]\times[0,b])$; $a>0$, then, since $u_n\to u$ as $n\to\infty$ and f is continuous, (2.40) gives

$$\|N(u_n)-N(u)\|_{BC}\to 0\quad\text{as } n\to\infty.$$

Case 2. If $(t,x)\in(a,\infty)\times[0,b];\ a>0$, then from (2.13.2) and (2.40), we get

$$\begin{aligned}
&|(Nu_n)(t,x)-(Nu)(t,x)|\\
\leq\ & \tfrac{2q^*}{\Gamma(r_1)\Gamma(r_2)}\int_0^t\int_0^x(t-\tau)^{r_1-1}(x-s)^{r_2-1}\left(\sum_{i=1}^m|u(\tau-\tau_i,s-\xi_i)|d_i(\tau,s)\right)\\
\times\ & \left(1+\sum_{i=1}^m|u_i(\tau-\tau_i,s-\xi_i)|\right)^{-1}dsd\tau\\
\leq\ & \sum_{i=1}^m\frac{2q^*}{\Gamma(r_1)\Gamma(r_2)}\int_0^t\int_0^x(t-\tau)^{r_1-1}(x-s)^{r_2-1}d_i(\tau,s)dsd\tau\\
\leq\ & 2q^*\sum_{i=1}^m I_\theta^r d_i(t,x).
\end{aligned}$$

Then

$$|(Nu_n)(t,x)-(Nu)(t,x)|\leq 2q^*\sum_{i=1}^m I_\theta^r d_i(t,x). \tag{2.41}$$

Thus (2.41) gives

$$\|N(u_n)-N(u)\|_{BC}\to 0 \quad \text{as } n\to\infty.$$

Step 2: *$N(B_\eta)$ is uniformly bounded.*
This is clear since $N(B_\eta)\subset B_\eta$ and B_η is bounded.

Step 3: *$N(B_\eta)$ is equicontinuous on every compact subset $[-T,a]\times[-\xi,b]$ of $[-T,a]\times[-\xi,\infty);\ a>0$.*
Let $(t_1,x_1),(t_2,x_2)\in[0,a]\times[0,b], t_1<t_2,\ x_1<x_2$ and let $u\in B_\eta$. Thus we have

$$\begin{aligned}
&|(Nu)(t_2,x_2)-(Nu)(t_1,x_1)|\leq|\varphi(t_2)-\varphi(t_1)|+|p(t_2,x_2)-p(t_1,x_1)|\\
+\ & \frac{1}{\Gamma(r_1)\Gamma(r_2)}\int_0^{t_1}\int_0^{x_1}[(t_2-\tau)^{r_1-1}(x_2-s)^{r_2-1}-(t_1-\tau)^{r_1-1}(x_1-s)^{r_2-1}]\\
&\times|q(\tau,s)|.|f(\tau,s,I_{0,s}^{r_2}u(\tau,s),u(\tau-\tau_1,s-\xi_1),\ldots,u(\tau-\tau_m,s-\xi_m))|dsd\tau\\
+\ & \frac{1}{\Gamma(r_1)\Gamma(r_2)}\int_{t_1}^{t_2}\int_{x_1}^{x_2}(t_2-\tau)^{r_1-1}(x_2-s)^{r_2-1}|q(\tau,s)|\\
&\times|f(\tau,s,I_{0,s}^{r_2}u(\tau,s),u(\tau-\tau_1,s-\xi_1),\ldots,u(\tau-\tau_m,s-\xi_m))|dsd\tau\\
+\ & \frac{1}{\Gamma(r_1)\Gamma(r_2)}\int_0^{t_1}\int_{x_1}^{x_2}(t_2-\tau)^{r_1-1}(x_2-s)^{r_2-1}|q(\tau,s)|\\
&\times|f(\tau,s,I_{0,s}^{r_2}u(\tau,s),u(\tau-\tau_1,s-\xi_1),\ldots,u(\tau-\tau_m,s-\xi_m))|dsd\tau\\
+\ & \frac{1}{\Gamma(r_1)\Gamma(r_2)}\int_{t_1}^{t_2}\int_0^{x_1}(t_2-\tau)^{r_1-1}(x_2-s)^{r_2-1}|q(\tau,s)|\\
&\times|f(\tau,s,I_{0,s}^{r_2}u(\tau,s),u(\tau-\tau_1,s-\xi_1),\ldots,u(\tau-\tau_m,s-\xi_m))|dsd\tau.
\end{aligned}$$

Thus

$$
\begin{aligned}
&|(Nu)(t_2,x_2)-(Nu)(t_1,x_1)| \leq |\varphi(t_2)-\varphi(t_1)|+|p(t_2,x_2)-p(t_1,x_1)| \\
+ &\frac{q^*}{\Gamma(r_1)\Gamma(r_2)}\int_0^{t_1}\int_0^{x_1}[(t_2-\tau)^{r_1-1}(x_2-s)^{r_2-1}-(t_1-\tau)^{r_1-1}(x_1-s)^{r_2-1}] \\
&\times\sum_{i=0}^{m} d_i(\tau,s)dsd\tau \\
+ &\frac{q^*}{\Gamma(r_1)\Gamma(r_2)}\int_{t_1}^{t_2}\int_{x_1}^{x_2}(t_2-\tau)^{r_1-1}(x_2-s)^{r_2-1}\sum_{i=0}^{m} d_i(\tau,s)dsd\tau \\
+ &\frac{q^*}{\Gamma(r_1)\Gamma(r_2)}\int_{0}^{t_1}\int_{x_1}^{x_2}(t_2-\tau)^{r_1-1}(x_2-s)^{r_2-1}\sum_{i=0}^{m} d_i(\tau,s)dsd\tau \\
+ &\frac{q^*}{\Gamma(r_1)\Gamma(r_2)}\int_{t_1}^{t_2}\int_{0}^{x_1}(t_2-\tau)^{r_1-1}(x_2-s)^{r_2-1}\sum_{i=0}^{m} d_i(\tau,s)dsd\tau.
\end{aligned}
$$

From continuity of φ, p, d_i; $i=0\ldots,m$ and as $t_1 \to t_2$ and $x_1 \to x_2$, the right-hand side of the above inequality tends to zero. The equicontinuity for the cases $t_1 < t_2 < 0$, $x_1 < x_2 < 0$ and $t_1 \leq 0 \leq t_2$, $x_1 \leq 0 \leq x_2$ is obvious.

Step 4: *$N(B_\eta)$ is equiconvergent.*
Let $(t,x) \in \mathbb{R}_+ \times [0,b]$ and $u \in B_\eta$, then we have

$$
\begin{aligned}
|(Nu)(t,x)| \;\leq\; & |\varphi(t)| + \left|\frac{1}{\Gamma(r_1)\Gamma(r_2)}\int_0^t\int_0^x (t-\tau)^{r_1-1}(x-s)^{r_2-1}[p(\tau,s)\right. \\
- \; & q(\tau,s)f(\tau,s,u(\tau-\tau_1,s-\xi_1),\ldots,u(\tau-\tau_m,s-\xi_m))]dsd\tau| \\
\leq \; & |\varphi(t)| + \frac{1}{\Gamma(r_1)\Gamma(r_2)}\int_0^t\int_0^x (t-\tau)^{r_1-1}(x-s)^{r_2-1}|p(\tau,s)|dsd\tau \\
+ \; & \frac{q^*}{\Gamma(r_1)\Gamma(r_2)}\int_0^t\int_0^x (t-\tau)^{r_1-1}(x-s)^{r_2-1}\sum_{i=0}^{m} d_i(\tau,s)dsd\tau \\
\leq \; & |\varphi(t)| + I_\theta^r p(t,x) + q^*\sum_{i=0}^{m} I_\theta^r d_i(t,x).
\end{aligned}
$$

Thus, for each $x \in [0,b]$, we get

$$|(Nu)(t,x)| \to 0, \; as\, t \to +\infty.$$

Hence,

$$|(Nu)(t,x)-(Nu)(+\infty,x)| \to 0, \; as\, t \to +\infty.$$

As a consequence of Steps 1 to 4 together with the Lemma 1.69, we can conclude that $N : B_\eta \to B_\eta$ is continuous and compact. From an application of Schauder's theorem [204], we deduce that N has a fixed point u which is a solution of the problem (2.35)-(2.37).

Step 5: *The global asymptotic stability of solutions.*
Now we investigate the global asymptotic stability of solutions of problem (2.35)-(2.37).

Let us assume that u and v are solutions of problem (2.35)-(2.37). Then for each $(t,x) \in J$

$$\begin{aligned} |u(t,x)-v(t,x)| &= |(Nu)(t,x)-(Nv)(t,x)| \\ &\leq \frac{q^*}{\Gamma(r_1)\Gamma(r_2)} \int_0^t \int_0^x (t-\tau)^{r_1-1}(x-s)^{r_2-1} \\ &\quad \times |f(\tau,s,u(\tau-\tau_1,s-\xi_1),\dots,u(\tau-\tau_m,s-\xi_m)) \\ &- \quad f(\tau,s,v(\tau-\tau_1,s-\xi_1),\dots,v(\tau-\tau_m,s-\xi_m))|dsd\tau \\ &\leq \frac{2q^*}{\Gamma(r_1)\Gamma(r_2)} \int_0^t \int_0^x (t-\tau)^{r_1-1}(x-s)^{r_2-1} \sum_{i=1}^m d_i(\tau,s)dsd\tau \\ &\leq 2q^* \sum_{i=1}^m I_\theta^r d_i(t,x). \end{aligned} \tag{2.42}$$

By using (2.42) and the fact that $I_\theta^r d_i(t,x) \to 0$ as $t \to \infty$; $i = 0 \dots m$ we deduce that

$$\lim_{t\to\infty} |u(t,x)-v(t,x)| = 0.$$

Consequently, all solutions of problem (2.35)-(2.37) are globally asymptotically stable. □

2.5.3. An Example

As an application and to illustrate our results, we consider the following system of delay differential equations of fractional order

$${}^cD_\theta^r u(t,x) = p(t,x) - q(t,x) f\left(t,x,u\left(t-1,x-\frac{1}{4}\right),u\left(t-\frac{2}{3},x-\frac{1}{5}\right)\right);$$

$$\text{for } (t,x) \in J := [0,\infty)\times[0,1], \tag{2.43}$$

$$u(t,x) = e^{-t}; \quad \text{for } (t,x) \in \tilde{J} := [-1,\infty)\times[-\frac{1}{4},1]\backslash(0,\infty)\times(0,1], \tag{2.44}$$

$$\begin{cases} u(t,0) = e^{-t}; & t \in [0,\infty), \\ u(0,x) = 1; & x \in [0,1], \end{cases} \tag{2.45}$$

where $r = (r_1,r_2) = (\frac{1}{4},\frac{1}{2})$, $q(t,x) = \frac{1}{1+t^2+x^2}$; $(t,x) \in (0,\infty)\times[0,1]$,

$$\begin{cases} p(t,x) = xt^{\frac{-3}{4}}\sin t; \ (t,x) \in (0,\infty)\times[0,1], \\ p(0,x) = 0; \qquad\qquad\qquad x \in [0,1], \end{cases}$$

and

$$\begin{cases} f(t,x,u,v) = \dfrac{xt^{\frac{-3}{4}}\left(|u|\sin t + |v|e^{-\frac{1}{t}}\right)}{2+|u|+|v|}; \\ \qquad (t,x) \in (0,\infty)\times[0,1] \text{ and } u,v \in \mathbb{R}, \\ f(0,x,u,v) = 0; \qquad\qquad x \in [0,1] \text{ and } u,v \in \mathbb{R}. \end{cases}$$

We have for each $(t,x) \in [0,\infty)\times[0,1]$, $\Phi(t,x) = e^{-t}$ and $\varphi(t) = e^{-t} \to 0 \text{ as } t \to \infty$, $\Phi^* = \varphi^* = 1$, $p^* = 1$, then the assumption (2.13.1) is satisfies. Let us notice that the function f satisfies assumption (2.13.2), where

$$\begin{cases} d_1(t,x) = xt^{\frac{-3}{4}}|\sin t|; \ (t,x) \in (0,\infty)\times[0,1], \\ d_1(0,x) = 0; \qquad\qquad\qquad x \in [0,1], \end{cases}$$

and

$$\begin{cases} d_2(t,x) = xt^{\frac{-3}{4}}e^{-\frac{1}{t}}; \ (t,x) \in (0,\infty)\times[0,1], \\ d_2(0,x) = 0; \qquad x \in [0,1]. \end{cases}$$

Also, for each $x \in [0,1]$, we get

$$\begin{aligned} I_\theta^r p(t,x) &= \frac{1}{\Gamma(r_1)\Gamma(r_2)}\int_0^t\int_0^x (t-\tau)^{r_1-1}(x-s)^{r_2-1}p(\tau,s)dsd\tau \\ &= \frac{1}{\Gamma(\frac{1}{4})\Gamma(\frac{1}{2})}\int_0^t\int_0^x (t-\tau)^{\frac{-3}{4}}(x-s)^{\frac{-1}{2}}s\tau^{\frac{-3}{4}}|\sin\tau|dsd\tau \\ &\leq \frac{1}{\Gamma(\frac{1}{4})\Gamma(\frac{1}{2})}\int_0^t\int_0^x (t-\tau)^{\frac{-3}{4}}(x-s)^{\frac{-1}{2}}s\tau^{\frac{-3}{4}}dsd\tau \\ &= \frac{\Gamma(\frac{1}{4})\Gamma(2)}{\Gamma(\frac{5}{4})\Gamma(\frac{5}{2})}t^{\frac{-1}{2}}x^{\frac{3}{2}} \to 0 \ as\ t\to\infty, \end{aligned}$$

$I_\theta^r d_1(t,x) = I_\theta^r p(t,x)$ and

$$\begin{aligned} I_\theta^r d_2(t,x) &= \frac{1}{\Gamma(r_1)\Gamma(r_2)}\int_0^t\int_0^x (t-\tau)^{r_1-1}(x-s)^{r_2-1}d_2(\tau,s)dsd\tau \\ &= \frac{1}{\Gamma(\frac{1}{4})\Gamma(\frac{1}{2})}\int_0^t\int_0^x (t-\tau)^{\frac{-3}{4}}(x-s)^{\frac{-1}{2}}s\tau^{\frac{-3}{4}}e^{-\frac{1}{\tau}}dsd\tau \end{aligned}$$

$$\begin{aligned} &\leq \frac{1}{\Gamma(\frac{1}{4})\Gamma(\frac{1}{2})}\int_0^t\int_0^x (t-\tau)^{\frac{-3}{4}}(x-s)^{\frac{-1}{2}}s\tau^{\frac{-3}{4}}dsd\tau \\ &= \frac{\Gamma(\frac{1}{4})\Gamma(2)}{\Gamma(\frac{5}{4})\Gamma(\frac{5}{2})}t^{\frac{-1}{2}}x^{\frac{3}{2}} \to 0 \ as\ t\to\infty. \end{aligned}$$

Thus

$$\lim_{t\to\infty} I_\theta^r p(t,x) = \lim_{t\to\infty} I_\theta^r d_i(t,x) = 0; \ x\in[0,1]; \ i=0,1,2.$$

Hence by Theorem 2.12, the problem (2.43)-(2.45) has a solution defined on $[-1,\infty)\times[-\frac{1}{4},1]$ and all solutions are globally asymptotically stable on $[-1,\infty)\times[-\frac{1}{4},1]$.

2.6. Notes and Remarks

The results of Chapter 2 are taken from Abbas *et al.* [14, 16, 24, 41]. Other results may be found in [319, 320, 321].

Chapter 3

Fractional Order Riemann-Liouville Integral Equations

3.1. Introduction

In this chapter, we shall present existence results for some classes of Riemann-Liouville integral equations of fractional order by using some fixed point theorems.

3.2. Partial Integral Equations

3.2.1. Introduction

In this section we present some results concerning the existence, the uniqueness and the attractivity of solutions for some functional integral equations of Riemann-Liouville fractional order, by using some fixed point theorems. First, we consider the following partial integral equation of Riemann-Liouville fractional order of the form

$$u(x,y) = \mu(x,y) + f(x,y,I_\theta^r u(x,y), u(x,y)); \; (x,y) \in J := [0,a] \times [0,b], \tag{3.1}$$

where $\theta = (0,0)$, $r = (r_1, r_2)$, $r_1, r_2 \in (0,\infty)$, $\mu : J \to \mathbb{R}^n$, $f : J \times \mathbb{R}^n \times \mathbb{R}^n \to \mathbb{R}^n$ are given continuous functions, I_θ^r is the left-sided mixed Riemann-Liouville integral of order r.
Next, we prove some results concerning the existence and the attractivity of solutions for the following partial Riemann-Liouville fractional order integral equation of the form

$$u(x,y) = \mu(x,y) + f(x,y,I_\theta^r u(x,y), u(x,y)); \; (x,y) \in J' := \mathbb{R}_+ \times [0,b], \tag{3.2}$$

where $b > 0$, $\mathbb{R}_+ = [0,\infty)$ and $\mu : J' \to \mathbb{R}^n$, $f : J' \times \mathbb{R}^n \times \mathbb{R}^n \to \mathbb{R}^n$ are given continuous functions. Our investigations are conducted in Banach spaces with an application of Banach's contraction principle and Schauder's fixed point theorem for the existence and uniqueness of solutions of the equation (3.1). We use the Schauder fixed point theorem for the existence of solutions of the equation (3.2), and we prove that all solutions are globally asymptotically stable. Also, we present some examples illustrating the applicability of the imposed conditions.

3.2.2. Main Results

Let us start by defining what we mean by a solution of of the equation (3.1).

Definition 3.1. *A function $u \in C(J)$ is said to be a solution of (3.1) if u satisfies equation (3.1) on J.*

Now, we shall prove the following theorem concerning the existence and uniqueness of a solution of equation (3.1).
Set

$$f^*(x,y) = f(x,y,0,0).$$

Theorem 3.2. *Assume that the following hypothesis holds*

(3.2.1) *There exist constants $L_1, L_2 > 0$ such that*

$$\|f(x,x,u,v) - f(x,y,\overline{u},\overline{v})\| \le L_2\|u-\overline{u}\| + L_1\|v-\overline{v}\|,$$

for each $(x,y) \in J$ and $u,v,\overline{u},\overline{v} \in \mathbb{R}^n$.

If

$$L_1 + \frac{L_2 a^{r_1} b^{r_2}}{\Gamma(1+r_1)\Gamma(1+r_2)} \le 1, \tag{3.3}$$

then the equation (3.1) has a unique solution on J.

Proof. Transform the equation (3.1) into a fixed point problem. Let $u \in C(J)$ and define the operator N by,

$$(Nu)(x,y) = \mu(x,y) + f(x,y,I_\theta^r u(x,y), u(x,y)). \tag{3.4}$$

Now, we verify that the operator N is a contraction map. Let $u,v \in C(J)$. Using (3.2.1), for each $(x,y) \in J$, we have

$$\begin{aligned}
& \|(Nu)(x,y) - (Nv)(x,y)\| \\
\le \; & \|f(x,y,I_\theta^r u(x,y), u(x,y)) - f(x,y,I_\theta^r v(x,y), v(x,y))\| \\
\le \; & L_1\|u(x,y) - v(x,y)\| \\
+ \; & \frac{L_2}{\Gamma(r_1)\Gamma(r_2)} \int_0^x \int_0^y (x-s)^{r_1-1}(y-t)^{r_2-1}\|u(s,t)-v(s,t)\|dtds \\
\le \; & \left(L_1 + \frac{L_2 a^{r_1} b^{r_2}}{\Gamma(1+r_1)\Gamma(1+r_2)}\right)\|u-v\|_{C(J)}.
\end{aligned}$$

Hence

$$\|N(u) - N(v)\|_{C(J)} \le \left(L_1 + \frac{L_2 a^{r_1} b^{r_2}}{\Gamma(1+r_1)\Gamma(1+r_2)}\right)\|u-v\|_{C(J)}.$$

From (3.3), it follows that N has a unique fixed point in $C(J)$ by Banach contraction principle. The fixed point of N is however a solution of equation (3.1).

For the existence of solutions for the equation (3.1) and by using Schauder's fixed point theorem [204], we prove the following theorem

Theorem 3.3. *Assume that the hypothesis* (3.2.1) *holds. If there exists a constant* $\eta > 0$ *such that*

$$\|\mu\| + \|f^*\| \leq \eta\left(1 - L_1 - \frac{L_2 a^{r_1} b^{r_2}}{\Gamma(1+r_1)\Gamma(1+r_2)}\right), \tag{3.5}$$

then the equation (3.1) has at least one solution in the space $C(J)$.

Proof. Consider the operator N defined in (3.4). We have

$$\begin{aligned}
&\|(Nu)(x,y)\| \leq \|\mu(x,y)\| + \|f(x,y,0,0)\| \\
+ \;& \|f(x,y,I_\theta^r u(x,y), u(x,y)) - f(x,y,0,0)\| \\
\leq \;& \|\mu(x,y)\| + \|f(x,y,0,0)\| + L_1\|u(x,y)\| \\
+ \;& \frac{L_2}{\Gamma(r_1)\Gamma(r_2)} \int_0^x \int_0^y (x-s)^{r_1-1}(y-t)^{r_2-1}\|u(s,t)\| dtds \\
\leq \;& \|\mu\| + \|f^*\| + \left(L_1 + \frac{L_2 a^{r_1} b^{r_2}}{\Gamma(1+r_1)\Gamma(1+r_2)}\right)\|u\|_{c(J)}.
\end{aligned}$$

Then,

$$\|(Nu)\|_{C(J)} \leq \|\mu\| + \|f^*\| + \eta\left(L_1 + \frac{L_2 a^{r_1} b^{r_2}}{\Gamma(1+r_1)\Gamma(1+r_2)}\right) \leq \eta.$$

Hence, (3.5) implies that N transforms the ball $B_\eta := \{u \in C(J) : \|u\|_{C(J)} \leq \eta\}$ into itself. We shall show that $N : B_\eta \to B_\eta$ satisfies the assumptions of Schauder's fixed point theorem [204]. The proof will be given in several steps.

Step 1: *N is continuous.*
Let $\{u_n\}_{n\in\mathbb{N}}$ be a sequence such that $u_n \to u$ in B_η. Then, for each $(x,y) \in J$, we have

$$\begin{aligned}
&\|(Nu_n)(x,y) - (Nu)(x,y)\| \\
\leq \;& \|f(x,y,I_\theta^r u_n(x,y), u_n(x,y)) - f(x,y,I_\theta^r u(x,y), u(x,y))\| \\
\leq \;& \left(L_1 + \frac{L_2 a^{r_1} b^{r_2}}{\Gamma(1+r_1)\Gamma(1+r_2)}\right)\|u_n - u\|_{C(J)}.
\end{aligned}$$

Hence

$$\|N(u_n) - N(u)\|_{C(J)} \leq \left(L_1 + \frac{L_2 a^{r_1} b^{r_2}}{\Gamma(1+r_1)\Gamma(1+r_2)}\right)\|u_n - u\|_{C(J)}. \tag{3.6}$$

Since $u_n \to u$ as $n \to \infty$, then (3.6) gives

$$\|N(u_n) - N(u)\|_{C(J)} \to 0 \quad \text{as } n \to \infty.$$

Step 2: *$N(B_\eta)$ is bounded.*
This is clear since $N(B_\eta) \subset B_\eta$ and B_η is bounded.

Step 3: *$N(B_\eta)$ is equicontinuous.*
Let $(x_1,y_1),\ (x_2,y_2) \in J,\ x_1 < x_2,\ y_1 < y_2$ and $u \in B_\eta$. Then

$$\begin{aligned}
&\|N(u)(x_2,y_2) - N(u)(x_1,y_1)\| \\
\leq \;& \|\mu(x_1,y_1) - \mu(x_2,y_2)\| + L_1\|u(x_1,y_1) - u(x_2,y_2)\|
\end{aligned}$$

$$
\begin{aligned}
&+\left\|\frac{L_2}{\Gamma(r_1)\Gamma(r_2)}\int_0^{x_1}\int_0^{y_1}[(x_2-s)^{r_1-1}(y_2-t)^{r_2-1}\right.\\
&-(x_1-s)^{r_1-1}(y_1-t)^{r_2-1}]u(s,t)dtds\\
&+\frac{L_2}{\Gamma(r_1)\Gamma(r_2)}\int_{x_1}^{x_2}\int_{y_1}^{y_2}(x_2-s)^{r_1-1}(y_2-t)^{r_2-1}u(s,t)dtds\\
&+\frac{L_2}{\Gamma(r_1)\Gamma(r_2)}\int_0^{x_1}\int_{y_1}^{y_2}(x_2-s)^{r_1-1}(y_2-t)^{r_2-1}u(s,t)dtds\\
&\left.+\frac{L_2}{\Gamma(r_1)\Gamma(r_2)}\int_{x_1}^{x_2}\int_0^{y_1}(x_2-s)^{r_1-1}(y_2-t)^{r_2-1}u(s,t)dtds\right\|\\
\leq\ &\|\mu(x_1,y_1)-\mu(x_2,y_2)\|+L_1\|u(x_1,y_1)-u(x_2,y_2)\|\\
&+\frac{L_2\eta}{\Gamma(1+r_1)\Gamma(1+r_2)}[2y_2^{r_2}(x_2-x_1)^{r_1}+2x_2^{r_1}(y_2-y_1)^{r_2}\\
&+x_1^{r_1}y_1^{r_2}-x_2^{r_1}y_2^{r_2}-2(x_2-x_1)^{r_1}(y_2-y_1)^{r_2}].
\end{aligned}
$$

As $x_1 \to x_2$, $y_1 \to y_2$ the right-hand side of the above inequality tends to zero. As a consequence of steps 1 to 3 together with the Arzela-Ascoli theorem, we can conclude that N is continuous and compact. From an application of Schauder's theorem [204], we deduce that N has a fixed point u which is a solution of the equation (3.1). □

Now, we are concerned with the existence and global asymptotic stability of solutions for the equation (3.2).

Theorem 3.4. *Assume that the following hypotheses hold.*

(3.5.1) *The function μ is continuous and bounded with*

$$\mu^* = \sup_{(x,y)\in\mathbb{R}_+\times[0,b]} \|\mu(x,y)\|,$$

and

$$\lim_{x\to\infty}\mu(x,y) = 0;\ \ for\ y\in[0,b],$$

(3.5.2) *The function f is continuous and there exist functions $p,q:\mathbb{R}_+\times[0,b]\to\mathbb{R}_+$ such that*

$$\|f(x,y,u,v)\| \leq \frac{p(x,y)\|u\|+q(x,y)\|v\|}{1+\|u\|+\|v\|},$$

for $(x,y)\in\mathbb{R}_+\times[0,b]$ and for $u,v\in\mathbb{R}^n$. Moreover, assume that

$$\lim_{x\to\infty}p(x,y) = \lim_{x\to\infty}q(x,y) = 0;\ \ for\ y\in[0,b].$$

Then the equation (3.2) has at least one solution in the space BC. Moreover, solutions of equation (3.2) are globally asymptotically stable.

Proof. Let us define the operator T, such that for any $u \in BC$

$$(Tu)(x,y) = \mu(x,y)+f(x,y,I_\theta^r u(x,y),u(x,y));\ (x,y)\in\mathbb{R}_+\times[0,b]. \tag{3.7}$$

By considering conditions of theorem we infer that $T(u)$ is continuous on $\mathbb{R}_+ \times [0,b]$. Now we prove that $N(u) \in BC$ for any $u \in BC$. For arbitrarily fixed $(x,y) \in \mathbb{R}_+ \times [0,b]$ we have

$$\begin{aligned} \|(Tu)(x,y)]\| &\leq \|\mu(x,y) + f(x,y,I_\theta^r u(x,y),u(x,y))\| \\ &\leq \|\mu(x,y)\| + (p(x,y)\|I_\theta^r u(x,y)\| + q(x,y)\|u(x,y))\|) \\ &\times (1 + \|I_\theta^r u(x,y)\| + \|u(x,y))\|)^{-1} \\ &\leq \|\mu(x,y)\| + p(x,y) + q(x,y) \\ &\leq \mu^* + p^* + q^*, \end{aligned}$$

where

$$p^* = \sup_{(x,y)\in\mathbb{R}_+\times[0,b]} p(x,y) \;\; and \;\; q^* = \sup_{(x,y)\in\mathbb{R}_+\times[0,b]} q(x,y).$$

Thus

$$\|T(u)\| \leq \mu^* + p^* + q^*. \tag{3.8}$$

Hence $T(u) \in BC$. Equation (3.8) yields that T transforms the ball $B_\eta := B(0,\eta)$ into itself where $\eta = \mu^* + p^* + q^*$. We shall show that $T : B_\eta \to B_\eta$ satisfies the assumptions of Schauder's fixed point theorem [204]. The proof will be given in several steps and cases.

Step 1: *T is continuous on B_η.*
Let $\{u_n\}_{n\in\mathbb{N}}$ be a sequence such that $u_n \to u$ in B_η. Then, for each $(x,y) \in \mathbb{R}_+ \times [0,b]$, we have

$$\|(Tu_n)(x,y) - (Tu)(x,y)\| \leq |f(x,y,I_\theta^r u_n(x,y),u_n(x,y)) - f(x,y,I_\theta^r u(x,y),u(x,y))|. \tag{3.9}$$

Case 1. If $(x,y) \in [0,a] \times [0,b]$; $a > 0$, then, since $u_n \to u$ as $n \to \infty$ and f is continuous, (3.9) gives

$$\|T(u_n) - T(u)\|_{BC} \to 0 \quad \text{as } n \to \infty.$$

Case 2. If $(x,y) \in (a,\infty) \times [0,b]$; $a > 0$, then from (3.5.2) and (3.9), for each $(x,y) \in \mathbb{R}_+ \times [0,b]$ we get

$$\|(Tu_n)(x,y) - (Tu)(x,y)\| \leq 2(p(x,y) + q(x,y)). \tag{3.10}$$

Since $u_n \to u$ as $n \to \infty$ and $x \to \infty$, then (3.10) gives

$$\|T(u_n) - T(u)\|_{BC} \to 0 \quad \text{as } n \to \infty.$$

Step 2: *$T(B_\eta)$ is uniformly bounded.*
This is clear since $T(B_\eta) \subset B_\eta$ and B_η is bounded.

Step 3: *$T(B_\eta)$ is equicontinuous on every compact $[0,a] \times [0,b]$; $a > 0$ of $(\mathbb{R}_+ \times [0,b]$.*
Let $(x_1,y_1),(x_2,y_2) \in [0,a] \times [0,b], x_1 < x_2,\ y_1 < y_2$ and let $u \in B_\eta$, thus we have

$$\|(Tu)(x_2,y_2) - (Tu)(x_1,y_1)\| \leq \|\mu(x_2,y_2) - \mu(x_1,y_1)\|$$

$$+\|f(x_2,y_2,I_\theta^r u(x_2,y_2),u(x_2,y_2)) - f(x_1,y_1,I_\theta^r u(x_1,y_1),u(x_1,y_1))\|$$

From continuity of μ, f, I_θ^r and as $x_1 \longrightarrow x_2,\ y_1 \longrightarrow y_2$, the right-hand side of the above inequality tends to zero.

Step 4: $T(B_\eta)$ *is equiconvergent.*

Let $(x,y) \in \mathbb{R}_+ \times [0,b]$ and $u \in B_\eta$, then we have

$$\|(Tu)(x,y)\| \leq \|\mu(x,y)\| + p(x,y) + q(x,y).$$

Thus, for each $y \in [0,b]$, we get

$$\|(Tu)(x,y)\| \longrightarrow 0, \ as\ x \longrightarrow +\infty.$$

As a consequence of Steps 1 to 4 together with the Lemma 1.69, we can conclude that T is continuous and compact. From an application of Schauder's theorem [204], we deduce that T has a fixed point u which is a solution of the equation (3.2).

Step 5: Finally, we show the global asymptotic stability of solutions of the equation (3.2). Let u and v be any two solutions of equation (3.2) in B_η, then for each $(x,y) \in \mathbb{R}_+ \times [0,b]$ we have

$$\begin{aligned}\|u(x,y) - v(x,y)\| &\leq \|f(x,y,I_\theta^r u(x,y), u(x,y)) - f(x,y,I_\theta^r v(x,y), v(x,y)))\| \\ &\leq 2(p(x,y) + q(x,y)). \end{aligned} \tag{3.11}$$

By using (3.11), we deduce that

$$\lim_{x \longrightarrow \infty} (u(x,y) - v(x,y)) = 0.$$

Consequently, equation (3.2) has a solution and all solutions are globally asymptotically stable. □

3.2.3. Examples

As applications and to illustrate our results, we present the following examples.

Example 1. Consider the following integral equation of fractional order

$$u(x,y) = x + y^2 + \frac{2 + |u(x,y)| + |I_\theta^r u(x,y)|}{e^{x+y+10}(1 + |u(x,y)|)(1 + |I_\theta^r u(x,y)|)};\ (x,y) \in [0,1] \times [0,1]. \tag{3.12}$$

Set $\mu(x,y) = x + y^2$ and

$$f(x,y,u,v) = \frac{e^{-(x+y+10)}}{1+|u|} + \frac{e^{-(x+y+10)}}{1+|v|};\ (x,y) \in [0,1] \times [0,1].$$

We have $f^*(x,y) = 2e^{-(x+y+10)}$.

For each $u, v \in \mathbb{R}$ and $(x,y) \in [0,1] \times [0,1]$, we have

$$|f(x,y,u(x,y)) - f(x,y,v(x,y))| \leq \frac{1}{e^{10}}|u - v|.$$

Hence condition (3.2.1) is satisfied with $L_1 = L_2 = \frac{1}{e^{10}}$. We shall show that condition (3.3) holds with $a = b = 1$. Indeed

$$L_1 + \frac{L_2 a^{r_1} b^{r_2}}{\Gamma(r_1+1)\Gamma(r_2+1)} = \frac{1}{e^{10}} + \frac{1}{e^{10}\Gamma(1+r_1)\Gamma(1+r_2)} < 1,$$

which is satisfied for each $(r_1, r_2) \in (0,1] \times (0,1]$. Consequently Theorem 3.2 implies that equation (3.12) has a unique solution defined on $[0,1] \times [0,1]$.

Example 2. Consider now the following fractional order integral equation of the form

$$u(x,y) = \frac{e^{3-2x+y}}{1+x+y^2}$$

$$+\frac{e^{2-x+y}|u(x,y)| + \frac{1}{1+x+y}|I_\theta^r u(x,y)|}{1+e^{2-x+y}|u(x,y)| + \frac{1}{1+x+y}|I_\theta^r u(x,y)|}; \ (x,y) \in \mathbb{R}_+ \times [0,1]. \quad (3.13)$$

Set $\mu(x,y) = \frac{e^{3-2x+y}}{1+x+y^2}$ and

$$f(x,y,u,v) = \frac{\frac{u}{1+x+y} + ve^{2-x+y}}{1 + \frac{u}{1+x+y} + ve^{2-x+y}}; \ (x,y) \in \mathbb{R}_+ \times [0,1].$$

We have $\mu^* = e^4$, then condition (3.5.1) is satisfied.
Also, the function f is continuous and satisfies assumption (3.5.2), with $p(x,y) = \frac{1}{1+x+y}$ and $q(x,y) = e^{2-x+y}$. Hence by Theorem 3.4, the equation (3.13) has a solution defined on $\mathbb{R}_+ \times [0,1]$ and all the solutions are globally asymptotically stable.

3.3. Qualitative Theory for Riemann-Liouville Integral Equations in Two Independent Variables

3.3.1. Introduction

In this section, by means of integral inequalities and the fixed point approach, we investigate the following partial integral equation of Riemann-Liouville fractional order of the form

$$u(t,x) = \mu(t,x) + \frac{1}{\Gamma(r_1)} \int_0^t (t-s)^{r_1-1} f(t,x,s,u(s,x))ds$$

$$+\frac{1}{\Gamma(r_1)\Gamma(r_2)} \int_0^t \int_0^b (t-s)^{r_1-1}(b-y)^{r_2-1} g(t,x,s,y,u(s,y))dyds; \ (t,x) \in J, \quad (3.14)$$

where $J = \mathbb{R}_+ \times [0,b]$, $b > 0$, $r = (r_1, r_2)$, $r_1, r_2 \in (0,\infty)$, $\mu : J \to \mathbb{R}$, $f : J_1 \times \mathbb{R} \to \mathbb{R}$, $g : J_2 \times \mathbb{R} \to \mathbb{R}$ are given continuous functions,

$$J_1 = \{(t,x,s) : 0 \le s \le t < \infty, \ x \in [0,b]\},$$

$$J_2 = \{(t,x,s,y) : 0 \le s \le t < \infty, \ x \in [0,b], \ y \in [0,b)\}.$$

3.3.2. Main Results

By $BC := BC(J)$ we denote the Banach space of all bounded and continuous functions from J into $\mathbb{R}$ equipped with the standard norm

$$\|u\|_{BC} = \sup_{(t,x)\in J} |u(t,x)|.$$

Denote by $D_1 := \frac{\partial}{\partial t}$, the partial derivative of a function defined on J_1 (or J_2) with respect to the first variable. In the sequel we will make use of the following Lemma due to Pachpatte.

Lemma 3.5. *[305] Let $u \in C(J)$, $q, D_1q \in C(J_1), k, D_1k \in C(J_2)$ be positive functions, and $c \geq 0$ is a constant. If*

$$u(t,x) \leq c + \int_0^t q(t,x,s)u(s,x)ds + \int_0^t \int_0^b k(t,x,s,y)u(s,y)dyds;\ (t,x) \in J, \tag{3.15}$$

then,

$$u(t,x) \leq cP(t,x)\exp\left(\int_0^t A(\sigma,x)d\sigma\right);\ (t,x) \in J, \tag{3.16}$$

where

$$P(t,x) = \exp(Q(t,x)), \tag{3.17}$$

in which

$$Q(t,x) = \int_0^t [q(\eta,x,\eta) + \int_0^\eta D_1q(\eta,x,\xi)d\xi]d\eta, \tag{3.18}$$

and

$$A(t,x) = \int_0^b k(t,x,t,y)P(t,y)dy + \int_0^t \int_0^b P(s,y)D_1k(t,x,s,y)dyds;\ (t,x) \in J. \tag{3.19}$$

Our first result is about the existence and uniqueness of the solution of equation (3.14).

Theorem 3.6. *Assume that following hypotheses hold*

(3.8.1) *The function μ is continuous and bounded with*

$$\mu^* = \sup_{(t,x)\in\mathbb{R}_+\times[0,b]} |\mu(t,x)|,$$

(3.8.2) *There exists a positive function $q \in BC(J_1)$ such that*

$$|f(t,x,s,u) - f(t,x,s,v)| \leq q(t,x,s)|u - v|,$$

for each $(t,x,s) \in J_1$ and $u, v \in \mathbb{R}$. Moreover, assume that the function $t \to \int_0^t (t-s)^{r_1-1} f(t,x,s,0)ds$ is bounded on J with

$$f^* = \sup_{(t,x)\in J} \frac{1}{\Gamma(r_1)} \int_0^t (t-s)^{r_1-1}|f(t,x,s,0)|ds,$$

(3.8.3) *There exists a positive function $k \in BC(J_2)$ such that*

$$|g(t,x,s,y,u) - g(t,x,s,y,v)| \leq k(t,x,s,y)|u-v|,$$

for each $(t,x,s,y) \in J_2$ and $u,v \in \mathbb{R}$. Moreover, assume that the function $t \to \int_0^b (t-s)^{r_1-1}(b-y)^{r_2-1}g(t,x,s,y,0)dyds$ is bounded on J with

$$g^* = \sup_{(t,x)\in J} \frac{1}{\Gamma(r_1)\Gamma(r_2)} \int_0^t \int_0^b (t-s)^{r_1-1}(b-y)^{r_2-1}$$

$$\times |g(t,x,s,y,0)|dyds.$$

If

$$q^* + k^* < 1, \tag{3.20}$$

where

$$p^* = \sup_{(t,x)\in J} \left[\frac{1}{\Gamma(r_1)} \int_0^t (t-s)^{r_1-1} q(t,x,s)ds \right],$$

and

$$k^* = \sup_{(t,x)\in J} \left[\frac{1}{\Gamma(r_1)\Gamma(r_2)} \int_0^t \int_0^b (t-s)^{r_1-1}(b-y)^{r_2-1}k(t,x,s,y)dyds \right],$$

then equation (3.14) has a unique solution on J.

Proof. Let us define the operator $N : BC \to BC$, such that for each $(t,x) \in J$,

$$(Nu)(t,x) = \mu(t,x) + \frac{1}{\Gamma(r_1)} \int_0^t (t-s)^{r_1-1} f(t,x,s,u(s,x))ds$$

$$+\frac{1}{\Gamma(r_1)\Gamma(r_2)} \int_0^t \int_0^b (t-s)^{r_1-1}(b-y)^{r_2-1}g(t,x,s,y,u(s,y))dyds; \; (t,x) \in J. \tag{3.21}$$

It is clear that the function $(t,x) \mapsto N(u)(t,x)$ is continuous on J. Now we prove that $N(u) \in BC$ for any $u \in BC$. For arbitrarily fixed $(t,x) \in J$ we have

$$\begin{aligned}
|(Nu)(t,x)| = & \Bigg| \mu(t,x) + \frac{1}{\Gamma(r_1)} \int_0^t (t-s)^{r_1-1} f(t,x,s,u(s,x))ds \\
& + \frac{1}{\Gamma(r_1)\Gamma(r_2)} \int_0^t \int_0^b (t-s)^{r_1-1}(b-y)^{r_2-1}g(t,x,s,y,u(s,y))dyds \Bigg| \\
& \leq |\mu(t,x)| + \frac{1}{\Gamma(r_1)} \int_0^t (t-s)^{r_1-1} |f(t,x,s,u(s,x)) - f(t,x,s,0)|ds \\
& + \frac{1}{\Gamma(r_1)} \int_0^t (t-s)^{r_1-1} |f(t,x,s,0)|ds \\
& + \frac{1}{\Gamma(r_1)\Gamma(r_2)} \int_0^t \int_0^b (t-s)^{r_1-1}(b-y)^{r_2-1} \\
& \times |g(t,x,s,y,u(s,y)) - g(t,x,s,y,0)|dyds
\end{aligned}$$

$$
\begin{aligned}
&+\frac{1}{\Gamma(r_1)\Gamma(r_2)}\int_0^t\int_0^b (t-s)^{r_1-1}(b-y)^{r_2-1}|g(t,x,s,y,0)|dyds\\
&\leq |\mu(t,x)|+\frac{1}{\Gamma(r_1)}\int_0^t (t-s)^{r_1-1}q(t,x,s)|u(s,x)|ds\\
&+\frac{1}{\Gamma(r_1)}\int_0^t (t-s)^{r_1-1}|f(t,x,s,0)|ds\\
&+\frac{1}{\Gamma(r_1)\Gamma(r_2)}\int_0^t\int_0^b (t-s)^{r_1-1}(b-y)^{r_2-1}k(t,x,s,y)|u(s,y)|dyds\\
&+\frac{1}{\Gamma(r_1)\Gamma(r_2)}\int_0^t\int_0^b (t-s)^{r_1-1}(b-y)^{r_2-1}|g(t,x,s,y,0)|dyds\\
&\leq \mu^*+f^*+g^*+(q^*+k^*)\|u\|_{BC}.
\end{aligned}
$$

Hence $N(u)\in BC$. Let $u,\ v\in BC$. Using the hypotheses, for each $(t,x)\in J$, we have

$$
\begin{aligned}
&|(Nu)(t,x)-(Nv)(t,x)|\leq \frac{1}{\Gamma(r_1)}\int_0^t (t-s)^{r_1-1}\\
&\times|f(t,x,s,u(s,x))-f(t,x,s,v(s,x))|ds\\
&+\frac{1}{\Gamma(r_1)\Gamma(r_2)}\int_0^t\int_0^b (t-s)^{r_1-1}(b-y)^{r_2-1}\\
&\times|g(t,x,s,y,u(s,y))-g(t,x,s,y,v(s,y))|dyds\\
&\leq \frac{1}{\Gamma(r_1)}\int_0^t (t-s)^{r_1-1}q(t,x,s)|u(s,x)-v(s,x)|ds\\
&+\frac{1}{\Gamma(r_1)\Gamma(r_2)}\int_0^t\int_0^b (t-s)^{r_1-1}(b-y)^{r_2-1}k(t,x,s,y)|u(s,y)-v(s,y)|dyds\\
&\leq \sup_{(t,x)\in J}\left[\frac{1}{\Gamma(r_1)}\int_0^t (t-s)^{r_1-1}q(t,x,s)ds\right.\\
&\left.+\frac{1}{\Gamma(r_1)\Gamma(r_2)}\int_0^t\int_0^b (t-s)^{r_1-1}(b-y)^{r_2-1}k(t,x,s,y)dyds\right]\|u-v\|_{BC}\\
&\leq (q^*+k^*)\|u-v\|_{BC}.
\end{aligned}
$$

From (3.20), it follows that N has a unique fixed point in BC by Banach contraction principle. The fixed point of N is however a solution of equation (3.14). □

Now, we shall prove the following theorem concerning the estimate on the solution of equation (3.14).

Theorem 3.7. *Set*

$$d=\mu^*+f^*+g^*. \tag{3.22}$$

Assume that (3.8.1) – (3.8.3) *and the following hypothesis holds*

(3.10.1) $q_1, D_1q_1\in BC(J_1)$ *and* $k_1, D_1k_1\in BC(J_2)$, *where*

$$q_1(t,x,s)=\frac{1}{\Gamma(r_1)}(t-s)^{r_1-1}q(t,x,s)$$

and

$$k_1(t,x,s,y) = \frac{1}{\Gamma(r_1)\Gamma(r_2)}(t-s)^{r_1-1}(b-y)^{r_2-1}k(t,x,s,y).$$

If u is any solution of (3.14) on J, then

$$|u(t,x)| \le dP_1(t,x)\exp\left(\int_0^t A_1(\sigma,x)d\sigma\right);\ (t,x)\in J, \tag{3.23}$$

where

$$P_1(t,x) = \exp(Q_1(t,x)), \tag{3.24}$$

in which

$$Q_1(t,x) = \int_0^t \left[q_1(\eta,x,\eta) + \int_0^\eta D_1q_1(\eta,x,\xi)d\xi\right]d\eta, \tag{3.25}$$

and

$$A_1(t,x) = \int_0^b k_1(t,x,t,y)P_1(t,y)dy + \int_0^t\int_0^b P_1(s,y)D_1k_1(t,x,s,y)dyds. \tag{3.26}$$

Proof. Using the fact that u is a solution of (3.14) and hypotheses, then for each $(t,x)\in J$, we have

$$\begin{aligned}
|u(t,x)| &\le |\mu(t,x)| + \frac{1}{\Gamma(r_1)}\int_0^t (t-s)^{r_1-1}|f(t,x,s,u(s,x)) - f(t,x,s,0)|ds \\
&+ \frac{1}{\Gamma(r_1)}\int_0^t (t-s)^{r_1-1}|f(t,x,s,0)|ds \\
&+ \frac{1}{\Gamma(r_1)\Gamma(r_2)}\int_0^t\int_0^b (t-s)^{r_1-1}(b-y)^{r_2-1} \\
&\times |g(t,x,s,y,u(s,y)) - g(t,x,s,y,0)|dyds \\
&+ \frac{1}{\Gamma(r_1)\Gamma(r_2)}\int_0^t\int_0^b (t-s)^{r_1-1}(b-y)^{r_2-1}|g(t,x,s,y,0)|dyds \\
&\le d + \frac{1}{\Gamma(r_1)}\int_0^t (t-s)^{r_1-1}|f(t,x,s,u(s,x)) - f(t,x,s,0)|ds \\
&+ \frac{1}{\Gamma(r_1)\Gamma(r_2)}\int_0^t\int_0^b (t-s)^{r_1-1}(b-y)^{r_2-1} \\
&\times |g(t,x,s,y,u(s,y)) - g(t,x,s,y,0)|dyds \\
&\le d + \frac{1}{\Gamma(r_1)}\int_0^t (t-s)^{r_1-1}q(t,x,s)|u(x,s)|ds
\end{aligned}$$

$$+\frac{1}{\Gamma(r_1)\Gamma(r_2)}\int_0^t\int_0^b (t-s)^{r_1-1}(b-y)^{r_2-1}k(t,x,s,y)|u(s,y)|dyds. \tag{3.27}$$

Now an application of Lemma 3.5, to (3.27) yields (3.23). □

Theorem 3.8. *Set*

$$\overline{d} := f^* + g^* + \mu^*(q^* + k^*). \tag{3.28}$$

Assume that (3.8.1) – (3.8.3) *hold. If u is any solution of (3.14) on J, then*

$$|u(t,x)-\mu(t,x)| \leq \bar{d}P_1(t,x)\exp\left(\int_0^t A_1(\sigma,x)d\sigma\right);\ (t,x)\in J, \tag{3.29}$$

where P_1 and A_1 are given by (3.24) and (3.26), respectively.

Proof. Let $h(t,x) = |u(t,x)-\mu(t,x)|$. Using the fact that u is a solution of (3.14) and hypotheses, for each $(t,x)\in J$, we have

$$\begin{aligned}
h(t,x) \leq & \frac{1}{\Gamma(r_1)}\int_0^t (t-s)^{r_1-1}|f(t,x,s,u(s,x))-f(t,x,s,\mu(s,x))|ds \\
& +\frac{1}{\Gamma(r_1)}\int_0^t (t-s)^{r_1-1}|f(t,x,s,\mu(s,x))|ds \\
& +\frac{1}{\Gamma(r_1)\Gamma(r_2)}\int_0^t\int_0^b (t-s)^{r_1-1}(b-y)^{r_2-1} \\
& \times |g(t,x,s,y,u(s,y))-g(t,x,s,y,\mu(s,x))|dyds \\
& +\frac{1}{\Gamma(r_1)\Gamma(r_2)}\int_0^t\int_0^b (t-s)^{r_1-1}(b-y)^{r_2-1}|g(t,x,s,y,\mu(s,x))|dyds \\
& \leq \bar{d}+\frac{1}{\Gamma(r_1)}\int_0^t (t-s)^{r_1-1}|f(t,x,s,u(s,x))-f(t,x,s,\mu(s,x))|ds \\
& +\frac{1}{\Gamma(r_1)\Gamma(r_2)}\int_0^t\int_0^b (t-s)^{r_1-1}(b-y)^{r_2-1} \\
& \times |g(t,x,s,y,u(s,y))-g(t,x,s,y,\mu(s,x))|dyds \\
& \leq \bar{d}+\frac{1}{\Gamma(r_1)}\int_0^t (t-s)^{r_1-1}q(t,x,s)h(x,s)ds \\
& +\frac{1}{\Gamma(r_1)\Gamma(r_2)}\int_0^t\int_0^b (t-s)^{r_1-1}(b-y)^{r_2-1}k(t,x,s,y)h(s,y)dyds.
\end{aligned} \tag{3.30}$$

Now an application of Lemma 3.5, to (3.30) yields (3.29). □

We next prove under more appropriate conditions on the functions involved in (3.14) that the solutions tends exponentially toward zero as $t\longrightarrow\infty$.

Theorem 3.9. *Assume that the following hypotheses hold*

(3.14.1) *There exist constants* $\alpha>0$ *and* $M\geq 0$ *such that*

$$|\mu(t,x)|\leq Me^{-\alpha t}; \tag{3.31}$$

$$|f(t,x,s,u)-f(t,x,s,v)|\leq q(t,x,s)e^{-\alpha(t-s)}|u-v|; \tag{3.32}$$

$$|g(t,x,s,y,u)-g(t,x,s,y,v)|\leq k(t,x,s,y)e^{-\alpha(t-s)}|u-v|; \tag{3.33}$$

and $f(t,x,s,0)=g(t,s,s,y,0)=0$; *for each* $(t,x)\in J$, $(t,x,s)\in J_1$, $(t,x,s,y)\in J_2$, $u,v\in\mathbb{R}$, *and the functions* q, k *be as in Theorem 3.6,*

(3.14.2) $\sup_{(t,x)\in J} Q_1(t,x) < \infty$, $\int_0^\infty A_1(\sigma,x)d\sigma < \infty$, *where* Q_1 *and* A_1 *are given by (3.25) and (3.26).*

If u *is any solution of (3.14) on* J, *then all solutions of equation (3.14) are uniformly globally attractive on* J.

Proof. From the hypotheses, for each $(t,x) \in J$, we have that

$$\begin{aligned}|u(t,x)| &\le |\mu(t,x)| + \frac{1}{\Gamma(r_1)}\int_0^t (t-s)^{r_1-1}|f(t,x,s,u(s,x)) - f(t,x,s,0)|ds \\ &+ \frac{1}{\Gamma(r_1)\Gamma(r_2)}\int_0^t\int_0^b (t-s)^{r_1-1}(b-y)^{r_2-1} \\ &\times |g(t,x,s,y,u(s,y)) - g(t,x,s,y,0)|dyds \\ &\le Me^{-\alpha t} + \frac{1}{\Gamma(r_1)}\int_0^t (t-s)^{r_1-1}q(t,x,s)e^{-\alpha(t-s)}|u(x,s)|ds\end{aligned}$$

$$+\frac{1}{\Gamma(r_1)\Gamma(r_2)}\int_0^t\int_0^b (t-s)^{r_1-1}(b-y)^{r_2-1}k(t,x,s,y)e^{-\alpha(t-s)}|u(s,y)|dyds. \tag{3.34}$$

From (3.34), we get

$$|u(t,x)|e^{\alpha t} \le M + \frac{1}{\Gamma(r_1)}\int_0^t (t-s)^{r_1-1}q(t,x,s)e^{\alpha s}|u(x,s)|ds$$

$$+\frac{1}{\Gamma(r_1)\Gamma(r_2)}\int_0^t\int_0^b (t-s)^{r_1-1}(b-y)^{r_2-1}k(t,x,s,y)e^{\alpha s}|u(s,y)|dyds. \tag{3.35}$$

Now an application of Lemma 3.5 to (3.35) yields

$$|u(t,x)|e^{\alpha t} \le MP_1(t,x)\exp\left(\int_0^t A_1(\sigma,x)d\sigma\right);\ (t,x) \in J. \tag{3.36}$$

Multiplying both sides of (3.36) by $e^{-\alpha t}$ and in view of (3.14.2), we get

$$\lim_{t\to\infty}|u(t,x)| \le \lim_{t\to\infty} MP_1(t,x)\exp\left(-\alpha t + \int_0^t A_1(\sigma,x)d\sigma\right) = 0.$$

Hence, the solution u tends to zero as $t \to \infty$. Consequently, all solutions of equation (3.14) are uniformly globally attractive on J. □

3.3.3. An Example

To illustrate our results, we consider the following partial integral equation of Riemann-Liouville fractional order of the form

$$u(t,x) = \frac{e^{x-t}}{1+t+x^2} + \frac{1}{\Gamma(r_1)}\int_0^t (t-s)^{r_1-1}f(t,x,s,u(s,x))ds$$

$$+\frac{1}{\Gamma(r_1)\Gamma(r_2)}\int_0^t\int_0^1 (t-s)^{r_1-1}(1-y)^{r_2-1}g(t,x,s,y,u(s,y))dyds; \tag{3.37}$$

$(t,x) \in \mathbb{R}_+ \times [0,1]$, where $r_1, r_2 \in (0,\infty)$,

$$\begin{cases} f(t,x,s,u) = \dfrac{x^2 t^{-r_1} s^{-\frac{1}{2}} \sin s \sin t}{2c(1+t^{-\frac{1}{2}})(1+|u|)}; \\ \quad \quad for\ \ (t,x,s) \in J_1,\ st \neq 0\ and\ \ u \in \mathbb{R}, \\ f(t,x,0,u) = f(0,x,0,u) = 0, \end{cases}$$

$$J_1 = \{(t,x,s) : 0 \le s \le t < \infty,\ x \in [0,1]\},$$

$$c := \frac{\Gamma(\frac{1}{2})}{\Gamma(\frac{1}{2}+r_1)} + \frac{\Gamma(\frac{1}{2})e}{\Gamma(\frac{1}{2}+r_1)\Gamma(1+r_2)},$$

$$\begin{cases} g(t,x,s,y,u) = \dfrac{t^{-r_1} s^{-\frac{1}{2}} e^{x-y-\frac{1}{s}-\frac{1}{t}}}{2c(1+t^{-\frac{1}{2}})(1+|u|)}; \\ \quad \quad for\ \ (t,x,s,y) \in J_2,\ st \neq 0\ and\ \ u \in \mathbb{R}, \\ g(t,x,0,y,u) = g(0,x,0,y,u) = 0, \end{cases}$$

and

$$J_2 = \{(t,x,s,y) : 0 \le s \le t < \infty,\ x \in [0,1],\ y \in [0,1)\}.$$

Set

$$\mu(t,x) = \frac{e^{x-t}}{1+t+x^2};\ (t,x) \in J.$$

We can see that the function μ is continuous and bounded with $\mu^* = e$.
For each $u, v \in \mathbb{R}$ and $(t,x,s) \in J_1$, we have

$$|f(t,x,s,u) - f(t,x,s,v)| \le \frac{1}{2c(1+t^{-\frac{1}{2}})} \left(x^2 t^{-r_1} s^{-\frac{1}{2}} |\sin s \sin t| \right) |u-v|,$$

and for each $u, v \in \mathbb{R}$ and $(t,x,s,y) \in J_2$, we have

$$|g(t,x,s,y,u) - g(t,x,s,y,v)| \le \frac{1}{2c(1+t^{-\frac{1}{2}})} \left(t^{-r_1} s^{-\frac{1}{2}} e^{x-y-t-\frac{1}{s}-\frac{1}{t}} \right) |u-v|.$$

Hence condition (3.8.2) is satisfied with

$$\begin{cases} q(t,x,s) = \dfrac{1}{2c(1+t^{-\frac{1}{2}})} \left(x^2 t^{-r_1} s^{-\frac{1}{2}} |\sin s \sin t| \right);\ st \neq 0, \\ \quad q(t,x,0) = q(0,x,0) = 0, \end{cases}$$

and condition (3.8.3) is satisfied with

$$\begin{cases} k(t,x,s,y) = \dfrac{1}{2c(1+t^{-\frac{1}{2}})}\left(t^{-r_1}s^{-\frac{1}{2}}e^{x-y-t-\frac{1}{s}-\frac{1}{t}}\right); \; st \neq 0, \\ k(t,x,0,y) = k(0,x,0,y) = 0. \end{cases}$$

We shall show that condition (3.20) holds with $b = 1$. Indeed

$$\begin{aligned} & \frac{1}{\Gamma(r_1)}\int_0^t (t-s)^{r_1-1}q(t,x,s)ds \\ \leq \;& \frac{1}{2c(1+t^{-\frac{1}{2}})\Gamma(r_1)}\int_0^t (t-s)^{r_1-1}x^2t^{-r_1}s^{-\frac{1}{2}}ds \\ = \;& x^2t^{-r_1}t^{-\frac{1}{2}+r_1}\frac{\Gamma(\frac{1}{2})}{2c(1+t^{-\frac{1}{2}})\Gamma(\frac{1}{2}+r_1)} \\ \leq \;& \frac{\Gamma(\frac{1}{2})}{2c(1+t^{-\frac{1}{2}})\Gamma(\frac{1}{2}+r_1)}t^{-\frac{1}{2}}, \end{aligned}$$

then

$$q^* \leq \frac{\Gamma(\frac{1}{2})}{2c\Gamma(\frac{1}{2}+r_1)}.$$

Also,

$$\begin{aligned} & \frac{1}{\Gamma(r_1)\Gamma(r_2)}\int_0^t\int_0^b (t-s)^{r_1-1}(b-y)^{r_2-1}k(t,x,s,y)dyds \\ \leq \;& \frac{1}{2c(1+t^{-\frac{1}{2}})\Gamma(r_1)\Gamma(r_2)}\int_0^t\int_0^1 (t-s)^{r_1-1}(1-y)^{r_2-1}t^{-r_1}s^{-\frac{1}{2}}e^x dyds \\ \leq \;& e^xt^{-r_1}t^{-\frac{1}{2}+r_1}\frac{\Gamma(\frac{1}{2})}{2c(1+t^{-\frac{1}{2}})\Gamma(\frac{1}{2}+r_1)\Gamma(1+r_2)} \\ \leq \;& \frac{\Gamma(\frac{1}{2})et^{-\frac{1}{2}}}{2c(1+t^{-\frac{1}{2}})\Gamma(\frac{1}{2}+r_1)\Gamma(1+r_2)}, \end{aligned}$$

then

$$k^* \leq \frac{e\Gamma(\frac{1}{2})}{2c\Gamma(\frac{1}{2}+r_1)\Gamma(1+r_2)}.$$

Thus,

$$q^* + k^* \leq \frac{1}{2c}\left(\frac{\Gamma(\frac{1}{2})}{\Gamma(\frac{1}{2}+r_1)} + \frac{\Gamma(\frac{1}{2})e}{\Gamma(\frac{1}{2}+r_1)\Gamma(1+r_2)}\right) = \frac{1}{2} < 1,$$

which is satisfied for each $r_1, r_2 \in (0,\infty)$. Consequently Theorem 3.6 implies that equation (3.37) has a unique solution defined on $\mathbb{R}_+ \times [0,1]$.

3.4. Attractivity Results for Functional Partial Integral Equations

3.4.1. Introduction

This section deals with the existence of solutions to the following partial integral equation of Riemann-Liouville fractional order,

$$\begin{aligned} u(t,x) &= f(t,x,u(t,x)) + \frac{g(t,x,u(t,x))}{\Gamma(r_1)\Gamma(r_2)}\int_0^t\int_0^x (t-s)^{r_1-1}(x-y)^{r_2-1} \\ &\quad \times h(t,x,s,y,u(s,y))dyds;\ (t,x)\in\mathbb{R}_+\times[0,b], \end{aligned} \tag{3.38}$$

where $b>0$, $\mathbb{R}_+ := [0,\infty)$, $r_1,r_2\in(0,\infty)$, $f,g: J\times\mathbb{R}\to\mathbb{R}$ and $h: J_1\times\mathbb{R}\to\mathbb{R}$ are given continuous functions $J_1 := \{(t,x,s,y) : 0\le s\le t<\infty,\ 0\le y\le x\le b]\}$.
Our investigations are conducted in a Banach space BC with an application of Schauder's fixed point theorem for the existence of solutions of the equation (3.38). Also, we obtain some results about the attractivity of solutions of equation (3.38). Finally, we present an example illustrating the applicability of the imposed conditions.

3.4.2. Main Results

Theorem 3.10. *Assume that the following hypotheses hold*

(3.16.1) *There exist constants $M,L>0$ with $M<L$ and a continuous function $\alpha: J\to\mathbb{R}_+$ with $\lim_{t\to\infty}\alpha(t,x)=\infty$; $x\in[0,b]$, such that*

$$|f(t,x,u)-f(t,x,v)| \le \frac{M|u-v|}{(1+\alpha(t,x))(L+|u-v|)},$$

for $(t,x)\in J$ and $u,v\in\mathbb{R}$. Moreover, the function $(t,x)\mapsto f(t,x,0)$ is bounded on J with

$$f^* = \sup_{(t,x)\in\mathbb{R}_+\times[0,b]} |f(t,x,0)|,\ \textit{and}\ \lim_{t\to\infty}|f(t,x,0)| = 0;\ x\in[0,b],$$

(3.16.2) *There exists an increasing function $\Phi:\mathbb{R}_+\to\mathbb{R}_+$ such that*

$$|g(t,x,u)| \le |u|\Phi(|u|),\ \textit{for}\ (t,x)\in J\ \textit{and}\ u\in\mathbb{R}.$$

(3.16.3) *There exist a continuous function $P: J_1\to\mathbb{R}_+$ and a function $\Psi:\mathbb{R}_+\to\mathbb{R}_+$ being continuous and increasing on $\mathbb{R}_+$ such that*

$$|h(t,x,s,y,u)| \le P(t,x,s,y)\Psi(|u|),\ \textit{for}\ (t,x,s,y)\in J_1\ \textit{and for}\ u\in\mathbb{R}.$$

Moreover, assume that

$$\lim_{t\to\infty}\int_0^t (t-s)^{r_1-1}P(t,x,s,y)ds = 0;\ \textit{for}\ x,y\in[0,b].$$

(3.16.4) *There exists a constant* $\eta > 0$, *such that*

$$M + f^* \leq \eta\left(1 - q^* \Phi(\eta)\Psi(\eta)\right),$$

where $q^* := \sup_{(t,x)\in J} Q(t,x)$ *and*

$$Q(t,x) = \frac{1}{\Gamma(r_1)\Gamma(r_2)} \int_0^t \int_0^x (t-s)^{r_1-1}(x-y)^{r_2-1} P(t,x,s,y)dyds.$$

Then the equation (3.38) has at least one solution in the space BC. Moreover, solutions of equation (3.38) are uniformly locally attractive.

Proof. Let us define the operator N such that, for any $u \in BC$,

$$\begin{aligned} (Nu)(t,x) &= f(t,x,u(t,x)) + \frac{g(t,x,u(t,x))}{\Gamma(r_1)\Gamma(r_2)} \int_0^t \int_0^x (t-s)^{r_1-1}(x-y)^{r_2-1} \\ &\quad \times h(t,x,s,y,u(s,y))dyds; \; (t,x) \in \mathbb{R}_+ \times [0,b]. \end{aligned} \tag{3.39}$$

By considering the assumptions, we infer that $N(u)$ is continuous on J. Now we prove that $N(u) \in BC$ for any $u \in BC$. For arbitrarily fixed $(t,x) \in J$ we have

$$\begin{aligned} |(Nu)(t,x)| &= |f(t,x,u(t,x)) \\ &\quad + \frac{g(t,x,u(t,x))}{\Gamma(r_1)\Gamma(r_2)} \int_0^t \int_0^x (t-s)^{r_1-1}(x-y)^{r_2-1} \\ &\quad \times h(t,x,s,y,u(s,y))dyds| \\ &\leq |f(t,x,u(t,x)) - f(t,x,0) + f(t,x,0)| \\ &\quad + \frac{\|u\|_{BC}\Phi(\|u\|_{BC})}{\Gamma(r_1)\Gamma(r_2)} \int_0^t \int_0^x (t-s)^{r_1-1}(x-y)^{r_2-1} \\ &\quad \times P(t,x,s,y)\Psi(\|u\|_{BC})dyds \\ &\leq \frac{M|u(t,x)|}{(1+\alpha(t,x))(L+|u(t,x)|)} + |f(t,x,0)| \\ &\quad + \|u\|_{BC}\Phi(\|u\|_{BC})\Psi(\|u\|_{BC})Q(t,x,s,y) \\ &\leq M + f^* + q^*\|u\|_{BC}\Phi(\|u\|_{BC})\Psi(\|u\|_{BC}). \end{aligned}$$

Thus

$$\|N(u)\|_{BC} \leq M + f^* + q^*\|u\|_{BC}\Phi(\|u\|_{BC})\Psi(\|u\|_{BC}). \tag{3.40}$$

Hence $N(u) \in BC$. Equation (3.40) and hypothesis (3.16.4) implies that N transforms the ball $B_\eta := B(0,\eta)$ into itself. We shall show that $N : B_\eta \to B_\eta$ satisfies the assumptions of Schauder's fixed point theorem [204]. The proof will be given in several steps and cases.

Step 1: *N is continuous.*
Let $\{u_n\}_{n\in\mathbb{N}}$ be a sequence such that $u_n \to u$ in B_η. Then, for each $(t,x) \in J$, we have

$$\begin{aligned}
|(Nu_n)(t,x)-(Nu)(t,x)| &\leq |f(t,x,u_n(t,x))-f(t,x,u(t,x))| \\
&+\left|\frac{g(t,x,u_n(t,x))}{\Gamma(r_1)\Gamma(r_2)}\int_0^t\int_0^x (t-s)^{r_1-1}(x-y)^{r_2-1}h(t,x,s,y,u_n(s,y))dyds\right. \\
&\left.-\frac{g(t,x,u(t,x))}{\Gamma(r_1)\Gamma(r_2)}\int_0^t\int_0^x (t-s)^{r_1-1}(x-y)^{r_2-1}h(t,x,s,y,u(s,y))dyds\right| \\
\leq\ & \frac{M}{L}\|u_n-u\|+\frac{|g(t,x,u_n(t,x))-g(t,x,u(t,x))|}{\Gamma(r_1)\Gamma(r_2)} \\
&\times\int_0^t\int_0^x (t-s)^{r_1-1}(x-y)^{r_2-1}|h(t,x,s,y,u_n(s,y))|dyds \\
&+\frac{|g(t,x,u(t,x))|}{\Gamma(r_1)\Gamma(r_2)}\int_0^t\int_0^x (t-s)^{r_1-1}(x-y)^{r_2-1} \\
&\times|h(t,x,s,y,u_n(s,y))-h(t,x,s,y,u(s,y))|dyds. \qquad (3.41)
\end{aligned}$$

Case 1. If $(t,x) \in [0,a]\times[0,b]$; $a>0$, then, since $u_n \to u$ as $n\to\infty$ and g,h are continuous, (3.41) gives

$$\|N(u_n)-N(u)\|_{BC} \to 0 \quad \text{as } n\to\infty.$$

Case 2. If $(t,x) \in (a,\infty)\times[0,b]$, $a>0$, then from (3.16.2), (3.16.3) and (3.41), for each $(t,x)\in J$, we get

$$\begin{aligned}
|(Nu_n)(t,x)-(Nu)(t,x)| &\leq \frac{M}{L}\|u_n-u\| \\
&+[\|u_n\|_{BC}\Phi(\|u_n\|_{BC})+\|u\|_{BC}\Phi(\|u\|_{BC})]Q(t,x)\Psi(\|u_n\|_{BC}) \\
&+\|u\|_{BC}\Phi(\|u\|_{BC})Q(t,x)[\Psi(\|u_n\|_{BC})+\Psi(\|u\|_{BC})] \\
&\leq \frac{M}{L}\|u_n-u\|+4\eta\Phi(\eta)\Psi(\eta)Q(t,x). \qquad (3.42)
\end{aligned}$$

Since $u_n \to u$ as $n\to\infty$ and $t\to\infty$, then (3.42) gives

$$\|N(u_n)-N(u)\|_{BC} \to 0 \quad \text{as } n\to\infty.$$

Step 2: *$N(B_\eta)$ is uniformly bounded.*
This is clear since $N(B_\eta)\subset B_\eta$ and B_η is bounded.

Step 3: *$N(B_\eta)$ is equicontinuous on every compact subset $[0,a]\times[0,b]$ of J, $a>0$.*
Let $(t_1,x_1),(t_2,x_2)\in[0,a]\times[0,b]$, $t_1<t_2$, $x_1<x_2$ and let $u\in B_\eta$. Thus we have

$$\begin{aligned}
|(Nu)(t_2,x_2)-(Nu)(t_1,x_1)| &\leq |f(t_2,x_2,u(t_2,x_2))-f(t_2,x_2,u(t_1,x_1))| \\
&+|f(t_2,x_2,u(t_1,x_1))-f(t_1,x_1,u(t_1,x_1))|
\end{aligned}$$

$$+\frac{|g(t_2,x_2,u(t_2,x_2))-g(t_1,x_1,u(t_1,x_1))|}{\Gamma(r_1)\Gamma(r_2)}$$
$$\times\int_0^{t_2}\int_0^{x_2}(t_2-s)^{r_1-1}(x_2-y)^{r_2-1}|h(t_2,x_2,s,y,u(s,y))|dyds$$
$$+\frac{|g(t_1,x_1,u(t_1,x_1))|}{\Gamma(r_1)\Gamma(r_2)}\int_0^{t_1}\int_0^{x_1}(t_1-s)^{r_1-1}(x_1-y)^{r_2-1}$$
$$\times|h(t_2,x_2,s,y,u(s,y))-h(t_1,x_1,s,y,u(s,y))|dyds$$
$$\leq\frac{M}{L}|u(t_2,x_2)-u(t_1,x_1)|+|f(t_2,x_2,u(t_1,x_1))-f(t_1,x_1,u(t_1,x_1))|$$
$$+|g(t_2,x_2,u(t_2,x_2))-g(t_1,x_1,u(t_1,x_1))|Q(t_2,x_2)\Psi(\eta)$$
$$+\frac{\eta\Phi(\eta)}{\Gamma(r_1)\Gamma(r_2)}\int_0^{t_1}\int_0^{x_1}(t_1-s)^{r_1-1}(x_1-y)^{r_2-1}$$
$$\times|h(t_2,x_2,s,y,u(s,y))-h(t_1,x_1,s,y,u(s,y))|dyds$$

From continuity of f,g,h and as $t_1\longrightarrow t_2$ and $x_1\longrightarrow x_2$, the right-hand side of the above inequality tends to zero.

Step 4: *$N(B_\eta)$ is equiconvergent.*
Let $(t,x)\in J$ and $u\in B_\eta$, then we have

$$\begin{aligned}|(Nu)(t,x)|&\leq|f(t,x,u(t,x))\\&\quad+\frac{g(t,x,u(t,x))}{\Gamma(r_1)\Gamma(r_2)}\int_0^t\int_0^x(t-s)^{r_1-1}(x-y)^{r_2-1}\\&\quad\times h(t,x,s,y,u(s,y))dyds|\\&\leq|f(t,x,u(t,x))-f(t,x,0)+f(t,x,0)|\\&\quad+\frac{\|u\|_{BC}\Phi(\|u\|_{BC})}{\Gamma(r_1)\Gamma(r_2)}\int_0^t\int_0^x(t-s)^{r_1-1}(x-y)^{r_2-1}\\&\quad\times P(t,x,s,y)\Psi(\|u\|_{BC})dyds\\&\leq\frac{M|u(t,x)|}{(1+\alpha(t,x))(L+|u(t,x)|)}+|f(t,x,0)|\\&\quad+\|u\|_{BC}\Phi(\|u\|_{BC})\Psi(\|u\|_{BC})Q(t,x)\\&\leq\frac{M}{1+\alpha(t,x)}+|f(t,x,0)|+\eta\Phi(\eta)\Psi(\eta)Q(t,x).\end{aligned}$$

Thus, for each $x\in[0,b]$, we get

$$|(Nu)(t,x)|\longrightarrow 0,\ as\, t\longrightarrow+\infty.$$

Hence,

$$|(Nu)(t,x)-(Nu)(+\infty,x)|\longrightarrow 0,\ as\, t\longrightarrow+\infty.$$

As a consequence of Steps 1 to 4 together with the Lemma 1.69, we can conclude that $N:B_\eta\rightarrow B_\eta$ is continuous and compact. From an application of Schauder's theorem [204], we deduce that N has a fixed point u which is a solution of the equation (3.38).

Step 5: Now we investigate the uniform local attractivity for solutions of equation (3.38). Let us assume that u_0 is a solution of equation (3.38) with the conditions of this theorem. Consider the ball $B(u_0,\eta^*)$ with $u_0 \in B(0,\eta)$, $\eta^* = \frac{LM^*}{L-M}$, where

$$M^* := \sup_{(t,x)\in J}\left\{\frac{1}{\Gamma(r_1)\Gamma(r_2)}\int_0^t\int_0^x (t-s)^{r_1-1}(x-y)^{r_2-1}\right.$$

$$\left.\times|g(t,x,u(t,x))h(t,x,s,y,u(s,y))-g(t,x,u_0(t,x))h(t,x,s,y,u_0(s,y))|dyds\right\}.$$

Taking $u \in B(u_0,\eta^*)$, we have

$$\begin{aligned}|(Nu)(t,x)-u_0(t,x)| &= |(Nu)(t,x)-(Nu_0)(t,x)|\\ &\leq \frac{M}{L}\|u-u_0\|+M^*\\ &\leq \frac{M\eta^*}{L}+M^*=\eta^*.\end{aligned}$$

Thus we observe that N is a continuous function such that $N(B(u_0,\eta^*)) \subset B(u_0,\eta^*)$. Moreover, if u is a solution of equation (3.38), then

$$\begin{aligned}|u(t,x)-u_0(t,x)| &= |(Nu)(t,x)-(Nu_0)(t,x)|\\ &\leq |f(t,x,u(t,x))-f(t,x,u_0(t,x))|\\ &\quad+\frac{1}{\Gamma(r_1)\Gamma(r_2)}\int_0^t\int_0^x (t-s)^{r_1-1}(x-y)^{r_2-1}\\ &\quad\times|g(t,x,u(t,x))h(t,x,s,y,u(s,y))\\ &\quad-g(t,x,u_0(t,x))h(t,x,s,y,u_0(s,y))|dyds\\ &\leq \frac{M}{L}|u(t,x)-u_0(t,x)|+Q(t,x)\\ &\quad\times[|u(t,x)|\Phi(|u(t,x)|)\Psi(|u(t,x)|)\\ &\quad+|u_0(t,x)|\Phi(|u_0(t,x)|)\Psi(|u_0(t,x)|)]\\ &\leq \frac{M}{L}|u(t,x)-u_0(t,x)|+Q(t,x)\end{aligned}$$

$$\times[(\eta^*+2|u_0(t,x)|)\Phi(\eta^*+|u_0(t,x)|)\Psi(\eta^*+|u_0(t,x)||). \tag{3.43}$$

By using (3.43) and the fact that $M < L$, we deduce that

$$|u(t,x)-u_0(t,x)| \leq \frac{LM_*}{L-M}Q(t,x),$$

where $M_* = (2\eta+\eta^*)\Phi(\eta+\eta^*)\Psi(\eta+\eta^*)$. Hence

$$\lim_{t\to\infty}|u(t,x)-u_0(t,x)| = 0.$$

Consequently, all solutions of equation (3.38) are uniformly locally attractive. □

3.4.3. An Example

As an application of our results we consider the following integral equation of fractional order

$$\begin{aligned} u(t,x) &= f(t,x,u(t,x)) + \tfrac{g(t,x,u(t,x))}{\Gamma(r_1)\Gamma(r_2)} \int_0^t \int_0^x (t-s)^{r_1-1}(x-y)^{r_2-1} \\ & \times h(t,x,s,y,u(s,y))dyds;\ (t,x) \in \mathbb{R}_+ \times [0,1], \end{aligned} \tag{3.44}$$

where $(r_1, r_2) = (\frac{1}{3}, 2)$,

$$f(t,x,u) = \frac{ctx\sin u}{1+x^2+t^2},\quad c := \frac{9}{8\Gamma^2(\frac{1}{3})},$$

$$g(t,x,u) = |u|^2 \ \ and \ \ h(t,x,s,y,u) = \frac{|u|\ln(1+sxy|u|)}{(1+t+|u|)^2(1+x^2+t^4)}.$$

First, we can see that $c < \frac{9}{8\pi} < \frac{2}{5}$. The function f is continuous and satisfies assumption (3.16.1), where

$$M = c < \frac{2}{5},\ L = 1,\ \alpha(t,x) = x^2+t^2 \ and \ f^* = 0.$$

The function g satisfies assumption (3.16.2), where $\Phi(w) = w$.
Next, let us notice that the function h satisfies assumption (3.16.3), where

$$P(t,x,s,y) = \frac{sxy}{1+x^2+t^4},\quad and \ \ \Psi(w) = w;$$

for $(t,x,s,y) \in J_1 := \{(t,x,s,y) : 0 \le s \le t < \infty,\ 0 \le y \le x \le 1]\}$, and $u \in \mathbb{R}$. Also, we have,

$$\begin{aligned} \int_0^t (t-s)^{r_1-1}P(t,x,s,y)ds &= \int_0^t (t-s)^{r_1-1}\frac{sxy}{1+x^2+t^4}ds \\ &\le \frac{xy}{1+t^4}\int_0^t (t-s)^{r_1-1}sds \\ &\le \frac{1}{1+t^4}\int_0^t (t-s)^{\frac{-2}{3}}sds \\ &= \frac{\Gamma(2)}{\Gamma(\frac{7}{3})}\frac{t^{\frac{4}{3}}}{1+t^4}. \end{aligned}$$

Thus

$$\lim_{t\to\infty}\int_0^t (t-s)^{r-1}P(t,x,s,y)ds = 0.$$

Finally, we can see that the hypothesis (3.16.4) is satisfied with $\eta = 1$. Indeed,

$$q^* \le \frac{\Gamma(2)}{\Gamma(\frac{1}{3})\Gamma(3)\Gamma(\frac{7}{3})} = c,\ \ M+f^* = c < \frac{2}{5},$$

and

$$\eta(1-q^*\Phi(\eta)\Psi(\eta)) \ge \eta\left(1-\frac{\Gamma(2)}{\Gamma(\frac{1}{3})\Gamma(3)\Gamma(\frac{7}{3})}\eta^2\right) = \eta(1-c\eta^2) > \eta(1-\frac{2}{5}\eta^2).$$

So, for $\eta = 1$, we obtain

$$M + f^* \leq \eta(1 - q^*\Phi(\eta)\Psi(\eta)).$$

Hence by Theorem 3.10, the equation (3.44) has a solution defined on $\mathbb{R}_+ \times [0,1]$ and solutions of this equation are uniformly locally attractive.

3.5. Notes and Remarks

The results of Chapter 3 are taken from Abbas *et al.* [20, 25, 26]. Other results may be found in [290, 291, 303].

Chapter 4

Fractional Order Riemann-Liouville Volterra-Stieltjes Quadratic Integral Equations

4.1. Introduction

In this chapter, we shall present existence results for some classes of Riemann-Liouville Volterra-Stieltjes quadratic integral equations of fractional order by using some fixed point theorems.

4.2. Existence and Stability Results for Nonlinear Riemann-Liouville Volterra-Stieltjes Quadratic Integral Equations

4.2.1. Introduction

In this section, we establish some sufficient conditions for the existence of solutions of the following fractional order Riemann-Liouville Volterra-Stieltjes quadratic integral equations of the form

$$u(t,x) = \mu(t,x) + \frac{f(t,x,u(t,x))}{\Gamma(r)} \int_0^t (t-s)^{r-1} h(t,x,s,u(s,x)) d_s g(t,s), \qquad (4.1)$$

for $(t,x) \in J := [0,a] \times [0,b], a,b > 0,\ r \in (0,\infty),\ \mu : J \to \mathbb{R},\ g : \mathbb{R}_+ \times \mathbb{R}_+ \to \mathbb{R},\ f : J \times \mathbb{R} \to \mathbb{R},\ h : J_1 \times \mathbb{R} \to \mathbb{R}$ are given continuous functions, $\mathbb{R}_+ = [0,\infty)$, $J_1 = \{(t,x,s) \in J \times [0,a] : s \le t\}$. We present two results for the existence of solutions of the equation (4.1). The first one is based on Shauder's fixed point theorem (Theorem 4.2) and the second one on the nonlinear alternative of Leray-Schauder type (Theorem 4.3).

Next, we establish a sufficient condition for the existence and the stability of solutions of the following fractional order Riemann-Liouville Volterra-Stieltjes quadratic integral equations of the form

$$u(t,x) = \mu(t,x) + \frac{f(t,x,u(t,x))}{\Gamma(r)} \int_0^t (t-s)^{r-1} h(t,x,s,u(s,x)) d_s g(t,s), \tag{4.2}$$

where $(t,x) \in J' := \mathbb{R}_+ \times [0,b], b > 0,\ r \in (0,\infty),\ \mu : J' \to \mathbb{R},\ g : \mathbb{R}_+ \times \mathbb{R}_+ \to \mathbb{R},\ f : J' \times \mathbb{R} \to \mathbb{R},\ h : J_1' \times \mathbb{R} \to \mathbb{R}$ are given continuous functions and $J_1' = \{(t,x,s) \in J' \times \mathbb{R}_+ : s \leq t\}$. We use the Schauder fixed point theorem for the existence of solutions of the equation (4.2), and we prove that all solutions are locally asymptotically stable (Theorem 4.6).

4.2.2. Main Results

Let $C := C(J)$ be the Banach space of all continuous functions from J into $\mathbb{R}$ endowed with the norm

$$\|u\|_C = \sup_{(t,x) \in J} |u(t,x)|.$$

By $BC := BC(J')$ we denote the Banach space of all bounded and continuous functions from J' into $\mathbb{R}$ equipped with the standard norm

$$\|u\|_{BC} = \sup_{(t,x) \in J'} |u(t,x)|.$$

For $u_0 \in C$ $(or\ u_0 \in BC)$ and $\eta \in (0,\infty)$, we denote by $B(u_0,\eta)$, the closed ball in C $(or\ BC)$ centered at u_0 with radius η.

Definition 4.1. *We mean by a solution of equation (4.1), every function $u \in C$ such that u satisfies equation (4.1) on J.*

The following hypotheses will be used in the sequel:

(4.1.1) The function f is a continuous and there exists a positive function $p \in C$ such that

$$|f(t,x,u) - f(t,x,v)| \leq p(t,x)|u-v|;\ (t,x) \in J,\ u,v \in \mathbb{R},$$

(4.1.2) For all $t_1,t_2 \in [0,a]$ such that $t_1 < t_2$ the function $s \mapsto g(t_2,s) - g(t_1,s)$ is nondecreasing on $[0,a]$,

(4.1.3) The function $s \mapsto g(0,s)$ is nondecreasing on $[0,a]$,

(4.1.4) The functions $s \mapsto g(t,s)$ and $t \mapsto g(t,s)$ are continuous on $[0,a]$ for each fixed $t \in [0,a]$ or $s \in [0,a]$, respectively,

(4.1.5) The function h is continuous and there exist continuous functions $q = q(t,x,s) : J_1 \to \mathbb{R}_+;\ \Phi : \mathbb{R}_+ \to \mathbb{R}_+$ such that Φ is nondecreasing and

$$|h(t,x,s,u)| \leq q(t,x,s)\Phi(|u|);\ (t,x,s) \in J_1,\ u \in \mathbb{R}.$$

Set

$$\mu^* := \sup_{(t,x)\in J} \mu(t,x),\ f^* := \sup_{(t,x)\in J} f(t,x,0),\ g^* = \sup_{t\in[0,a]} \bigvee_{k=0}^{t} g(t,k),$$

$$p^* := \sup_{(t,x)\in J} p(t,x) \ \ and \ \ q^* := \sup_{(t,x,s)\in J_1} \frac{(t-s)^{r-1}q(t,x,s)}{\Gamma(r)}.$$

Now, we shall prove the following theorem concerning the existence of a solution of the equation (4.1), based on Schauder's fixed point theorem [204].

Theorem 4.2. *Assume that the hypotheses* (4.1.1) – (4.1.5) *and the following hypothesis hold*

(4.2.1) *There exists a constant* $\eta > 0$, *such that* $\mu^* + q^* g^* \Phi(\eta)(f^* + p^*\eta) \leq \eta$.

Then the equation (4.1) *has at least one solution in the space* C.

Proof. Let us define the operator $N : C \to C$ such that, for any $u \in C$,

$$(Nu)(t,x) = \mu(t,x)$$
$$+ \frac{f(t,x,u(t,x))}{\Gamma(r)} \int_0^t (t-s)^{r-1} h(t,x,s,u(s,x)) d_s g(t,s);\ (t,x) \in J. \tag{4.3}$$

It is clear that the operator N maps C into C. The problem of finding the solutions of the equation (4.1) is reduced to finding the solutions of the operator equation $N(u) = u$. Hypothesis (4.2.1) implies that N transforms the ball $B_\eta := B(0,\eta)$ into itself. Indeed, for any $u \in B_\eta$, and for each $(t,x) \in J$ we have

$$\begin{aligned}
&|(Nu)(t,x)| \leq |\mu(t,x)| \\
&+ \frac{|f(t,x,0)|}{\Gamma(r)} \int_0^t (t-s)^{r-1} |h(t,x,s,u(s,x))||d_s g(t,s)| \\
&+ \frac{|f(t,x,u(t,x)) - f(t,x,0)|}{\Gamma(r)} \int_0^t (t-s)^{r-1} \\
&\quad \times |h(t,x,s,u(s,x))| d_s \left(\bigvee_{k=0}^{s} g(t,k) \right) \\
&\leq \mu^* + \frac{f^*}{\Gamma(r)} \int_0^t (t-s)^{r-1} q(t,x,s) \Phi(|u(s,x)|) d_s \left(\bigvee_{k=0}^{s} g(t,k) \right) \\
&+ \frac{p(t,x)|u(t,x)|}{\Gamma(r)} \int_0^t (t-s)^{r-1} q(t,x,s) \Phi(|u(s,x)|) d_s \left(\bigvee_{k=0}^{s} g(t,k) \right) \\
&\leq \mu^* + q^* g^* \Phi(\|u\|_C)(f^* + p^* \|u\|_C) \\
&\leq \mu^* + q^* g^* \Phi(\eta)(f^* + p^*\eta) \\
&\leq \eta.
\end{aligned}$$

Hence, $N(u) \in B_\eta$. We shall show that $N : B_\eta \to B_\eta$ satisfies the assumptions of Schauder's fixed point theorem [204]. The proof will be given in several steps.

Step 1: *N is continuous.*
Let $\{u_n\}_{n\in\mathbb{N}}$ be a sequence such that $u_n \to u$ in B_η. Then, for each $(t,x)\in J$, we have

$$\begin{aligned}
&|(Nu_n)(t,x)-(Nu)(t,x)|\\
\le\ & |f(t,x,u_n(t,x))-f(t,x,u(t,x))|\int_0^t \frac{(t-s)^{r-1}}{\Gamma(r)}\\
&\times|h(t,x,s,u_n(s,x))|d_sg(t,s)|\\
+\ & |f(t,x,u(t,x))|\int_0^t \frac{(t-s)^{r-1}}{\Gamma(r)}|h(t,x,s,u_n(s,x))\\
-\ & h(t,x,s,u(s,x))||d_sg(t,s)|\\
\le\ & p^*q^*g^*\Phi(\|u_n\|_C)|u_n(t,x)-u(t,x)|\\
+\ & \frac{f^*+p^*\|u\|_C}{\Gamma(r)}\int_0^t (t-s)^{r-1}|h(t,x,s,u_n(s,x))\\
-\ & h(t,x,s,u(s,x))|d_s\left(\bigvee_{k=0}^{s} g(t,k)\right)\\
\le\ & p^*q^*g^*\Phi(\eta)|u_n(t,x)-u(t,x)|
\end{aligned}$$

$$+\frac{f^*+p^*\eta}{\Gamma(r)}\int_0^t (t-s)^{r-1}|h(t,x,s,u_n(s,x))-h(t,x,s,u(s,x))|d_s\left(\bigvee_{k=0}^{s} g(t,k)\right). \tag{4.4}$$

Since $u_n \to u$ as $n\to\infty$ and h is continuous, (4.4) gives

$$\|N(u_n)-N(u)\|_C \to 0 \quad \text{as } n\to\infty.$$

Step 2: *$N(B_\eta)$ is bounded.*
This is clear since $N(B_\eta)\subset B_\eta$ and B_η is bounded.

Step 3: *$N(B_\eta)$ is equicontinuous.*
Let $(t_1,x_1),(t_2,x_2)\in J$, $t_1<t_2$, $x_1<x_2$ and let $u\in B_\eta$. Thus we have

$$\begin{aligned}
&|(Nu)(t_2,x_2)-(Nu)(t_1,x_1)| \le |\mu(t_2,x_2)-\mu(t_1,x_1)|\\
+\ & |f(t_2,x_2,u(t_2,x_2))-f(t_1,x_1,u(t_1,x_1))|\\
\times\ & \int_0^{t_2}\frac{(t_2-s)^{r-1}}{\Gamma(r)}|h(t_2,x_2,s,u(s,x_2))|d_s\left(\bigvee_{k=0}^{s} g(t_2,k)\right)\\
+\ & |f(t_1,x_1,u(t_1,x_1))|\left|\int_0^{t_2}\frac{(t_2-s)^{r-1}}{\Gamma(r)}h(t_2,x_2,s,u(s,x_2))d_sg(t_2,s)\right.\\
-\ & \left.\int_0^{t_1}\frac{(t_1-s)^{r-1}}{\Gamma(r)}h(t_1,x_1,s,u(s,x_1))d_sg(t_1,s)\right|.
\end{aligned}$$

Thus

$$\begin{aligned}
&|(Nu)(t_2,x_2)-(Nu)(t_1,x_1)| \le |\mu(t_2,x_2)-\mu(t_1,x_1)|\\
+\ & q^*g^*\Phi(\eta)|f(t_2,x_2,u(t_2,x_2))-f(t_1,x_1,u(t_1,x_1))|
\end{aligned}$$

$$+ \quad p^*\eta\left|\int_0^{t_2}\frac{(t_2-s)^{r-1}}{\Gamma(r)}h(t_2,x_2,s,u(s,x_2))d_sg(t_2,s)\right.$$
$$\left. - \quad \int_0^{t_1}\frac{(t_1-s)^{r-1}}{\Gamma(r)}h(t_1,x_1,s,u(s,x_1))d_sg(t_1,s)\right|.$$

Hence

$$|(Nu)(t_2,x_2)-(Nu)(t_1,x_1)| \leq |\mu(t_2,x_2)-\mu(t_1,x_1)|$$
$$+ \quad p^*q^*g^*\Phi(\eta)|u(t_2,x_2)-u(t_1,x_1)|+\frac{p^*\eta}{\Gamma(r)}\int_0^{t_1}|(t_2-s)^{r-1}$$
$$\times \quad h(t_2,x_2,s,u(s,x_2))-(t_1-s)^{r-1}h(t_1,x_1,s,u(s,x_1))|\,d_s\left(\bigvee_{k=0}^{s}g(t_1,k)\right)$$
$$+ \quad \frac{p^*\eta}{\Gamma(r)}\int_0^{t_1}|(t_2-s)^{r-1}h(t_2,x_2,s,u(s,x_2))||d_s(g(t_2,s)-g(t_1,s))|$$
$$+ \quad p^*q^*\eta\Phi(\eta)\bigvee_{k=t_1}^{t_2}g(t_2,k).$$

From continuity of μ, g, h and as $t_1 \to t_2$ and $x_1 \to x_2$, the right-hand side of the above inequality tends to zero. As a consequence of Steps 1 to 3 together with the Arzela-Ascoli theorem, we can conclude that $N : B_\eta \to B_\eta$ is continuous and compact. From an application of Schauder's theorem [204], we deduce that N has a fixed point u which is a solution of the equation (4.1). □

Now, we present another result based on the nonlinear alternative of Leray-Schauder type (Theorem 1.57).

Theorem 4.3. *Assume that the hypotheses* (3.1.1) – (3.1.4) *and the following hypothesis hold*

(4.4.1) *The function h is continuous and there exist continuous functions $k_1, k_2 : J_1 \to \mathbb{R}_+$ such that*

$$|h(t,x,s,u)| \leq k_1(t,x,s)+\frac{k_2(t,x,s)}{1+|u|};\ (t,x,s)\in J_1,\ u\in\mathbb{R},$$

with

$$k_i^* := \sup_{(t,x,s)\in J_1}\frac{1}{\Gamma(r)}(t-s)^{r-1}k_i(t,x,s);\ i=1,2.$$

If

$$p^*g^*k_1^* < 1, \tag{4.5}$$

then the equation (4.1) *has at least one solution in the space C.*

Proof. We shall show that the operator N defined in (4.3) satisfies all the conditions of Theorem 1.57. As in Theorem 4.2, we can show that N is completely continuous.

A priori bounds.

We shall show there exists an open set $U \subseteq C$ with $u \neq \lambda N(u)$, for $\lambda \in (0,1)$ and $u \in \partial U$. Let $u \in C$ be such that $u = \lambda N(u)$ for some $0 < \lambda < 1$. Thus for each $(x,y) \in J$, we have

$$
\begin{aligned}
|u(t,x)| &\leq |\lambda\mu(t,x)| \\
&+ \frac{|\lambda f(t,x,0)|}{\Gamma(r)} \int_0^t (t-s)^{r-1} |h(t,x,s,u(s,x))| |d_s g(t,s)| \\
&+ \frac{|\lambda||f(t,x,u(t,x)) - f(t,x,0)|}{\Gamma(r)} \int_0^t (t-s)^{r-1} \\
&\quad \times |h(t,x,s,u(s,x))| d_s \left(\bigvee_{k=0}^{s} g(t,k) \right) \\
&\leq \mu^* + \frac{f^*}{\Gamma(r)} \int_0^t (t-s)^{r-1} \left(k_1(t,x,s) + \frac{k_2(t,x,s)}{1+|u(s,x)|} \right) d_s \left(\bigvee_{k=0}^{s} g(t,k) \right) \\
&+ \frac{p(t,x)|u(t,x)|}{\Gamma(r)} \int_0^t (t-s)^{r-1} \left(k_1(t,x,s) + \frac{k_2(t,x,s)}{1+|u(s,x)|} \right) d_s \left(\bigvee_{k=0}^{s} g(t,k) \right) \\
&\leq \mu^* + f^* g^* (k_1^* + k_2^*) + p^* g^* k_2^* + p^* g^* k_1^* \|u\|_C.
\end{aligned}
$$

Then,

$$\|u\|_C \leq \mu^* + f^* g^* (k_1^* + k_2^*) + p^* g^* k_2^* + p^* g^* k_1^* \|u\|_C.$$

Thus, by (4.5) we get

$$\|u\|_C \leq \frac{\mu^* + f^* g^* (k_1^* + k_2^*) + p^* g^* k_2^*}{1 - p^* g^* k_1^*} := M.$$

Set

$$U = \{u \in C : \|u\|_C < M+1\}.$$

By our choice of U, there is no $u \in \partial U$ such that $u = \lambda N(u)$, for $\lambda \in (0,1)$.
From Theorem 1.57, we deduce that N has a fixed point u in $\overline{U}$ which is a solution to the equation (4.1). □

4.2.3. Global Existence and Stability of Solutions

Definition 4.4. *We mean by a solution of equation* (4.2)*, every function $u \in BC$ such that u satisfies equation* (4.2) *on J'.*

The following hypotheses will be used in the sequel:

(4.6.1) The function μ is in BC. Moreover, assume that $\lim_{t\to\infty} \mu(t,x) = 0$; $x \in [0,b]$,

(4.6.2) The function f is a continuous and there exists a positive function $p \in BC$ such that

$$|f(t,x,u) - f(t,x,v)| \leq p(t,x)|u-v|; \; (t,x) \in J, \; u,v \in \mathbb{R}.$$

Moreover, assume that $\lim_{t\to\infty} f(t,x,0) = 0$; $x \in [0,b]$,

(4.6.3) For all $t_1, t_2 \in \mathbb{R}_+$ such that $t_1 < t_2$ the function $s \mapsto g(t_2, s) - g(t_1, s)$ is nondecreasing on $\mathbb{R}_+$,

(4.6.4) The function $s \mapsto g(0, s)$ is nondecreasing on $\mathbb{R}_+$,

(4.6.5) The functions $s \mapsto g(t, s)$ and $t \mapsto g(t, s)$ are continuous on $\mathbb{R}_+$ for each fixed $t \in \mathbb{R}_+$ or $s \in \mathbb{R}_+$, respectively,

(4.6.6) The function h is continuous and there exist continuous functions $q : J_1' \to \mathbb{R}_+$, $\Phi : \mathbb{R}_+ \to \mathbb{R}_+$ such that Φ is nondecreasing and

$$|h(t,x,s,u)| \leq q(t,x,s)\Phi(|u|); \; (t,x,s) \in J_1', \; u \in \mathbb{R}.$$

Moreover, assume that $\lim_{t\to\infty} \int_0^t (t-s)^{r-1} q(t,x,s) d_s g(t,s) = 0.$

Remark 4.5. *Set* $\mu^* := \sup_{(t,x)\in J'} \mu(t,x)$, $f^* := \sup_{(t,x)\in J'} f(t,x,0)$, $p^* := \sup_{(t,x)\in J'} p(t,x)$ *and* $q^* :=$ $\sup_{(t,x)\in J'} \frac{1}{\Gamma(r)} \int_0^t (t-s)^{r-1} q(t,x,s) d_s \left(\bigvee_{k=0}^{s} g(t,k) \right)$. *From hypotheses, we infer that* μ^*, f^*, p^* *and* q^* *are finite.*

Now, we shall prove the following theorem concerning the existence and the stability of a solution of the equation (4.2).

Theorem 4.6. *Assume that the hypotheses* (4.6.1) – (4.6.6) *and the following hypothesis hold*

(4.8.1) *There exists a constant* $\eta > 0$, *such that* $\mu^* + q^*\Phi(\eta)(f^* + p^*\eta) \leq \eta$.

Then the equation (4.2) *has at least one solution in the space BC. Moreover, if there exists a constant* $\eta^* > 0$, *such that*

$$q^*(f^* + p^*\eta)\Phi(\eta) + [q^* f^* + p^* q^*(\eta + \eta^*)]\Phi(\eta + \eta^*) \leq \eta^*,$$

then solutions of the equation (4.2) *are locally asymptotically stable.*

Proof. Let us define the operator N' such that, for any $u \in BC$,

$$(N'u)(t,x) = \mu(t,x) + \frac{f(t,x,u(t,x))}{\Gamma(r)} \int_0^t (t-s)^{r-1} h(t,x,s,u(s,x)) d_s g(t,s); (t,x) \in J'. \tag{4.6}$$

The operator N' maps BC into BC. Indeed the map $N'(u)$ is continuous on J' for any $u \in BC$, and for each $(t,x) \in J'$ we have

$$|(N'u)(t,x)| \leq |\mu(t,x)| + \frac{|f(t,x,0)|}{\Gamma(r)} \int_0^t (t-s)^{r-1} |h(t,x,s,u(s,x))||d_s g(t,s)|$$
$$+ \frac{|f(t,x,u(t,x)) - f(t,x,0)|}{\Gamma(r)} \int_0^t (t-s)^{r-1} |h(t,x,s,u(s,x))| d_s \left(\bigvee_{k=0}^{s} g(t,k) \right)$$

$$\begin{aligned}
&\leq \mu^* + \frac{f^*}{\Gamma(r)} \int_0^t (t-s)^{r-1} q(t,x,s)\Phi(|u(s,x)|) d_s \left(\bigvee_{k=0}^{s} g(t,k) \right) \\
&+ \frac{p(t,x)|u(t,x)|}{\Gamma(r)} \int_0^t (t-s)^{r-1} q(t,x,s)\Phi(|u(s,x)|) d_s \left(\bigvee_{k=0}^{s} g(t,k) \right) \\
&\leq \mu^* + \frac{f^*\Phi(\|u\|_{BC})}{\Gamma(r)} \int_0^t (t-s)^{r-1} q(t,x,s) d_s \left(\bigvee_{k=0}^{s} g(t,k) \right) \\
&+ \frac{p^*\|u\|_{BC}\Phi(\|u\|_{BC})}{\Gamma(r)} \int_0^t (t-s)^{r-1} q(t,x,s) d_s \left(\bigvee_{k=0}^{s} g(t,k) \right) \\
&\leq \mu^* + q^*\Phi(\|u\|_{BC})(f^* + p^*\|u\|_{BC}).
\end{aligned}$$

Thus,

$$\|N'(u)\|_{BC} \leq \mu^* + q^*\Phi(\|u\|_{BC})(f^* + p^*\|u\|_{BC}). \tag{4.7}$$

Hence, $N'(u) \in BC$. This proves that the operator N' maps BC into itself.
The problem of finding the solutions of the equation (4.2) is reduced to finding the solutions of the operator equation $N'(u) = u$. Hypothesis (4.8.1) implies that N' transforms the ball $B_\eta := B(0,\eta)$ into itself. We shall show that $N' : B_\eta \to B_\eta$ satisfies the assumptions of Schauder's fixed point theorem [204]. The proof will be given in several steps and cases.

Step 1: *N' is continuous.*
Let $\{u_n\}_{n\in\mathbb{N}}$ be a sequence such that $u_n \to u$ in B_η. Then, for each $(t,x) \in J'$, we have

$$\begin{aligned}
& |(N'u_n)(t,x) - (N'u)(t,x)| \\
&\leq |f(t,x,u_n(t,x)) - f(t,x,u(t,x))| \int_0^t \frac{(t-s)^{r-1}}{\Gamma(r)} |h(t,x,s,u_n(s,x))| |d_s g(t,s)| \\
&+ |f(t,x,u(t,x))| \int_0^t \frac{(t-s)^{r-1}}{\Gamma(r)} |h(t,x,s,u_n(s,x)) - h(t,x,s,u(s,x))| |d_s g(t,s)| \\
&\leq p^*q^*\Phi(\|u_n\|_{BC})|u_n(t,x) - u(t,x)| \\
&+ \frac{f^* + p^*\|u\|_{BC}}{\Gamma(r)} \int_0^t (t-s)^{r-1} |h(t,x,s,u_n(s,x)) \\
&- h(t,x,s,u(s,x))| d_s \left(\bigvee_{k=0}^{s} g(t,k) \right) \\
&\leq p^*q^*\Phi(\eta)|u_n(t,x) - u(t,x)|
\end{aligned}$$

$$+ \frac{f^* + p^*\eta}{\Gamma(r)} \int_0^t (t-s)^{r-1} |h(t,x,s,u_n(s,x)) - h(t,x,s,u(s,x))| d_s \left(\bigvee_{k=0}^{s} g(t,k) \right). \tag{4.8}$$

Case 1. If $(t,x) \in [0,a] \times [0,b]$; $a > 0$, then, since $u_n \to u$ as $n \to \infty$ and h is continuous, (4.8) gives

$$\|N'(u_n) - N'(u)\|_{BC} \to 0 \quad \text{as } n \to \infty.$$

Case 2. If $(t,x) \in (a,\infty) \times [0,b]$; $a > 0$, then from (4.6.4) and (4.8), we get

$$\begin{aligned} & |(N'u_n)(t,x) - (N'u)(t,x)| \\ \leq\ & p^*q^*\Phi(\eta)|u_n(t,x) - u(t,x)| \\ +\ & \frac{f^* + p^*\eta}{\Gamma(r)} \int_0^t (t-s)^{r-1}|h(t,x,s,u_n(s,x)) - h(t,x,s,u(s,x))|d_s\left(\bigvee_{k=0}^{s} g(t,k)\right) \end{aligned}$$

$$\begin{aligned} \leq\ & p^*q^*\Phi(\eta)|u_n(t,x) - u(t,x)| \\ +\ & \frac{f^* + p^*\eta}{\Gamma(r)} \int_0^t (t-s)^{r-1}q(t,x,s)(\Phi(|u(s,x)|) + \Phi(|u_n(s,x)|))d_s\left(\bigvee_{k=0}^{s} g(t,k)\right) \\ \leq\ & p^*q^*\Phi(\eta)|u_n(t,x) - u(t,x)| \\ +\ & \frac{2\Phi(\eta)(f^* + p^*\eta)}{\Gamma(r)} \int_0^t (t-s)^{r-1}q(t,x,s)d_s\left(\bigvee_{k=0}^{s} g(t,k)\right). \end{aligned}$$

Then

$$\begin{aligned} & |(N'u_n)(t,x) - (N'u)(t,x)| \leq p^*q^*\Phi(\eta)|u_n(t,x) - u(t,x)| \\ & + \frac{2\Lambda(\eta)(f^* + p^*\eta)}{\Gamma(r)} \int_0^t (t-s)^{r-1}q(t,x,s)d_s\left(\bigvee_{k=0}^{s} g(t,k)\right). \end{aligned} \tag{4.9}$$

Thus (4.9) gives

$$\|N'(u_n) - N'(u)\|_{BC} \to 0 \quad \text{as } n \to \infty.$$

Step 2: *$N'(B_\eta)$ is uniformly bounded.*
This is clear since $N'(B_\eta) \subset B_\eta$ and B_η is bounded.

Step 3: *$N'(B_\eta)$ is equicontinuous on every compact subset $[0,a] \times [0,b]$ of J'; $a > 0$.*
Let $(t_1,x_1), (t_2,x_2) \in [0,a] \times [0,b]$, $t_1 < t_2$, $x_1 < x_2$ and let $u \in B_\eta$. Thus we have

$$\begin{aligned} & |(N'u)(t_2,x_2) - (N'u)(t_1,x_1)| \leq |\mu(t_2,x_2) - \mu(t_1,x_1)| \\ +\ & |f(t_2,x_2,u(t_2,x_2)) - f(t_1,x_1,u(t_1,x_1))| \\ \times\ & \int_0^{t_2} \frac{(t_2-s)^{r-1}}{\Gamma(r)}|h(t_2,x_2,s,u(s,x_2))|d_s\left(\bigvee_{k=0}^{s} g(t_2,k)\right) \\ +\ & |f(t_1,x_1,u(t_1,x_1))|\left|\int_0^{t_2} \frac{(t_2-s)^{r-1}}{\Gamma(r)}h(t_2,x_2,s,u(s,x_2))d_s g(t_2,s)\right. \\ -\ & \left.\int_0^{t_1} \frac{(t_1-s)^{r-1}}{\Gamma(r)}h(t_1,x_1,s,u(s,x_1))d_s g(t_1,s)\right|. \end{aligned}$$

Thus

$$\begin{aligned} & |(N'u)(t_2,x_2) - (N'u)(t_1,x_1)| \leq |\mu(t_2,x_2) - \mu(t_1,x_1)| \\ +\ & q^*\Phi(\eta)|f(t_2,x_2,u(t_2,x_2)) - f(t_1,x_1,u(t_1,x_1))| \end{aligned}$$

$$+ \quad p^*\eta\left|\int_0^{t_2}\frac{(t_2-s)^{r-1}}{\Gamma(r)}h(t_2,x_2,s,u(s,x_2))d_sg(t_2,s)\right.$$
$$- \quad \left.\int_0^{t_1}\frac{(t_1-s)^{r-1}}{\Gamma(r)}h(t_1,x_1,s,u(s,x_1))d_sg(t_1,s)\right|.$$

Hence

$$|(N'u)(t_2,x_2)-(N'u)(t_1,x_1)| \leq |\mu(t_2,x_2)-\mu(t_1,x_1)|$$
$$+ \quad q^*\Phi(\eta)|f(t_2,x_2,u(t_2,x_2))-f(t_1,x_1,u(t_1,x_1))|$$
$$+ \quad \frac{p^*\eta}{\Gamma(r)}\int_0^{t_1}|(t_2-s)^{r-1}h(t_2,x_2,s,u(s,x_2))$$

$$- \quad (t_1-s)^{r-1}h(t_1,x_1,s,u(s,x_1))|d_s\left(\bigvee_{k=0}^{s}g(t_1,k)\right)$$
$$+ \quad \frac{p^*\eta}{\Gamma(r)}\int_0^{t_1}|(t_2-s)^{r-1}h(t_2,x_2,s,u(s,x_2))||d_s(g(t_2,s)-g(t_1,s))|$$
$$+ \quad \frac{p^*\eta}{\Gamma(r)}\int_{t_1}^{t_2}(t_2-s)^{r-1}|h(t_2,x_2,s,u(s,x_2))|d_s\left(\bigvee_{k=0}^{s}g(t_2,k)\right).$$

From continuity of μ, f, g, h and as $t_1 \to t_2$ and $x_1 \to x_2$, the right-hand side of the above inequality tends to zero.

Step 4: *$N'(B_\eta)$ is equiconvergent.*
Let $(t,x) \in \mathbb{R}_+ \times [0,b]$ and $u \in B_\eta$, then we have

$$|(Nu)(t,x)| \leq |\mu(t,x)| + \frac{|f(t,x,0)|}{\Gamma(r)}\int_0^t (t-s)^{r-1}|h(t,x,s,u(s,x))||d_sg(t,s)|$$
$$+ \quad \frac{|f(t,x,u(t,x))-f(t,x,0)|}{\Gamma(r)}\int_0^t (t-s)^{r-1}|h(t,x,s,u(s,x))|d_s\left(\bigvee_{k=0}^{s}g(t,k)\right)$$
$$\leq \quad |\mu(t,x)| + \frac{f^*}{\Gamma(r)}\int_0^t (t-s)^{r-1}q(t,x,s)\Phi(|u(s,x)|)d_s\left(\bigvee_{k=0}^{s}g(t,k)\right)$$
$$+ \quad \frac{p(t,x)|u(t,x)|}{\Gamma(r)}\int_0^t (t-s)^{r-1}q(t,x,s)\Phi(|u(s,x)|)d_s\left(\bigvee_{k=0}^{s}g(t,k)\right)$$
$$\leq \quad |\mu(t,x)| + \frac{\Phi(\eta)(f^*+p^*\eta)}{\Gamma(r)}\int_0^t (t-s)^{r-1}q(t,x,s)d_s\left(\bigvee_{k=0}^{s}g(t,k)\right).$$

Thus, for each $x \in [0,b]$, we get

$$|(Nu)(t,x)| \to 0, \ as\, t \to +\infty.$$

Hence,

$$|(Nu)(t,x)-(Nu)(+\infty,x)| \to 0, \ as\, t \to +\infty.$$

As a consequence of Steps 1 to 4 together with the Lemma 1.69, we can conclude that $N' : B_\eta \to B_\eta$ is continuous and compact. From an application of Schauder's theorem [204], we deduce that N' has a fixed point u which is a solution of the equation (4.2).

Step 5: *The local asymptotic stability of solutions.*
Now we investigate the local asymptotic stability of solutions of the equation (4.2). Let us assume that u_0 is a solution of the equation(4.2) with the conditions of this theorem. Taking $u \in B(u_0, \eta^*)$, we have

$$\begin{aligned}
& |(N'u)(t,x) - u_0(t,x)| = |(N'u)(t,x) - (N'u_0)(t,x)| \\
\leq\ & p^*q^*\Phi(\eta+\eta^*)|u(t,x) - u_0(t,x)| \\
+\ & \frac{f^*+p^*\eta}{\Gamma(r)} \int_0^t (t-s)^{r-1}|h(t,x,s,u(s,x)) - h(t,x,s,u_0(s,x))|d_s\left(\bigvee_{k=0}^{s} g(t,k)\right) \\
\leq\ & p^*q^*\Phi(\eta+\eta^*)|u(t,x) - u_0(t,x)| \\
+\ & \frac{f^*+p^*\eta}{\Gamma(r)} \int_0^t (t-s)^{r-1}q(t,x,s)(\Phi(|u(s,x)|) + \Phi(|u_0(s,x)|))d_s\left(\bigvee_{k=0}^{s} g(t,k)\right) \\
\leq\ & p^*q^*\eta^*\Phi(\eta+\eta^*) + q^*(f^*+p^*\eta)[\Phi(\eta) + \Phi(\eta+\eta^*)] \\
\leq\ & \eta^*.
\end{aligned}$$

Thus we observe that N' is a continuous function such that

$$N'(B(u_0, \eta^*)) \subset B(u_0, \eta^*).$$

Moreover, if u is a solution of the equation (4.2), then

$$\begin{aligned}
& |u(t,x) - u_0(t,x)| = |(N'u)(t,x) - (N'u_0)(t,x)| \\
\leq\ & p^*\Phi(\eta+\eta^*)|u(t,x) - u_0(t,x)| \\
\times\ & \frac{1}{\Gamma(r)} \int_0^t (t-s)^{r-1}q(t,x,s)d_s\left(\bigvee_{k=0}^{s} g(t,k)\right) \\
+\ & \frac{f^*+p^*\eta}{\Gamma(r)} \int_0^t (t-s)^{r-1}|h(t,x,s,u(s,x)) - h(t,x,s,u_0(s,x))|d_s\left(\bigvee_{k=0}^{s} g(t,k)\right) \\
\leq\ & \frac{p^*\eta^*\Phi(\eta+\eta^*)}{\Gamma(r)} \int_0^t (t-s)^{r-1}q(t,x,s)d_s\left(\bigvee_{k=0}^{s} g(t,k)\right) \\
+\ & \frac{f^*+p^*\eta}{\Gamma(r)} \int_0^t (t-s)^{r-1}q(t,x,s)(\Phi(|u(s,x)|) + \Phi(|u_0(s,x)|))d_s\left(\bigvee_{k=0}^{s} g(t,k)\right) \\
\leq\ & (p^*\eta^*\Phi(\eta+\eta^*) + (f^*+p^*\eta)[\Phi(\eta) + \Phi(\eta+\eta^*)]) \\
\times\ & \int_0^t \frac{(t-s)^{r-1}}{\Gamma(r)} q(t,x,s)d_s\left(\bigvee_{k=0}^{s} g(t,k)\right).
\end{aligned}$$

Thus

$$|u(t,x) - u_0(t,x)| \leq (p^*\eta^*\Phi(\eta+\eta^*) + (f^*+p^*\eta)[\Phi(\eta) + \Phi(\eta+\eta^*)])$$

$$\times \int_0^t \frac{(t-s)^{r-1}}{\Gamma(r)} q(t,x,s)d_s\left(\bigvee_{k=0}^{s} g(t,k)\right). \tag{4.10}$$

By using (4.10) we deduce that

$$\lim_{t\to\infty} |u(t,x) - u_0(t,x)| = 0.$$

Consequently, all solutions of the equation (4.2) are locally asymptotically stable.

4.2.4. Examples

As applications and to illustrate our results, we present two examples.

Example 1. Consider the following fractional order Riemann-Liouville Volterra-Stieltjes quadratic integral equations

$$u(t,x) = \mu(t,x) + \frac{f(t,x,u(t,x))}{\Gamma(r)} \int_0^t (t-s)^{r-1} h(t,x,s,u(s,x))d_s g(t,s);$$

$$\text{for } (t,x) \in J := [0,1] \times [0,1], \tag{4.11}$$

where $r = \frac{1}{4}$, $\mu(t,x) = \frac{1}{2+t^2}$; $t,x \in [0,1]$,

$$f(t,x,u) = \frac{e^{x-t}|u|}{1+|u|};\ t,x \in [0,1],\ u \in \mathbb{R},$$

$$g(t,s) = s;\ t,s \in [0,1],$$

$$h(t,x,s,u) = \frac{cx|u|}{1+|u|}(t-s)^{\frac{3}{4}} \sin t \sin s;\ (t,x,s) \in J_1,$$

$$c = \frac{\Gamma(\frac{1}{4})}{8e} \text{ and } J_1 = \{(t,x,s) \in [0,1]^3 : s \le t\}.$$

The function f is a continuous, $f^* = 0$, and

$$|f(t,x,u) - f(t,x,v)| \le e^{x-t}|u-v|;\ t,x \in [0,1],\ u,v \in \mathbb{R}.$$

Then, the assumption (4.1.1) is satisfies with $p(t,x) = e^{x-t}$; $t,x \in [0,1]$, and then $p^* = e$. Also, we can easily see that the function g satisfies the hypotheses $(4.1.2)-(4.1.4)$. The function h satisfies the assumption $(4.1.5)$. Indeed, h is continuous and

$$|h(t,x,s,u)| \le q(t,x,s)\Phi(|u|);\ (t,x,s) \in J_1,\ u \in \mathbb{R},$$

where $\Phi : \mathbb{R}_+ \to \mathbb{R}_+ :\ \Phi(w) = w$, and

$$q(t,x,s) = cx(t-s)^{\frac{3}{4}} \sin t \sin s;\ (t,x,s) \in J_1.$$

Then,

$$q^* = \sup_{(t,x,s)\in J_1} \frac{(t-s)^{\frac{-3}{4}} q(t,x,s)}{\Gamma(\frac{1}{4})} = \frac{1}{8e}.$$

Finally, we can see that the hypothesis (4.2.1) is satisfies with $\eta = 1$. Indeed, we have $\mu^* = \frac{1}{2}$, $g^* = 1$ and the inequality

$$\mu^* + q^* g^* \Phi(\eta)(f^* + p^* \eta) \leq \eta$$

implies

$$\frac{1}{2} + \frac{1}{8}\eta^2 \leq \eta,$$

which is satisfies for $\eta = 1$. Consequently, by Theorem 4.2, the equation (4.11) has a solution defined on $[0,1] \times [0,1]$.

Example 2. Consider now the following fractional order Riemann-Liouville Volterra-Stieltjes quadratic integral equations

$$u(t,x) = \mu(t,x) + \frac{f(t,x,u(t,x))}{\Gamma(r)} \int_0^t (t-s)^{r-1} h(t,x,s,u(s,x)) d_s g(t,s);$$

$$\text{for } (t,x) \in J' := \mathbb{R}_+ \times [0,1], \tag{4.12}$$

where $r = \frac{1}{4}$, $\mu(t,x) = \frac{1}{2+t^2}$; $(t,x) \in J'$,

$$f(t,x,u) = \frac{e^{x-t}|u|}{1+|u|}; \ (t,x) \in J',\ u \in \mathbb{R},\ \ g(t,s) = s;\ (t,s) \in \mathbb{R}_+^2,$$

$$\begin{cases} h(t,x,s,u) = \dfrac{cxs^{\frac{-3}{4}}|u| \sin\sqrt{t} \sin s}{(1+t^2)(2+|u|)}; & (t,x,s) \in J_1',\ s \neq 0 \textit{ and } u \in \mathbb{R}, \\ h(t,x,0,u) = 0; & (t,x) \in J \textit{ and } u \in \mathbb{R}, \end{cases}$$

$$c = \frac{\sqrt{\pi}}{8e\Gamma(\frac{1}{4})} \ \textit{and}\ J_1' = \{(t,x,s) \in J \times \mathbb{R}_+ : s \leq t,\ x \in [0,1]\}.$$

First, we can see that $\lim\limits_{t \to 0} \dfrac{1}{2+t^2} = 0$, then the assumption (4.6.1) is satisfies and $\mu^* = \frac{1}{2}$. Next, the function f is a continuous, $f^* = 0$, and

$$|f(t,x,u) - f(t,x,v)| \leq e^{x-t}|u-v|;\ (t,x) \in J',\ u,v \in \mathbb{R}.$$

Then, the assumption (4.6.2) is satisfies with $p(t,x) = e^{x-t}$; $(t,x) \in J$, and then $p^* = e$. Also, we can easily see that the function g satisfies the hypotheses $(4.6.3)-(4.6.5)$. The function h satisfies the assumption (4.6.6). Indeed, h is continuous and

$$|h(t,x,s,u)| \leq q(t,x,s)\Phi(|u|);\ (t,x,s) \in J_1',\ u \in \mathbb{R},$$

where $\Phi : \mathbb{R}_+ \to \mathbb{R}_+ :\ \Phi(w) = w$, and

$$\begin{cases} q(t,x,s) = \dfrac{cxs^{\frac{-3}{4}} \sin\sqrt{t} \sin s}{1+t^2}; & (t,x,s) \in J_1',\ s \neq 0, \\ q(t,x,0) = 0; & (t,x) \in J'. \end{cases}$$

Then,

$$\begin{aligned}
\left|\int_0^t (t-s)^{r-1}q(t,x,s)d_sg(t,s)\right| &\leq \int_0^t (t-s)^{\frac{-3}{4}} cxs^{\frac{-3}{4}} |\sin\sqrt{t}\sin s| d_s\left(\bigvee_{k=0}^{s} g(t,k)\right)\\
&\leq cx|\sin\sqrt{t}|\int_0^t (t-s)^{\frac{-3}{4}} s^{\frac{-3}{4}} ds\\
&\leq \frac{cx\Gamma^2(\frac{1}{4})}{\sqrt{\pi}}\left|\frac{\sin\sqrt{t}}{\sqrt{t}}\right|\\
&\leq \frac{cx\Gamma^2(\frac{1}{4})}{\sqrt{\pi t}} \longrightarrow 0 \; as \; t\to\infty,
\end{aligned}$$

and

$$\begin{aligned}
q^* &:= \sup_{(t,x)\in J}\frac{1}{\Gamma(r)}\int_0^t (t-s)^{r-1}q(t,x,s)d_s\left(\bigvee_{k=0}^{s} g(t,k)\right)\\
&\leq \sup_{(t,x)\in J}\frac{cx\Gamma(\frac{1}{4})}{\sqrt{\pi}}\left|\frac{\sin\sqrt{t}}{\sqrt{t}}\right|\\
&= \frac{c\Gamma(\frac{1}{4})}{\sqrt{\pi}} = \frac{1}{8e}.
\end{aligned}$$

Finally, we can see that hypothesis (4.8.1) is satisfies with $\eta = 1$. Indeed, the inequality

$$\mu^* + q^*\Phi(\eta)(f^* + p^*\eta) \leq \eta$$

implies $\frac{1}{2} + \frac{1}{8}\eta^2 \leq \eta$, which is satisfies for $\eta = 1$. Moreover, the inequality

$$q^*(f^* + p^*\eta)\Phi(\eta) + [q^*f^* + p^*q^*(\eta + \eta^*)]\Phi(\eta + \eta^*) \leq \eta^*$$

is satisfies with $\eta = \eta^* = 1$ and gives $\frac{5}{8} \leq 1$. Consequently, by Theorem 4.6, equation (4.12) has a solution defined on $\mathbb{R}_+ \times [0,1]$ and all solutions are locally asymptotically stable on $\mathbb{R}_+ \times [0,1]$. □

4.3. Asymptotic Behavior of Solutions of Nonlinear Riemann-Liouville Volterra-Stieltjes Quadratic Integral Equations

4.3.1. Introduction

This section deals with the existence and the stability of solutions to the following nonlinear fractional order Riemann-Liouville Volterra-Stieltjes quadratic integral equations of the form,

$$\begin{aligned}
u(t,x) &= f(t,x,u(t,x),u(\alpha(t),x)) + \tfrac{1}{\Gamma(r)}\int_0^{\beta(t)}(\beta(t)-s)^{r-1}\\
&\quad \times h(t,x,s,u(s,x),u(\gamma(s),x))d_sg(t,s);\; (t,x)\in J := \mathbb{R}_+\times[0,b],
\end{aligned} \tag{4.13}$$

where $b > 0$, $r \in (0,\infty)$, α, β, $\gamma : \mathbb{R}_+ \to \mathbb{R}_+$, $f : J \times \mathbb{R} \times \mathbb{R} \to \mathbb{R}$, $g : \mathbb{R}_+ \times \mathbb{R}_+ \to \mathbb{R}$, $h : J_1 \times \mathbb{R} \times \mathbb{R} \to \mathbb{R}$ are given continuous functions and $J_1 = \{(t,x,s) \in J \times \mathbb{R}_+ : s \le t\}$.

Our investigations are conducted in Banach spaces with an application of Schauder's fixed point theorem for the existence of solutions of the equation (4.13). Also, we prove that all solutions are locally asymptotically stable. Finally, we present an example illustrating the applicability of the imposed conditions.

4.3.2. Main Results

By $BC := BC(J)$ we denote the Banach space of all bounded and continuous functions from J into $\mathbb{R}$ equipped with the standard norm

$$\|u\|_{BC} = \sup_{(t,x)\in J} |u(t,x)|.$$

Definition 4.7. *We mean by a solution of equation (4.13), every function $u \in BC$ such that u satisfies equation (4.13) on J.*

The following hypotheses will be used in the sequel.

(4.10.1) The functions $\alpha, \beta, \gamma : \mathbb{R}_+ \to \mathbb{R}_+$ are continuous and $\lim_{t\to\infty} \alpha(t) = \infty$.

(4.10.2) The function f is continuous and there exist positive constants M and L such that $M + L < 1$ and

$$(1+\alpha(t))|f(t,x,u_1,u_2) - f(t,x,v_1,v_2)| \le M|u_1 - v_1| + L|u_2 - v_2|,$$

for $(t,x) \in J$ and $u_1, u_2, v_1, v_2 \in \mathbb{R}$.

(4.10.3) The function $t \to f(t,x,0,0)$ is bounded on J with

$$f^* = \sup_{(t,x)\in\mathbb{R}_+\times[0,b]} f(t,x,0,0) \ \textit{and} \ \lim_{t\to\infty} |f(t,x,0,0)| = 0,\ x \in [0,b].$$

(4.10.4) For all $t_1, t_2 \in \mathbb{R}_+$ such that $t_1 < t_2$ the function $s \mapsto g(t_2,s) - g(t_1,s)$ is nondecreasing on $\mathbb{R}_+$.

(4.10.5) The function $s \mapsto g(0,s)$ is nondecreasing on $\mathbb{R}_+$.

(4.10.6) The functions $s \mapsto g(t,s)$ and $t \mapsto g(t,s)$ are continuous on $\mathbb{R}_+$ for each fixed $t \in \mathbb{R}_+$ or $s \in \mathbb{R}_+$, respectively.

(4.10.7) The function h is continuous and there exist continuous functions $p_1, p_2 : J_1 \to \mathbb{R}_+$, $\Phi, \Psi : \mathbb{R}_+ \to \mathbb{R}_+$ such that Φ and Ψ are nondecreasing and

$$|h(t,x,s,u,v)| \le p_1(t,x,s)\Phi(|u|) + p_2(t,x,s)\Psi(|v|);$$

$(t,x,s) \in J_1$, $u,v \in \mathbb{R}$. Moreover, assume that $\lim_{t\to\infty} \int_0^{\beta(t)} (\beta(t) - s)^{r-1} p_i(t,x,s) d_s g(t,s) = 0$; $i = 1,2$.

Remark 4.8. *Set*

$$p_i^* := \sup_{(t,x)\in J'} \frac{1}{\Gamma(r)} \int_0^{\beta(t)} (\beta(t)-s)^{r-1} p_i(t,x,s) d_s \left(\bigvee_{k=0}^{s} g(t,k) \right); \; i = 1,2.$$

From hypothesis (4.10.7), *we infer that* p_i^* *is finite, for* $i = 1,2$.

Theorem 4.9. *Assume that hypotheses* (4.10.1) – (4.10.7) *hold. If there exists a constant* $\eta > 0$, *such that*

$$f^* + p_1^*\Phi(\eta) + p_2^*\Psi(\eta) \le \eta(1-M-L), \tag{4.14}$$

then the equation (4.13) has at least one solution in the space BC. Moreover, solutions of equation (4.13) are locally asymptotically stable.

Proof. Let us define the operator N such that, for any $u \in BC$,

$$\begin{aligned} (Nu)(t,x) \; &= \; f(t,x,u(t,x),u(\alpha(t),x)) + \tfrac{1}{\Gamma(r)} \textstyle\int_0^{\beta(t)} (\beta(t)-s)^{r-1} \\ &\quad \times h(t,x,s,u(s,x),u(\gamma(s),x)) d_s g(t,s); \; (t,x) \in J. \end{aligned} \tag{4.15}$$

From the assumptions of this theorem, we infer that $N(u)$ is continuous on J. Now we prove that $N(u) \in BC$ for any $u \in BC$. For arbitrarily fixed $(t,x) \in J$, we have

$$\begin{aligned} |(Nu)(t,x)| &= \Big| f(t,x,u(t,x),u(\alpha(t),x)) \\ &\quad + \frac{1}{\Gamma(r)} \int_0^{\beta(t)} (\beta(t)-s)^{r-1} h(t,x,s,u(s,x),u(\gamma(s),x)) d_s g(t,s) \Big| \\ &\le |f(t,x,u(t,x),u(\alpha(t),x)) - f(t,x,0,0) + f(t,x,0,0)| \\ &\quad + \left| \frac{1}{\Gamma(r)} \int_0^{\beta(t)} (\beta(t)-s)^{r-1} h(t,x,s,u(s,x),u(\gamma(s),x)) d_s g(t,s) \right| \\ &\le M|u(t,x)| + L|u(\alpha(t),x)| + |f(t,x,0,0)| + \int_0^{\beta(t)} \frac{(\beta(t)-s)^{r-1}}{\Gamma(r)} \\ &\quad \times (p_1(t,x,s)\Phi(|u(s,x)|) \\ &\quad + p_2(t,x,s)\Psi(|u(\gamma(s),x)|)) d_s \left(\bigvee_{k=0}^{s} g(t,k) \right) \\ &\le f^* + (M+L)\|u\|_{BC} + p_1^*\Phi(\|u\|_{BC}) + p_2^*\Psi(\|u\|_{BC}). \end{aligned}$$

Thus

$$\|N(u)\| \le f^* + (M+L)\|u\|_{BC} + p_1^*\Phi(\|u\|_{BC}) + p_2^*\Psi(\|u\|_{BC}). \tag{4.16}$$

Hence $N(u) \in BC$. From (4.14) and (4.16), we infer that N transforms the ball $B_\eta := B(0,\eta)$ into itself. We shall show that $N : B_\eta \to B_\eta$ satisfies the assumptions of Schauder's fixed point theorem [204]. The proof will be given in several steps and cases.

Step 1: *N is continuous.*
Let $\{u_n\}_{n\in\mathbb{N}}$ be a sequence such that $u_n \to u$ in B_η. Then, for each $(t,x)\in J$, we have

$$\begin{aligned}
|(Nu_n)(t,x)-(Nu)(t,x)| &\leq |f(t,x,u_n(t,x),u_n(\alpha(t),x))\\
&-f(t,x,u(t,x),u(\alpha(t),x))|\\
&+\frac{1}{\Gamma(r)}\int_0^{\beta(t)}(\beta(t)-s)^{r-1}|h(t,x,s,u_n(s,x),u_n(\gamma(s),x))\\
&-h(t,x,s,u(s,x),u(\gamma(s),x))|d_sg(t,s)\\
\leq\ & (M+L)\|u_n-u\|_{BC}\\
&+\frac{1}{\Gamma(r)}\int_0^{\beta(t)}(\beta(t)-s)^{r-1}|h(t,x,s,u_n(s,x),u_n(\gamma(s),x))\\
&-h(t,x,s,u(s,x),u(\gamma(s),x))|d_s(\textstyle\bigvee_{k=0}^{s}g(t,k)).
\end{aligned} \tag{4.17}$$

Case 1. If $(t,x)\in[0,a]\times[0,b]$; $a>0$, then, since $u_n\to u$ as $n\to\infty$ and g,h are continuous, (4.17) gives

$$\|N(u_n)-N(u)\|_{BC}\to 0 \quad \text{as } n\to\infty.$$

Case 2. If $(t,x)\in(a,\infty)\times[0,b]$; $a>0$, then from (4.10.4) and (4.17), for each $(t,x)\in\mathbb{R}_+\times[0,b]$, we get

$$\begin{aligned}
|(Nu_n)(t,x)-(Nu)(t,x)| &\leq (M+L)\|u_n-u\|\\
&+\int_0^{\beta(t)}\frac{(\beta(t)-s)^{r-1}}{\Gamma(r)}(p_1(t,x,s)(\Phi(|u_n(s,x)|))\\
&+\Phi(|u(s,x)|))+p_2(t,x,s)(\Psi(|u_n(\gamma(s),x)|))\\
&+\Psi(|u(\gamma(s),x)|))d_s(\textstyle\bigvee_{k=0}^{s}g(t,k))\\
\leq\ & (M+L)\|u_n-u\|\\
&+\frac{\Phi(\eta)}{\Gamma(r)}\int_0^{\beta(t)}(\beta(t)-s)^{r-1}p_1(t,x,s)d_s(\textstyle\bigvee_{k=0}^{s}g(t,k))\\
&+\frac{\Psi(\eta)}{\Gamma(r)}\int_0^{\beta(t)}(\beta(t)-s)^{r-1}p_2(t,x,s)d_s(\textstyle\bigvee_{k=0}^{s}g(t,k)).
\end{aligned} \tag{4.18}$$

From (4.10.7) and since $u_n\to u$ as $n\to\infty$ and $t\to\infty$, then (4.18) gives

$$\|N(u_n)-N(u)\|_{BC}\to 0 \quad \text{as } n\to\infty.$$

Step 2: *$N(B_\eta)$ is uniformly bounded.*
This is clear since $N(B_\eta)\subset B_\eta$ and B_η is bounded.

Step 3: *$N(B_\eta)$ is equicontinuous on every compact subset $[0,a]\times[0,b]$ of J, $a>0$.*
Let $(t_1,x_1),(t_2,x_2)\in[0,a]\times[0,b]$, $t_1<t_2$, $x_1<x_2$ and let $u\in B_\eta$. Also without loss of generality, suppose that $\beta(t_1)\leq\beta(t_2)$. Then we have

$$\begin{aligned}
&|(Nu)(t_2,x_2)-(Nu)(t_1,x_1)|\\
&\leq |f(t_2,x_2,u(t_2,x_2),u(\alpha(t_2),x_2))-f(t_2,x_2,u(t_1,x_1),u(\alpha(t_1),x_1))|\\
&+|f(t_2,x_2,u(t_1,x_1),u(\alpha(t_1),x_1))-f(t_1,x_1,u(t_1,x_1),u(\alpha(t_1),x_1))|\\
&+\frac{1}{\Gamma(r)}\left|\int_0^{\beta(t_2)}(\beta(t_2)-s)^{r-1}\right.
\end{aligned}$$

$$\times[h(t_2,x_2,s,u(t_2,x_2),u(\gamma(s),x_2))$$
$$-h(t_1,x_1,s,u(t_1,x_1),u(\gamma(s),x_1))]d_sg(t,s)|$$
$$+\frac{1}{\Gamma(r)}\left|\int_0^{\beta(t_2)}(\beta(t_2)-s)^{r-1}h(t_1,x_1,s,u(t_1,x_1),u(\gamma(s),x_1))d_sg(t,s)\right.$$
$$\left.-\int_0^{\beta(t_1)}(\beta(t_2)-s)^{r-1}h(t_1,x_1,s,u(t_1,x_1),u(\gamma(s),x_1))d_sg(t,s)\right|$$
$$+\frac{1}{\Gamma(r)}\left|\int_0^{\beta(t_1)}(\beta(t_2)-s)^{r-1}h(t_1,x_1,s,u(t_1,x_1),u(\gamma(s),x_1))d_sg(t,s)\right.$$
$$\left.-\int_0^{\beta(t_1)}(\beta(t_1)-s)^{r-1}h(t_1,x_1,s,u(t_1,x_1),u(\gamma(s),x_1))d_sg(t,s)\right|$$
$$\leq M|u(\alpha(t_2),x_2)-u(\alpha(t_1),x_1)|+L|u(t_2,x_2)-u(t_1,x_1)|$$
$$+|f(t_2,x_2,u(t_1,x_1),u(\alpha(t_1),x_1))-f(t_1,x_1,u(t_1,x_1),u(\alpha(t_1),x_1))|$$
$$+\frac{1}{\Gamma(r)}\int_0^{\beta(t_2)}(\beta(t_2)-s)^{r-1}$$
$$\times|h(t_2,x_2,s,u(t_2,x_2),u(\gamma(s),x_2))$$
$$-h(t_1,x_1,s,u(t_1,x_1),u(\gamma(s),x_1))|d_s\left(\bigvee_{k=0}^{s}g(t,k)\right)$$
$$+\frac{1}{\Gamma(r)}\int_{\beta(t_1)}^{\beta(t_2)}(\beta(t_2)-s)^{r-1}|h(t_1,x_1,s,u(t_1,x_1),u(\gamma(s),x_1))|d_s\left(\bigvee_{k=0}^{s}g(t,k)\right)$$
$$+\frac{1}{\Gamma(r)}\int_0^{\beta(t_1)}|(\beta(t_2)-s)^{r-1}-(\beta(t_1)-s)^{r-1}|$$
$$\times|h(t_1,x_1,s,u(t_1,x_1),u(\gamma(s),x_1))|d_s\left(\bigvee_{k=0}^{s}g(t,k)\right).$$

Thus we have

$$|(Nu)(t_2,x_2)-(Nu)(t_1,x_1)|$$
$$\leq M|u(\alpha(t_2),x_2)-u(\alpha(t_1),x_1)|+L|u(t_2,x_2)-u(t_1,x_1)|$$
$$+|f(t_2,x_2,u(\alpha(t_1),x_1))-f(t_1,x_1,u(\alpha(t_1),x_1))|$$
$$+\frac{1}{\Gamma(r)}\int_0^{\beta(t_2)}(\beta(t_2)-s)^{r-1}$$
$$\times|h(t_2,x_2,s,u(s,x_2),u(\gamma(s),x_2))$$
$$-h(t_1,x_1,s,u(s,x_1),u(\gamma(s),x_1))|d_s\left(\bigvee_{k=0}^{s}g(t,k)\right)$$
$$+\frac{\Phi(\eta)}{\Gamma(r)}\int_{\beta(t_1)}^{\beta(t_2)}(\beta(t_2)-s)^{r-1}p_1(t_1,x_1,s)d_s\left(\bigvee_{k=0}^{s}g(t,k)\right)$$
$$+\frac{\Psi(\eta)}{\Gamma(r)}\int_{\beta(t_1)}^{\beta(t_2)}(\beta(t_2)-s)^{r-1}p_2(t_1,x_1,s)d_sg(t_1,s)$$

$$+\frac{\Phi(\eta)}{\Gamma(r)}\int_0^{\beta(t_1)}|(\beta(t_2)-s)^{r-1}-(\beta(t_1)-s)^{r-1}|p_1(t_1,x_1,s)d_s\left(\bigvee_{k=0}^{s}g(t,k)\right)$$
$$+\frac{\Psi(\eta)}{\Gamma(r)}\int_0^{\beta(t_1)}|(\beta(t_2)-s)^{r-1}-(\beta(t_1)-s)^{r-1}|p_2(t_1,x_1,s)d_s\left(\bigvee_{k=0}^{s}g(t,k)\right).$$

From continuity of α,β,f,g,h and as $t_1\longrightarrow t_2$ and $x_1\longrightarrow x_2$, the right-hand side of the above inequality tends to zero.

Step 4: *$N(B_\eta)$ is equiconvergent.*

Let $(t,x)\in J$ and $u\in B_\eta$, then we have

$$\begin{aligned}|(Nu)(t,x)|&\leq|f(t,x,u(t,x),u(\alpha(t),x))-f(t,x,0,0)+f(t,x,0,0)|\\&+\left|\frac{1}{\Gamma(r)}\int_0^{\beta(t)}(\beta(t)-s)^{r-1}g(t,x,s,u(s,x),u(\gamma(s),x))d_sg(t,s)\right|\\&\leq\frac{M}{1+\alpha(t)}|u(t,x)|+\frac{L}{1+\alpha(t)}|u(\alpha(t),x)|+|f(t,x,0,0)|\end{aligned}$$

$$+\frac{\Phi(\eta)}{\Gamma(r)}\int_0^{\beta(t)}(\beta(t)-s)^{r-1}p_1(t,x,s)d_s\left(\bigvee_{k=0}^{s}g(t,k)\right)$$
$$+\frac{\Psi(\eta)}{\Gamma(r)}\int_0^{\beta(t)}(\beta(t)-s)^{r-1}p_2(t,x,s)d_s\left(\bigvee_{k=0}^{s}g(t,k)\right).$$

Then,

$$\begin{aligned}\lim_{t\longrightarrow\infty}|(Nu)(t,x)|&\leq\lim_{t\longrightarrow\infty}\frac{2M\eta}{1+\alpha(t)}+f(t,x,0,0)\\&+\frac{\Phi(\eta)}{\Gamma(r)}\int_0^{\beta(t)}(\beta(t)-s)^{r-1}p_1(t,x,s)d_s\left(\bigvee_{k=0}^{s}g(t,k)\right)\\&+\frac{\Psi(\eta)}{\Gamma(r)}\int_0^{\beta(t)}(\beta(t)-s)^{r-1}p_2(t,x,s)d_s\left(\bigvee_{k=0}^{s}g(t,k)\right).\end{aligned}$$

Thus, for each $x\in[0,b]$, we get

$$|(Nu)(t,x)|\longrightarrow 0,\ as\,t\longrightarrow+\infty.$$

Hence,

$$|(Nu)(t,x)-(Nu)(+\infty,x)|\longrightarrow 0,\ as\,t\longrightarrow+\infty.$$

As a consequence of Steps 1 to 4 together with the Lemma 1.69, we can conclude that $N:B_\eta\rightarrow B_\eta$ is continuous and compact. From an application of Schauder's theorem [204], we deduce that N has a fixed point u which is a solution of the equation (4.13).

Step 5: *The local asymptotic stability of solutions.*
Now we investigate the stability for solutions of equation (4.13). Let us assume that u_0 is a solution of equation (4.13) with the conditions of this theorem. Consider the ball $B(u_0,\eta^*)$ with $\eta^* = \frac{M^*}{1-M-L}$, where

$$M^* := \frac{1}{\Gamma(r)} \sup_{(t,x)\in J} \left\{ \int_0^{\beta(t)} (\beta(t)-s)^{r-1} |h(t,x,s,u(s,x),u(\gamma(s),x)) \right.$$
$$\left. - h(t,x,s,u_0(s,x),u_0(\gamma(s),x))| d_s\left(\bigvee_{k=0}^{s} g(t,k)\right);\ u\in BC \right\}.$$

Taking $u \in B(u_0,\eta^*)$, we have

$$\begin{aligned}
& |(Nu)(t,x)-u_0(t,x)| = |(Nu)(t,x)-(Nu_0)(t,x)| \\
\leq\ & |f(t,x,u(t,x),u(\alpha(t),x)) - f(t,x,u_0(t,x),u_0(\alpha(t),x))| \\
& + \frac{1}{\Gamma(r)} \int_0^{\beta(t)} (\beta(t)-s)^{r-1} |h(t,x,s,u(s,x),u(\gamma(s),x)) \\
& - h(t,x,s,u_0(s,x),u_0(\gamma(s),x))| d_s g(t,s) \\
\leq\ & (M+L)\|u-u_0\| \\
& + \frac{1}{\Gamma(r)} \int_0^{\beta(t)} (\beta(t)-s)^{r-1} |h(t,x,s,u(s,x),u(\gamma(s),x)) \\
& - h(t,x,s,u_0(s,x),u_0(\gamma(s,x)))| d_s\left(\bigvee_{k=0}^{s} g(t,k)\right) \\
\leq\ & \eta^*(M+L)+M^* = \eta^*.
\end{aligned}$$

Thus we observe that N is a continuous function such that $N(B(u_0,\eta^*)) \subset B(u_0,\eta^*)$. Moreover, if u is a solution of equation (4.13), then

$$\begin{aligned}
& |u(t,x)-u_0(t,x)| = |(Nu)(t,x)-(Nu_0)(t,x)| \\
\leq\ & |f(t,x,u(t,x),u(\alpha(t),x)) - f(t,x,u_0(t,x),u_0(\alpha(t),x))| \\
& + \tfrac{1}{\Gamma(r)} \int_0^{\beta(t)} (\beta(t)-s)^{r-1} |h(t,x,s,u(s,x),u(\gamma(s),x)) \\
& - h(t,x,s,u_0(s,x),u_0(\gamma(s),x))| d_s(\textstyle\bigvee_{k=0}^{s} g(t,k)) \\
\leq\ & M|u(t,x)-u_0(t,x)| + L|u(\alpha(t),x)-u_0(\alpha(t),x)| \\
& + \tfrac{\Phi(\eta)+\Phi(\eta+\eta^*)}{\Gamma(r)} \int_0^{\beta(t)} (\beta(t)-s)^{r-1} p_1(t,x,s) d_s(\textstyle\bigvee_{k=0}^{s} g(t,k)) \\
& + \tfrac{\Psi(\eta)+\Psi(\eta+\eta^*)}{\Gamma(r)} \int_0^{\beta(t)} (\beta(t)-s)^{r-1} p_2(t,x,s) d_s(\textstyle\bigvee_{k=0}^{s} g(t,k)).
\end{aligned} \tag{4.19}$$

By using (4.19) and the fact that $\alpha(t) \longrightarrow \infty$ as $t \longrightarrow \infty$, we deduce that

$$\lim_{t\to\infty} |u(t,x)-u_0(t,x)| \leq \lim_{t\to\infty} \frac{1}{(1-M-L)\Gamma(r)}$$
$$\times \left[(\Phi(\eta)+\Phi(\eta+\eta^*)) \int_0^{\beta(t)} (\beta(t)-s)^{r-1} p_1(t,x,s) d_s\left(\bigvee_{k=0}^{s} g(t,k)\right) \right.$$
$$\left. + (\Psi(\eta)+\Psi(\eta+\eta^*)) \int_0^{\beta(t)} (\beta(t)-s)^{r-1} p_2(t,x,s) d_s\left(\bigvee_{k=0}^{s} g(t,k)\right) \right]$$

$$= \quad 0.$$

Hence,

$$\lim_{t\to\infty} |u(t,x) - u_0(t,x)| = 0.$$

Consequently, all solutions of equation (4.13) are locally asymptotically stable. □

4.3.3. An Example

As an application of our results we consider the following nonlinear fractional order Riemann-Liouville Volterra-Stieltjes quadratic integral equation of the form

$$\begin{aligned} u(t,x) &= f(t,x,u(t,x),u(\alpha(t),x)) + \tfrac{1}{\Gamma(r)} \int_0^{\beta(t)} (\beta(t)-s)^{r-1} \\ &\quad \times h(t,x,s,u(s,x),u(\gamma(s),x))d_s g(t,s),\ (t,x) \in J := \mathbb{R}_+ \times [0,1], \end{aligned} \tag{4.20}$$

where $r = \frac{1}{4}$, $\alpha(t) = \beta(t) = \gamma(t) = t$; $t \in \mathbb{R}_+$,

$$f(t,x,u,v) = \frac{xe^{-t}|uv|}{8(1+t^2)(1+|u|+2|v|)},\ (t,x) \in J \text{ and } u,v \in \mathbb{R},$$

$$g(t,s) = s,\ (t,s) \in \mathbb{R}_+^2,$$

$$\begin{cases} h(t,x,s,u,v) = \dfrac{cxs^{\frac{-3}{4}}(1+|u|)\sin\sqrt{t}\sin s}{(1+t^2)(2+|u|)}; \\ \qquad (t,x,s) \in J_1,\ s \neq 0 \textit{ and } u,v \in \mathbb{R}, \\ h(t,x,0,u,v) = 0; \qquad (t,x) \in J \textit{ and } u,v \in \mathbb{R}, \end{cases}$$

$$c = \frac{\sqrt{\pi}}{8e\Gamma(\frac{1}{4})} \textit{ and } J_1 = \{(t,x,s) \in J \times \mathbb{R}_+ : s \leq t\}.$$

First, we can see that $\lim_{t\to 0} \alpha(t) = 0$. Then the assumption (4.10.1) is satisfied. Next, the function f is a continuous, and

$$|f(t,x,u_1,u_2) - f(t,x,v_1,v_2)| \leq \frac{1}{8}(|u_1 - v_1| + 2|u_2 - v_2|);\ (t,x) \in J,\ u,v \in \mathbb{R}.$$

Then, the assumption (4.10.2) is satisfied with $M = \frac{1}{8}$, $L = \frac{1}{4}$, and (4.10.3) is satisfied with $f^* = 0$. Also, we can easily see that the function g satisfies the hypotheses (4.10.4) – (4.10.6).

The function h satisfies the assumption (4.10.7). Indeed, h is continuous and

$$|h(t,x,s,u,v)| \leq p_1(t,x,s)\Phi(|u|) + p_2(t,x,s)\Psi(|v|);\ (t,x,s) \in J_1,\ u,v \in \mathbb{R},$$

where $\Phi, \Psi : \mathbb{R}_+ \to \mathbb{R}_+ :\ \Phi(w) = w,\ \Psi(w) = 1$, and

$$\begin{cases} p_i(t,x,s) = \dfrac{cxs^{\frac{-3}{4}}\sin\sqrt{t}\sin s}{1+t^2};\ (t,x,s) \in J_1,\ s \neq 0,\ i = 1,2, \\ p_i(t,x,0) = 0; \qquad (t,x) \in J,\ i = 1,2. \end{cases}$$

Then, for $i = 1,2$, we have

$$\begin{aligned}\left|\int_0^t (t-s)^{r-1} p_i(t,x,s) d_s g(t,s)\right| &\leq \int_0^t (t-s)^{\frac{-3}{4}} cxs^{\frac{-3}{4}} \\ &\times |\sin\sqrt{t}\sin s| d_s\left(\bigvee_{k=0}^{s} g(t,k)\right) \\ &\leq cx|\sin\sqrt{t}|\int_0^t (t-s)^{\frac{-3}{4}} s^{\frac{-3}{4}} ds \\ &\leq \frac{cx\Gamma^2(\frac{1}{4})}{\sqrt{\pi}}\left|\frac{\sin\sqrt{t}}{\sqrt{t}}\right| \\ &\leq \frac{cx\Gamma^2(\frac{1}{4})}{\sqrt{\pi t}} \longrightarrow 0 \; as \; t \to \infty,\end{aligned}$$

and

$$\begin{aligned}p_i^* &:= \sup_{(t,x)\in J} \frac{1}{\Gamma(r)} \int_0^t (t-s)^{r-1} p_i(t,x,s) d_s\left(\bigvee_{k=0}^{s} g(t,k)\right) \\ &\leq \sup_{(t,x)\in J} \frac{cx\Gamma(\frac{1}{4})}{\sqrt{\pi}}\left|\frac{\sin\sqrt{t}}{\sqrt{t}}\right| \\ &= \frac{c\Gamma(\frac{1}{4})}{\sqrt{\pi}} = \frac{1}{8e}.\end{aligned}$$

Finally, let us notice that the condition (4.14) is satisfied with $\eta = 1$. Indeed, the inequality

$$f^* + p_1^*\Phi(\eta) + p_2^*\Psi(\eta) \leq \eta(1 - M - L),$$

implies $\eta \geq \frac{1}{5e-1}$, which is satisfied for $\eta = 1$. Hence by Theorem 4.9, the equation (4.20) has a solution defined on $\mathbb{R}_+ \times [0,1]$ and solutions of this equation are locally asymptotically stable.

4.4. Global Stability Results for Nonlinear Partial Riemann-Liouville Volterra-Stieltjes Functional Integral Equations

4.4.1. Introduction

This section deals with the existence of solutions to the following partial integral equation of Riemann-Liouville fractional order,

$$\begin{aligned}u(t,x) &= f(t,x,u(t,x)) + \frac{1}{\Gamma(r_1)\Gamma(r_2)}\int_0^t\int_0^x (t-s)^{r_1-1}(x-y)^{r_2-1} \\ &\quad \times h(t,x,s,y,u(s,y))dyd_sg(t,s); \; (t,x) \in J := \mathbb{R}_+ \times [0,b],\end{aligned} \tag{4.21}$$

where $b > 0$, $\mathbb{R}_+ := [0,\infty)$, $r_1, r_2 \in (0,\infty)$, $f : J \times \mathbb{R} \to \mathbb{R}$, $g : \mathbb{R}_+ \times \mathbb{R}_+ \to \mathbb{R}$ and $h : J_1 \times \mathbb{R} \to \mathbb{R}$ are given continuous functions $J_1 := \{(t,x,s,y) : 0 \leq s \leq t < \infty, \; 0 \leq y \leq x \leq b\}$. Our investigations are conducted with an application of Schauder's fixed point theorem for the existence of solutions of the equation (4.21). Also, we obtain some results about the global asymptotic stability of solutions of our equation. Finally, we present an example illustrating the applicability of the imposed conditions.

4.4.2. Main Results

Definition 4.10. *We mean by a solution of equation (4.21), every function $u \in BC$ such that u satisfies equation (4.21) on J.*

The following hypotheses will be used in the sequel.

(4.14.1) There exist constants $M, L > 0$ with $M < L$ and a continuous function $\alpha : J \to \mathbb{R}_+$ with $\lim_{t\to\infty} \alpha(t,x) = \infty$; $x \in [0,b]$, such that

$$|f(t,x,u) - f(t,x,v)| \leq \frac{M|u-v|}{(1+\alpha(t,x))(L+|u-v|)},$$

for $(t,x) \in J$ and $u, v \in \mathbb{R}$. Moreover, the function $(t,x) \mapsto f(t,x,0)$ is bounded on J with

$$f^* = \sup_{(t,x)\in\mathbb{R}_+\times[0,b]} |f(t,x,0)|, \ \ and \ \lim_{t\to\infty} |f(t,x,0)| = 0; \ x \in [0,b],$$

(4.14.2) For all $t_1, t_2 \in \mathbb{R}_+$ such that $t_1 < t_2$ the function $s \mapsto g(t_2,s) - g(t_1,s)$ is nondecreasing on $\mathbb{R}_+$,

(4.14.3) The function $s \mapsto g(0,s)$ is nondecreasing on $\mathbb{R}_+$,

(4.14.4) The functions $s \mapsto g(t,s)$ and $t \mapsto g(t,s)$ are continuous on $\mathbb{R}_+$ for each fixed $t \in \mathbb{R}_+$ or $s \in \mathbb{R}_+$, respectively,

(4.14.5) There exist a continuous function $P : J_1 \to \mathbb{R}_+$ and a function $\Psi : \mathbb{R}_+ \to \mathbb{R}_+$ being continuous and increasing on $\mathbb{R}_+$ such that

$$|h(t,x,s,y,u)| \leq P(t,x,s,y)\Psi(|u|); \text{ for } (t,x,s,y) \in J_1 \text{ and for } u \in \mathbb{R}.$$

Moreover, assume that

$$\lim_{t\to\infty} \int_0^t (t-s)^{r_1-1} P(t,x,s,y) d_s g(t,s) = 0; \ for \ x, y \in [0,b].$$

Remark 4.11. *Set* $q^* := \sup_{(t,x)\in J} Q(t,x)$ *where*

$$Q(t,x) = \frac{1}{\Gamma(r_1)\Gamma(r_2)} \int_0^t \int_0^x (t-s)^{r_1-1} (x-y)^{r_2-1} P(t,x,s,y) dy d_s \left(\bigvee_{k=0}^{s} g(t,k) \right).$$

From hypothesis (4.14.5), *we infer that q^* is finite and* $\lim_{t\to\infty} Q(t,x) = 0$.

Theorem 4.12. *Assume that* (4.14.1) – (4.14.5) *and the following hypothesis hold*

(4.16.1) *There exists a constant* $\eta > 0$, *such that*

$$M + f^* + q^*\Psi(\eta) \leq \eta.$$

Then the equation (4.21) has at least one solution in the space BC. Moreover, solutions of equation (4.21) are globally asymptotically stable.

Proof. Let us define the operator N such that, for any $u \in BC$,

$$\begin{aligned}(Nu)(t,x) &= f(t,x,u(t,x)) + \frac{1}{\Gamma(r_1)\Gamma(r_2)}\int_0^t\int_0^x (t-s)^{r_1-1}(x-y)^{r_2-1} \\ &\quad \times h(t,x,s,y,u(s,y))dyd_sg(t,s);\ (t,x) \in J.\end{aligned} \tag{4.22}$$

By considering the assumptions, we infer that $N(u)$ is continuous on J. Now we prove that $N(u) \in BC$ for any $u \in BC$. For arbitrarily fixed $(t,x) \in J$ we have

$$\begin{aligned}|(Nu)(t,x)| &= |f(t,x,u(t,x)) \\ &\quad + \frac{1}{\Gamma(r_1)\Gamma(r_2)}\int_0^t\int_0^x (t-s)^{r_1-1}(x-y)^{r_2-1} \\ &\quad \times h(t,x,s,y,u(s,y))dyd_sg(t,s)| \\ &\leq |f(t,x,u(t,x)) - f(t,x,0) + f(t,x,0)| \\ &\quad + \frac{1}{\Gamma(r_1)\Gamma(r_2)}\int_0^t\int_0^x (t-s)^{r_1-1}(x-y)^{r_2-1} \\ &\quad \times P(t,x,s,y)\Psi(\|u\|_{BC})dyd_s\left(\bigvee_{k=0}^{s} g(t,k)\right) \\ &\leq \frac{M|u(t,x)|}{(1+\alpha(t,x))(L+|u(t,x)|)} + |f(t,x,0)| + \Psi(\|u\|_{BC})Q(t,x) \\ &\leq M + f^* + q^*\Psi(\|u\|_{BC}).\end{aligned}$$

Thus

$$\|N(u)\|_{BC} \leq M + f^* + q^*\Psi(\|u\|_{BC}). \tag{4.23}$$

Hence $N(u) \in BC$. Equation (4.23) and hypothesis (4.16.1) implies that N transforms the ball $B_\eta := B(0,\eta)$ into itself. We shall show that $N : B_\eta \to B_\eta$ satisfies the assumptions of Schauder's fixed point theorem [204]. The proof will be given in several steps and cases.

Step 1: *N is continuous.*
Let $\{u_n\}_{n\in\mathbb{N}}$ be a sequence such that $u_n \to u$ in B_η. Then, for each $(t,x) \in J$, we have

$$\begin{aligned}&|(Nu_n)(t,x) - (Nu)(t,x)| \leq |f(t,x,u_n(t,x)) - f(t,x,u(t,x))| \\ &+ \left|\frac{1}{\Gamma(r_1)\Gamma(r_2)}\int_0^t\int_0^x (t-s)^{r_1-1}(x-y)^{r_2-1}|h(t,x,s,y,u_n(s,y))\right. \\ &-h(t,x,s,y,u(s,y))|dyd_sg(t,s)| \\ \leq\ & \frac{M}{L}\|u_n - u\| + \frac{1}{\Gamma(r_1)\Gamma(r_2)}\int_0^t\int_0^x (t-s)^{r_1-1}(x-y)^{r_2-1}\end{aligned}$$

$$\times|h(t,x,s,y,u_n(s,y))-h(t,x,s,y,u(s,y))|dy\left(\bigvee_{k=0}^{s}g(t,k)\right). \tag{4.24}$$

Case 1. If $(t,x)\in[0,a]\times[0,b]$; $a>0$, then, since $u_n\to u$ as $n\to\infty$ and g,h are continuous, (4.24) gives

$$\|N(u_n)-N(u)\|_{BC}\to 0 \quad \text{as } n\to\infty.$$

Case 2. If $(t,x)\in(a,\infty)\times[0,b]$, $a>0$, then from (4.14.5) and (4.24), for each $(t,x)\in J$, we get

$$|(Nu_n)(t,x)-(Nu)(t,x)|\le\frac{M}{L}\|u_n-u\|+(\Phi(\|u_n\|_{BC})+\Phi(\|u\|_{BC}))Q(t,x).$$

Then

$$|(Nu_n)(t,x)-(Nu)(t,x)|\le\frac{M}{L}\|u_n-u\|+(\Phi(\eta)+\Phi(\eta))Q(t,x). \tag{4.25}$$

Since $u_n\to u$ as $n\to\infty$ and $Q(t,x)\to 0$ as $t\to\infty$, then (4.25) gives

$$\|N(u_n)-N(u)\|_{BC}\to 0 \quad \text{as } n\to\infty.$$

Step 2: *$N(B_\eta)$ is uniformly bounded.*
This is clear since $N(B_\eta)\subset B_\eta$ and B_η is bounded.

Step 3: *$N(B_\eta)$ is equicontinuous on every compact subset $[0,a]\times[0,b]$ of J, $a>0$.*
Let $(t_1,x_1),(t_2,x_2)\in[0,a]\times[0,b]$, $t_1<t_2$, $x_1<x_2$ and let $u\in B_\eta$. Thus we have

$$\begin{aligned}
&|(Nu)(t_2,x_2)-(Nu)(t_1,x_1)|\le|f(t_2,x_2,u(t_2,x_2))-f(t_2,x_2,u(t_1,x_1))|\\
&+|f(t_2,x_2,u(t_1,x_1))-f(t_1,x_1,u(t_1,x_1))|\\
&+\frac{1}{\Gamma(r_1)\Gamma(r_2)}\int_0^{t_1}\int_0^{x_1}(t_1-s)^{r_1-1}(x_1-y)^{r_2-1}\\
&\times|h(t_2,x_2,s,y,u(s,y))-h(t_1,x_1,s,y,u(s,y))|dyd_s\left(\bigvee_{k=0}^{s}g(t,k)\right)\\
&+\frac{1}{\Gamma(r_1)\Gamma(r_2)}\int_0^{t_1}\int_{x_1}^{x_2}(t_1-s)^{r_1-1}(x_1-y)^{r_2-1}\\
&\times|h(t_1,x_1,s,y,u(s,y))|dyd_s\left(\bigvee_{k=0}^{s}g(t,k)\right)\\
&+\frac{1}{\Gamma(r_1)\Gamma(r_2)}\int_{t_1}^{t_2}\int_0^{x_1}(t_1-s)^{r_1-1}(x_1-y)^{r_2-1}\\
&\times|h(t_1,x_1,s,y,u(s,y))|dyd_s\left(\bigvee_{k=0}^{s}g(t,k)\right)\\
&+\frac{1}{\Gamma(r_1)\Gamma(r_2)}\int_{t_1}^{t_2}\int_{x_1}^{x_2}(t_1-s)^{r_1-1}(x_1-y)^{r_2-1}\\
&\times|h(t_1,x_1,s,y,u(s,y))|dyd_s\left(\bigvee_{k=0}^{s}g(t,k)\right)
\end{aligned}$$

$$\le \frac{M}{L}|u(t_2,x_2)-u(t_1,x_1)|+|f(t_2,x_2,u(t_1,x_1))-f(t_1,x_1,u(t_1,x_1))|$$
$$+\frac{1}{\Gamma(r_1)\Gamma(r_2)}\int_0^{t_1}\int_0^{x_1}(t_1-s)^{r_1-1}(x_1-y)^{r_2-1}$$
$$\times|h(t_2,x_2,s,y,u(s,y))-h(t_1,x_1,s,y,u(s,y))|dyd_s\left(\bigvee_{k=0}^{s}g(t,k)\right)$$
$$+\frac{\Psi(\eta)}{\Gamma(r_1)\Gamma(r_2)}\int_0^{t_1}\int_{x_1}^{x_2}(t_1-s)^{r_1-1}(x_1-y)^{r_2-1}$$
$$\times P(t_1,x_1,s,y)dyd_s\left(\bigvee_{k=0}^{s}g(t,k)\right)$$
$$+\frac{\Psi(\eta)}{\Gamma(r_1)\Gamma(r_2)}\int_{t_1}^{t_2}\int_0^{x_1}(t_1-s)^{r_1-1}(x_1-y)^{r_2-1}$$
$$\times P(t_1,x_1,s,y)dyd_s\left(\bigvee_{k=0}^{s}g(t,k)\right)$$
$$+\frac{\Psi(\eta)}{\Gamma(r_1)\Gamma(r_2)}\int_{t_1}^{t_2}\int_{x_1}^{x_2}(t_1-s)^{r_1-1}(x_1-y)^{r_2-1}$$
$$\times P(t_1,x_1,s,y)dyd_s\left(\bigvee_{k=0}^{s}g(t,k)\right).$$

From continuity of f,g,h,P and as $t_1\longrightarrow t_2$ and $x_1\longrightarrow x_2$, the right-hand side of the above inequality tends to zero.

Step 4: *$N(B_\eta)$ is equiconvergent.*
Let $(t,x)\in J$ and $u\in B_\eta$, then we have

$$\begin{aligned}|(Nu)(t,x)|\le{}&|f(t,x,u(t,x))\\&+\frac{1}{\Gamma(r_1)\Gamma(r_2)}\int_0^{t}\int_0^{x}(t-s)^{r_1-1}(x-y)^{r_2-1}\\&\times h(t,x,s,y,u(s,y))dyd_sg(t,s)|\\&\le|f(t,x,u(t,x))-f(t,x,0)+f(t,x,0)|\\&+\frac{1}{\Gamma(r_1)\Gamma(r_2)}\int_0^{t}\int_0^{x}(t-s)^{r_1-1}(x-y)^{r_2-1}P(t,x,s,y)\\&\times\Psi(\|u\|_{BC})dyd_s\left(\bigvee_{k=0}^{s}g(t,k)\right)\\&\le\frac{M|u(t,x)|}{(1+\alpha(t,x))(L+|u(t,x)|)}+|f(t,x,0)|\\&+\Psi(\|u\|_{BC})Q(t,x)\\&\le\frac{M}{1+\alpha(t,x)}+|f(t,x,0)|+\Psi(\eta)Q(t,x).\end{aligned}$$

Thus, for each $x \in [0,b]$, we get

$$|(Nu)(t,x)| \longrightarrow 0, \ as\, t \longrightarrow +\infty.$$

Hence,

$$|(Nu)(t,x) - (Nu)(+\infty,x)| \longrightarrow 0, \ as\, t \longrightarrow +\infty.$$

As a consequence of Steps 1 to 4 together with the Lemma 1.69, we can conclude that $N : B_\eta \to B_\eta$ is continuous and compact. From an application of Schauder's theorem [204], we deduce that N has a fixed point u which is a solution of the equation (4.21).

Step 5: *The global asymptotic stability of solutions.*
Now we investigate the stability for solutions of equation (4.21). Let us assume that u and v are two solutions of the equation (4.21). Then for each $(t,x) \in J$, we have

$$\begin{aligned}
&|u(t,x) - v(t,x)| = |(Nu)(t,x) - (Nv)(t,x)| \\
\leq\ & |f(t,x,u(t,x)) - f(t,x,v(t,x))| \\
&+ \frac{1}{\Gamma(r_1)\Gamma(r_2)} \int_0^t \int_0^x (t-s)^{r_1-1}(x-y)^{r_2-1} \\
&\times |h(t,x,s,y,u(s,y)) - h(t,x,s,y,v(s,y))| dy d_s \left(\bigvee_{k=0}^{s} g(t,k) \right) \\
\leq\ & \frac{M}{L}|u(t,x) - v(t,x)| + (\Psi(\|u\|_{BC}) + \Psi(\|v\|_{BC}))Q(t,x). \qquad (4.26)
\end{aligned}$$

By using (4.26) and the fact that $M < L$, we deduce that

$$|u(t,x) - v(t,x)| \leq \frac{L(\Psi(\|u\|_{BC}) + \Psi(\|v\|_{BC}))}{L-M} Q(t,x).$$

Hence, for each $x \in [0,b]$, we get

$$\lim_{t \longrightarrow \infty} |u(t,x) - v(t,x)| = 0.$$

Consequently, all solutions of equation (4.21) are globally asymptotically stable. □

4.4.3. An Example

As an application of our results we consider the following integral equation of fractional order

$$\begin{aligned}
u(t,x) \ &= \ f(t,x,u(t,x)) + \tfrac{1}{\Gamma(r_1)\Gamma(r_2)} \textstyle\int_0^t \int_0^x (t-s)^{r_1-1}(x-y)^{r_2-1} \\
&\quad \times h(t,x,s,y,u(s,y)) dy d_s g(t,s); \ (t,x) \in J := \mathbb{R}_+ \times [0,1],
\end{aligned} \qquad (4.27)$$

where

$$(r_1, r_2) = \left(\frac{1}{3}, 2\right), \ f(t,x,u) = \frac{ctx\sin u}{1+x^2+t^2}; \ (t,x) \in J, \ \ c := \frac{9}{8\Gamma^2(\frac{1}{3})},$$

$$g(t,s)=s;\ (t,s)\in\mathbb{R}_+^2,\quad h(t,x,s,y,u)=\frac{sxye^{-t}|u|}{1+x^2+t^4};\ (t,x,s,y)\in J_1,\ u\in\mathbb{R},$$

and $J_1:=\{(t,x,s,y):0\le s\le t<\infty,\ 0\le y\le x\le 1\}$. First, we can see that $c<\frac{2}{5}$. The function f is continuous and satisfies assumption (4.14.1), where

$$M=c<\frac{2}{5},\ L=1,\ \alpha(t,x)=x^2+t^2\ and\ f^*=0.$$

Also, we can easily see that the function g satisfies the hypotheses (4.14.2) – (4.14.4). Next, let us notice that the function h satisfies assumption (4.14.5), where

$$P(t,x,s,y)=\frac{sxye^{-t}}{1+x^2+t^4};\ (t,x,s,y)\in J_1,\ \ and\ \ \Psi(w)=w;\ w\in\mathbb{R}_+.$$

Indeed, we have,

$$\begin{aligned}\int_0^t(t-s)^{r_1-1}P(t,x,s,y)d_sg(t,s) &\le \int_0^t(t-s)^{r_1-1}\frac{sxy}{1+x^2+t^4}d_s\left(\bigvee_{k=0}^{s}g(t,k)\right)\\ &\le \frac{xy}{1+t^4}\int_0^t(t-s)^{r_1-1}sds\\ &\le \frac{1}{1+t^4}\int_0^t(t-s)^{\frac{-2}{3}}sds\\ &= \frac{1}{\Gamma(\frac{7}{3})}\frac{t^{\frac{4}{3}}}{1+t^4}.\end{aligned}$$

Thus

$$\lim_{t\to\infty}\int_0^t(t-s)^{r-1}P(t,x,s,y)d_sg(t,s)=0.$$

Finally, we can see that the hypothesis (4.16.1) is satisfied with $\eta=1$. Indeed,

$$q^*\le\frac{\Gamma(2)}{\Gamma(\frac{1}{3})\Gamma(3)\Gamma(\frac{7}{3})}=c,\ and\ M+f^*+q^*\Psi(\eta)<\frac{2}{5}(1+\eta).$$

So, for $\eta=1$, we have $\frac{2}{5}(1+\eta)\le\eta$, witch gives immediately $M+f^*+q^*\Psi(\eta)\le\eta$. Hence by Theorem 4.12, the equation (4.27) has a solution defined on $\mathbb{R}_+\times[0,1]$ and solutions of this equation are globally asymptotically stable.

4.5. On the Existence and Global Asymptotic Stability of Solutions of Riemann-Liouville Volterra-Stieltjes Integral Equations

4.5.1. Introduction

This section deals with the existence of solutions to the following nonlinear quadratic Volterra functional Riemann-Liouville Volterra-Stieltjes integral equations of fractional order

$$u(t)=f(t,u(\alpha(t))),$$

$$\frac{1}{\Gamma(r)}\int_0^{\beta(t)}(\beta(t)-s)^{r-1}h(t,s,u(\gamma(s)))d_sg(t,s)\Bigg);\ if\ t\in\mathbb{R}_+, \quad (4.28)$$

where $r\in(0,\infty)$, α, β, $\gamma:\mathbb{R}_+\to\mathbb{R}_+$, $f:\mathbb{R}_+\times\mathbb{R}\times\mathbb{R}\to\mathbb{R}$, $g:\mathbb{R}_+\times\mathbb{R}_+\to\mathbb{R}$, $h:J_1\times\mathbb{R}\to\mathbb{R}$ are given continuous functions, $\lim_{t\to\infty}\alpha(t)=\infty$, and $J_1:=\{(t,s)\in\mathbb{R}_+^2:s\le t\}$. We prove the existence of solutions of equation (4.28) by using Schauder's fixed point theorem, and we obtain some results about the global asymptotic stability of solutions.

4.5.2. Main Results

By $BC:=BC(\mathbb{R}_+)$ we denote the Banach space of all bounded and continuous functions from $\mathbb{R}_+$ into $\mathbb{R}$ equipped with the standard norm

$$\|u\|_{BC}=\sup_{t\in\mathbb{R}_+}|u(t)|.$$

For $u_0\in BC$ and $\eta\in(0,\infty)$, we denote by $B(u_0,\eta)$, the closed ball in BC centered at u_0 with radius η.

Definition 4.13. *We mean by a solution of equation (4.28), every function $u\in BC$ such that u satisfies equation (4.28) on $\mathbb{R}_+$.*

The following hypotheses will be used in the sequel.

(4.18.1) There exist constants $0<M<1$ and $L>0$ such that

$$|f(t,u_1,u_2)-f(t,v_1,v_2)|\le\frac{M|u_1-v_1|+L|u_2-v_2|}{1+\alpha(t)+|u_1-v_1|},$$

for $t\in\mathbb{R}_+$ and for $u_1,v_1,u_2,v_2\in\mathbb{R}$,

(4.18.2) The function $t\to f(t,0,0)$ is bounded on $\mathbb{R}_+$ with $f^*=\sup_{t\in\mathbb{R}_+}f(t,0,0)$ and $\lim_{t\to\infty}|f(t,0,0)|=0$,

(4.18.3) For all $t_1,t_2\in\mathbb{R}_+$ such that $t_1<t_2$ the function $s\mapsto g(t_2,s)-g(t_1,s)$ is nondecreasing on $\mathbb{R}_+$,

(4.18.4) The function $s\mapsto g(0,s)$ is nondecreasing on $\mathbb{R}_+$,

(4.18.5) The functions $s\mapsto g(t,s)$ and $t\mapsto g(t,s)$ are continuous on $\mathbb{R}_+$ for each fixed $t\in\mathbb{R}_+$ or $s\in\mathbb{R}_+$, respectively,

(4.18.6) There exists a continuous function $p:J_1\to\mathbb{R}_+$ such that

$$|h(t,s,u)|\le\frac{p(t,s)}{1+\alpha(t)+|u|};\ \text{for } (t,s)\in J_1 \text{ and for } u\in\mathbb{R}.$$

Moreover, assume that

$$\lim_{t\to\infty}\int_0^{\beta(t)}(\beta(t)-s)^{r_1-1}p(t,s)d_sg(t,s)=0.$$

Remark 4.14. *Set* $q^* := \sup_{t\in\mathbb{R}_+} q(t)$ *where*

$$q(t) = \frac{1}{\Gamma(r)}\int_0^{\beta(t)} (\beta(t)-s)^{r_1-1}p(t)d_s\left(\bigvee_{k=0}^{s} g(t,k)\right).$$

From the hypothesis $(4.18.6)$, *we infer that* q^* *is finite and* $\lim_{t\to\infty} q(t) = 0$.

Theorem 4.15. *Assume that hypotheses* $(4.18.1)-(4.18.6)$ *hold. Then the equation (4.28) has at least one solution in the space BC. Moreover, solutions of equation (4.28) are globally asymptotically stable.*

Proof. Let us define the operator N, such that for any $u \in BC$

$$(Nu)(t) = f(t,u(\alpha(t)),$$

$$\frac{1}{\Gamma(r)}\int_0^{\beta(t)} (\beta(t)-s)^{r-1}h(t,s,u(\gamma(s)))d_s g(t,s)\Big);\ t\in\mathbb{R}_+. \tag{4.29}$$

By considering conditions of theorem we infer that $N(u)$ is continuous on $\mathbb{R}_+$. Now we prove that $N(u) \in BC$ for any $u \in BC$. For arbitrarily fixed $t \in \mathbb{R}_+$ we have

$$\begin{aligned}
|(Nu)(t)| &= \left|f\left(t,u(\alpha(t)),\frac{1}{\Gamma(r)}\int_0^{\beta(t)} (\beta(t)-s)^{r-1}h(t,s,u(\gamma(s)))d_s g(t,s)\right)\right| \\
&\le |f(t,0,0)| + |f(t,0,0) \\
&- f\left(t,u(\alpha(t)),\frac{1}{\Gamma(r)}\int_0^{\beta(t)} (\beta(t)-s)^{r-1}h(t,s,u(\gamma(s)))d_s g(t,s)\right)\Bigg| \\
&\le |f(t,0,0)| + \frac{M|u(\alpha(t))|}{1+\alpha(t)+|u(\alpha(t))|} \\
&+ \frac{L}{\Gamma(r)(1+\alpha(t)+|u(\alpha(t))|)} \\
&\times \int_0^{\beta(t)} \frac{(\beta(t)-s)^{r-1}p(t,s)}{1+\alpha(t)+|u(\gamma(s))|}d_s\left(\bigvee_{k=0}^{s} g(t,k)\right) \\
&\le |f(t,0,0)| + M + Lq(t) \\
&\le f^* + M + Lq^*.
\end{aligned}$$

Thus

$$\|N(u)\|_{BC} \le f^* + M + Lq^*. \tag{4.30}$$

Hence $N(u) \in BC$. Equation (4.30) yields that N transforms the ball $B_\eta := B(0,\eta)$ into itself where $\eta = f^* + M + Lq^*$. We shall show that $N : B_\eta \to B_\eta$ satisfies the assumptions of Schauder's fixed point theorem [204]. The proof will be given in several steps.

Step 1: *N is continuous.*
Let $\{u_n\}_{n\in\mathbb{N}}$ be a sequence such that $u_n \to u$ in B_η. Then, for each $t \in \mathbb{R}_+$, we have

$$\begin{aligned}
& |(Nu_n)(t) - (Nu)(t)| \\
\leq & \left| f\left(t, u_n(\alpha(t)), \frac{1}{\Gamma(r)} \int_0^{\beta(t)} (\beta(t)-s)^{r-1} h(t,s,u_n(\gamma(s)))d_s g(t,s)\right) \right. \\
- & \left. f\left(t, u(\alpha(t)), \frac{1}{\Gamma(r)} \int_0^{\beta(t)} (\beta(t)-s)^{r-1} h(t,s,u(\gamma(s)))d_s g(t,s)\right) \right| \\
\leq & \frac{M|u_n(\alpha(t)) - u(\alpha(t))|}{1+\alpha(t)+|u_n(\alpha(t)-u(\alpha(t)|} \\
+ & \frac{L}{\Gamma(r)(1+\alpha(t)+|u_n(\alpha(t))-u(\alpha(t))|)} \\
\times & \int_0^{\beta(t)} (\beta(t)-s)^{r-1} |h(t,s,u_n(\gamma(s))) - h(t,s,u(\gamma(s)))| d_s \left(\bigvee_{k=0}^{s} g(t,k) \right) \\
\leq & M|u_n(\alpha(t)) - u(\alpha(t))|
\end{aligned}$$

$$+\frac{L}{\Gamma(r)} \int_0^{\beta(t)} (\beta(t)-s)^{r-1} |h(t,s,u_n(\gamma(s))) - h(t,s,u(\gamma(s)))| d_s \left(\bigvee_{k=0}^{s} g(t,k) \right). \tag{4.31}$$

Case 1. If $t \in [0,T]$; $T > 0$, then, since $u_n \to u$ as $n \to \infty$ and g is continuous, (4.31) gives

$$\|N(u_n) - N(u)\|_{BC} \to 0 \quad \text{as } n \to \infty.$$

Case 2. If $t > T$; $T > 0$, then from (4.18.6) and (4.31) we get

$$\begin{aligned}
& |(Nu_n)(t) - (Nu)(t)| \leq M|u_n(\alpha(t)) - u(\alpha(t))| \\
+ & \frac{L}{\Gamma(r)} \int_0^{\beta(t)} (\beta(t)-s)^{r-1} (|h(t,s,u_n(\gamma(s)))| \\
+ & |h(t,s,u(\gamma(s)))|) d_s \left(\bigvee_{k=0}^{s} g(t,k) \right) \\
\leq & M|u_n(\alpha(t)) - u(\alpha(t))| + \frac{L}{\Gamma(r)} \int_0^{\beta(t)} (\beta(t)-s)^{r-1} p(t,s) \\
\times & \left(\frac{1}{1+\alpha(t)+|u_n(\gamma(s))|} + \frac{1}{1+\alpha(t)+|u(\gamma(s))|} \right) d_s \left(\bigvee_{k=0}^{s} g(t,k) \right)
\end{aligned}$$

$$\leq M|u_n(\alpha(t)) - u(\alpha(t))| + 2Lq(t). \tag{4.32}$$

Since $u_n \to u$ as $n \to \infty$ and $q(t) \to 0$ as $t \to \infty$, then (4.32) gives

$$\|N(u_n) - N(u)\|_{BC} \to 0 \quad \text{as } n \to \infty.$$

Step 2: *$N(B_\eta)$ is uniformly bounded.*
This is clear since $N(B_\eta) \subset B_\eta$ and B_η is bounded.

Step 3: *$N(B_\eta)$ is equicontinuous on every compact interval $[0,a]$; $a>0$ of $\mathbb{R}_+$.*
Let $t_1, t_2 \in [0,a], t_1 < t_2$ and let $u \in B_\eta$. Also without lose of generality suppose that $\beta(t_1) \leq \beta(t_2)$, then we have

$$\begin{aligned}
& |(Nu)(t_2) - (Nu)(t_1)| \\
\leq & \left| f\left(t_2, u(\alpha(t_2)), \frac{1}{\Gamma(r)} \int_0^{\beta(t_2)} (\beta(t_2)-s)^{r-1} h(t_2,s,u(\gamma(s))) d_s g(t_2,s)\right)\right. \\
- & \left. f\left(t_2, u(\alpha(t_1)), \frac{1}{\Gamma(r)} \int_0^{\beta(t_1)} (\beta(t_1)-s)^{r-1} h(t_1,s,u(\gamma(s))) d_s g(t_1,s)\right)\right| \\
+ & \left| f\left(t_1, u(\alpha(t_2)), \frac{1}{\Gamma(r)} \int_0^{\beta(t_2)} (\beta(t_2)-s)^{r-1} h(t_2,s,u(\gamma(s))) d_s g(t_2,s)\right)\right. \\
- & \left. f\left(t_1, u(\alpha(t_1)), \frac{1}{\Gamma(r)} \int_0^{\beta(t_1)} (\beta(t_1)-s)^{r-1} h(t_1,s,u(\gamma(s))) d_s g(t_1,s)\right)\right|.
\end{aligned}$$

Thus, we get

$$\begin{aligned}
& |(Nu)(t_2) - (Nu)(t_1)| \\
\leq & \frac{M|u(\alpha(t_2)) - u(\alpha(t_1))|}{1+\alpha(t_2) + |u(\alpha(t_2) - u(\alpha(t_1))|} \\
+ & \frac{L}{\Gamma(r)(1+\alpha(t_2) + |u(\alpha(t_2)) - u(\alpha(t_1))|)} \\
\times & \left[\left| \int_0^{\beta(t_2)} (\beta(t_2)-s)^{r-1} [h(t_2,s,u(\gamma(s))) - h(t_1,s,u(\gamma(s)))] d_s g(t,s)\right|\right. \\
& + \left| \int_0^{\beta(t_2)} (\beta(t_2)-s)^{r-1} h(t_1,s,u(\gamma(s))) d_s g(t,s)\right. \\
& \left. - \int_0^{\beta(t_1)} (\beta(t_2)-s)^{r-1} h(t_1,s,u(\gamma(s))) d_s g(t,s)\right| \\
& + \left| \int_0^{\beta(t_1)} (\beta(t_2)-s)^{r-1} h(t_1,s,u(\gamma(s))) d_s g(t,s)\right. \\
& \left. \left[- \int_0^{\beta(t_1)} (\beta(t_1)-s)^{r-1} h(t_1,s,u(\gamma(s))) d_s g(t,s)\right| \right] \\
\leq & M|u(\alpha(t_2)) - u(\alpha(t_1))| + \frac{L}{\Gamma(r)} \\
\times & \left[\int_0^{\beta(t_2)} (\beta(t_2)-s)^{r-1} |h(t_2,s,u(\gamma(s))) - h(t_1,s,u(\gamma(s)))| d_s \left(\bigvee_{k=0}^{s} g(t,k)\right)\right. \\
& + \int_{\beta(t_1)}^{\beta(t_2)} (\beta(t_2)-s)^{r-1} |h(t_1,s,u(\gamma(s)))| d_s \left(\bigvee_{k=0}^{s} g(t,k)\right) \\
& + \int_0^{\beta(t_1)} |(\beta(t_2)-s)^{r-1} - (\beta(t_1)-s)^{r-1}|
\end{aligned}$$

$$\times \quad |h(t_1,s,u(\gamma(s)))|d_s\left(\bigvee_{k=0}^{s} g(t,k)\right)\Bigg].$$

Hence,

$$\begin{aligned}
&|(Nu)(t_2)-(Nu)(t_1)| \\
\leq\ & M|u(\alpha(t_2))-u(\alpha(t_1))|+\frac{L}{\Gamma(r)} \\
\times\ & \Bigg[\int_0^{\beta(t_2)}(\beta(t_2)-s)^{r-1}|h(t_2,s,u(\gamma(s))) \\
& -h(t_1,s,u(\gamma(s)))|d_s\left(\bigvee_{k=0}^{s} g(t,k)\right) \\
& +\int_{\beta(t_1)}^{\beta(t_2)}(\beta(t_2)-s)^{r-1}p(t_1,s)d_s\left(\bigvee_{k=0}^{s} g(t,k)\right) \\
& +\int_0^{\beta(t_1)}|(\beta(t_2)-s)^{r-1}-(\beta(t_1)-s)^{r-1}|\times p(t_1,s)d_s\left(\bigvee_{k=0}^{s} g(t,k)\right)\Bigg].
\end{aligned}$$

From continuity of α,β,f,g,h,p and as $t_1\to t_2$, the right-hand side of the above inequality tends to zero.

Step 4: *$N(B_\eta)$ is equiconvergent.*
Let $t\in\mathbb{R}_+$ and $u\in B_\eta$, then we have

$$\begin{aligned}
|(Nu)(t)| &= \left|f\left(t,u(\alpha(t)),\frac{1}{\Gamma(r)}\int_0^{\beta(t)}(\beta(t)-s)^{r-1}h(t,s,u(\gamma(s)))d_sg(t,s)\right)\right| \\
&\leq |f(t,0,0)| \\
&+\left|f\left(t,u(\alpha(t)),\frac{1}{\Gamma(r)}\int_0^{\beta(t)}(\beta(t)-s)^{r-1}\right.\right. \\
&\times h(t,s,u(\gamma(s)))d_sg(t,s)\Big)-f(t,0,0)\Big| \\
&\leq |f(t,0,0)|+\frac{M|u(\alpha(t))|}{1+\alpha(t)+|u(\alpha(t))|} \\
&+\frac{L}{\Gamma(r)(1+\alpha(t)+|u(\alpha(t))|)}\int_0^{\beta(t)}\frac{(\beta(t)-s)^{r-1}p(t,s)}{1+\alpha(t)+|u(\gamma(s))|}d_s\left(\bigvee_{k=0}^{s} g(t,k)\right) \\
&\leq |f(t,0,0)|+\frac{M\eta}{1+\alpha(t)}+Lq(t).
\end{aligned}$$

Thus

$$|(Nu)(t)|\to 0,\ \text{as}\ t\to+\infty.$$

Hence, we get

$$|(Nu)(t)-(Nu)(+\infty)|\to 0,\ \text{as}\ t\to+\infty.$$

As a consequence of Steps 1 to 4 together with the Lemma 1.69, we can conclude that $N : B_\eta \to B_\eta$ is continuous and compact. From an application of Schauder's theorem [204], we deduce that N has a fixed point u which is a solution of the equation (4.28).

Step 5: *The global asymptotic stability of solutions.*
Now we investigate the stability for solutions of equation (4.28). Let us assume that u and v are two solutions of the equation (4.28). Then for each $t \in \mathbb{R}_+$, we have

$$\begin{aligned}
&|u(t) - v(t)| = |(Nu)(t) - (Nv)(t)| \\
\leq\ & M|u(\alpha(t)) - v(\alpha(t))| \\
+\ & \frac{L}{\Gamma(r)} \int_0^{\beta(t)} (\beta(t) - s)^{r-1}(|h(t,s,u(\gamma(s)))| \\
+\ & |h(t,s,v(\gamma(s)))|)d_s\left(\bigvee_{k=0}^{s} g(t,k)\right) \\
\leq\ & M|u(\alpha(t)) - v(\alpha(t))| + \frac{L}{\Gamma(r)} \int_0^{\beta(t)} (\beta(t) - s)^{r-1} p(t,s) \\
\times\ & \left(\frac{1}{1+\alpha(t)+|u(\gamma(s))|} + \frac{1}{1+\alpha(t)+|v(\gamma(s))|}\right) d_s\left(\bigvee_{k=0}^{s} g(t,k)\right)
\end{aligned}$$

$$\leq M|u(\alpha(t)) - v(\alpha(t))| + 2Lq(t). \tag{4.33}$$

Since $\alpha(t) \to \infty$ as $t \to \infty$, then $\lim_{t\to\infty} |u(\alpha(t)) - v(\alpha(t))| = \lim_{t\to\infty} |u(t) - v(t)|$. Thus, by using (4.33), we deduce that

$$\lim_{t\to\infty} |u(t) - v(t)| \leq \lim_{t\to\infty} \frac{2Lq(t)}{1-M} = 0.$$

Consequently, all solutions of equation (4.28) are globally asymptotically stable. □

4.5.3. An Example

As an application of our results we consider the following nonlinear quadratic Volterra functional Riemann-Liouville Volterra-Stieltjes integral equation of fractional order

$$u(t) = f(t, u(\alpha(t)),$$

$$\frac{1}{\Gamma(r)} \int_0^{\beta(t)} (\beta(t) - s)^{r-1} h(t,s,u(\gamma(s)))d_s g(t,s)\Bigg); \ if\ t \in \mathbb{R}_+, \tag{4.34}$$

where $r = \frac{2}{3}$, $\alpha(t) = \beta(t) = \gamma(t) = t,\ g(t,s) = s;\ (t,s) \in \mathbb{R}_+^2$,

$$f(t,u,v) = \frac{2}{2(1+t)+|u|} + |v|;\ t \in \mathbb{R}_+,\ u,v \in \mathbb{R},$$

and

$$h(t,s,u) = \frac{s|u|}{(1+t+|u|)(1+t^4)};\ (t,s) \in J_1 := \{(t,s) \in \mathbb{R}_+^2 : s \leq t\},\ u \in \mathbb{R}.$$

For each $t \in \mathbb{R}_+$ and $u_1, u_2, v_1, v_2 \in \mathbb{R}$, we have

$$|f(t,u_1,u_2) - f(t,v_1,v_2)| \leq \frac{\frac{1}{2}|u_1 - v_1| + |u_2 - v_2|}{1 + t + |u_1 - v_1|},$$

then, we have that the function f is continuous and satisfies assumption (4.18.1), with $M = \frac{1}{2}$ and $L = 1$. Also f satisfies assumption (4.18.2) with $f^* = 1$. Next, we can easily see that the function g satisfies the hypotheses (4.18.3) – (4.18.5).
Finally, let us notice that the function h satisfies the assumption (4.18.6), with

$$p(t,s) = \frac{s}{1+t^4}; \ (t,s) \in J_1.$$

Indeed, we have,

$$\begin{aligned}\int_0^{\beta(t)} (\beta(t) - s)^{r-1} p(t,s) d_s g(t,s) &\leq \int_0^t (t-s)^{-\frac{1}{3}} \frac{s}{1+t^4} ds \\ &= \frac{9}{10(1+t^4)} t^{\frac{5}{3}} \to 0, \ as\ t \to \infty.\end{aligned}$$

Hence by Theorem 4.15, equation (4.34) has a solution defined on $\mathbb{R}_+$ and solutions of this equation are globally asymptotically stable.

4.6. Global Attractivity of Solutions for Nonlinear Riemann-Liouville Volterra-Stieltjes Partial Integral Equations

4.6.1. Introduction

This section deals with the existence and the attractivity of solutions to the following nonlinear fractional order Riemann-Liouville Volterra-Stieltjes quadratic partial integral equations of the form,

$$\begin{aligned} u(t,x) &= f(t,x,u(t,x),u(\alpha(t),x)) \\ &+\tfrac{1}{\Gamma(r_1)\Gamma(r_2)} \int_0^{\beta(t)} \int_0^x (\beta(t)-s)^{r_1-1}(x-y)^{r_2-1} \\ &\times h(t,x,s,y,u(s,y),u(\gamma(s),y)) dy d_s g(t,s); \ (t,x) \in J := \mathbb{R}_+ \times [0,b], \end{aligned} \tag{4.35}$$

where $b > 0$, $r_1, r_2 \in (0,\infty)$, $\mathbb{R}_+ = [0,\infty)$, α, β, $\gamma : \mathbb{R}_+ \to \mathbb{R}_+$, $f : J \times \mathbb{R} \times \mathbb{R} \to \mathbb{R}$, $g : \mathbb{R}_+ \times \mathbb{R}_+ \to \mathbb{R}$, $h : J' \times \mathbb{R} \times \mathbb{R} \to \mathbb{R}$ are given continuous functions, $\lim_{t\to\infty} \alpha(t) = \infty$, $J' = \{(t,x,s,y) \in J^2 : s \leq t, \ y \leq x\}$. We use the Schauder fixed point theorem for the existence of solutions of the equation (4.35), and we prove that all solutions are uniformly globally attractive. Finally, we present an example illustrating the applicability of the imposed conditions.

4.6.2. Main Results

For $u_0 \in BC$ and $\eta \in (0,\infty)$, we denote by $B(u_0,\eta)$, the closed ball in BC centered at u_0 with radius η.

Definition 4.16. *We mean by a solution of equation (4.35), every function $u \in BC$ such that u satisfies equation (4.35) on J.*

The following hypotheses will be used in the sequel.

(4.22.1) There exist positive constants M and L such that

$$|f(t,x,u_1,u_2)-f(t,x,v_1,v_2)| \le \frac{M|u_1-v_1|+L|u_2-v_2|}{1+\alpha(t)},$$

for $(t,x) \in J$ and $u_1,u_2,v_1,v_2 \in \mathbb{R}$.

(4.22.2) The function $t \to f(t,x,0,0)$ is bounded on J with

$$f^* = \sup_{(t,x)\in\mathbb{R}_+\times[0,b]} f(t,x,0,0) \ \ and \ \ \lim_{t\to\infty}|f(t,x,0,0)| = 0, \ x \in [0,b].$$

(4.22.3) For all $t_1,t_2 \in \mathbb{R}_+$ such that $t_1 < t_2$ the function $s \mapsto g(t_2,s)-g(t_1,s)$ is nondecreasing on $\mathbb{R}_+$.

(4.22.4) The function $s \mapsto g(0,s)$ is nondecreasing on $\mathbb{R}_+$.

(4.22.5) The functions $s \mapsto g(t,s)$ and $t \mapsto g(t,s)$ are continuous on $\mathbb{R}_+$ for each fixed $t \in \mathbb{R}_+$ or $s \in \mathbb{R}_+$, respectively.

(4.22.6) There exist continuous functions $p_1,p_2 : J' \to \mathbb{R}_+$ such that

$$|h(t,x,s,y,u,v)| \le \frac{p_1(t,x,s,y)|u|+p_2(t,x,s,y)|v|}{1+|u|+|v|};$$

for $(t,x,s,y) \in J'$, $u,v \in \mathbb{R}$. Moreover, assume that

$$\lim_{t\to\infty}\int_0^{\beta(t)}(\beta(t)-s)^{r-1}p_i(t,x,s,y)d_sg(t,s) = 0; \ i = 1,2.$$

Remark 4.17. *Set* $p_i^* := \sup_{(t,x)\in J}\frac{1}{\Gamma(r_1)\Gamma(r_2)}\int_0^{\beta(t)}\int_0^x(\beta(t)-s)^{r_1-1}(x-y)^{r_2-1}p_i(t,x,s,y)dyd_s\left(\bigvee_{k=0}^{s}g(t,k)\right)$; *for* $i=1,2$. *From hypothesis* (4.22.6), *we infer that* p_i^* *is finite, for* $i=1,2$.

Theorem 4.18. *Assume that hypotheses* (4.22.1) – (4.22.6) *hold. If*

$$M+L+p_1^*+p_2^* < 1, \tag{4.36}$$

then the equation (4.35) has at least one solution in the space BC. Moreover, solutions of equation (4.35) are uniformly globally attractive.

Proof. Let us define the operator N such that, for any $u \in BC$,

$$
\begin{aligned}
(Nu)(t,x) &= f(t,x,u(t,x),u(\alpha(t),x)) \\
&\quad + \frac{1}{\Gamma(r_1)\Gamma(r_2)} \int_0^{\beta(t)} \int_0^x (\beta(t)-s)^{r_1-1}(x-y)^{r_2-1} \\
&\quad \times h(t,x,s,y,u(s,x),u(\gamma(s),x))dyd_sg(t,s); \ (t,x) \in J.
\end{aligned}
\tag{4.37}
$$

From the assumptions of this theorem, we infer that $N(u)$ is continuous on J. Now we prove that $N(u) \in BC$ for any $u \in BC$. For arbitrarily fixed $(t,x) \in J$, we have

$$
\begin{aligned}
|(Nu)(t,x)| &= |f(t,x,u(t,x),u(\alpha(t),x)) \\
&\quad + \frac{1}{\Gamma(r_1)\Gamma(r_2)} \int_0^{\beta(t)} \int_0^x (\beta(t)-s)^{r_1-1}(x-y)^{r_2-1} \\
&\quad \times h(t,x,s,y,u(s,y),u(\gamma(s),y))dyd_sg(t,s)| \\
&\le |f(t,x,u(t,x),u(\alpha(t),x)) - f(t,x,0,0) + f(t,x,0,0)| \\
&\quad + |\frac{1}{\Gamma(r_1)\Gamma(r_2)} \int_0^{\beta(t)} \int_0^x (\beta(t)-s)^{r_1-1}(x-y)^{r_2-1} \\
&\quad \times h(t,x,s,y,u(s,y),u(\gamma(s),y))dyd_sg(t,s)| \\
&\le \frac{M}{1+\alpha(t)}|u(t,x)| + \frac{L}{1+\alpha(t)}|u(\alpha(t),x)| + |f(t,x,0,0)| \\
&\quad + \frac{1}{\Gamma(r_1)\Gamma(r_2)} \int_0^{\beta(t)} \int_0^x (\beta(t)-s)^{r_1-1}(x-y)^{r_2-1} \\
&\quad \times \frac{p_1(t,x,s,y)|u(s,y)| + p_2(t,x,s,y)|u(\gamma(s),y)|}{1+|u(s,y)|+|u(\gamma(s),y)|} dyd_s\left(\bigvee_{k=0}^{s} g(t,k)\right) \\
&\le |f(t,x,0,0)| + M|u(t,x)| + L|u(\alpha(t),x)| \\
&\quad + \frac{1}{\Gamma(r_1)\Gamma(r_2)} \int_0^{\beta(t)} \int_0^x (\beta(t)-s)^{r_1-1}(x-y)^{r_2-1} \\
&\quad \times (p_1(t,x,s,y) + p_2(t,x,s,y))dyd_s\left(\bigvee_{k=0}^{s} g(t,k)\right) \\
&\le f^* + (M+L+p_1^*+p_2^*)\|u\|_{BC}.
\end{aligned}
$$

Thus

$$
\|N(u)\| \le f^* + (M+L+p_1^*+p_2^*)\|u\|_{BC}. \tag{4.38}
$$

Hence $N(u) \in BC$. From (4.36) and (4.38), we infer that N transforms the ball $B_\eta := B(0,\eta)$ into itself, where

$$
\eta \ge \frac{f^*}{1-M-L-p_1^*-p_2^*}.
$$

We shall show that $N : B_\eta \to B_\eta$ satisfies the assumptions of Schauder's fixed point theorem [204]. The proof will be given in several steps and cases.

Step 1: *N is continuous.*
Let $\{u_n\}_{n\in\mathbb{N}}$ be a sequence such that $u_n \longrightarrow u$ in B_η. Then, for each $(t,x)\in J$, we have

$$\begin{aligned}
& |(Nu_n)(t,x)-(Nu)(t,x)| \\
\leq\ & |f(t,x,u_n(t,x),u_n(\alpha(t),x))-f(t,x,u(t,x),u(\alpha(t),x))| \\
& +\tfrac{1}{\Gamma(r_1)\Gamma(r_2)}\int_0^{\beta(t)}\int_0^x(\beta(t)-s)^{r_1-1}(x-y)^{r_2-1} \\
\times\ & |h(t,x,s,y,u_n(s,y),u_n(\gamma(s),y))-h(t,x,s,y,u(s,y),u(\gamma(s),y))|dyd_sg(t,s) \qquad (4.39)\\
\leq\ & (M+L)\|u_n-u\|_{BC}+\tfrac{1}{\Gamma(r_1)\Gamma(r_2)}\int_0^{\beta(t)}\int_0^x(\beta(t)-s)^{r_1-1}(x-y)^{r_2-1} \\
\times\ & |h(t,x,s,y,u_n(s,y),u_n(\gamma(s),y)) \\
& -h(t,x,s,y,u(s,y),u(\gamma(s),y))|dyd_s\left(\vee_{k=0}^{s}g(t,k)\right).
\end{aligned}$$

Case 1. If $(t,x)\in[0,a]\times[0,b]$; $a>0$, then, since $u_n\longrightarrow u$ as $n\longrightarrow\infty$ and g,h are continuous, (4.39) gives

$$\|N(u_n)-N(u)\|_{BC}\to 0 \quad \text{as } n\longrightarrow\infty.$$

Case 2. If $(t,x)\in(a,\infty)\times[0,b]$; $a>0$, then from (4.22.3) and (4.39), for each $(t,x)\in J$, we get

$$\begin{aligned}
& |(Nu_n)(t,x)-(Nu)(t,x)| \\
\leq\ & (M+L)\|u_n-u\|+\tfrac{1}{\Gamma(r_1)\Gamma(r_2)}\int_0^{\beta(t)}\int_0^x(\beta(t)-s)^{r_1-1}(x-y)^{r_2-1} \\
\times\ & (p_1(t,x,s,y)(|u_n(s,x)|+|u(s,x)|) \\
+\ & p_2(t,x,s,y)(|u_n(\gamma(s),x)|+|u(\gamma(s),x)|))dyd_s\left(\vee_{k=0}^{s}g(t,k)\right) \qquad (4.40)\\
\leq\ & (M+L)\|u_n-u\|+\tfrac{2\eta}{\Gamma(r_1)\Gamma(r_2)}\int_0^{\beta(t)}\int_0^x(\beta(t)-s)^{r_1-1}(x-y)^{r_2-1} \\
\times\ & (p_1(t,x,s,y)+p_2(t,x,s,y))dyd_s\left(\vee_{k=0}^{s}g(t,k)\right)
\end{aligned}$$

From (4.22.6) and since $u_n\longrightarrow u$ as $n\longrightarrow\infty$ and $t\longrightarrow\infty$, then (4.40) gives

$$\|N(u_n)-N(u)\|_{BC}\to 0 \quad \text{as } n\longrightarrow\infty.$$

Step 2: *$N(B_\eta)$ is uniformly bounded.*
This is clear since $N(B_\eta)\subset B_\eta$ and B_η is bounded.

Step 3: *$N(B_\eta)$ is equicontinuous on every compact subset $[0,a]\times[0,b]$ of J, $a>0$.*
Let $(t_1,x_1),(t_2,x_2)\in[0,a]\times[0,b]$, $t_1<t_2$, $x_1<x_2$ and let $u\in B_\eta$. Also without loss of generality, suppose that $\beta(t_1)\leq\beta(t_2)$. Then we have

$$\begin{aligned}
& |(Nu)(t_2,x_2)-(Nu)(t_1,x_1)| \\
& \leq|f(t_2,x_2,u(t_2,x_2),u(\alpha(t_2),x_2))-f(t_2,x_2,u(t_1,x_1),u(\alpha(t_1),x_1))| \\
& +|f(t_2,x_2,u(t_1,x_1),u(\alpha(t_1),x_1))-f(t_1,x_1,u(t_1,x_1),u(\alpha(t_1),x_1))| \\
& +\frac{1}{\Gamma(r_1)\Gamma(r_2)}\left|\int_0^{\beta(t_2)}\int_0^{x_2}(\beta(t_2)-s)^{r_1-1}(x_2-y)^{r_2-1}\right. \\
& \times[h(t_2,x_2,s,y,u(s,y),u(\gamma(s),y)) \\
& -h(t_1,x_1,s,y,u(s,y),u(\gamma(s),y))]dyd_sg(t,s)\Big|
\end{aligned}$$

$$+\frac{1}{\Gamma(r_1)\Gamma(r_2)}|\int_0^{\beta(t_2)}\int_0^{x_2}(\beta(t_2)-s)^{r_1-1}(x_2-y)^{r_2-1}$$
$$\times h(t_1,x_1,s,y,u(s,y),u(\gamma(s),y))dyd_sg(t,s)$$
$$-\int_0^{\beta(t_1)}\int_0^{x_2}(\beta(t_2)-s)^{r_1-1}(x_2-y)^{r_2-1}$$
$$\times h(t_1,x_1,s,y,u(s,y),u(\gamma(s),y))dyd_sg(t,s)|$$
$$+\frac{1}{\Gamma(r_1)\Gamma(r_2)}|\int_0^{\beta(t_1)}\int_0^{x_2}(\beta(t_2)-s)^{r_1-1}(x_2-y)^{r_2-1}$$
$$\times h(t_1,x_1,s,y,u(s,y),u(\gamma(s),y))dyd_sg(t,s)$$
$$-\int_0^{\beta(t_1)}\int_0^{x_1}(\beta(t_1)-s)^{r_1-1}(x_1-y)^{r_2-1}$$
$$\times h(t_1,x_1,s,y,u(s,y),u(\gamma(s),y))dyd_sg(t,s)|.$$

Thus we get

$$|(Nu)(t_2,x_2)-(Nu)(t_1,x_1)|$$
$$\leq M|u(\alpha(t_2),x_2)-u(\alpha(t_1),x_1)|+L|u(t_2,x_2)-u(t_1,x_1)|$$
$$+|f(t_2,x_2,u(t_1,x_1),u(\alpha(t_1),x_1))-f(t_1,x_1,u(t_1,x_1),u(\alpha(t_1),x_1))|$$
$$+\frac{1}{\Gamma(r_1)\Gamma(r_2)}\int_0^{\beta(t_2)}\int_0^{x_2}(\beta(t_2)-s)^{r_1-1}(x_2-y)^{r_2-1}$$
$$\times|h(t_2,x_2,s,y,u(s,y),u(\gamma(s),y))$$
$$-h(t_1,x_1,s,y,u(s,y),u(\gamma(s),y))|dyd_s\left(\bigvee_{k=0}^{s}g(t,k)\right)$$
$$+\frac{1}{\Gamma(r_1)\Gamma(r_2)}\int_{\beta(t_1)}^{\beta(t_2)}\int_0^{x_2}(\beta(t_2)-s)^{r_1-1}(x_2-y)^{r_2-1}$$
$$\times|h(t_1,x_1,s,y,u(s,y),u(\gamma(s),y))|dyd_s\left(\bigvee_{k=0}^{s}g(t,k)\right)$$
$$+\frac{1}{\Gamma(r_1)\Gamma(r_2)}\int_0^{\beta(t_1)}\int_0^{x_1}|(\beta(t_2)-s)^{r_1-1}(x_2-y)^{r_2-1}$$
$$-(\beta(t_1)-s)^{r_1-1}(x_1-y)^{r_2-1}||h(t_1,x_1,s,y,u(s,y),u(\gamma(s),y))|dyd_s\left(\bigvee_{k=0}^{s}g(t,k)\right)$$
$$+\frac{1}{\Gamma(r_1)\Gamma(r_2)}\int_0^{\beta(t_1)}\int_{x_1}^{x_2}|(\beta(t_2)-s)^{r_1-1}(x_2-y)^{r_2-1}|$$
$$\times|h(t_1,x_1,s,y,u(s,y),u(\gamma(s),y))|dyd_s\left(\bigvee_{k=0}^{s}g(t,k)\right).$$

Hence

$$
\begin{aligned}
&|(Nu)(t_2,x_2)-(Nu)(t_1,x_1)|\\
&\leq M|u(\alpha(t_2),x_2)-u(\alpha(t_1),x_1)|+L|u(t_2,x_2)-u(t_1,x_1)|\\
&+|f(t_2,x_2,u(t_1,x_1),u(\alpha(t_1),x_1))-f(t_1,x_1,u(t_1,x_1),u(\alpha(t_1),x_1))|\\
&+\frac{1}{\Gamma(r_1)\Gamma(r_2)}\int_0^{\beta(t_2)}\int_0^{x_2}(\beta(t_2)-s)^{r_1-1}(x_2-y)^{r_2-1}\\
&\times|h(t_2,x_2,s,y,u(s,y),u(\gamma(s),y))\\
&-h(t_1,x_1,s,y,u(s,y),u(\gamma(s),y))|dyd_s\left(\bigvee_{k=0}^{s}g(t,k)\right)\\
&+\frac{1}{\Gamma(r_1)\Gamma(r_2)}\int_{\beta(t_1)}^{\beta(t_2)}\int_0^{x_2}(\beta(t_2)-s)^{r_1-1}(x_2-y)^{r_2-1}\\
&\times(p_1(t_1,x_1,s,y)+p_2(t_1,x_1,s,y))dyd_s\left(\bigvee_{k=0}^{s}g(t,k)\right)\\
&+\frac{1}{\Gamma(r_1)\Gamma(r_2)}\int_0^{\beta(t_1)}\int_0^{x_1}|(\beta(t_2)-s)^{r_1-1}(x_2-y)^{r_2-1}\\
&-(\beta(t_1)-s)^{r_1-1}(x_1-y)^{r_2-1}|(p_1(t_1,x_1,s,y)+p_2(t_1,x_1,s,y))dyd_s\left(\bigvee_{k=0}^{s}g(t,k)\right)\\
&+\frac{1}{\Gamma(r_1)\Gamma(r_2)}\int_0^{\beta(t_1)}\int_{x_1}^{x_2}|(\beta(t_2)-s)^{r_1-1}(x_2-y)^{r_2-1}|\\
&\times(p_1(t_1,x_1,s,y)+p_2(t_1,x_1,s,y))dyd_s\left(\bigvee_{k=0}^{s}g(t,k)\right)
\end{aligned}
$$

From continuity of $\alpha,\beta,f,g,h,p_1,p_2$ and as $t_1\longrightarrow t_2$ and $x_1\longrightarrow x_2$, the right-hand side of the above inequality tends to zero.

Step 4: *$N(B_\eta)$ is equiconvergent.*

Let $(t,x)\in J$ and $u\in B_\eta$, then we have

$$
\begin{aligned}
|(Nu)(t,x)|&\leq|f(t,x,u(t,x),u(\alpha(t),x))-f(t,x,0,0)+f(t,x,0,0)|\\
&+|\frac{1}{\Gamma(r_1)\Gamma(r_2)}\int_0^{\beta(t)}\int_0^{x}(\beta(t)-s)^{r_1-1}(x-y)^{r_2-1}\\
&\times h(t,x,s,y,u(s,y),u(\gamma(s),y))dyd_sg(t,s)|\\
&\leq|f(t,x,0,0)|+\frac{M}{1+\alpha(t)}|u(t,x)|+\frac{L}{1+\alpha(t)}|u(\alpha(t),x)|\\
&+\frac{1}{\Gamma(r_1)\Gamma(r_2)}\int_0^{\beta(t)}\int_0^{x}(\beta(t)-s)^{r_1-1}(x-y)^{r_2-1}\\
&\times(p_1(t,x,s,y)+p_2(t,x,s,y))d_s\left(\bigvee_{k=0}^{s}g(t,k)\right)
\end{aligned}
$$

$$\leq |f(t,x,0,0)| + \frac{\eta(M+L)}{1+\alpha(t)}$$
$$+ \frac{1}{\Gamma(r_1)\Gamma(r_2)} \int_0^{\beta(t)} \int_0^x (\beta(t)-s)^{r_1-1}(x-y)^{r_2-1}$$
$$\times (p_1(t,x,s,y) + p_2(t,x,s,y))d_s \left(\bigvee_{k=0}^{s} g(t,k) \right).$$

Then, since $\alpha(t) \longrightarrow \infty$ as $t \longrightarrow \infty$, we deduce that, for each $x \in [0,b]$, we get

$$|(Nu)(t,x)| \longrightarrow 0, \ as\ t \longrightarrow +\infty.$$

Hence,

$$|(Nu)(t,x) - (Nu)(+\infty,x)| \longrightarrow 0, \ as\ t \longrightarrow +\infty.$$

As a consequence of Steps 1 to 4 together with the Lemma 1.69, we can conclude that $N : B_\eta \to B_\eta$ is continuous and compact. From an application of Schauder's theorem [204], we deduce that N has a fixed point u which is a solution of the equation (4.35).

Step 5: *The uniform global attractivity of solutions.*
Now we investigate the stability of solutions of equation (4.35). Let us assume that u and v are two solutions of the equation (4.35) with the conditions of this theorem. Then, for each $(t,x) \in J$, we have

$$\begin{aligned}
&|u(t,x) - v(t,x)| = |(Nu)(t,x) - (Nv)(t,x)| \\
\leq\ & |f(t,x,u(t,x),u(\alpha(t),x)) - f(t,x,v(t,x),v(\alpha(t),x))| \\
&+ \tfrac{1}{\Gamma(r_1)\Gamma(r_2)} \textstyle\int_0^{\beta(t)} \int_0^x (\beta(t)-s)^{r_1-1}(x-y)^{r_2-1} \\
&\times |h(t,x,s,y,u(s,y),u(\gamma(s),y)) - h(t,x,s,y,v(s,y),v(\gamma(s),y))| dyd_sg(t,s) \\
\leq\ & M\|u(t,x) - v(t,x)\| + L\|u(\alpha(t),x) - v(\alpha(t),x)\| \\
&+ \tfrac{1}{\Gamma(r_1)\Gamma(r_2)} \textstyle\int_0^{\beta(t)} \int_0^x (\beta(t)-s)^{r_1-1}(x-y)^{r_2-1} \\
&\times |h(t,x,s,y,u(s,y),u(\gamma(s),y)) \\
&- h(t,x,s,y,v(s,y),v(\gamma(s,y)))| dyd_s(\textstyle\bigvee_{k=0}^{s} g(t,k)) \\
\leq\ & M\|u(t,x) - v(t,x)\| + L\|u(\alpha(t),x) - v(\alpha(t),x)\| \\
&+ \tfrac{1}{\Gamma(r_1)\Gamma(r_2)} \textstyle\int_0^{\beta(t)} \int_0^x (\beta(t)-s)^{r_1-1}(x-y)^{r_2-1} \\
&\times (p_1(t,x,s,y) + p_2(t,x,s,y)) dyd_s(\textstyle\bigvee_{k=0}^{s} g(t,k)).
\end{aligned} \tag{4.41}$$

By using (4.36), (4.41) and the fact that $\alpha(t) \longrightarrow \infty$ as $t \longrightarrow \infty$, we deduce that

$$\begin{aligned}
& \lim_{t\to\infty} |u(t,x) - v(t,x)| \leq \lim_{t\to\infty} \frac{1}{(1-M-L)\Gamma(r_1)\Gamma(r_2)} \\
\times\ & \int_0^{\beta(t)} \int_0^x (\beta(t)-s)^{r_1-1}(x-y)^{r_2-1} \\
& \times (p_1(t,x,s,y) + p_2(t,x,s,y)) dyd_s \left(\bigvee_{k=0}^{s} g(t,k) \right) \\
=\ & 0.
\end{aligned}$$

Hence,

$$\lim_{t\longrightarrow\infty} |u(t,x)-v(t,x)| = 0.$$

Consequently, all solutions of equation (4.35) are uniformly globally attractive.

As a consequence of Theorem 4.18, we prove the following result.

Theorem 4.19. *Assume that hypotheses* (4.22.1) – (4.22.6) *and the equation (4.36) hold. If*

$$f(t,x,0,0)+\int_0^{\beta(t)}\int_0^x \frac{(\beta(t)-s)^{r_1-1}(x-y)^{r_2-1}}{\Gamma(r_1)\Gamma(r_2)} h(t,x,s,y,0;0)dyd_sg(t,s) = 0; \quad (4.42)$$

for $(t,x)\in J$, *then the zero solution of equation (4.35) is globally attractive.*

Proof. From the equation (4.42), it is clear that $v(t,x)=0$ is a solution of our equation (4.35). Let us assume that u is any solution of the equation (4.35). Then, for each $(t,x)\in J$, we have

$$\begin{aligned}
&|u(t,x)| = |u(t,x)-v(t,x)| = |(Nu)(t,x)-(Nv)(t,x)|\\
\leq\ & |f(t,x,u(t,x),u(\alpha(t),x)) - f(t,x,0,0)|\\
&+\tfrac{1}{\Gamma(r_1)\Gamma(r_2)}\int_0^{\beta(t)}\int_0^x(\beta(t)-s)^{r_1-1}(x-y)^{r_2-1}\\
&\times|h(t,x,s,y,u(s,y),u(\gamma(s),y)) - h(t,x,s,y,0,0)|dyd_sg(t,s)\\
\leq\ & M\|u(t,x)\| + L\|u(\alpha(t),x)\| \qquad (4.43)\\
&+\tfrac{1}{\Gamma(r_1)\Gamma(r_2)}\int_0^{\beta(t)}\int_0^x(\beta(t)-s)^{r_1-1}(x-y)^{r_2-1}\\
&\times|h(t,x,s,y,u(s,y),u(\gamma(s),y)) - h(t,x,s,y,0,0)|dyd_s(\vee_{k=0}^{s}g(t,k))\\
\leq\ & M\|u(t,x)\| + L\|u(\alpha(t),x)\|\\
&+\tfrac{1}{\Gamma(r_1)\Gamma(r_2)}\int_0^{\beta(t)}\int_0^x(\beta(t)-s)^{r_1-1}(x-y)^{r_2-1}p_1(t,x,s,y)dyd_s(\vee_{k=0}^{s}g(t,k)).
\end{aligned}$$

By using (4.36), (4.43) and the fact that $\alpha(t)\longrightarrow\infty$ as $t\longrightarrow\infty$, we deduce that

$$\begin{aligned}
\lim_{t\longrightarrow\infty}|u(t,x)| &\leq \lim_{t\longrightarrow\infty}\frac{\int_0^{\beta(t)}\int_0^x(\beta(t)-s)^{r_1-1}(x-y)^{r_2-1}}{(1-M-L)\Gamma(r_1)\Gamma(r_2)}\\
&\times\ p_1(t,x,s,y)dyd_s\left(\bigvee_{k=0}^{s}g(t,k)\right) = 0.
\end{aligned}$$

Thus, $\lim_{t\longrightarrow\infty}|u(t,x)| = 0$. Hence, the zero solution of equation (4.35) is globally attractive. □

4.6.3. An Example

As an application of our results we consider the following nonlinear fractional order Riemann-Liouville Volterra-Stieltjes quadratic partial integral equation of the form

$$\begin{aligned}
u(t,x)\ =\ & f(t,x,u(t,x),u(\alpha(t),x))\\
&+\tfrac{1}{\Gamma(r_1)\Gamma(r_2)}\int_0^{\beta(t)}\int_0^x(\beta(t)-s)^{r_1-1}(x-y)^{r_2-1} \qquad (4.44)\\
&\times h(t,x,s,y,u(s,y),u(\gamma(s),y))dyd_sg(t,s),\ (t,x)\in J := \mathbb{R}_+\times[0,1],
\end{aligned}$$

where $r_1 = \frac{1}{4}$, $r_2 = \frac{1}{2}$, $\alpha(t) = \beta(t) = \gamma(t) = t$; $t \in \mathbb{R}_+$,

$$f(t,x,u,v) = \frac{xe^{-t}|uv|}{8(1+t)(1+|u|+2|v|)}, \ (t,x) \in J \text{ and } u,v \in \mathbb{R},$$

$$g(t,s) = s, \ (t,s) \in \mathbb{R}_+^2,$$

$$\begin{cases} h(t,x,s,y,u,v) = \dfrac{cxs^{\frac{-3}{4}}(1+|u|)\sin\sqrt{t}\sin s}{(1+y^2+t^2)(1+|u|+|v|)}; \\ \quad if \ (t,x,s,y) \in J', \ s \neq 0, \ y \in [0,1] \ and \ u,v \in \mathbb{R}, \\ h(t,x,0,y,u,v) = 0; \ if \ (t,x) \in J, \ y \in [0,1] \ and \ u,v \in \mathbb{R}, \end{cases}$$

$$c = \frac{\pi}{8e\Gamma(\frac{1}{4})} \ and \ J' = \{(t,x,s,y) \in J^2 : s \leq t \ and \ x \leq y\}.$$

First, we can see that $\lim_{t\to 0} \alpha(t) = 0$. Next, the function f is a continuous, and

$$|f(t,x,u_1,u_2) - f(t,x,v_1,v_2)| \leq \frac{1}{8(1+t)}(|u_1 - v_1| + 2|u_2 - v_2|); \ (t,x) \in J, \ u,v \in \mathbb{R}.$$

Then, the assumption $(4.22.1)$ is satisfied with $M = \frac{1}{8}$, $L = \frac{1}{4}$, and $(4.22.2)$ is satisfied with $f^* = 0$. Also, we can easily see that the function g satisfies the hypotheses $(4.22.3)$ – $(4.22.5)$.
The function h satisfies the assumption $(4.22.6)$. Indeed, h is continuous and

$$|h(t,x,s,y,u,v)| \leq p_1(t,x,s,y)|u| + p_2(t,x,s,y)|v|; \ (t,x,s,y) \in J', \ u,v \in \mathbb{R},$$

and for $i = 1,2$,

$$\begin{cases} p_i(t,x,s,y) = \dfrac{cxs^{\frac{-3}{4}}\sin\sqrt{t}\sin s}{1+y^2+t^2}; \ (t,x,s,y) \in J', \ y \in [0,1], \ s \neq 0, \\ p_i(t,x,0,y) = 0; \qquad (t,x) \in J, \ y \in [0,1]. \end{cases}$$

Then, for $i = 1,2$, we have

$$\begin{aligned} \left| \int_0^t (t-s)^{r-1} p_i(t,x,s,y) d_s g(t,s) \right| &\leq \int_0^t (t-s)^{\frac{-3}{4}} cxs^{\frac{-3}{4}} | \\ &\times \sin\sqrt{t}\sin s| d_s \left(\bigvee_{k=0}^{s} g(t,k) \right) \\ &\leq cx|\sin\sqrt{t}| \int_0^t (t-s)^{\frac{-3}{4}} s^{\frac{-3}{4}} ds \\ &\leq \frac{cx\Gamma^2(\frac{1}{4})}{\sqrt{\pi}} \left| \frac{\sin\sqrt{t}}{\sqrt{t}} \right| \\ &\leq \frac{cx\Gamma^2(\frac{1}{4})}{\sqrt{\pi t}} \longrightarrow 0 \ as \ t \to \infty, \end{aligned}$$

and

$$\begin{aligned} p_i^* &:= \sup_{(t,x)\in J} \frac{1}{\Gamma(r_1)\Gamma(r_2)} \int_0^t \int_0^x (t-s)^{r_1-1}(x-y)^{r_2-1} \\ &\times \ p_i(t,x,s,y)dyd_s \left(\bigvee_{k=0}^{s} g(t,k) \right) \\ &\leq \ \sup_{(t,x)\in J} \frac{cx\Gamma(\frac{1}{4})}{\sqrt{\pi}} \left| \frac{\sin\sqrt{t}}{\sqrt{t}} \right| \\ &= \ \frac{c\Gamma(\frac{1}{4})}{\pi} = \frac{1}{8e}. \end{aligned}$$

Finally, we can see that $M+L+p_1^*+p_2^* \leq \frac{2+3e}{8e} < 1$. Hence the condition (4.36) is satisfied. Consequently, by Theorem 4.18, the equation (4.44) has a solution defined on $\mathbb{R}_+ \times [0,1]$ and solutions of this equation are uniformly globally attractive.

4.7. Notes and Remarks

The results of Chapter 4 are taken from Abbas *et al.* [19, 28, 45, 47]. Other results may be found in [140, 327].

Chapter 5

Fractional Order Riemann-Liouville Volterra-Stieltjes Quadratic Delay Integral Equations

5.1. Introduction

In this chapter, we shall present existence results for some classes of Riemann-Liouville Volterra-Stieltjes quadratic delay integral equations of fractional order by using some fixed point theorems.

5.2. Asymptotic Behavior of Solutions of Nonlinear Riemann-Liouville Volterra-Stieltjes Multi-delay Integral Equations

5.2.1. Introduction

In this section we establish some sufficient conditions for the existence and the stability of solutions of the following system of delay fractional order Riemann-Liouville Volterra-Stieltjes quadratic integral equations of the form

$$u(t,x) = \mu(t,x) + \frac{f(t,x,u(t,x))}{\Gamma(r_1)\Gamma(r_2)} \int_0^t \int_0^x (t-s)^{r_1-1}(x-y)^{r_2-1}$$

$$\times h(t,x,s,y,u(s-\tau_1,y-\xi_1),\ldots,u(s-\tau_m,y-\xi_m))dyd_sg(t,s); \tag{5.1}$$

$$if \ \ (t,x) \in J := \mathbb{R}_+ \times [0,b],$$

$$u(t,x) = \Phi(t,x); \ if \ \ (t,x) \in \tilde{J} := [-T,\infty) \times [-\xi,b] \backslash (0,\infty) \times (0,b], \tag{5.2}$$

where $J = \mathbb{R}_+ \times [0,b]$, $b > 0$, $r_1, r_2 \in (0,\infty)$, $\tau_i, \xi_i \geq 0$; $i = 1\ldots,m$, $T = \max_{i=1\ldots,m} \{\tau_i\}$, $\xi = \max_{i=1\ldots,m} \{\xi_i\}$, $\mu : J \to \mathbb{R}$, $g : \mathbb{R}_+ \times \mathbb{R}_+ \to \mathbb{R}$, $f : J \times \mathbb{R} \to \mathbb{R}$, $h : J' \times \mathbb{R}^m \to \mathbb{R}$ are given continuous functions, $J' = \{(t,x,s,y) \in J^2 : s \leq t, \ y \leq x\}$, and $\Phi : \tilde{J} \to \mathbb{R}$ is continuous with $\mu(t,0) = \Phi(t,0)$ for each $t \in \mathbb{R}_+$ and $\mu(0,x) = \Phi(0,x)$; for each $x \in [0,b]$. We use the Schauder fixed point theorem for the existence of solutions of the problem (5.1)-(5.2), and we prove that all solutions are globally asymptotically stable.

5.2.2. Main Results

By $BC := BC([-T,\infty) \times [-\xi,b])$ we denote the Banach space of all bounded and continuous functions from $[-T,\infty) \times [-\xi,b]$ into $\mathbb{R}$ equipped with the standard norm

$$\|u\|_{BC} = \sup_{(t,x)\in[-T,\infty)\times[-\xi,b]} |u(t,x)|.$$

For $u_0 \in BC$ and $\eta \in (0,\infty)$, we denote by $B(u_0,\eta)$, the closed ball in BC centered at u_0 with radius η.

Definition 5.1. *We mean by a solution of problem (5.1)-(5.2), every function $u \in BC$ such that u satisfies equation (5.1) on J and equation (5.2) on $\tilde{J}$.*

The following hypotheses will be used in the sequel:

(5.1.1) The functions μ and Φ are in BC. Moreover, assume that

$$\lim_{t\longrightarrow\infty} \mu(t,x) = 0;\ x \in [0,b],$$

(5.1.2) There exists a positive function $P \in BC$ such that

$$|f(t,x,u) - f(t,x,v)| \le P(t,x)|u-v|;\ (t,x) \in J,\ u,v \in \mathbb{R}.$$

Moreover, assume that $\lim_{t\longrightarrow\infty} f(t,x,0) = 0;\ x \in [0,b]$,

(5.1.3) For all $t_1,t_2 \in \mathbb{R}_+$ such that $t_1 < t_2$ the function $s \mapsto g(t_2,s) - g(t_1,s)$ is nondecreasing on $\mathbb{R}_+$,

(5.1.4) The function $s \mapsto g(0,s)$ is nondecreasing on $\mathbb{R}_+$,

(5.1.5) The functions $s \mapsto g(t,s)$ and $t \mapsto g(t,s)$ are continuous on $\mathbb{R}_+$ for each fixed $t \in \mathbb{R}_+$ or $s \in \mathbb{R}_+$, respectively,

(5.1.6) There exists a continuous function $Q : J' \to \mathbb{R}_+$ such that

$$\left(1 + \sum_{i=1}^{m} |u_i|\right) |h(t,x,s,y,u_1,\ldots,u_m)| \le Q(t,x,s,y);$$

for each $(t,x,s,y) \in J'$, and all $u_i \in \mathbb{R};\ i = 1\ldots m$. Moreover, assume that

$$\lim_{t\longrightarrow\infty} \int_0^t (t-s)^{r-1} Q(t,x,s,y)\, d_s g(t,s) = 0.$$

Remark 5.2. *Set* $\mu^* := \sup_{(t,x)\in J} \mu(t,x)$, $\Phi^* := \sup_{(t,x)\in \tilde{J}} \Phi(t,x)$, $f^* := \sup_{(t,x)\in J} f(t,x,0)$, $p^* := \sup_{(t,x)\in J} P(t,x)$ *and*

$$q^* := \sup_{(t,x)\in J} \frac{1}{\Gamma(r_1)\Gamma(r_2)} \int_0^t \int_0^x (t-s)^{r_1-1}(x-y)^{r_2-1} Q(t,x,s,y)dyd_s\left(\bigvee_{k=0}^{s} g(t,k)\right).$$

From hypotheses, we infer that μ^*, Φ^*, f^*, p^* *and* q^* *are finite.*

Now, we shall prove the following theorem concerning the existence and the stability of a solution of the problem (5.1)-(5.2).

Theorem 5.3. *Assume that the hypotheses* $(5.1.1)-(5.1.6)$ *hold. If*

$$p^*q^* < 1, \tag{5.3}$$

then the problem (5.1)-(5.2) has at least one solution in the space BC. Moreover, solutions of the problem (5.1)-(5.2) are globally asymptotically stable.

Proof. Let us define the operator N such that, for any $u \in BC$,

$$(Nu)(t,x) = \begin{cases} \Phi(t,x); & (t,x)\in \tilde{J}, \\ \mu(t,x) + \frac{f(t,x,u(t,x))}{\Gamma(r_1)\Gamma(r_2)} \int_0^t \int_0^x (t-s)^{r_1-1}(x-y)^{r_2-1} & \\ \times h(t,x,s,y,u(s-\tau_1,y-\xi_1),\dots, & \\ u(s-\tau_m,y-\xi_m))dyd_s g(t,s); & (t,x)\in J. \end{cases} \tag{5.4}$$

The operator N maps BC into BC. Indeed the map $N(u)$ is continuous on J for any $u \in BC$, and for each $(t,x)\in J$, we have

$$\begin{aligned} |(Nu)(t,x)| \leq\ & |\mu(t,x)| + \frac{f(t,x,u(t,x))}{\Gamma(r_1)\Gamma(r_2)} \int_0^t \int_0^x (t-s)^{r_1-1}(x-y)^{r_2-1} \\ & \times |h(t,x,s,y,u(s-\tau_1,y-\xi_1),\dots,u(s-\tau_m,y-\xi_m))|dyd_s g(t,s) \\ + & \frac{|f(t,x,u(t,x))-f(t,x,0)|}{\Gamma(r_1)\Gamma(r_2)} \int_0^t \int_0^x (t-s)^{r_1-1}(x-y)^{r_2-1} \\ & \times |h(t,x,s,u(s,x))|dyd_s\left(\bigvee_{k=0}^{s} g(t,k)\right) \\ \leq\ & \mu^* + \frac{f^*}{\Gamma(r_1)\Gamma(r_2)} \int_0^t \int_0^x (t-s)^{r_1-1}(x-y)^{r_2-1} \\ & \times \frac{Q(t,x,s,y)}{\left(1+\sum_{i=1}^{m} |u_i(s-\tau_i,y-\xi_i)|\right)} dyd_s\left(\bigvee_{k=0}^{s} g(t,k)\right) \\ + & \frac{P(t,x)|u(t,x)|}{\Gamma(r_1)\Gamma(r_2)} \int_0^t \int_0^x (t-s)^{r_1-1}(x-y)^{r_2-1} \end{aligned}$$

$$
\begin{aligned}
& \times Q(t,x,s,y)dyd_s\left(\bigvee_{k=0}^{s} g(t,k)\right) \\
\leq\ & \mu^* + \frac{f^*}{\Gamma(r_1)\Gamma(r_2)}\int_0^t\int_0^x (t-s)^{r_1-1}(x-y)^{r_2-1} \\
& \times Q(t,x,s,y)dyd_s\left(\bigvee_{k=0}^{s} g(t,k)\right) \\
+\ & \frac{p^*\|u\|_{BC}}{\Gamma(r_1)\Gamma(r_2)}\int_0^t\int_0^x (t-s)^{r_1-1}(x-y)^{r_2-1} \\
\times\ & Q(t,x,s,y)dyd_s\left(\bigvee_{k=0}^{s} g(t,k)\right) \\
\leq\ & \mu^* + q^*(f^* + p^*\|u\|_{BC}).
\end{aligned}
$$

Thus, for each $(t,x) \in J$, we have

$$|(Nu)(t,x)| \leq \mu^* + q^*(f^* + p^*\|u\|_{BC}). \tag{5.5}$$

Also, for $(t,x) \in \tilde{J}$, we have

$$|(Nu)(t,x)| = |\Phi(t,x)| \leq \Phi^*.$$

Thus,

$$\|Nu\|_{BC} \leq \max\{\Phi^*, \mu^* + q^*(f^* + p^*\|u\|_{BC})\}.$$

Hence, $N(u) \in BC$. This proves that the operator N maps BC into itself.
The problem of finding the solutions of the problem (5.1)-(5.2) is reduced to finding the solutions of the operator equation $N(u) = u$. From the inequalities (5.3) and (5.5), we can show that N transforms the ball $B_\eta := B(0,\eta)$ into itself, where $\eta = \max\{\Phi^*, \eta^*\}$ and $\eta^* > \frac{\mu^* + q^* f^*}{1 - p^* q^*}$. We shall show that $N : B_\eta \to B_\eta$ satisfies the assumptions of Schauder's fixed point theorem [204]. The proof will be given in several steps and cases.

Step 1: *N is continuous.*
Let $\{u_n\}_{n\in\mathbb{N}}$ be a sequence such that $u_n \to u$ in B_η. Then, for each $(t,x) \in [-T,\infty) \times [-\xi, b]$, we have

$$
\begin{aligned}
& |(Nu_n)(t,x) - (Nu)(t,x)| \\
\leq\ & |f(t,x,u_n(t,x)) - f(t,x,u(t,x))| \int_0^t\int_0^x \frac{(t-s)^{r_1-1}(x-y)^{r_2-1}}{\Gamma(r_1)\Gamma(r_2)} \\
& \times |h(t,x,s,y,u_n(s-\tau_1,y-\xi_1),\dots,u_n(s-\tau_m,y-\xi_m))|dyd_s g(t,s) \\
+\ & |f(t,x,u(t,x))| \int_0^t\int_0^x \frac{(t-s)^{r_1-1}(x-y)^{r_2-1}}{\Gamma(r_1)\Gamma(r_2)} \\
& \times |h(t,x,s,y,u_n(s-\tau_1,y-\xi_1),\dots,u_n(s-\tau_m,y-\xi_m)) \\
& -h(t,x,s,y,u(s-\tau_1,y-\xi_1),\dots,u(s-\tau_m,y-\xi_m))|dyd_s g(t,s) \\
\leq\ & p^*q^*|u_n(t,x) - u(t,x)| + \frac{f^* + p^*\|u\|_{BC}}{\Gamma(r_1)\Gamma(r_2)}\int_0^t\int_0^x (t-s)^{r_1-1}(x-y)^{r_2-1}
\end{aligned}
$$

$$\times|h(t,x,s,y,u_n(s-\tau_1,y-\xi_1),\ldots,u_n(s-\tau_m,y-\xi_m))$$
$$-h(t,x,s,y,u(s-\tau_1,y-\xi_1),\ldots,u(s-\tau_m,y-\xi_m))|dyd_s\left(\bigvee_{k=0}^{s}g(t,k)\right)$$
$$\leq p^*q^*|u_n(t,x)-u(t,x)|+\frac{f^*+p^*\eta}{\Gamma(r_1)\Gamma(r_2)}\int_0^t\int_0^x(t-s)^{r_1-1}(x-y)^{r_2-1}$$
$$\times|h(t,x,s,y,u_n(s-\tau_1,y-\xi_1),\ldots,u_n(s-\tau_m,y-\xi_m))$$
$$-h(t,x,s,y,u(s-\tau_1,y-\xi_1),\ldots,u(s-\tau_m,y-\xi_m))|dyd_s\left(\bigvee_{k=0}^{s}g(t,k)\right). \tag{5.6}$$

Case 1. If $(t,x)\in\tilde{J}\cup[0,a]\times[0,b]$; $a>0$, then, since $u_n\to u$ as $n\to\infty$ and h is continuous, (5.6) gives

$$\|N(u_n)-N(u)\|_{BC}\to 0 \quad \text{as } n\to\infty.$$

Case 2. If $(t,x)\in(a,\infty)\times[0,b]$; $a>0$, then from (5.1.6) and (5.6), we get

$$|(Nu_n)(t,x)-(Nu)(t,x)|$$
$$\leq p^*q^*|u_n(t,x)-u(t,x)|+\frac{f^*+p^*\eta}{\Gamma(r_1)\Gamma(r_2)}\int_0^t\int_0^x(t-s)^{r_1-1}(x-y)^{r_2-1}$$
$$\times 2Q(t,x,s,y)dyd_s\left(\bigvee_{k=0}^{s}g(t,k)\right)$$
$$\leq p^*q^*|u_n(t,x)-u(t,x)|+\frac{2(f^*+p^*\eta)}{\Gamma(r_1)\Gamma(r_2)}\int_0^x(x-y)^{r_2-1}\left[\int_0^t(t-s)^{r_1-1}\right.$$
$$\left.\times Q(t,x,s,y)d_s\left(\bigvee_{k=0}^{s}g(t,k)\right)\right]dy. \tag{5.7}$$

Thus, (5.7) gives

$$\|N(u_n)-N(u)\|_{BC}\to 0 \quad \text{as } n\to\infty.$$

Step 2: *$N(B_\eta)$ is uniformly bounded.*
This is clear since $N(B_\eta)\subset B_\eta$ and B_η is bounded.

Step 3: *$N(B_\eta)$ is equicontinuous on every compact subset $[-T,a]\times[-\xi,b]$ of $[-T,\infty)\times[-\xi,b]$; $a>0$.*
Let $(t_1,x_1),(t_2,x_2)\in[0,a]\times[0,b]$, $t_1<t_2$, $x_1<x_2$ and let $u\in B_\eta$. Thus we have

$$|(Nu)(t_2,x_2)-(Nu)(t_1,x_1)|\leq|\mu(t_2,x_2)-\mu(t_1,x_1)|$$
$$+\ |f(t_2,x_2,u(t_2,x_2))-f(t_1,x_1,u(t_1,x_1))|\int_0^{t_2}\int_0^{x_2}\frac{(t_2-s)^{r_1-1}(x_2-y)^{r_2-1}}{\Gamma(r_1)\Gamma(r_2)}$$
$$\times|h(t_2,x_2,s,y,u(s-\tau_1,x_2-\xi_1),\ldots,u(s-\tau_m,x_2-\xi_m))|dyd_s\left(\bigvee_{k=0}^{s}g(t_2,k)\right)$$
$$+\ |f(t_1,x_1,u(t_1,x_1))|\left|\int_0^{t_2}\int_0^{x_2}\frac{(t_2-s)^{r_1-1}(x_2-y)^{r_2-1}}{\Gamma(r_1)\Gamma(r_2)}\right.$$

$$\begin{aligned}
&\times h(t_2,x_2,s,y,u(s-\tau_1,x_2-\xi_1),\dots,u(s-\tau_m,x_2-\xi_m))dyd_sg(t_2,s)\\
-&\int_0^{t_1}\int_0^{x_1}\frac{(t_1-s)^{r_1-1}(x_1-y)^{r_2-1}}{\Gamma(r_1)\Gamma(r_2)}\\
&\times h(t_1,x_1,s,y,u(s-\tau_1,x_1-\xi_1),\dots,u(s-\tau_m,x_1-\xi_m))dyd_sg(t_1,s)\big|.
\end{aligned}$$

Thus,

$$\begin{aligned}
&|(Nu)(t_2,x_2)-(Nu)(t_1,x_1)|\le|\mu(t_2,x_2)-\mu(t_1,x_1)|\\
+&\ q^*|f(t_2,x_2,u(t_2,x_2))-f(t_1,x_1,u(t_1,x_1))|\\
+&\ \frac{p^*\eta}{\Gamma(r_1)\Gamma(r_2)}\left|\int_0^{t_1}\int_0^{x_1}(t_2-s)^{r_1-1}(x_2-y)^{r_2-1}\right.\\
\times&\ h(t_2,x_2,s,y,u(s-\tau_1,x_2-\xi_1),\dots,u(s-\tau_m,x_2-\xi_m))dyd_sg(t_2,s)\\
-&\ (t_1-s)^{r_1-1}(x_1-y)^{r_2-1}\\
\times&\ h(t_1,x_1,s,y,u(s-\tau_1,x_1-\xi_1),\dots,u(s-\tau_m,x_1-\xi_m))dyd_sg(t_1,s)\big|\\
+&\ \frac{p^*\eta}{\Gamma(r_1)\Gamma(r_2)}\left|\int_{t_1}^{t_2}\int_{x_1}^{x_2}(t_2-s)^{r_1-1}(x_2-y)^{r_2-1}\right.\\
\times&\ h(t_2,x_2,s,y,u(s-\tau_1,x_2-\xi_1),\dots,u(s-\tau_m,x_2-\xi_m))dyd_sg(t_2,s)\big|\\
+&\ \frac{p^*\eta}{\Gamma(r_1)\Gamma(r_2)}\left|\int_0^{t_1}\int_{x_1}^{x_2}(t_2-s)^{r_1-1}(x_2-y)^{r_2-1}\right.\\
\times&\ h(t_2,x_2,s,y,u(s-\tau_1,x_2-\xi_1),\dots,u(s-\tau_m,x_2-\xi_m))dyd_sg(t_2,s)\big|\\
+&\ \frac{p^*\eta}{\Gamma(r_1)\Gamma(r_2)}\left|\int_{t_1}^{t_2}\int_0^{x_1}(t_2-s)^{r_1-1}(x_2-y)^{r_2-1}\right.\\
\times&\ h(t_2,x_2,s,y,u(s-\tau_1,x_2-\xi_1),\dots,u(s-\tau_m,x_2-\xi_m))dyd_sg(t_2,s)\big|.
\end{aligned}$$

From continuity of μ, f, g, h and as $t_1 \to t_2$ and $x_1 \to x_2$, the right-hand side of the above inequality tends to zero. The equicontinuity for the cases $t_1 < t_2 < 0$, $x_1 < x_2 < 0$ and $t_1 \le 0 \le t_2$, $x_1 \le 0 \le x_2$ is obvious.

Step 4: *$N(B_\eta)$ is equiconvergent.*
Let $(t,x) \in J$ and $u \in B_\eta$, then we have

$$\begin{aligned}
|(Nu)(t,x)|\ &\le\ |\mu(t,x)|+\frac{|f(t,x,u(t,x))|}{\Gamma(r_1)\Gamma(r_2)}\int_0^t\int_0^x(t-s)^{r_1-1}(x-y)^{r_2-1}\\
&\quad\times|h(t,x,s,y,u(s-\tau_1,y-\xi_1),\dots,u(s-\tau_m,y-\xi_m))|dyd_sg(t,s)\\
&+\ \frac{|f(t,x,u(t,x))-f(t,x,0)|}{\Gamma(r_1)\Gamma(r_2)}\int_0^t\int_0^x(t-s)^{r_1-1}(x-y)^{r_2-1}\\
&\quad\times|h(t,x,s,u(s,x))|dyd_s\left(\bigvee_{k=0}^{s}g(t,k)\right)\\
&\le\ |\mu(t,x)|+\frac{f^*}{\Gamma(r_1)\Gamma(r_2)}\int_0^t\int_0^x(t-s)^{r_1-1}(x-y)^{r_2-1}
\end{aligned}$$

$$\times \frac{Q(t,x,s,y)}{\left(1+\sum_{i=1}^{m}|u_i(s-\tau_i,y-\xi_i)|\right)} dyd_s\left(\bigvee_{k=0}^{s} g(t,k)\right)$$

$$+ \frac{P(t,x)|u(t,x)|}{\Gamma(r_1)\Gamma(r_2)}\int_0^t\int_0^x (t-s)^{r_1-1}(x-y)^{r_2-1}$$

$$\times Q(t,x,s,y)dyd_s\left(\bigvee_{k=0}^{s} g(t,k)\right)$$

$$\leq |\mu(t,x)| + \frac{f^*+p^*\eta}{\Gamma(r_1)\Gamma(r_2)}\int_0^t\int_0^x (t-s)^{r_1-1}(x-y)^{r_2-1}$$

$$\times Q(t,x,s,y)dyd_s\left(\bigvee_{k=0}^{s} g(t,k)\right).$$

Thus, for each $x \in [0,b]$, we get

$$|(Nu)(t,x)| \to 0, \ as\, t \to +\infty.$$

Hence,

$$|(Nu)(t,x)-(Nu)(+\infty,x)| \to 0, \ as\, t \to +\infty.$$

As a consequence of Steps 1 to 4 together with the Lemma 1.69, we can conclude that $N: B_\eta \to B_\eta$ is continuous and compact. From an application of Schauder's theorem [204], we deduce that N has a fixed point u which is a solution of the problem (5.1)-(5.2).

Step 5: *The Global asymptotic stability of solutions.*
Now we investigate the global asymptotic stability of solutions of the the problem (5.1)-(5.2). Let us assume that u and v are two solutions of the problem (5.1)-(5.2). Then, for each $(t,x) \in J$, we have

$$|u(t,x)-v(t,x)| = |(Nu)(t,x)-(Nv)(t,x)|$$

$$\leq p^*q^*|u(t,x)-v(t,x)| + \frac{f^*+p^*\|u\|_{BC}}{\Gamma(r_1)\Gamma(r_2)}\int_0^t\int_0^x (t-s)^{r_1-1}(x-y)^{r_2-1}$$

$$\times 2Q(t,x,s,y)dyd_s\left(\bigvee_{k=0}^{s} g(t,k)\right)$$

$$\leq p^*q^*|u(t,x)-v(t,x)| + \frac{2(f^*+p^*\|u\|_{BC})}{\Gamma(r_1)\Gamma(r_2)}\int_0^x (x-y)^{r_2-1}\left[\int_0^t (t-s)^{r_1-1}\right.$$

$$\left.\times Q(t,x,s,y)d_s\left(\bigvee_{k=0}^{s} g(t,k)\right)\right] dy.$$

Thus, for each $(t,x) \in J$, we get

$$|u(t,x)-v(t,x)| \leq \frac{2(f^*+p^*\|u\|_{BC})}{(1-p^*q^*)\Gamma(r_1)\Gamma(r_2)}\int_0^x (x-y)^{r_2-1}\left[\int_0^t (t-s)^{r_1-1}\right.$$

$$\times Q(t,x,s,y)d_s\left(\bigvee_{k=0}^{s} g(t,k)\right)\Bigg] dy. \quad (5.8)$$

By using (5.8), we deduce that

$$\lim_{t\to\infty} |u(t,x)-v(t,x)| = 0.$$

Consequently, all solutions of the problem (5.1)-(5.2) are globally asymptotically stable. □

5.2.3. An Example

As an application and to illustrate our results, we consider the following problem of fractional order Riemann-Liouville Volterra-Stieltjes quadratic multi-delay integral equations

$$\begin{aligned} u(t,x) &= \mu(t,x) + \tfrac{f(t,x,u(t,x))}{\Gamma(r_1)\Gamma(r_2)} \int_0^t \int_0^x (t-s)^{r_1-1}(x-y)^{r_2-1} \\ &\quad \times h(t,x,s,y,u(s-1,y-2),u(s-\tfrac{1}{2},y-\tfrac{2}{5}))dyd_sg(t,s); \quad (5.9) \\ &\quad if\ \ (t,x)\in J := \mathbb{R}_+ \times [0,1], \end{aligned}$$

$$u(t,x) = \frac{2}{(2+t^2)(2+x^2)};\ if\ \ (t,x)\in \tilde{J} := [-1,\infty)\times[-2,1]\backslash(0,\infty)\times(0,1], \quad (5.10)$$

where

$$r_1 = \frac{1}{4},\ r_2 = \frac{1}{2},\ \mu(t,x) = \frac{1}{2+t^2+x^2};\ (t,x)\in J,$$

$$f(t,x,u) = \frac{e^{x-t}|u|}{1+|u|};\ (t,x)\in J,\ u\in\mathbb{R},\ \ g(t,s) = s;\ (t,s)\in\mathbb{R}_+^2,$$

$$\begin{cases} h(t,x,s,y,u_1,u_2) = \dfrac{cxs^{\frac{-3}{4}}|u|\sin\sqrt{t}\sin s}{(1+t^2+y^2)(2+|u_1|+|u_2|)}; \\ \qquad (t,x,s,y)\in J',\ s\neq 0\ and\ u_1,u_2\in\mathbb{R}, \\ h(t,x,0,y,u_1,u_2) = 0; \qquad (t,x)\in J,\ y\in[0,1]\ and\ u_1,u_2\in\mathbb{R}, \end{cases}$$

$$c = \frac{\pi}{16e\Gamma(\frac{1}{4})}\ and\ \ J' = \{(t,x,s,y)\in J^2 : s\le t,\ y\le x\}.$$

Set

$$\tau_1 = 1, \tau_2 = \frac{1}{2},\ \xi_1 = 2,\ \xi_2 = \frac{2}{5}\ \ and\ \ \Phi(t,x) = \frac{2}{(2+t^2)(2+x^2)};\ (t,x)\in\tilde{J}.$$

Then $T = 1$ and $\xi = 2$.

First, we can see that for each $x\in[0,1]$, we have, $\lim_{t\to 0}\frac{1}{2+t^2+x^2} = 0$. Then the assumption (5.1.1) is satisfies and $\mu^* = \Phi^* = \frac{1}{2}$.

Next, the function f is a continuous, $f^* = 0$, and

$$|f(t,x,u)-f(t,x,v)| \le e^{x-t}|u-v|;\ (t,x)\in J,\ u,v\in\mathbb{R}.$$

Then, the assumption (5.1.2) is satisfies with $P(t,x) = e^{x-t}$; $(t,x) \in J$, and then $p^* = e$. Also, we can easily see that the function g satisfies the hypotheses $(5.1.3) - (5.1.5)$.
The function h satisfies the assumption (5.1.6). Indeed, h is continuous and

$$|h(t,x,s,y,u_1,u_2)| \leq \frac{Q(t,x,s,y)}{1+|u_1|+|u_2|};\ (t,x,s,y) \in J',\ u_1,u_2 \in \mathbb{R},$$

where

$$\begin{cases} Q(t,x,s,y) = \dfrac{cxs^{\frac{-3}{4}} \sin\sqrt{t} \sin s}{1+t^2+y^2};\ (t,x,s,y) \in J',\ s \neq 0, \\ Q(t,x,0,y) = 0; \qquad (t,x) \in J,\ y \in [0,1]. \end{cases}$$

Then,

$$\begin{aligned} \left| \int_0^t (t-s)^{r-1} Q(t,x,s,y) d_s g(t,s) \right| &\leq \int_0^t (t-s)^{\frac{-3}{4}} cxs^{\frac{-3}{4}} \\ &\quad \times |\sin\sqrt{t} \sin s| d_s \left(\bigvee_{k=0}^{s} g(t,k) \right) \\ &\leq cx|\sin\sqrt{t}| \int_0^t (t-s)^{\frac{-3}{4}} s^{\frac{-3}{4}} ds \\ &\leq \frac{cx\Gamma^2(\frac{1}{4})}{\sqrt{\pi}} \left| \frac{\sin\sqrt{t}}{\sqrt{t}} \right| \\ &\leq \frac{cx\Gamma^2(\frac{1}{4})}{\sqrt{\pi t}} \longrightarrow 0\ as\ t \to \infty, \end{aligned}$$

and

$$\begin{aligned} q^* &:= \sup_{(t,x)\in J} \frac{1}{\Gamma(r_1)\Gamma(r_2)} \int_0^t \int_0^x (t-s)^{r_1-1} (x-y)^{r_2-1} \\ &\quad \times Q(t,x,s,y) dy d_s \left(\bigvee_{k=0}^{s} g(t,k) \right) \\ &\leq \sup_{(t,x)\in J} \frac{1}{\Gamma(\frac{3}{2})} \frac{cx\Gamma(\frac{1}{4})}{\sqrt{\pi}} \left| \frac{\sin\sqrt{t}}{\sqrt{t}} \right| = \frac{2c\Gamma(\frac{1}{4})}{\pi} = \frac{1}{8e}. \end{aligned}$$

Finally, we can see that $p^* q^* \leq \frac{1}{8} < 1$. Hence the inequality (5.3) is satisfied.
Consequently, by Theorem 5.3, the problem (5.9)-(5.10) has at least one solution in the space $BC([-1,\infty) \times [-2,1])$, and all solutions of (5.9)-(5.10) are globally asymptotically stable.

5.3. Existence and Attractivity of Solutions for Nonlinear Riemann-Liouville Volterra-Stieltjes Multi-delay Integral Equations

5.3.1. Introduction

This section deals with the existence and the attractivity of solutions to the following system of nonlinear fractional order Riemann-Liouville Volterra-Stieltjes quadratic multi-delay partial integral equations of the form,

$$\begin{aligned} u(t,x) &= \mu(\alpha(t),x)+\frac{1}{\Gamma(r_1)\Gamma(r_2)}\int_0^{\beta(t)}\int_0^x(\beta(t)-s)^{r_1-1}(x-y)^{r_2-1} \\ &\quad \times f(t,x,s,y,u(\gamma(s)-\tau_1,y-\xi_1),\dots, \qquad (5.11) \\ &\quad u(\gamma(s)-\tau_m,y-\xi_m))dyd_sg(t,s); \qquad if\ \ (t,x)\in J:=\mathbb{R}_+\times[0,b], \end{aligned}$$

$$u(t,x)=\Phi(t,x);\ if\ \ (t,x)\in\tilde{J}:=[-T,\infty)\times[-\xi,b]\backslash(0,\infty)\times(0,b], \qquad (5.12)$$

where $b>0,\ r_1,r_2\in(0,\infty),\ \mathbb{R}_+=[0,\infty),\ \tau_i,\xi_i\geq 0;\ i=1\dots,m,\ T=\max_{i=1\dots,m}\{\tau_i\},\ \xi=\max_{i=1\dots,m}\{\xi_i\},\ \alpha,\ \beta,\ \gamma\colon\mathbb{R}_+\to\mathbb{R}_+,\ \mu:J\to\mathbb{R},\ f:J'\times\mathbb{R}^m\to\mathbb{R},\ g:\mathbb{R}_+\times\mathbb{R}_+\to\mathbb{R}$ are given continuous functions, $\lim_{t\to\infty}\alpha(t)=\infty$, μ is bounded, $J'=\{(t,x,s,y)\in J^2: s\leq t,\ y\leq x\}$, $\Phi:\tilde{J}\to\mathbb{R}$ is continuous and bounded with $\lim_{t\to\infty}\Phi(t,x)=0;\ x\in[-\xi,a],\ \mu(\alpha(t),0)=\Phi(t,0)$ for each $t\in\mathbb{R}_+$ and $\mu(\alpha(0),x)=\Phi(0,x)$; for each $x\in[0,b]$. We use the Schauder fixed point theorem for the existence of solutions of the problem (5.11)-(5.12), and we prove that all solutions are uniformly globally attractive. Finally, we present an example illustrating the applicability of the imposed conditions.

5.3.2. Main Results

Definition 5.4. *We mean by a solution of problem (5.11)-(5.12), every function $u\in BC$ such that u satisfies equation (5.11) on J and equation (5.12) on $\tilde{J}$.*

The following hypotheses will be used in the sequel.

(5.5.1) There exists a positive function $p\in BC(J)$ such that

$$(1+\alpha(t))|\mu(\alpha(t),x)|\leq p(t,x);\ (t,x)\in J.$$

(5.5.2) For all $t_1,t_2\in\mathbb{R}_+$ such that $t_1<t_2$ the function $s\mapsto g(t_2,s)-g(t_1,s)$ is nondecreasing on $\mathbb{R}_+$.

(5.5.3) The function $s\mapsto g(0,s)$ is nondecreasing on $\mathbb{R}_+$.

(5.5.4) The functions $s\mapsto g(t,s)$ and $t\mapsto g(t,s)$ are continuous on $\mathbb{R}_+$ for each fixed $t\in\mathbb{R}_+$ or $s\in\mathbb{R}_+$, respectively.

(5.5.5) There exist continuous functions $q_i : J' \to \mathbb{R}_+$; $i = 1,\ldots,2m$, such that

$$\left(1+\sum_{i=1}^{m}|u_i|\right)|h(t,x,s,y,u_1,...,u_m)| \leq \sum_{i=1}^{m} q_i(t,x,s,y)|u_i|;$$

for $(t,x,s,y) \in J'$, $u_i \in \mathbb{R}$; $i = 1,\ldots,2m$. Moreover, assume that

$$\lim_{t\to\infty}\int_0^{\beta(t)}(\beta(t)-s)^{r-1}q_i(t,x,s,y)d_sg(t,s) = 0;\ i = 1,\ldots,m.$$

Remark 5.5. *Set* $\Phi^* := \sup_{(t,x)\in\tilde{J}} \Phi(t,x)$, $p^* := \sup_{(t,x)\in J} p(t,x)$,

$$q_i^* := \sup_{(t,x)\in J}\frac{1}{\Gamma(r_1)\Gamma(r_2)}\int_0^{\beta(t)}\int_0^x(\beta(t)-s)^{r_1-1}(x-y)^{r_2-1}$$

$q_i(t,x,s,y)dyd_s(\bigvee_{k=0}^{s}g(t,k))$; *for* $i = 1,\ldots,m$. *From hypotheses, we infer that* Φ^*, p^*, q_i^*; $i = 1,\ldots,m$ *are finite.*

Theorem 5.6. *Assume that hypotheses* $(5.5.1)-(5.5.5)$ *hold. Then the problem (5.11)-(5.12) has at least one solution in the space BC. Moreover, solutions of problem (5.11)-(5.12) are uniformly globally attractive.*

Proof. Let us define the operator N such that, for any $u \in BC$,

$$(Nu)(t,x) = \begin{cases} \Phi(t,x); & (t,x)\in\tilde{J}, \\ \mu(\alpha(t),x) & \\ +\frac{1}{\Gamma(r_1)\Gamma(r_2)}\int_0^{\beta(t)}\int_0^x(\beta(t)-s)^{r_1-1}(x-y)^{r_2-1} & \\ \times f(t,x,s,y,u(\gamma(s)-\tau_1,y-\xi_1),..., & \\ u(\gamma(s)-\tau_m,y-\xi_m))dyd_sg(t,s); & (t,x)\in J. \end{cases} \tag{5.13}$$

From the assumptions of this theorem, we infer that $N(u)$ is continuous on $[-T,\infty)\times[-\xi,b]$. Now we prove that $N(u)\in BC$ for any $u\in BC$. For arbitrarily fixed $(t,x)\in J$, we have

$$\begin{aligned}|(Nu)(t,x)| &= |\mu(\alpha(t),x)+\frac{1}{\Gamma(r_1)\Gamma(r_2)}\int_0^{\beta(t)}\int_0^x(\beta(t)-s)^{r_1-1}(x-y)^{r_2-1} \\ &\quad\times f(t,x,s,y,u(\gamma(s)-\tau_1,y-\xi_1),...,u(\gamma(s)-\tau_m,y-\xi_m))dyd_sg(t,s)| \\ &\leq \frac{p(t,x)}{1+\alpha(t))}+|\frac{1}{\Gamma(r_1)\Gamma(r_2)}\int_0^{\beta(t)}\int_0^x(\beta(t)-s)^{r_1-1}(x-y)^{r_2-1} \\ &\quad\times\left.\frac{\sum_{i=1}^{m}q_i(t,x,s,y)|u(\gamma(s)-\tau_i,y-\xi_i)|}{\left(1+\sum_{i=1}^{m}|u(\gamma(s)-\tau_i,y-\xi_i)|\right)}dyd_sg(t,s)\right| \\ &\leq p(t,x)+\frac{1}{\Gamma(r_1)\Gamma(r_2)}\int_0^{\beta(t)}\int_0^x(\beta(t)-s)^{r_1-1}(x-y)^{r_2-1}\end{aligned}$$

$$\times \sum_{i=1}^{m} q_i(t,x,s,y)dyd_s\left(\bigvee_{k=0}^{s} g(t,k)\right)$$
$$\leq p^* + \sum_{i=1}^{m} q_i^*,$$

and for $(t,x) \in \tilde{J}$, we have

$$|(Nu)(t,x)| = |\Phi(t,x)| \leq \Phi^*.$$

Thus,

$$\|N(u)\|_{BC} \leq \max\left\{\Phi^*, p^* + \sum_{i=1}^{m} q_i^*\right\} := \eta. \tag{5.14}$$

Hence $N(u) \in BC$. The problem of finding the solutions of the problem (5.11)-(5.12) is reduced to finding the solutions of the operator equation $N(u) = u$. From (5.14), we infer that N transforms the ball $B_\eta := B(0,\eta)$ into itself. We shall show that $N : B_\eta \to B_\eta$ satisfies the assumptions of Schauder's fixed point theorem [204]. The proof will be given in several steps and cases.

Step 1: *N is continuous.*
Let $\{u_n\}_{n\in\mathbb{N}}$ be a sequence such that $u_n \to u$ in B_η. Then, for each $(t,x) \in [-T,\infty)\times[-\xi,b]$, we have

$$\begin{aligned}
&|(Nu_n)(t,x) - (Nu)(t,x)| \\
\leq\; & \tfrac{1}{\Gamma(r_1)\Gamma(r_2)} \textstyle\int_0^{\beta(t)} \int_0^x (\beta(t)-s)^{r_1-1}(x-y)^{r_2-1} \\
&\times|f(t,x,s,y,u_n(\gamma(s)-\tau_1,y-\xi_1),\dots,u_n(\gamma(s)-\tau_m,y-\xi_m)) \\
&-f(t,x,s,y,u(\gamma(s)-\tau_1,y-\xi_1),\dots,u(\gamma(s)-\tau_m,y-\xi_m))|dyd_s\,(\textstyle\bigvee_{k=0}^{s} g(t,k)).
\end{aligned} \tag{5.15}$$

Case 1. If $(t,x) \in \tilde{J}\cup([0,a]\times[0,b])$; $a > 0$, then, since $u_n \to u$ as $n \to \infty$ and f,g,γ are continuous, (5.15) gives

$$\|N(u_n) - N(u)\|_{BC} \to 0 \quad \text{as } n \to \infty.$$

Case 2. If $(t,x) \in (a,\infty)\times[0,b]$; $a > 0$, then from (5.5.5) and (5.15), for each $(t,x) \in J$, we get

$$\begin{aligned}
&|(Nu_n)(t,x) - (Nu)(t,x)| \\
\leq\; & \tfrac{2}{\Gamma(r_1)\Gamma(r_2)} \textstyle\int_0^{\beta(t)} \int_0^x (\beta(t)-s)^{r_1-1}(x-y)^{r_2-1} \\
&\times \sum_{i=1}^{m} q_i(t,x,s,y)dyd_s\left(\bigvee_{k=0}^{s} g(t,k)\right) \\
\leq\; & \sum_{i=1}^{m} \frac{2}{\Gamma(r_1)\Gamma(r_2)} \int_0^{\beta(t)} \int_0^x (\beta(t)-s)^{r_1-1}(x-y)^{r_2-1} \\
&\times q_i(t,x,s,y)dyd_s(\textstyle\bigvee_{k=0}^{s} g(t,k)).
\end{aligned} \tag{5.16}$$

Since $t \to \infty$, then (5.16) gives

$$\|N(u_n) - N(u)\|_{BC} \to 0 \quad \text{as } n \to \infty.$$

Step 2: $N(B_\eta)$ *is uniformly bounded.*
This is clear since $N(B_\eta) \subset B_\eta$ and B_η is bounded.

Step 3: $N(B_\eta)$ *is equicontinuous on every compact subset* $[0,a]\times[0,b]$ *of* J, $a>0$.
Let $(t_1,x_1),(t_2,x_2)\in[0,a]\times[0,b]$, $t_1<t_2$, $x_1<x_2$ and let $u\in B_\eta$. Also without loss of generality, suppose that $\beta(t_1)\le\beta(t_2)$. Thus we have

$$
\begin{aligned}
&|(Nu)(t_2,x_2)-(Nu)(t_1,x_1)| \le |\mu(\alpha(t_2),x_2)-\mu(\alpha(t_1),x_1)|\\
&+\frac{1}{\Gamma(r_1)\Gamma(r_2)}\Big|\int_0^{\beta(t_2)}\int_0^{x_2}(\beta(t_2)-s)^{r_1-1}(x_2-y)^{r_2-1}\\
&\times[f(t_2,x_2,s,y,u(\gamma(s)-\tau_1,y-\xi_1),\dots,u(\gamma(s)-\tau_m,y-\xi_m))\\
&-f(t_1,x_1,s,y,u(\gamma(s)-\tau_1,y-\xi_1),\dots,u(\gamma(s)-\tau_m,y-\xi_m))]dyd_sg(t,s)\Big|\\
&+\frac{1}{\Gamma(r_1)\Gamma(r_2)}\Big|\int_0^{\beta(t_2)}\int_0^{x_2}(\beta(t_2)-s)^{r_1-1}(x_2-y)^{r_2-1}\\
&\times f(t_1,x_1,s,y,u(\gamma(s)-\tau_1,y-\xi_1),\dots,u(\gamma(s)-\tau_m,y-\xi_m))dyd_sg(t,s)\\
&-\int_0^{\beta(t_1)}\int_0^{x_2}(\beta(t_2)-s)^{r_1-1}(x_2-y)^{r_2-1}\\
&\times f(t_1,x_1,s,y,u(\gamma(s)-\tau_1,y-\xi_1),\dots,u(\gamma(s)-\tau_m,y-\xi_m))dyd_sg(t,s)\Big|\\
&+\frac{1}{\Gamma(r_1)\Gamma(r_2)}\Big|\int_0^{\beta(t_1)}\int_0^{x_2}(\beta(t_2)-s)^{r_1-1}(x_2-y)^{r_2-1}\\
&\times f(t_1,x_1,s,y,u(\gamma(s)-\tau_1,y-\xi_1),\dots,u(\gamma(s)-\tau_m,y-\xi_m))dyd_sg(t,s)\\
&-\int_0^{\beta(t_1)}\int_0^{x_1}(\beta(t_1)-s)^{r_1-1}(x_1-y)^{r_2-1}\\
&\times f(t_1,x_1,s,y,u(\gamma(s)-\tau_1,y-\xi_1),\dots,u(\gamma(s)-\tau_m,y-\xi_m))dyd_sg(t,s)\Big|\\
&\le |\mu(\alpha(t_2),x_2)-\mu(\alpha(t_1),x_1)|\\
&+\frac{1}{\Gamma(r_1)\Gamma(r_2)}\int_0^{\beta(t_2)}\int_0^{x_2}(\beta(t_2)-s)^{r_1-1}(x_2-y)^{r_2-1}\\
&\times|f(t_2,x_2,s,y,u(\gamma(s)-\tau_1,y-\xi_1),\dots,u(\gamma(s)-\tau_m,y-\xi_m))\\
&-f(t_1,x_1,s,y,u(\gamma(s)-\tau_1,y-\xi_1),\dots,u(\gamma(s)-\tau_m,y-\xi_m))|dyd_s\left(\bigvee_{k=0}^{s}g(t,k)\right)\\
&+\frac{1}{\Gamma(r_1)\Gamma(r_2)}\int_{\beta(t_1)}^{\beta(t_2)}\int_0^{x_2}(\beta(t_2)-s)^{r_1-1}(x_2-y)^{r_2-1}\\
&\times|f(t_1,x_1,s,y,u(\gamma(s)-\tau_1,y-\xi_1),\dots,u(\gamma(s)-\tau_m,y-\xi_m))|dyd_s\left(\bigvee_{k=0}^{s}g(t,k)\right)\\
&+\frac{1}{\Gamma(r_1)\Gamma(r_2)}\int_0^{\beta(t_1)}\int_0^{x_1}|(\beta(t_2)-s)^{r_1-1}(x_2-y)^{r_2-1}\\
&-(\beta(t_1)-s)^{r_1-1}(x_1-y)^{r_2-1}|\\
&\times|f(t_1,x_1,s,y,u(\gamma(s)-\tau_1,y-\xi_1),\dots,u(\gamma(s)-\tau_m,y-\xi_m))|dyd_s\left(\bigvee_{k=0}^{s}g(t,k)\right)
\end{aligned}
$$

$$+\frac{1}{\Gamma(r_1)\Gamma(r_2)}\int_0^{\beta(t_1)}\int_{x_1}^{x_2}|(\beta(t_2)-s)^{r_1-1}(x_2-y)^{r_2-1}|$$
$$\times|f(t_1,x_1,s,y,u(\gamma(s)-\tau_1,y-\xi_1),...,u(\gamma(s)-\tau_m,y-\xi_m))|dyd_s\left(\bigvee_{k=0}^{s}g(t,k)\right).$$

Then

$$|(Nu)(t_2,x_2)-(Nu)(t_1,x_1)|$$
$$\leq|\mu(\alpha(t_2),x_2)-\mu(\alpha(t_1),x_1)|$$
$$+\frac{1}{\Gamma(r_1)\Gamma(r_2)}\int_0^{\beta(t_2)}\int_0^{x_2}(\beta(t_2)-s)^{r_1-1}(x_2-y)^{r_2-1}$$
$$\times|f(t_2,x_2,s,y,u(\gamma(s)-\tau_1,y-\xi_1),...,u(\gamma(s)-\tau_m,y-\xi_m))$$
$$-f(t_1,x_1,s,y,u(\gamma(s)-\tau_1,y-\xi_1),...,u(\gamma(s)-\tau_m,y-\xi_m))|dyd_s\left(\bigvee_{k=0}^{s}g(t,k)\right)$$
$$+\frac{1}{\Gamma(r_1)\Gamma(r_2)}\int_{\beta(t_1)}^{\beta(t_2)}\int_0^{x_2}(\beta(t_2)-s)^{r_1-1}(x_2-y)^{r_2-1}$$
$$\times\sum_{i=1}^{m}q_i(t,x,s,y)dyd_s\left(\bigvee_{k=0}^{s}g(t,k)\right)$$
$$+\frac{1}{\Gamma(r_1)\Gamma(r_2)}\int_0^{\beta(t_1)}\int_0^{x_1}|(\beta(t_2)-s)^{r_1-1}(x_2-y)^{r_2-1}$$
$$-(\beta(t_1)-s)^{r_1-1}(x_1-y)^{r_2-1}|$$
$$\times\sum_{i=1}^{m}q_i(t,x,s,y)dyd_s\left(\bigvee_{k=0}^{s}g(t,k)\right)$$
$$+\frac{1}{\Gamma(r_1)\Gamma(r_2)}\int_0^{\beta(t_1)}\int_{x_1}^{x_2}|(\beta(t_2)-s)^{r_1-1}(x_2-y)^{r_2-1}|$$
$$\times\sum_{i=1}^{m}q_i(t,x,s,y)dyd_sd_s\left(\bigvee_{k=0}^{s}g(t,k)\right)$$

From continuity of $\mu,\alpha,\beta,\gamma,f,g,h,q_i$; $i=1\ldots m$ and as $t_1\longrightarrow t_2$ and $x_1\longrightarrow x_2$, the right-hand side of the above inequality tends to zero. The equicontinuity for the cases $t_1<t_2<0$, $x_1<x_2<0$ and $t_1\leq 0\leq t_2$, $x_1\leq 0\leq x_2$ is obvious.

Step 4: *$N(B_\eta)$ is equiconvergent.*
Let $(t,x)\in J$ and $u\in B_\eta$, then we have

$$|(Nu)(t,x)|\leq\mu(\alpha(t),x)|+|\frac{1}{\Gamma(r_1)\Gamma(r_2)}\int_0^{\beta(t)}\int_0^{x}(\beta(t)-s)^{r_1-1}(x-y)^{r_2-1}$$
$$\times f(t,x,s,y,u(\gamma(s)-\tau_1,y-\xi_1),...,u(\gamma(s)-\tau_m,y-\xi_m))dyd_sg(t,s)|$$
$$\leq\frac{p(t,x)}{1+\alpha(t)}+\frac{1}{\Gamma(r_1)\Gamma(r_2)}\int_0^{\beta(t)}\int_0^{x}(\beta(t)-s)^{r_1-1}(x-y)^{r_2-1}$$

$$\times \sum_{i=1}^{m} q_i(t,x,s,y)d_s\left(\bigvee_{k=0}^{s} g(t,k)\right)$$
$$\leq \frac{p^*}{1+\alpha(t)} + \sum_{i=1}^{m} \frac{1}{\Gamma(r_1)\Gamma(r_2)} \int_0^{\beta(t)} \int_0^x (\beta(t)-s)^{r_1-1}(x-y)^{r_2-1}$$
$$\times q_i(t,x,s,y)d_s\left(\bigvee_{k=0}^{s} g(t,k)\right).$$

Then, since $\alpha(t) \longrightarrow \infty$ as $t \longrightarrow \infty$, we deduce that, for each $x \in [0,b]$, we get

$$|(Nu)(t,x)| \longrightarrow 0, \; as\, t \longrightarrow +\infty.$$

Also, for each $x \in [-\xi, 0]$, we get

$$|(Nu)(t,x)| = |\Phi(t,x)| \longrightarrow 0, \; as\, t \longrightarrow +\infty.$$

Then, for each $x \in [-\xi, b]$, we get

$$|(Nu)(t,x)| \longrightarrow 0, \; as\, t \longrightarrow +\infty.$$

Hence,

$$|(Nu)(t,x) - (Nu)(+\infty,x)| \longrightarrow 0, \; as\, t \longrightarrow +\infty.$$

As a consequence of Steps 1 to 4 together with the Lemma 1.69, we can conclude that $N : B_\eta \to B_\eta$ is continuous and compact. From an application of Schauder's theorem [204], we deduce that N has a fixed point u which is a solution of the problem (5.11)-(5.12).

Step 5: *The uniform global attractivity of solutions.*
Now we investigate the stability of solutions of problem (5.11)-(5.12). Let us assume that u and v are two solutions of the problem (5.11)-(5.12) with the conditions of this theorem. Then, for each $(t,x) \in [-T,\infty) \times [-\xi,b]$, we have

$$\begin{aligned}
& |u(t,x)-v(t,x)| = |(Nu)(t,x)-(Nv)(t,x)| \\
\leq \; & \tfrac{1}{\Gamma(r_1)\Gamma(r_2)} \int_0^{\beta(t)} \int_0^x (\beta(t)-s)^{r_1-1}(x-y)^{r_2-1} \\
& \times |f(t,x,s,y,u(\gamma(s)-\tau_1,y-\xi_1),\ldots,u(\gamma(s)-\tau_m,y-\xi_m)) \\
& -f(t,x,s,y,v(\gamma(s)-\tau_1,y-\xi_1),\ldots,v(\gamma(s)-\tau_m,y-\xi_m))|dyd_sg(t,s) \\
\leq \; & \tfrac{1}{\Gamma(r_1)\Gamma(r_2)} \int_0^{\beta(t)} \int_0^x (\beta(t)-s)^{r_1-1}(x-y)^{r_2-1} \\
& \times |f(t,x,s,y,u(\gamma(s)-\tau_1,y-\xi_1),\ldots,u(\gamma(s)-\tau_m,y-\xi_m)) \\
& -f(t,x,s,y,v(\gamma(s)-\tau_1,y-\xi_1),\ldots,v(\gamma(s)-\tau_m,y-\xi_m))|dyd_s(\textstyle\bigvee_{k=0}^{s} g(t,k)) \\
\leq \; & \tfrac{1}{\Gamma(r_1)\Gamma(r_2)} \int_0^{\beta(t)} \int_0^x (\beta(t)-s)^{r_1-1}(x-y)^{r_2-1} \\
& \times \sum_{i=1}^{m} q_i(t,x,s,y)dyd_s\left(\bigvee_{k=0}^{s} g(t,k)\right) \\
\leq \; & \tfrac{2}{\Gamma(r_1)\Gamma(r_2)} \sum_{i=1}^{m} \int_0^{\beta(t)} \int_0^x (\beta(t)-s)^{r_1-1}(x-y)^{r_2-1} \\
& \times q_i(t,x,s,y)dyd_s(\textstyle\bigvee_{k=0}^{s} g(t,k)).
\end{aligned} \tag{5.17}$$

By using (5.17) and the fact that $\alpha(t) \longrightarrow \infty$ as $t \longrightarrow \infty$, we deduce that

$$\lim_{t\to\infty} |u(t,x) - v(t,x)| = 0.$$

Consequently, all solutions of the problem (5.11)-(5.12) are uniformly globally attractive. □

5.3.3. An Example

As an application of our results we consider the following system of nonlinear fractional order Riemann-Liouville Volterra-Stieltjes quadratic multi-delay partial integral equation of the form,

$$\begin{aligned} u(t,x) &= \mu(\alpha(t),x) + \frac{1}{\Gamma(r_1)\Gamma(r_2)} \int_0^{\beta(t)} \int_0^x (\beta(t)-s)^{r_1-1}(x-y)^{r_2-1} \\ &\quad \times f(t,x,s,y,u(\gamma(s)-\tfrac{1}{2},y-1),u(\gamma(s)-\tfrac{1}{4},y-2))dyd_sg(t,s); \qquad (5.18) \\ &\quad for\ \ (t,x) \in J := \mathbb{R}_+ \times [0,1], \end{aligned}$$

$$u(t,x) = \Phi(t,x); \quad \text{for } (t,x) \in \tilde{J} := [-\frac{1}{2},\infty) \times [-2,1] \backslash (0,\infty) \times (0,1], \qquad (5.19)$$

where $r_1 = \frac{1}{4}$, $r_2 = \frac{1}{2}$, $\alpha(t) = \beta(t) = \gamma(t) = t$; $t \in \mathbb{R}_+$,

$$\mu(\alpha(t),x) = \frac{x^2 e^{-t}}{1+t}; \ (t,x) \in \mathbb{R}_+ \times [0,1], \ \ \Phi(t,x) = x^2 e^{-t}; \ (t,x) \in \tilde{J},$$

$$\begin{cases} f(t,x,s,y,u,v) = \dfrac{xs^{\frac{-3}{4}}(|u_1|+|u_2|)\sin\sqrt{t}\sin s}{(1+y^2+t^2)(1+|u_1|+|u_2|)}; \\ \qquad if\ (t,x,s,y) \in J',\ s \neq 0,\ y \in [0,1]\ and\ u_1,u_2 \in \mathbb{R}, \\ f(t,x,0,y,u,v) = 0;\ if\ (t,x) \in J,\ y \in [0,1]\ and\ u_1,u_2 \in \mathbb{R}, \end{cases}$$

$$J' = \{(t,x,s,y) \in J^2 : s \le t\ and\ x \le y\}, \ \ and\ \ g(t,s) = s, \ (t,s) \in \mathbb{R}_+^2.$$

First, we can see that $\lim_{t\to\infty} \alpha(t) = \infty$ and $\lim_{t\to\infty} \Phi(t,x) = 0$. Next, the assumption (5.5.1) is satisfied with $p(t,x) = x^2 e^{-t}$ and then $p^* = 1$. Also, we can easily see that the function g satisfies the hypotheses $(5.5.2)-(5.5.4)$.

Finally, the function f satisfies the assumption $(5.5.5)$. Indeed, f is continuous and

$$|f(t,x,s,y,u_1,u_2)| \le \frac{q_1(t,x,s,y)|u_1| + q_2(t,x,s,y)|u_2|}{1+|u_1|+|u_2|};$$

$(t,x,s,y) \in J'$, $u_1,u_2 \in \mathbb{R}$. For $i = 1,2$, we have

$$\begin{cases} q_i(t,x,s,y) = \dfrac{xs^{\frac{-3}{4}}\sin\sqrt{t}\sin s}{1+y^2+t^2}; \ (t,x,s,y) \in J',\ y \in [0,1],\ s \neq 0, \\ q_i(t,x,0,y) = 0; \qquad\qquad (t,x) \in J,\ y \in [0,1]. \end{cases}$$

Then, for $i = 1,2$, we have

$$\begin{aligned}
\left|\int_0^t (t-s)^{r-1} q_i(t,x,s,y) d_s g(t,s)\right| &\leq \int_0^t (t-s)^{\frac{-3}{4}} x s^{\frac{-3}{4}} \\
&\quad \times |\sin\sqrt{t}\sin s| d_s\left(\bigvee_{k=0}^{s} g(t,k)\right) \\
&\leq x|\sin\sqrt{t}| \int_0^t (t-s)^{\frac{-3}{4}} s^{\frac{-3}{4}} ds \\
&\leq \frac{x\Gamma^2(\frac{1}{4})}{\sqrt{\pi}} \left|\frac{\sin\sqrt{t}}{\sqrt{t}}\right| \\
&\leq \frac{x\Gamma^2(\frac{1}{4})}{\sqrt{\pi t}} \longrightarrow 0 \; as \; t \to \infty,
\end{aligned}$$

and

$$\begin{aligned}
q_i^* &:= \sup_{(t,x)\in J} \frac{1}{\Gamma(r_1)\Gamma(r_2)} \int_0^t \int_0^x (t-s)^{r_1-1}(x-y)^{r_2-1} \\
&\quad \times q_i(t,x,s,y) dy d_s\left(\bigvee_{k=0}^{s} g(t,k)\right) \\
&\leq \sup_{(t,x)\in J} \frac{x\Gamma(\frac{1}{4})}{\pi} \left|\frac{\sin\sqrt{t}}{\sqrt{t}}\right| = \frac{\Gamma(\frac{1}{4})}{\pi}.
\end{aligned}$$

Consequently, by Theorem 5.6, the problem (5.18)-(5.19) has a solution defined on $[-\frac{1}{2},\infty) \times [-2,1]$ and solutions of this equation are uniformly globally attractive.

5.4. Notes and Remarks

The results of Chapter 5 are taken from Abbas *et al.* [12, 28, 37]. Other results may be found in [84, 87, 88].

Chapter 6

Fractional Order Riemann-Liouville Integro-Differential Equations

6.1. Introduction

In this chapter, we shall present existence results for some classes of Riemann-Liouville integro-differential equations of fractional order by using some fixed point theorems.

6.2. Integro-Differential Equations

6.2.1. Introduction

In this section we investigate the existence and uniqueness of solutions for for the following integro-differential equation of fractional order of the form

$$^cD^r_\theta u(x,y) = f(x,y,I^r_\theta u(x,y),u(x,y)); \; (x,y) \in J := [0,a]\times[0,b], \tag{6.1}$$

$$\begin{cases} u(x,0) = \varphi(x); \; x \in [0,a], \\ u(0,y) = \psi(y); \; y \in [0,b], \\ \varphi(0) = \psi(0), \end{cases} \tag{6.2}$$

where $\theta = (0,0)$, $r = (r_1,r_2)$, $r_1,r_2 \in (0,\infty)$, I^r_θ is the left-sided mixed Riemann-Liouville integral of order r, $^cD^r_\theta$ is the fractional Caputo derivative of order r, $f : J\times\mathbb{R}^n\times\mathbb{R}^n \to \mathbb{R}^n$ is a given continuous function, and $\varphi : [0,a] \to \mathbb{R}^n$, $\psi : [0,b] \to \mathbb{R}^n$ are absolutely continuous functions. Our investigations are conducted with an application of Banach's contraction principle, Schauder's fixed point theorem, and the nonlinear alternative of Leray-Schauder type for the existence and uniqueness of solutions of our problem.

6.2.2. Main Results

Definition 6.1. *A function $u \in C := C(J,\mathbb{R}^n)$ is said to be a solution of (6.1)-(6.2) if u satisfies equation (6.1) on J and conditions (6.2) are satisfied.*

Now, we shall prove the following theorem concerning the existence and uniqueness of a solution of (6.1)-(6.2). Set

$$f^*(x,y) = f(x,y,0,0).$$

Theorem 6.2. *Assume that the following hypothesis holds*

(6.2.1) *There exist constants* $L_1, L_2 > 0$ *such that*

$$\|f(x,x,u,v) - f(x,y,\overline{u},\overline{v})\| \leq L_2\|u-\overline{u}\| + L_1\|v-\overline{v}\|,$$

for each $(x,y) \in J$ *and* $u, v, \overline{u}, \overline{v} \in \mathbb{R}^n$.

If

$$L_3(L_1 + L_2L_3) \leq 1, \tag{6.3}$$

where

$$L_3 := \frac{a^{r_1}b^{r_2}}{\Gamma(1+r_1)\Gamma(1+r_2)},$$

then the problem (6.1)-(6.2) has a unique solution on J.

Proof. Transform the problem (6.1)-(6.2) into a fixed point problem. Let $u \in C$ and define the operator N by,

$$(Nu)(x,y) = \mu(x,y) + I_\theta^r f(x,y,I_\theta^r u(x,y), u(x,y)). \tag{6.4}$$

It is clear that N maps C into itself. By Lemma 1.28, the problem of finding the solutions of the problem (6.1)-(6.2) is reduced to finding the solutions of the operator equation $N(u) = u$. Now, we verify that the operator N is a contraction map. Let $u, v \in C$. Using (6.2.1), for each $(x,y) \in J$, we have

$$\begin{aligned}
&\|(Nu)(x,y) - (Nv)(x,y)\| \\
\leq\ & \|I_\theta^r (f(x,y,I_\theta^r u(x,y), u(x,y)) - f(x,y,I_\theta^r v(x,y), v(x,y)))\| \\
\leq\ & I_\theta^r (L_1\|u(x,y) - v(x,y)\| \\
+\ & \left.\frac{L_2}{\Gamma(r_1)\Gamma(r_2)} \int_0^x \int_0^y (x-s)^{r_1-1}(y-t)^{r_2-1}\|u(s,t) - v(s,t)\| dt ds\right) \\
\leq\ & \left(\frac{L_1 a^{r_1} b^{r_2}}{\Gamma(1+r_1)\Gamma(1+r_2)} + \frac{L_2 a^{2r_1} b^{2r_2}}{\Gamma^2(1+r_1)\Gamma^2(1+r_2)} \right) \|u - v\|_\infty.
\end{aligned}$$

Hence

$$\|N(u) - N(v)\|_\infty \leq L_3(L_1 + L_2L_3)\|u - v\|_\infty.$$

From (6.3), it follows that N has a unique fixed point in C by Banach contraction principle. The fixed point of N is however a solution of problem (6.1)-(6.2). □

For the existence of solutions for the problem (6.1)-(6.2) and by using Schauder's fixed point theorem [204], we prove the following theorem

Theorem 6.3. *Assume that hypothesis* (6.2.1) *holds. If there exists a constant* $\eta > 0$ *such that*

$$\|\mu\|_\infty + L_3\|f^*\|_\infty \leq \eta(1 - L_1L_3 - L_2L_3^2), \tag{6.5}$$

then the problem (6.1)-(6.2) has at least one solution in the space C.

Proof. Consider the operator N defined in (6.4). By (6.5), we get

$$\|(Nu)\|_\infty \leq \|\mu\|_\infty + L_3(\|f^*\|_\infty + L_1\eta + L_2L_3\eta) \leq \eta.$$

This implies that N transforms the ball $B_\eta := \{u \in C : \|u\|_\infty \leq \eta\}$ into itself. We shall show that $N : B_\eta \to B_\eta$ satisfies the assumptions of Schauder's fixed point theorem [204]. The proof will be given in several steps.

Step 1: *N is continuous.*
Let $\{u_n\}_{n\in\mathbb{N}}$ be a sequence such that $u_n \to u$ in B_η. Then, for each $(x,y) \in J$, we have

$$\begin{aligned} & \|(Nu_n)(x,y) - (Nu)(x,y)\| \\ \leq\ & \|I_\theta^r(f(x,y,I_\theta^r u_n(x,y),u_n(x,y)) - f(x,y,I_\theta^r u(x,y),u(x,y)))\| \\ \leq\ & L_3(L_1 + L_2L_3)\|u_n - u\|_\infty. \end{aligned}$$

Hence

$$\|N(u_n) - N(u)\|_\infty \leq L_3(L_1 + L_2L_3)\|u_n - u\|_\infty. \tag{6.6}$$

Since $u_n \to u$ as $n \to \infty$, then (6.6) gives

$$\|N(u_n) - N(u)\|_\infty \to 0 \quad \text{as } n \to \infty.$$

Step 2: *$N(B_\eta)$ is bounded.*
This is clear since $N(B_\eta) \subset B_\eta$ and B_η is bounded.

Step 3: *$N(B_\eta)$ is equicontinuous.*
Let $(x_1,y_1),\ (x_2,y_2) \in J,\ x_1 < x_2,\ y_1 < y_2$ and $u \in B_\eta$. Then

$$\begin{aligned} & \|N(u)(x_2,y_2) - N(u)(x_1,y_1)\| \\ \leq\ & \|\mu(x_1,y_1) - \mu(x_2,y_2)\| \\ & +\|\frac{L_1+L_2L_3}{\Gamma(r_1)\Gamma(r_2)}\int_0^{x_1}\int_0^{y_1}[(x_2-s)^{r_1-1}(y_2-t)^{r_2-1} - (x_1-s)^{r_1-1}(y_1-t)^{r_2-1}] \\ & \times u(s,t)dtds + \frac{L_1+L_2L_3}{\Gamma(r_1)\Gamma(r_2)}\int_{x_1}^{x_2}\int_{y_1}^{y_2}(x_2-s)^{r_1-1}(y_2-t)^{r_2-1}u(s,t)dtds \\ & +\frac{L_1+L_2L_3}{\Gamma(r_1)\Gamma(r_2)}\int_0^{x_1}\int_{y_1}^{y_2}(x_2-s)^{r_1-1}(y_2-t)^{r_2-1}u(s,t)dtds \\ & +\frac{L_1+L_2L_3}{\Gamma(r_1)\Gamma(r_2)}\int_{x_1}^{x_2}\int_0^{y_1}(x_2-s)^{r_1-1}(y_2-t)^{r_2-1}u(s,t)dtds\| \\ \leq\ & \|\mu(x_1,y_1) - \mu(x_2,y_2)\| + L_1\|u(x_1,y_1) - u(x_2,y_2)\| \\ & +\frac{\eta(L_1+L_2L_3)}{\Gamma(1+r_1)\Gamma(1+r_2)}[2y_2^{r_2}(x_2-x_1)^{r_1} + 2x_2^{r_1}(y_2-y_1)^{r_2} \\ & +x_1^{r_1}y_1^{r_2} - x_2^{r_1}y_2^{r_2} - 2(x_2-x_1)^{r_1}(y_2-y_1)^{r_2}]. \end{aligned}$$

As $x_1 \to x_2,\ y_1 \to y_2$ the right-hand side of the above inequality tends to zero. As a consequence of steps 1 to 3 together with the Arzela-Ascoli theorem, we can conclude that N is continuous and compact. From an application of Schauder's theorem [204], we deduce that N has a fixed point u which is a solution of the problem (6.1)-(6.2). □

Now, we present an existence result for the problem (6.1)-(6.2) based on the nonlinear alternative of Leray-Schauder type

Theorem 6.4. *Assume that the following hypothesis holds*

(6.6.1) *There exist positive functions $p, q, d \in C(J)$ such that*

$$\|f(x,y,u,v)\| \leq p(x,y) + q(x,y)\|u\| + d(x,y)\|v\|;$$

for all $(x,y) \in J$ and $u, v \in \mathbb{R}^n$.

Then problem (6.1)-(6.2) has at least one solution on J.

Proof. Consider the operator N defined in (6.4). We shall show that the operator N is completely continuous. By the continuity of f and the Arzela-Ascoli theorem, we can easily obtain that N is completely continuous.

A priori bounds. We shall show there exists an open set $U \subseteq C$ with $u \neq \lambda N(u)$, for $\lambda \in (0,1)$ and $u \in \partial U$. Let $u \in E$ and $u = \lambda N(u)$ for some $0 < \lambda < 1$. Thus for each $(x,y) \in J$, we have

$$u(x,y) = \lambda \mu(x,u) + \lambda I_\theta^r f(x,y,I_\theta^r u(x,y), u(x,y)).$$

This implies by (6.6.1) that, for each $(x,y) \in J$, we have

$$\begin{aligned}
\|u(x,y)\| &\leq \|\mu(x,y)\| + \frac{p^* a^{r_1} b^{r_2}}{\Gamma(1+r_1)\Gamma(1+r_2)} \\
&+ \frac{q^* a^{r_1} b^{r_2}}{\Gamma(r_1)\Gamma(r_2)\Gamma(1+r_1)\Gamma(1+r_2)} \int_0^x \int_0^y (x-s)^{r_1-1}(y-t)^{r_2-1}\|u(s,t)\| dt ds \\
&+ \frac{d^*}{\Gamma(r_1)\Gamma(r_2)} \int_0^x \int_0^y (x-s)^{r_1-1}(y-t)^{r_2-1} \\
&\times \|u(s,t)\| dt ds \\
&\leq \|\mu(x,y)\| + \frac{p^* a^{r_1} b^{r_2}}{\Gamma(1+r_1)\Gamma(1+r_2)} \\
&+ \left[\frac{q^* a^{r_1} b^{r_2}}{\Gamma(r_1)\Gamma(r_2)\Gamma(1+r_1)\Gamma(1+r_2)} + \frac{d^*}{\Gamma(r_1)\Gamma(r_2)}\right] \\
&\times \int_0^x \int_0^y (x-s)^{r_1-1}(y-t)^{r_2-1}\|u(s,t)\| dt ds,
\end{aligned}$$

where $p^* = \|p\|_\infty$, $q^* = \|q\|_\infty$ and $d^* = \|d\|_\infty$. Thus, for each $(x,y) \in J$, we get

$$\begin{aligned}
\|u(x,y)\| &\leq \|\mu\|_\infty + \frac{p^* a^{r_1} b^{r_2}}{\Gamma(1+r_1)\Gamma(1+r_2)} \\
&+ \left[\frac{q^* a^{r_1} b^{r_2}}{\Gamma(r_1)\Gamma(r_2)\Gamma(1+r_1)\Gamma(1+r_2)} + \frac{d^*}{\Gamma(r_1)\Gamma(r_2)}\right] \\
&\times \int_0^x \int_0^y (x-s)^{r_1-1}(y-t)^{r_2-1}\|u(s,t)\| dt ds \\
&\leq w + c\int_0^x \int_0^y (x-s)^{r_1-1}(y-t)^{r_2-1}\|u(s,t)\| dt ds
\end{aligned}$$

where

$$w := \|\mu\|_\infty + \frac{p^* a^{r_1} b^{r_2}}{\Gamma(1+r_1)\Gamma(1+r_2)}$$

and

$$c := \frac{q^* a^{r_1} b^{r_2}}{\Gamma(r_1)\Gamma(r_2)\Gamma(1+r_1)\Gamma(1+r_2)} + \frac{d^*}{\Gamma(r_1)\Gamma(r_2)}.$$

From Lemma 1.68, there exists $\delta := \delta(r_1, r_2) > 0$ such that, for each $(x, y) \in J$, we get

$$\begin{aligned} \|u\|_\infty &\leq w\left(1 + c\delta \int_0^x \int_0^y (x-s)^{r_1-1}(y-t)^{r_2-1} dt ds\right) \\ &\leq w\left(1 + \frac{c\delta a^{r_1} b^{r_2}}{r_1 r_2}\right) := M^*. \end{aligned}$$

Set

$$U = \{u \in C : \|u\|_\infty < 1 + M^*\}.$$

By our choice of U, there is no $u \in \partial U$ such that $u = \lambda N(u)$, for $\lambda \in (0,1)$. As a consequence of Theorem 1.56, we deduce that N has a fixed point u in $\overline{U}$ which is a solution to problem (6.1)-(6.2). □

6.2.3. Examples

As applications and to illustrate our results, we present the following examples.

Example 1. Consider the following system of integro-differential equations of fractional order

$$^c D_\theta^r u(x,y) = \frac{2 + |u(x,y)| + |I_\theta^r u(x,y)|}{e^{x+y+10}(1+|u(x,y)|)(1+|I_\theta^r u(x,y)|)}; \ (x,y) \in [0,1] \times [0,1], \tag{6.7}$$

$$\begin{cases} \varphi(x) = x; \ x \in [0,1], \\ \psi(y) = y^2; \ y \in [0,1]. \end{cases} \tag{6.8}$$

Set

$$f(x,y,u,v) = \frac{e^{-(x+y+10)}}{1+|u|} + \frac{e^{-(x+y+10)}}{1+|v|}; \ (x,y) \in [0,1] \times [0,1].$$

We have

$$f^*(x,y) = 2e^{-(x+y+10)}.$$

For each $u_1,\ u_2,\ v_1,\ v_2 \in \mathbb{R}$, we have

$$|f(x,y,u_1,v_1) - f(x,y,u_2,v_2)| \leq e^{-10}(|u_1 - u_2| + |v_1 - v_2|).$$

Hence condition (6.2.1) is satisfied with $L_1 = L_2 = e^{-10}$. We shall show that condition (6.3) holds with $a = b = 1$. Indeed

$$L_3 = \frac{a^{r_1} b^{r_2}}{\Gamma(1+r_1)\Gamma(1+r_2)} = \frac{1}{\Gamma(1+r_1)\Gamma(1+r_2)} < \frac{25}{16},$$

and

$$L_3(L_1 + L_2L_3) < \frac{25}{16e^{10}}\left(1 + \frac{25}{16}\right) = \frac{1025}{256e^{10}} < 1,$$

which is satisfied for each $(r_1, r_2) \in (0,1] \times (0,1]$. Consequently Theorem 6.2 implies that problem (6.7)-(6.8) has a unique solution defined on $[0,1] \times [0,1]$.

Example 2. Consider now the following system of integro-differential equations of fractional order

$$^cD_\theta^r u(x,y) = \frac{1}{e^{x+y+10}(1 + |u(x,y) + |I_\theta^r u(x,y)||)}; \ (x,y) \in [0,1] \times [0,1], \tag{6.9}$$

$$\begin{cases} \varphi(x) = xe^{-10};\ x \in [0,1], \\ \psi(y) = y^2 e^{-10};\ y \in [0,1]. \end{cases} \tag{6.10}$$

Set

$$f(x,y,u,v) = \frac{e^{-(x+y+10)}}{1 + |u| + |v|};\ (x,y) \in [0,1] \times [0,1].$$

We have

$$f^*(x,y) = e^{-(x+y+10)}.$$

For each $u_1,\ u_2,\ v_1,\ v_2 \in \mathbb{R}$, we have

$$|f(x,y,u_1,v_1) - f(x,y,u_2,v_2)| \leq e^{-10}(|u_1 - v_1| + |u_2 - v_2|).$$

Hence condition (6.2.1) is satisfied with $L_1 = L_2 = e^{-10}$. We shall show that condition (6.5) holds with $a = b = 1$ and $\eta > \frac{57e^{-10}}{8}$. Indeed, for each $(r_1, r_2) \in (0,1] \times (0,1]$, we get

$$L_3 = \frac{a^{r_1} b^{r_2}}{\Gamma(1+r_1)\Gamma(1+r_2)} = \frac{1}{\Gamma(1+r_1)\Gamma(1+r_2)} < \frac{25}{16},$$

also, we have

$$\|\mu\|_\infty < 2e^{-10},\ \|f^*\|_\infty < e^{-10},$$

$$\|\mu\|_\infty + L_3\|f^*\|_\infty \leq e^{-10}\left(2 + \frac{25}{16}\right) = \frac{57e^{-10}}{16},$$

and

$$\eta(1 - L_1L_3 - L_2L_3^2) \geq \eta\left(1 - \frac{1025}{256e^{10}}\right) > \frac{\eta}{2}.$$

So, for $\eta > \frac{57e^{-10}}{8}$, we obtain (6.5). Consequently Theorem 6.3 implies that problem (6.9)-(6.10) has at least one solution defined on $[0,1] \times [0,1]$.

6.3. Perturbed Integro-Differential Equations

6.3.1. Introduction

In this section, we investigate the following system of perturbed integro-differential equations of fractional order of the form

$$^cD^r_\theta u(x,y) = f(x,y,I^r_\theta u(x,y),u(x,y)) + g(x,y,I^r_\theta u(x,y),u(x,y));\ (x,y) \in J, \tag{6.11}$$

$$\begin{cases} u(x,0) = \varphi(x);\ x \in [0,a], \\ u(0,y) = \psi(y);\ y \in [0,b], \\ \varphi(0) = \psi(0), \end{cases} \tag{6.12}$$

where $J := [0,a] \times [0,b];\ a,b > 0,\ \theta = (0,0),\ r = (r_1,r_2),\ r_1,r_2 \in (0,\infty),\ I^r_\theta$ is the left-sided mixed Riemann-Liouville integral of order r, $^cD^r_\theta$ is the fractional Caputo derivative of order r, $f,g : J \times \mathbb{R}^n \times \mathbb{R}^n \to \mathbb{R}^n$ are given continuous functions, and $\varphi : [0,a] \to \mathbb{R}^n$, $\psi : [0,b] \to \mathbb{R}^n$ are absolutely continuous functions. We present existence results for the problem (6.11)-(6.12) based on a fixed point theorem due to Burton and Kirk [123] for the sum of contraction and completely continuous operators.

6.3.2. Main Results

Definition 6.5. *A function $u \in C$ is said to be a solution of (6.11)-(6.12) if u satisfies equation (6.11) on J and conditions (6.12) are satisfied.*

Theorem 6.6. *Assume that the following hypotheses hold*

(6.9.1) *There exist positive functions $p,q,d \in C(J)$ such that*

$$\|f(x,y,u,v)\| \leq p(x,y) + q(x,y)\|u\| + d(x,y)\|v\|,$$

for all $(x,y) \in J$ and $u,v \in \mathbb{R}^n$,

(6.9.2) *There exist constants $L_1, L_2 > 0$ such that*

$$\|g(x,x,u,v) - g(x,y,\overline{u},\overline{v})\| \leq L_2\|u-\overline{u}\| + L_1\|v-\overline{v}\|,$$

for each $(x,y) \in J$ and $u,v,\overline{u},\overline{v} \in \mathbb{R}^n$.

If

$$L_3(L_1 + L_2L_3) \leq 1, \tag{6.13}$$

where

$$L_3 = \frac{a^{r_1}b^{r_2}}{\Gamma(1+r_1)\Gamma(1+r_2)},\ p^* = \|p\|_\infty,\ q^* = \|q\|_\infty \text{ and } d^* = \|d\|_\infty,$$

then problem (6.11)-(6.12) has at least one solution on J.

Proof. Transform the problem (6.11)-(6.12) into a fixed point problem. Let $u \in C$ and define the operator N by,

$$(Nu)(x,y) = \mu(x,y) + I_\theta^r f(x,y,I_\theta^r u(x,y),u(x,y)) + I_\theta^r g(x,y,I_\theta^r u(x,y),u(x,y)). \qquad (6.14)$$

We can see that operator N maps C into itself. By Lemma 1.28, the problem of finding the solutions of the problem (6.11)-(6.12) is reduced to finding the solutions of the operator equation $N(u) = u$.
Consider the operators $F, G : C \to C$, defined by,

$$(Fu)(x,y) = \mu(x,y) + I_\theta^r f(x,y,I_\theta^r u(x,y),u(x,y)); \ (x,y) \in J,$$

and

$$(Gu)(x,y) = I_\theta^r g(x,y,I_\theta^r u(x,y),u(x,y)); \ (x,y) \in J.$$

Thus $N = F + G$. We shall show that the operators F and G satisfy all the conditions of Theorem 1.59. The proof will be given in several steps.

Step 1: *F is continuous.*
Let $\{u_n\}_{n\in\mathbb{N}}$ be a sequence such that $u_n \to u$ in C. Let $0 < \eta \le M$ such that $\|u\|_\infty, \|u_n\|_\infty \in B_\eta := \{u \in C : \|u\|_\infty \le \eta\}$. Then, for each $(x,y) \in J$, we have

$$\begin{aligned}
& \|(Fu_n)(x,y) - (Fu)(x,y)\| \\
\le\ & \|I_\theta^r (f(x,y,I_\theta^r u_n(x,y),u_n(x,y)) - f(x,y,I_\theta^r u(x,y),u(x,y)))\| \\
\le\ & \frac{1}{\Gamma(r_1)\Gamma(r_2)} \int_0^x \int_0^y (x-s)^{r_1-1}(y-t)^{r_2-1} \\
\times\ & \|f(s,t,I_\theta^r u_n(s,t),u_n(s,t)) - f(s,t,I_\theta^r u(s,t),u(s,t))\| dtds.
\end{aligned}$$

Hence, for each $(x,y) \in J$, we get

$$\|F(u_n) - F(u)\|_\infty \le \frac{1}{\Gamma(r_1)\Gamma(r_2)} \int_0^x \int_0^y (x-s)^{r_1-1}(y-t)^{r_2-1}$$

$$\times \|f(.,.,I_\theta^r u_n(.,.),u_n(.,.)) - f(.,.,I_\theta^r u(.,.),u(.,.))\|_\infty dtds.$$

Since $u_n \to u$ as $n \to \infty$, and f, I_θ^r are continuous, then

$$\|F(u_n) - F(u)\|_\infty \to 0 \quad \text{as } n \to \infty.$$

Step 2: *F maps bounded sets into bounded sets in C.*
Indeed, it is enough show that, for any $\eta > 0$, there exists a positive constant $\ell > 0$ such that, for each $u \in B_\eta := \{u \in C : \|u\|_\infty \le \eta\}$ we have $\|F(u)\|_C \le \ell$. By (6.9.1) we have for each $(x,y) \in J$,

$$\begin{aligned}
\|(Fu)(x,y)\| \ \le\ & \|\mu(x,y)\| + \frac{p^* a^{r_1} b^{r_2}}{\Gamma(1+r_1)\Gamma(1+r_2)} \\
& + \frac{q^* a^{r_1} b^{r_2}}{\Gamma(r_1)\Gamma(r_2)\Gamma(1+r_1)\Gamma(1+r_2)}
\end{aligned}$$

$$
\begin{aligned}
&\times \int_0^x \int_0^y (x-s)^{r_1-1}(y-t)^{r_2-1}\|u(s,t)\|dtds \\
&+ \frac{d^*}{\Gamma(r_1)\Gamma(r_2)} \int_0^x \int_0^y (x-s)^{r_1-1}(y-t)^{r_2-1}\|u(s,t)\|dtds \\
&\leq \|\mu(x,y)\| + \frac{p^* a^{r_1} b^{r_2}}{\Gamma(1+r_1)\Gamma(1+r_2)} \\
&+ \left(\frac{q^* a^{r_1} b^{r_2}}{\Gamma(r_1)\Gamma(r_2)\Gamma(1+r_1)\Gamma(1+r_2)} + \frac{d^*}{\Gamma(r_1)\Gamma(r_2)}\right) \\
&\times \int_0^x \int_0^y (x-s)^{r_1-1}(y-t)^{r_2-1}\|u(s,t)\|dtds \\
&\leq \|\mu\|_\infty + \frac{p^* a^{r_1} b^{r_2}}{\Gamma(1+r_1)\Gamma(1+r_2)} + \frac{\eta a^{r_1} b^{r_2}}{\Gamma(1+r_1)\Gamma(1+r_2)} \\
&\times \left(\frac{q^* a^{r_1} b^{r_2}}{\Gamma(1+r_1)\Gamma(1+r_2)} + d^*\right).
\end{aligned}
$$

Thus,

$$\|F(u)\|_\infty \leq \|\mu\|_\infty + L_3 p^* + \eta L_3 (q^* L_3 + d^*) := \ell.$$

Step 3: *F maps bounded sets into equicontinuous sets in C.*
Let $(x_1,y_1),(x_2,y_2) \in J$, $x_1 < x_2, y_1 < y_2$, B_η be a bounded set of C as in Step 2, and let $u \in B_\eta$. Then

$$
\begin{aligned}
& \|F(u)(x_2,y_2) - F(u)(x_1,y_1)\| \\
\leq\; & \|\mu(x_1,y_1) - \mu(x_2,y_2)\| \\
& + \left\| \frac{p^* + q^* L_3 \eta + d^* \eta}{\Gamma(r_1)\Gamma(r_2)} \int_0^{x_1} \int_0^{y_1} [(x_2-s)^{r_1-1}(y_2-t)^{r_2-1} \right. \\
- \; & (x_1-s)^{r_1-1}(y_1-t)^{r_2-1}]dtds \\
+ \; & \frac{p^* + q^* L_3 \eta + d^* \eta}{\Gamma(r_1)\Gamma(r_2)} \int_{x_1}^{x_2} \int_{y_1}^{y_2} (x_2-s)^{r_1-1}(y_2-t)^{r_2-1}dtds \\
& + \frac{p^* + q^* L_3 \eta + d^* \eta}{\Gamma(r_1)\Gamma(r_2)} \int_0^{x_1} \int_{y_1}^{y_2} (x_2-s)^{r_1-1}(y_2-t)^{r_2-1}dtds \\
& \left. + \frac{p^* + q^* L_3 \eta + d^* \eta}{\Gamma(r_1)\Gamma(r_2)} \int_{x_1}^{x_2} \int_0^{y_1} (x_2-s)^{r_1-1}(y_2-t)^{r_2-1}u(s,t)dtds \right\| \\
\leq\; & \|\mu(x_1,y_1) - \mu(x_2,y_2)\| + \frac{p^* + q^* L_3 \eta + d^* \eta}{\Gamma(1+r_1)\Gamma(1+r_2)} \\
& \times [2y_2^{r_2}(x_2-x_1)^{r_1} + 2x_2^{r_1}(y_2-y_1)^{r_2} + x_1^{r_1}y_1^{r_2} - x_2^{r_1}y_2^{r_2} \\
& -2(x_2-x_1)^{r_1}(y_2-y_1)^{r_2}].
\end{aligned}
$$

As $x_1 \to x_2$, $y_1 \to y_2$ the right-hand side of the above inequality tends to zero. As a consequence of steps 1 to 3 together with the Arzela-Ascoli theorem, we can conclude that F is continuous and completely continuous.

Step 4: *G is a contraction.*

Let $u,v\in C$. Using (6.9.2), for each $(x,y)\in J$, we have

$$\begin{aligned}
&\|(Gu)(x,y)-(Gv)(x,y)\|\\
\leq\ & \|I_\theta^r(g(x,y,I_\theta^r u(x,y),u(x,y))-g(x,y,I_\theta^r v(x,y),v(x,y)))\|\\
\leq\ & I_\theta^r(L_1\|u(x,y)-v(x,y)\|\\
+\ & \left.\frac{L_2}{\Gamma(r_1)\Gamma(r_2)}\int_0^x\int_0^y (x-s)^{r_1-1}(y-t)^{r_2-1}\|u(s,t)-v(s,t)\|dtds\right)\\
\leq\ & \left(\frac{L_1a^{r_1}b^{r_2}}{\Gamma(1+r_1)\Gamma(1+r_2)}+\frac{L_2a^{2r_1}b^{2r_2}}{\Gamma^2(1+r_1)\Gamma^2(1+r_2)}\right)\|u-v\|_\infty.
\end{aligned}$$

Hence

$$\|G(u)-G(v)\|_\infty\leq L_3(L_1+L_2L_3)\|u-v\|_\infty.$$

From (6.13), it follows that G is a contraction.

Step 5: *A priori bounds.*

Now it remains to show that the set

$$\mathcal{E}=\{u\in C: u=\lambda F(u)+\lambda G(\frac{u}{\lambda}),\ \text{ for some } 0<\lambda<1\}$$

is bounded. Let $u\in\mathcal{E}$, then $u=\lambda F(u)+\lambda G(\frac{u}{\lambda})$; for some $0<\lambda<1$. Thus, for each $(x,y)\in J$ we have

$$\begin{aligned}
u(x,y)\ =\ & \lambda\mu(x,u)+\lambda I_\theta^r f(x,y,I_\theta^r u(x,y),u(x,y))\\
+\ & \lambda I_\theta^r g\left(x,y,I_\theta^r\frac{u(x,y)}{\lambda},\frac{u(x,y)}{\lambda}\right).
\end{aligned}$$

This implies by (6.9.1) and (6.9.2) that, for each $(x,y)\in J$, we have

$$\begin{aligned}
\|u(x,y)\|\ \leq\ & \|\mu(x,y)\|+\frac{p^*a^{r_1}b^{r_2}}{\Gamma(1+r_1)\Gamma(1+r_2)}\\
+\ & \frac{q^*a^{r_1}b^{r_2}}{\Gamma(r_1)\Gamma(r_2)\Gamma(1+r_1)\Gamma(1+r_2)}\int_0^x\int_0^y(x-s)^{r_1-1}(y-t)^{r_2-1}\|u(s,t)\|dtds\\
+\ & \frac{d^*}{\Gamma(r_1)\Gamma(r_2)}\int_0^x\int_0^y(x-s)^{r_1-1}(y-t)^{r_2-1}\\
\times\ & \|u(s,t)\|dtds\\
+\ & \frac{1}{\Gamma(r_1)\Gamma(r_2)}\int_0^x\int_0^y(x-s)^{r_1-1}(y-t)^{r_2-1}\|g(s,t,0,0)\|dtds\\
+\ & \frac{L_1+L_2L_3}{\Gamma(r_1)\Gamma(r_2)}\int_0^x\int_0^y(x-s)^{r_1-1}(y-t)^{r_2-1}\left\|\frac{u(s,t)}{\lambda}\right\|dtds\\
\leq\ & \|\mu\|_\infty+L_3p^*\\
+\ & \frac{L_3q^*}{\Gamma(r_1)\Gamma(r_2)}\int_0^x\int_0^y(x-s)^{r_1-1}(y-t)^{r_2-1}\|u(s,t)\|dtds\\
+\ & \frac{d^*}{\Gamma(r_1)\Gamma(r_2)}\int_0^x\int_0^y(x-s)^{r_1-1}(y-t)^{r_2-1}\|u(s,t)\|dtds\\
+\ & g^*L_3+\frac{L_1+L_2L_3}{\Gamma(r_1)\Gamma(r_2)}\int_0^x\int_0^y(x-s)^{r_1-1}(y-t)^{r_2-1}\left\|\frac{u(s,t)}{\lambda}\right\|dtds
\end{aligned}$$

where $g^* = \sup_{(x,y)\in J} g(x,y,0,0)$. Thus, for each $(x,y) \in J$, we get

$$\begin{aligned}
\|u(x,y)\| &\leq \|\mu\|_\infty + L_3(p^* + g^*) \\
&+ \frac{\lambda(q^* L_3 + d^*) + L_1 + L_2 L_3}{\lambda\Gamma(r_1)\Gamma(r_2)} \int_0^x \int_0^y (x-s)^{r_1-1}(y-t)^{r_2-1} \\
&\times \|u(s,t)\| dt ds \\
&\leq w + c\int_0^x \int_0^y (x-s)^{r_1-1}(y-t)^{r_2-1}\|u(s,t)\| dt ds,
\end{aligned}$$

where

$$w := \|\mu\|_\infty + L_3(p^* + g^*)$$

and

$$c := \frac{\lambda(q^* L_3 + d^*) + L_1 + L_2 L_3}{\lambda\Gamma(r_1)\Gamma(r_2)}.$$

From Lemma 1.68, there exists $\delta := \delta(r_1, r_2) > 0$ such that, for each $(x,y) \in J$, we get

$$\begin{aligned}
\|u\|_\infty &\leq w\left(1 + c\delta \int_0^x \int_0^y (x-s)^{r_1-1}(y-t)^{r_2-1} dt ds\right) \\
&\leq w\left(1 + \frac{c\delta a^{r_1} b^{r_2}}{r_1 r_2}\right) := M^*.
\end{aligned}$$

This shows that the set $\mathcal{E}$ is bounded. As a consequence of Theorem 1.59, we deduce that N has a fixed point u which is a solution to problem (6.11)-(6.12). □

6.3.3. An Example

To illustrate our result, we consider the following system of integro-differential equations of fractional order

$$\begin{aligned}
{}^c D_\theta^r u(x,y) &= \frac{3 + (x+y)|u(x,y)| + xy|I_\theta^r u(x,y)|}{e^{10}(1 + |u(x,y)| + |I_\theta^r u(x,y)|)} \\
&+ \frac{2 + |u(x,y)| + |I_\theta^r u(x,y)|}{e^{x+y+10}(1 + |u(x,y)|)(1 + |I_\theta^r u(x,y)|)}; \ (x,y) \in [0,1]\times[0,1], \qquad (6.15)
\end{aligned}$$

$$\begin{cases} \varphi(x) = x; \ x \in [0,1], \\ \psi(y) = y^2; \ y \in [0,1]. \end{cases} \qquad (6.16)$$

Set

$$f(x,y,u,v) = \frac{3 + (x+y)|u| + xy|v|}{e^{10}(1 + |u| + |v|)}; \ (x,y) \in [0,1]\times[0,1],$$

and

$$g(x,y,u,v) = \frac{e^{-(x+y+10)}}{1+|u|} + \frac{e^{-(x+y+10)}}{1+|v|}; \ (x,y) \in [0,1]\times[0,1].$$

For each $u, u_1, u_2, v, v_1, v_2 \in \mathbb{R}$, we have

$$|f(x,y,u,v)| \leq e^{-10}(3 + (x+y)|u| + xy|v|),$$

and

$$|g(x,y,u_1,v_1)-g(x,y,u_2,v_2)| \le e^{-10}(|u_1-u_2|+|v_1-v_2|).$$

Hence condition (6.9.1) is satisfied with $p^* = 3e^{-10}, q^* = 2e^{-10}$ and $d^* = e^{-10}$.
Also,condition (6.9.2) is satisfied with $L_1 = L_2 = e^{-10}$. We shall show that the condition (6.13) holds for each $(r_1,r_2) \in (0,1] \times (0,1]$ with $a = b = 1$. Indeed,

$$L_3 = \frac{a^{r_1}b^{r_2}}{\Gamma(1+r_1)\Gamma(1+r_2)} = \frac{1}{\Gamma(1+r_1)\Gamma(1+r_2)} < \frac{25}{16},$$

and,

$$L_3(L_1+L_2L_3) < \frac{25}{16e^{10}}\left(1+\frac{25}{16}\right) = \frac{1025}{256e^{10}} < 1.$$

Consequently Theorem 6.6 implies that problem (6.15)-(6.16) has at least one solution defined on $[0,1] \times [0,1]$.

6.4. Impulsive Partial Integro-Differential Equations

6.4.1. Introduction

In this section, we prove some results for the following system of impulsive integro-differential equations of fractional order of the form

$$^cD^r_{x_k}u(x,y) = f(x,y,I^r_\theta u(x,y),u(x,y)), \quad \text{for } (x,y) \in J_k,\ k = 0,\dots,m, \tag{6.17}$$

$$u(x_k^+,y) = u(x_k^-,y) + I_k(u(x_k^-,y)), \quad \text{for } y \in [0,b],\ k = 1,\dots,m, \tag{6.18}$$

$$\begin{cases} u(x,0) = \varphi(x), & x \in [0,a], \\ u(0,y) = \psi(y), & y \in [0,b], \\ \varphi(0) = \psi(0), \end{cases} \tag{6.19}$$

where $a,b > 0$, $J_0 := [0,x_1] \times [0,b]$, $J_k := (x_k,x_{k+1}] \times [0,b]$, $k = 1,\dots,m$, $0 = x_0 < x_1 < \cdots < x_m < x_{m+1} = a$, $r = (r_1,r_2)$, $r_1,r_2 \in (0,\infty)$, I^r_θ is the left-sided mixed Riemann-Liouville integral of order r, $^cD^r_\theta$ is the fractional Caputo derivative of order r, $I_k : \mathbb{R}^n \to \mathbb{R}^n$, $k = 1,\dots,m$, $f : J \times \mathbb{R}^n \times \mathbb{R}^n \to \mathbb{R}^n$ are given continuous functions, and $\varphi : [0,a] \to \mathbb{R}^n$, $\psi : [0,b] \to \mathbb{R}^n$ are absolutely continuous functions. Our investigations are conducted with an application of Banach's contraction principle, Schauder's fixed point theorem, and the nonlinear alternative of Leray-Schauder type for the existence and uniqueness of solutions of our problem.

6.4.2. Main Results

Consider the space

$$\begin{aligned} PC = \{ & u : J \to \mathbb{R}^n : u \in C(J_k),\ k = 0,1,\dots,m, \text{ and} \\ & \text{there exist } u(x_k^-,y) \text{ and } u(x_k^+,y),\ k = 1,\dots,m, \\ & \text{with } u(x_k^-,y) = u(x_k,y) \text{ for each } y \in [0,b]\}. \end{aligned}$$

This set is a Banach space with the norm

$$\|u\|_{PC} = \sup_{(x,y)\in J} \|u(x,y)\|.$$

Definition 6.7. *A function $u \in PC$ is said to be a solution of (6.17)-(6.19) if u satisfies equation (6.17) on J_k, $k = 0,\dots,m$ and conditions (6.18),(6.19) are satisfied.*

Now, we shall prove the following theorem concerning the existence and uniqueness of a solution of (6.17)-(6.19). Set

$$f^*(x,y) = f(x,y,0,0).$$

Theorem 6.8. *Assume that the following hypotheses hold:*

(6.12.1) *There exist constants $L_2, L > 0$ such that*

$$\|f(x,x,u,v) - f(x,y,\overline{u},\overline{v})\| \le L_2\|u-\overline{u}\| + L_1\|v-\overline{v}\|,$$

for each $(x,y) \in J$ and $u,v,\overline{u},\overline{v} \in \mathbb{R}^n$;

(6.12.2) *There exists a constant $l > 0$ such that*

$$\|I_k(u) - I_k(\overline{u})\| \le l\|u-\overline{u}\|, \text{ for each } u,\overline{u} \in \mathbb{R}^n,\ k = 1,\dots,m.$$

If

$$2ml + 2L_1L_3 + 2L_2L_3^2 < 1, \tag{6.20}$$

where

$$L_3 := \frac{a^{r_1}b^{r_2}}{\Gamma(1+r_1)\Gamma(1+r_2)},$$

then the problem (6.17)-(6.19) has a unique solution on J.

Proof. We transform the problem (6.17)-(6.19) into a fixed point problem. Define the operator $N : PC \to PC$ by

$$\begin{aligned}(Nu)(x,y) &= \mu(x,y) + \sum_{0<x_k<x} (I_k(u(x_k^-,y)) - I_k(u(x_k^-,0)))\\ &+ \frac{1}{\Gamma(r_1)\Gamma(r_2)} \sum_{0<x_k<x} \int_{x_{k-1}}^{x_k}\int_0^y (x_k-s)^{r_1-1}(y-t)^{r_2-1}\\ &\times f(s,t,I_\theta^r u(s,t),u(s,t))dtds\\ &+ \frac{1}{\Gamma(r_1)\Gamma(r_2)}\int_{x_k}^{x}\int_0^y (x-s)^{r_1-1}(y-t)^{r_2-1} f(s,t,I_\theta^r u(s,t),u(s,t))dtds.\end{aligned} \tag{6.21}$$

It is clear that N maps the space PC into itself. The problem of finding the solutions of the problem (6.17)-(6.19) is reduced to finding the solutions of the operator equation $N(u) = u$.

To this end, we shall verify that the operator N is a contraction map. Let $u,v\in PC$. Using (6.12.1) and (6.12.2), for each $(x,y)\in J$, we have

$$\begin{aligned}
&\|(Nu)(x,y)-(Nv)(x,y)\|\\
\leq\;& \sum_{0<x_k<x}\|I_k(u(x_k^-,y))-I_k(v(x_k^-,y))\|+\sum_{0<x_k<x}\|I_k(u(x_k^-,0))-I_k(v(x_k^-,0))\|\\
+\;& 2\|I_\theta^r(f(x,y,I_\theta^r u(x,y),u(x,y))-f(x,y,I_\theta^r v(x,y),v(x,y)))\|\\
\leq\;& 2ml\|u(x,y)-v(x,y)\|+2I_\theta^r(L_1\|u(x,y)-v(x,y)\|\\
+\;& \frac{L_2}{\Gamma(r_1)\Gamma(r_2)}\int_0^x\int_0^y (x-s)^{r_1-1}(y-t)^{r_2-1}\|u(s,t)-v(s,t)\|dtds\Big)\\
\leq\;& \left(2ml+\frac{2L_1a^{r_1}b^{r_2}}{\Gamma(1+r_1)\Gamma(1+r_2)}+\frac{2L_2a^{2r_1}b^{2r_2}}{\Gamma^2(1+r_1)\Gamma^2(1+r_2)}\right)\|u-v\|_{PC}.
\end{aligned}$$

Hence,

$$\|N(u)-N(v)\|_{PC}\leq(2ml+2L_1L_3+2L_2L_3^2)\|u-v\|_{PC}.$$

From (6.20), it follows that N has a unique fixed point in PC by the Banach contraction principle. The fixed point of N is however a solution of problem (6.17)-(6.19). □

For the existence of solutions for the problem (6.17)-(6.19) and by using Schauder's fixed point theorem [204], we prove the following theorem:

Theorem 6.9. *Assume that hypotheses* (6.12.1), (6.12.2) *and the condition (6.20) hold. Then the problem (6.17)-(6.19) has at least one solution in the space PC.*

Proof. Consider the operator N defined in (6.21). For $u\in PC$ and for each $(x,y)\in J$, we have

$$\begin{aligned}
&\|(Nu)(x,y)\|\leq\|\mu(x,y)\|\\
&+\sum_{0<x_k<x}\|I_k(u(x_k^-,y))-I_k(0)\|+\sum_{0<x_k<x}\|I_k(0)-I_k(u(x_k^-,0))\|\\
&+\frac{1}{\Gamma(r_1)\Gamma(r_2)}\sum_{0<x_k<x}\int_{x_{k-1}}^{x_k}\int_0^y (x_k-s)^{r_1-1}(y-t)^{r_2-1}\|f(s,t,0,0)\|dtds\\
&+\frac{1}{\Gamma(r_1)\Gamma(r_2)}\int_{x_k}^{x}\int_0^y (x-s)^{r_1-1}(y-t)^{r_2-1}\|f(s,t,0,0)\|dtds\\
&+\frac{1}{\Gamma(r_1)\Gamma(r_2)}\sum_{0<x_k<x}\int_{x_{k-1}}^{x_k}\int_0^y (x_k-s)^{r_1-1}(y-t)^{r_2-1}\\
&\times\|f(s,t,I_\theta^r u(s,t),u(s,t))-f(s,t,0,0\|dtds\\
&+\frac{1}{\Gamma(r_1)\Gamma(r_2)}\int_{x_k}^{x}\int_0^y (x-s)^{r_1-1}(y-t)^{r_2-1}\\
&\times\|f(s,t,I_\theta^r u(s,t),u(s,t))-f(s,t,0,0\|dtds\\
\leq\;& \|\mu\|_{PC}+2L_3\|f^*\|_{PC}+(2lm+2L_1L_3+2L_2L_3^2)\|u\|_{PC}.
\end{aligned}$$

Thus,

$$\|N(u)\|_{PC}\leq\|\mu\|_{PC}+2L_3\|f^*\|_{PC}+(2lm+2L_1L_3+2L_2L_3^2)\|u\|_{PC}. \tag{6.22}$$

From (6.20) and (6.22), we infer that N transforms the ball B_η into itself, where

$$B_\eta := \{u \in PC : \|u\|_{PC} \leq \eta\}, \ and \ \ \eta > \frac{\|\mu\|_{PC} + 2L_3\|f^*\|_{PC}}{1 - 2lm - 2L_1L_3 - 2L_2L_3^2}.$$

We shall show that $N : B_\eta \to B_\eta$ satisfies the assumptions of Schauder's fixed point theorem [204]. The proof will be given in several steps.

Step 1: *N is continuous.*
Let $\{u_n\}_{n\in\mathbb{N}}$ be a sequence such that $u_n \to u$ in B_η. Then, for each $(x,y) \in J$, we have

$$\begin{aligned}
& \|(Nu_n)(x,y) - (Nu)(x,y)\| \\
\leq & \sum_{0<x_k<x} \|I_k(u_n(x_k^-,y)) - I_k(u(x_k^-,y))\| \\
& + \sum_{0<x_k<x} \|I_k(u_n(x_k^-,0)) - I_k(u(x_k^-,0))\| \\
& + \frac{1}{\Gamma(r_1)\Gamma(r_2)} \sum_{k=1}^{m} \int_{x_{k-1}}^{x_k} \int_0^y (x_k - s)^{r_1-1}(y-t)^{r_2-1} \\
& \times \|f(s,t,I_\theta^r u_n(s,t), u_n(s,t)) - f(s,t,I_\theta^r u(s,t), u(s,t))\| dtds \\
& + \frac{1}{\Gamma(r_1)\Gamma(r_2)} \int_{x_k}^{x} \int_0^y (x-s)^{r_1-1}(y-t)^{r_2-1} \\
& \times \|f(s,t,I_\theta^r u_n(s,t), u_n(s,t)) - f(s,t,I_\theta^r u(s,t), u(s,t))\| dtds \\
\leq & (2ml + 2L_1L_3 + 2L_2L_3^2)\|u_n - u\|_{PC}.
\end{aligned}$$

Hence,

$$\|N(u_n) - N(u)\|_{PC} \leq (2ml + 2L_1L_3 + 2L_2L_3^2)\|u_n - u\|_{PC}. \tag{6.23}$$

Since $u_n \to u$ as $n \to \infty$, then (6.23) gives

$$\|N(u_n) - N(u)\|_{PC} \to 0 \quad \text{as } n \to \infty.$$

Step 2: *$N(B_\eta)$ is bounded.*
This is clear since $N(B_\eta) \subset B_\eta$ and B_η is bounded.

Step 3: *$N(B_\eta)$ is equicontinuous.*
Let $(x_1,y_1),\ (x_2,y_2) \in J,\ x_1 < x_2,\ y_1 < y_2$ and $u \in B_\eta$. Then

$$\begin{aligned}
& \|N(u)(x_2,y_2) - N(u)(x_1,y_1)\| \\
\leq & \|\mu(x_1,y_1) - \mu(x_2,y_2)\| \\
& + \sum_{0<x_k<x} \|I_k(u(x_k^-,y_2)) - I_k(u(x_k^-,y_1))\| \\
& + \| \frac{2(L_1 + L_2L_3)}{\Gamma(r_1)\Gamma(r_2)} \int_0^{x_1} \int_0^{y_1} [(x_2 - s)^{r_1-1}(y_2 - t)^{r_2-1}
\end{aligned}$$

$$
\begin{aligned}
& -(x_1-s)^{r_1-1}(y_1-t)^{r_2-1}]u(s,t)dtds \\
& +\frac{2(L_1+L_2L_3)}{\Gamma(r_1)\Gamma(r_2)}\int_{x_1}^{x_2}\int_{y_1}^{y_2}(x_2-s)^{r_1-1}(y_2-t)^{r_2-1}u(s,t)dtds \\
& +\frac{2(L_1+L_2L_3)}{\Gamma(r_1)\Gamma(r_2)}\int_{0}^{x_1}\int_{y_1}^{y_2}(x_2-s)^{r_1-1}(y_2-t)^{r_2-1}u(s,t)dtds \\
& +\frac{2(L_1+L_2L_3)}{\Gamma(r_1)\Gamma(r_2)}\int_{x_1}^{x_2}\int_{0}^{y_1}(x_2-s)^{r_1-1}(y_2-t)^{r_2-1}u(s,t)dtds\| \\
\le\ & \|\mu(x_1,y_1)-\mu(x_2,y_2)\| \\
& +\sum_{0<x_k<x}\|I_k(u(x_k^-,y_2))-I_k(u(x_k^-,y_1))\| \\
& +\frac{2\eta(L_1+L_2L_3)}{\Gamma(1+r_1)\Gamma(1+r_2)}[2y_2^{r_2}(x_2-x_1)^{r_1}+2x_2^{r_1}(y_2-y_1)^{r_2} \\
& +x_1^{r_1}y_1^{r_2}-x_2^{r_1}y_2^{r_2}-2(x_2-x_1)^{r_1}(y_2-y_1)^{r_2}].
\end{aligned}
$$

As $x_1 \to x_2$, $y_1 \to y_2$ the right-hand side of the above inequality tends to zero. As a consequence of steps 1 to 3 together with the Arzela-Ascoli theorem, we can conclude that N is continuous and compact. From an application of Schauder's theorem [204], we deduce that N has a fixed point u which is a solution of the problem (6.17)-(6.19). □

Now, we present an existence result for the problem (6.17)-(6.19) based on the nonlinear alternative of Leray-Schauder type.

Theorem 6.10. *Assume that the following hypotheses hold:*

(6.16.1) *There exist positive functions $p, q, d \in C(J)$ such that*

$$\|f(x,y,u,v)\| \le p(x,y)+q(x,y)\|u\|+d(x,y)\|v\|;$$

for all $(x,y) \in J$ and $u, v \in \mathbb{R}^n$,

(6.16.2) *There exist a constant $L > 0$, and a continuous function $\psi^* : [0,\infty) \to (0,\infty)$ such that*

$$\|I_k(u)\| \le \frac{L}{1+\psi^*(\|u\|_{PC})},\quad k=1,\ldots,m,\quad \text{for all } u \in PC.$$

Then problem (6.17)-(6.19) has at least one solution defined on J.

Proof. Consider the operator N defined in (6.21). By the continuity of f and the Arzela-Ascoli theorem, we can easily obtain that N is completely continuous.

A priori bounds. We shall show there exists an open set $U \subseteq PC$ with $u \neq \lambda N(u)$, for $\lambda \in (0,1)$ and $u \in \partial U$. Let $u \in PC$ and $u = \lambda N(u)$ for some $0 < \lambda < 1$, then for each $(x,y) \in J$, we have

$$u(x,y) = \lambda\Bigg(\mu(x,u)+\sum_{0<x_k<x}(I_k(u(x_k^-,y))-I_k(u(x_k^-,0)))$$

$$+\frac{1}{\Gamma(r_1)\Gamma(r_2)}\sum_{k=1}^{m}\int_{x_{k-1}}^{x_k}\int_0^y (x_k-s)^{r_1-1}(y-t)^{r_2-1}$$
$$\times f(s,t,I_\theta^r u(s,t),u(s,t))dtds$$
$$+\frac{1}{\Gamma(r_1)\Gamma(r_2)}\int_{x_k}^{x}\int_0^y (x-s)^{r_1-1}(y-t)^{r_2-1}$$
$$\times f(s,t,I_\theta^r u(s,t),u(s,t))dtds).$$

This implies, by (6.16.1) and (6.16.2), that for each $(x,y)\in J$, we have

$$\begin{aligned}
\|u(x,y)\| &\leq \|\mu(x,y)\|+2mL+\frac{2p^*a^{r_1}b^{r_2}}{\Gamma(1+r_1)\Gamma(1+r_2)}\\
&+ \frac{2q^*a^{r_1}b^{r_2}}{\Gamma(r_1)\Gamma(r_2)\Gamma(1+r_1)\Gamma(1+r_2)}\int_0^x\int_0^y (x-s)^{r_1-1}(y-t)^{r_2-1}\|u(s,t)\|dtds\\
&+ \frac{2d^*}{\Gamma(r_1)\Gamma(r_2)}\int_0^x\int_0^y (x-s)^{r_1-1}(y-t)^{r_2-1}\|u(s,t)\|dtds\\
&\leq \|\mu(x,y)\|+2mL+\frac{2p^*a^{r_1}b^{r_2}}{\Gamma(1+r_1)\Gamma(1+r_2)}\\
&+ \left[\frac{2q^*a^{r_1}b^{r_2}}{\Gamma(r_1)\Gamma(r_2)\Gamma(1+r_1)\Gamma(1+r_2)}+\frac{2d^*}{\Gamma(r_1)\Gamma(r_2)}\right]\\
&\times \int_0^x\int_0^y (x-s)^{r_1-1}(y-t)^{r_2-1}\|u(s,t)\|dtds,
\end{aligned}$$

where $p^*=\|p\|_\infty$, $q^*=\|q\|_\infty$ and $d^*=\|d\|_\infty$. Thus, for each $(x,y)\in J$, we get

$$\begin{aligned}
\|u(x,y)\| &\leq \|\mu\|_{PC}+2mL+\frac{2p^*a^{r_1}b^{r_2}}{\Gamma(1+r_1)\Gamma(1+r_2)}\\
&+ \left[\frac{2q^*a^{r_1}b^{r_2}}{\Gamma(r_1)\Gamma(r_2)\Gamma(1+r_1)\Gamma(1+r_2)}+\frac{2d^*}{\Gamma(r_1)\Gamma(r_2)}\right]\\
&\times \int_0^x\int_0^y (x-s)^{r_1-1}(y-t)^{r_2-1}\|u(s,t)\|dtds\\
&\leq w+c\int_0^x\int_0^y (x-s)^{r_1-1}(y-t)^{r_2-1}\|u(s,t)\|dtds
\end{aligned}$$

where

$$w:=\|\mu\|_{PC}+2mL+\frac{2p^*a^{r_1}b^{r_2}}{\Gamma(1+r_1)\Gamma(1+r_2)}$$

and

$$c:=\frac{2q^*a^{r_1}b^{r_2}}{\Gamma(r_1)\Gamma(r_2)\Gamma(1+r_1)\Gamma(1+r_2)}+\frac{2d^*}{\Gamma(r_1)\Gamma(r_2)}.$$

From Lemma 1.68, there exists $\delta:=\delta(r_1,r_2)>0$ such that, for each $(x,y)\in J$, we obtain

$$\begin{aligned}
\|u\|_{PC} &\leq w\left(1+c\delta\int_0^x\int_0^y (x-s)^{r_1-1}(y-t)^{r_2-1}dtds\right)\\
&\leq w\left(1+\frac{c\delta a^{r_1}b^{r_2}}{r_1r_2}\right):=M_*.
\end{aligned}$$

Set

$$U = \{u \in PC : \|u\|_{PC} < M_* + 1\}.$$

By our choice of U, there is no $u \in \partial U$ such that $u = \lambda N(u)$, for $\lambda \in (0,1)$. As a consequence of Theorem 1.56, we deduce that N has a fixed point u in $\overline{U}$ which is a solution to problem (6.17)-(6.19). □

6.4.3. An Example

As an application and to illustrate our results, we consider the following system of impulsive integro-differential equations of fractional order

$$^{c}D^{r}_{x_k} u(x,y) = \frac{1}{e^{x+y+10}(1+|u(x,y)+|I^{r}_{\theta}u(x,y)|)}, \ (x,y) \in J_k, \ k = 0,\ldots,m, \tag{6.24}$$

$$u(x_k^+,y) = u(x_k^-,y) + \frac{1}{6e^{x+y+10}(1+|u(x_k^-,y)|)}, \ \text{for } y \in [0,1], \ k = 0,\ldots,m, \tag{6.25}$$

$$\varphi(x) = x, \ \psi(y) = y^2, \ x,y \in [0,1], \tag{6.26}$$

where $J_0 = [0,x_1] \times [0,1]$, $J_k = (x_k,x_{k+1}] \times [0,1]$, $k = 0,\ldots,m$, $0 = x_0 < x_1 < \cdots < x_m < x_{m+1} = 1$, $r_1,r_2 \in (0,1]$. Set

$$f(x,y,u,v) = \frac{e^{-(x+y+10)}}{1+|u|+|v|}, \ (x,y) \in [0,1] \times [0,1],$$

and

$$I_k(u(x_k^-,y)) = \frac{1}{6e^{x+y+10}(1+|u(x_k^-,y)|)}, \quad y \in [0,1].$$

We have $\mu(x,y) = x + y^2$ and $f^*(x,y) = e^{-(x+y+10)}$.
For each $u_1,\ u_2,\ v_1,\ v_2 \in \mathbb{R}$, we have

$$|f(x,y,u_1,v_1) - f(x,y,u_2,v_2)| \leq e^{-10}(|u_1 - v_1| + |u_2 - v_2|)$$

and

$$|I_k(u) - I_k(\overline{u})| \leq \frac{1}{6e^{10}}|u - \overline{u}|.$$

Hence conditions (6.12.1) and (6.12.2) are satisfied with $L_1 = L_2 = e^{-10}$ and $l = \frac{1}{6e^{10}}$
We shall show that condition (6.20) holds with $m = 3$ and $a = b = 1$. Indeed, for each $(r_1,r_2) \in (0,1] \times (0,1]$, we get

$$L_3 = \frac{a^{r_1}b^{r_2}}{\Gamma(1+r_1)\Gamma(1+r_2)} = \frac{1}{\Gamma(1+r_1)\Gamma(1+r_2)} < \frac{25}{16},$$

and

$$2ml + 2L_3(L_1 + L_2L_3) \leq \frac{1}{e^{10}} + \frac{25}{8e^{10}}\left(1 + \frac{25}{16}\right) = \frac{1153}{128e^{10}} < 1.$$

Consequently Theorem 6.8 implies that problem (6.24)-(6.26) has at least one solution defined on $[0,1] \times [0,1]$.

6.5. Notes and Remarks

The results of Chapter 6 are taken from Abbas *et al.* [18, 35]. Other results may be found in [186, 187, 188, 365].

Chapter 7

Fractional Order Riemann-Liouville Delay Integro-Differential Equations

7.1. Introduction

In this chapter, we shall present existence results for some classes of Riemann-Liouville integro-differential equations of fractional order with delay by using some fixed point theorems.

7.2. Partial Neutral Functional Integro-Differential Equations with Delay

7.2.1. Introduction

This section deals with the existence of solutions for two systems of neutral integro-differential equations of fractional order with delay. First, we consider the following system of fractional order neutral integro-differential equations with finite delay of the form

$$^{c}D^{r}_{\theta}[u(x,y)-g(x,y,u_{(x,y)})]=f(x,y,I^{r}_{\theta}u(x,y),u_{(x,y)});\ (x,y)\in J, \tag{7.1}$$

$$u(x,y)=\Phi(x,y);\ \text{if}\ (x,y)\in\tilde{J}:=([-\alpha,a]\times[-\beta,b])\backslash((0,a]\times(0,b]), \tag{7.2}$$

$$\begin{cases} u(x,0)=\varphi(x);\ x\in[0,a], \\ u(0,y)=\psi(y);\ y\in[0,b], \\ \varphi(0)=\psi(0), \end{cases} \tag{7.3}$$

where $J:=[0,a]\times[0,b];\ a,b,\alpha,\beta>0,\ \theta=(0,0),\ r=(r_1,r_2),\ r_1,r_2\in(0,\infty),\ I^{r}_{\theta}$ is the left-sided mixed Riemann-Liouville integral of order r, $^{c}D^{r}_{\theta}$ is the fractional Caputo derivative of order r, $f:J\times\mathbb{R}^n\times C\to\mathbb{R}^n,\ g:J\times C\to\mathbb{R}^n$ are given continuous functions, $\phi\in C(\tilde{J}),\ \varphi:[0,a]\to\mathbb{R}^n,\ \psi:[0,b]\to\mathbb{R}^n$ are given absolutely continuous functions with

$\varphi(x) = \phi(x,0)$, $\psi(y) = \phi(0,y)$ for each $x \in [0,a]$, $y \in [0,b]$, and $C := C([-\alpha,0] \times [-\beta,0])$ is the Banach space of continuous functions on $[-\alpha,0] \times [-\beta,0]$, coupled with the norm

$$\|w\|_C = \sup_{(x,y)\in[-\alpha,0]\times[-\beta,0]} \|w(x,y)\|.$$

We denote by $u_{(x,y)}$ the element of C defined by

$$u_{(x,y)}(s,t) = u(x+s,y+t);\ (s,t) \in [-\alpha,0] \times [-\beta,0],$$

here $u_{(x,y)}(\cdot,\cdot)$ represents the history of the state from time $(x-\alpha, y-\beta)$ up to the present time (x,y).

Next, we consider the following system of fractional order neutral integro-differential equations with infinite delay of the form

$$^cD^r_\theta[u(x,y) - g(x,y,u_{(x,y)})] = f(x,y,I^r_\theta u(x,y), u_{(x,y)});\ (x,y) \in J, \tag{7.4}$$

$$u(x,y) = \phi(x,y),\ \text{if } (x,y) \in \tilde{J}' := ((-\infty,a] \times (-\infty,b]) \backslash ((0,a] \times (0,b]), \tag{7.5}$$

$$\begin{cases} u(x,0) = \varphi(x);\ x \in [0,a], \\ u(0,y) = \psi(y);\ y \in [0,b], \\ \varphi(0) = \psi(0), \end{cases} \tag{7.6}$$

where J, φ, ψ are as in problem (7.1)-(7.3) and $\phi \in C(\tilde{J}')$, $f : J \times \mathbb{R}^n \times \mathcal{B} \to \mathbb{R}^n$, $g : J \times \mathcal{B} \to \mathbb{R}^n$ are given continuous functions, and $\mathcal{B}$ is called a phase space that will be specified later.

7.2.2. Existence Results with Finite Delay

Let $E := C([-\alpha,a] \times [-\beta,b])$, be the Banach space with the norm

$$\|w\|_E = \sup_{(x,y)\in[-\alpha,a]\times[-\beta,b]} \|w(x,y)\|.$$

Let us start by defining what we mean by a solution of of the problem (7.1)-(7.3).

Definition 7.1. *A function $u \in E$ is said to be a solution of the problem (7.1)-(7.3) if u satisfies equations (7.1),(7.3) on J and the condition (7.5) on $\tilde{J}$.*

Further, we present conditions for the existence of a solution of problem (7.1)-(7.3).

(7.1.1) There exist nonnegative functions $p, q, d \in C(J)$ such that

$$\|f(x,y,u,v)\| \le p(x,y) + q(x,y)\|u\| + d(x,y)\|v\|_C,$$

for all $(x,y) \in J$, $u \in \mathbb{R}^n$, and $v \in C$.

(7.1.2) For any bounded set B in E, the set $\{(x,y) \to g(x,y,u_{(x,y)}) : u \in B\}$ is equicontinuous in E, and there exist constants $L_1,\ L_2 \ge 0$ such that

$$\|g(x,y,u)\| \le L_1 + L_2\|u\|_C;\ (x,y) \in J \text{ and } u \in C.$$

Set

$$L = \frac{a^{r_1} b^{r_2}}{\Gamma(1+r_1)\Gamma(1+r_2)},\ p^* = \|p\|_J,\ q^* = \|q\|_J \text{ and } d^* = \|d\|_J.$$

Theorem 7.2. *Assume that the hypotheses* (7.1.1) *and* (7.1.2) *hold. Then, if*

$$4L_2 + q^* L^2 + d^* L < 1, \tag{7.7}$$

problem (7.1)-(7.3) has at least one solution $u \in AC([-\alpha, a] \times [-\beta, b])$.

Proof. Transform the problem (7.1)-(7.3) into a fixed point problem. Define the operator $N : E \to E$ by,

$$(Nu)(x,y) = \begin{cases} \Phi(x,y);\ (x,y) \in \tilde{J}, \\ \mu(x,y) + g(x,y,u_{(x,y)}) + g(0,0,u_{(0,0)}) \\ -g(x,0,u_{(x,0)}) - g(0,y,u_{(0,y)}) \\ +I_\theta^r f(x,y,I_\theta^r u(x,y),u_{(x,y)});\ (x,y) \in J. \end{cases} \tag{7.8}$$

It is clear that N maps E into itself. The problem of finding the solutions of the problem (7.1)-(7.3) is reduced to finding the solutions of the operator equation $N(u) = u$. We shall show that the operator N satisfies all the conditions of Theorem 1.56. The proof will be given in two steps.

Step 1: *N is continuous and completely continuous.*
Using (7.1.2) we deduce that g is a complete continuous operator from E to $\mathbb{R}^n$, so it suffices to show that the operator $N_1 : E \to E$ defined by,

$$(N_1 u)(x,y) = \begin{cases} \Phi(x,y);\ (x,y) \in \tilde{J}, \\ \mu(x,y) + I_\theta^r f(x,y,I_\theta^r u(x,y),u_{(x,y)});\ (x,y) \in J. \end{cases} \tag{7.9}$$

is continuous and completely continuous. The proof will be given in several claims.

Claim 1: N_1 *is continuous.*
Let $\{u_n\}_{n\in\mathbb{N}}$ be a sequence such that $u_n \to u$ in E. Then, for each $(x,y) \in [-\alpha, a] \times [-\beta, b]$, we have

$$\begin{aligned} & \|(N_1 u_n)(x,y) - (N_1 u)(x,y)\| \\ \leq\ & \|I_\theta^r \left(f(x,y,I_\theta^r u_n(x,y),u_{n(x,y)}) - f(x,y,I_\theta^r u(x,y),u_{(x,y)}) \right)\| \\ \leq\ & \frac{1}{\Gamma(r_1)\Gamma(r_2)} \int_0^x \int_0^y (x-s)^{r_1-1}(y-t)^{r_2-1} \\ \times\ & \|f(s,t,I_\theta^r u_n(s,t),u_{n(s,t)}) - f(s,t,I_\theta^r u(s,t),u_{(s,t)})\| dt ds. \end{aligned}$$

Hence, for each $(x,y) \in J$, we get

$$\|N_1(u_n) - N_1(u)\|_E \leq \frac{1}{\Gamma(r_1)\Gamma(r_2)} \int_0^a \int_0^b (x-s)^{r_1-1}(y-t)^{r_2-1}$$

$$\times \|f(\cdot,\cdot,I_\theta^r u_n(\cdot,\cdot),u_{n(\cdot,\cdot)}) - f(\cdot,\cdot,I_\theta^r u(\cdot,\cdot),u_{(\cdot,\cdot)})\|_E\, dt ds.$$

Since $u_n \to u$ as $n \to \infty$, and f, I_θ^r are continuous, then

$$\|N_1(u_n) - N_1(u)\|_E \to 0 \quad \text{as } n \to \infty.$$

Claim 2: *N_1 maps bounded sets into bounded sets in E.*
Indeed, it is enough to show that, for any $\eta > 0$, there exists a positive constant $\ell > 0$ such that, if $\|u\|_E \leq \eta$, we have that $\|N_1(u)\|_E \leq \ell$. By (7.1.1) we have that, for each $(x,y) \in J$ and $\|u\|_E \leq \eta$,

$$\begin{aligned}
\|(N_1 u)(x,y)\| &\leq \|\mu(x,y)\| + \frac{p^* a^{r_1} b^{r_2}}{\Gamma(1+r_1)\Gamma(1+r_2)} \\
&+ \frac{q^* a^{r_1} b^{r_2}}{\Gamma(r_1)\Gamma(r_2)\Gamma(1+r_1)\Gamma(1+r_2)} \\
&\times \int_0^x \int_0^y (x-s)^{r_1-1}(y-t)^{r_2-1}\|u(s,t)\| dt ds \\
&+ \frac{d^*}{\Gamma(r_1)\Gamma(r_2)} \int_0^x \int_0^y (x-s)^{r_1-1}(y-t)^{r_2-1}\|u_{(s,t)}\|_C dt ds \\
&\leq \|\mu(x,y)\| + \frac{p^* a^{r_1} b^{r_2}}{\Gamma(1+r_1)\Gamma(1+r_2)} \\
&+ \frac{q^* a^{r_1} b^{r_2}}{\Gamma(r_1)\Gamma(r_2)\Gamma(1+r_1)\Gamma(1+r_2)} \\
&\times \int_0^x \int_0^y (x-s)^{r_1-1}(y-t)^{r_2-1}\|u\|_J dt ds \\
&+ \frac{d^*}{\Gamma(r_1)\Gamma(r_2)} \int_0^x \int_0^y (x-s)^{r_1-1}(y-t)^{r_2-1}\|u\|_E dt ds \\
&\leq \|\mu\|_J + p^* L + (q^* L^2 + d^* L)\, \|u\|_E \\
&\leq \|\mu\|_J + p^* L + (q^* L + d^*) L\eta. := \ell_1.
\end{aligned}$$

Thus,

$$\|N_1(u)\|_E \leq \max\{\ell_1, \|\Phi\|_C\} := \ell.$$

Claim 3: *N_1 maps bounded sets in E into equicontinuous sets in E.*
Let $(x_1,y_1),(x_2,y_2) \in J$, $x_1 < x_2, y_1 < y_2$, $\eta > 0$ and let $u \in E$ be such that $\|u\|_E \leq \eta$. Then

$$\begin{aligned}
&\|(N_1 u)(x_2,y_2) - (N_1 u)(x_1,y_1)\| \\
\leq\ &\|\mu(x_2,y_2) - \mu(x_1,y_1)\| \\
&+ \frac{p^* + q^* L\eta + d^*\eta}{\Gamma(r_1)\Gamma(r_2)} \Bigg[\int_0^{x_1} \int_0^{y_1} [(x_2-s)^{r_1-1}(y_2-t)^{r_2-1} \\
&-(x_1-s)^{r_1-1}(y_1-t)^{r_2-1}] dt ds \\
&+ \int_{x_1}^{x_2} \int_{y_1}^{y_2} (x_2-s)^{r_1-1}(y_2-t)^{r_2-1} dt ds \\
&+ \int_0^{x_1} \int_{y_1}^{y_2} (x_2-s)^{r_1-1}(y_2-t)^{r_2-1} dt ds \\
&+ \int_{x_1}^{x_2} \int_0^{y_1} (x_2-s)^{r_1-1}(y_2-t)^{r_2-1} dt ds]
\end{aligned}$$

$$= \|\mu(x_1,y_1)-\mu(x_2,y_2)\| + \frac{p^*+q^*L\eta+d^*\eta}{\Gamma(1+r_1)\Gamma(1+r_2)}$$
$$\times[2y_2^{r_2}(x_2-x_1)^{r_1}+2x_2^{r_1}(y_2-y_1)^{r_2}+x_1^{r_1}y_1^{r_2}-x_2^{r_1}y_2^{r_2}-2(x_2-x_1)^{r_1}(y_2-y_1)^{r_2}].$$

As $x_1 \to x_2$, $y_1 \to y_2$, the right-hand side of the above inequality tends to zero with the same rate of convergence for all $u \in E$, with $\|u\|_E \leq \eta$. The equicontinuity for the cases $x_1 < x_2 < 0, y_1 < y_2 < 0$ and $x_1 \leq 0 \leq x_2, y_1 \leq 0 \leq y_2$ is obvious. As a consequence of claims 1 to 3 together with the Arzela-Ascoli theorem, we can conclude that N_1 is continuous and completely continuous.

Step 2: *A priori bounds.*

We shall show that there exists an open set $U \subseteq E$ with $u \neq \lambda N(u)$, for all $\lambda \in (0,1)$ and all $u \in \partial U$.

Let $u \in E$ be such that $u = \lambda N(u)$ for some $0 < \lambda < 1$. Thus, for each $(x,y) \in J$, we have

$$\begin{aligned} u(x,y) &= \lambda(\mu(x,y)+g(x,y,u_{(x,y)})+g(0,0,u_{(0,0)}) \\ &- g(x,0,u_{(x,0)})-g(0,y,u_{(0,y)})) \\ &+ \lambda I_\theta^r f(x,y,I_\theta^r u(x,y),u_{(x,y)}). \end{aligned}$$

Then, for $(x,y) \in J$, we have

$$\begin{aligned} & \|u(x,y)\| \leq \|\mu(x,y)\| \\ &+ 4L_1+L_2(\|u_{(x,y)}\|_C+\|u_{(x,0)}\|_C+\|u_{(0,y)}\|_C+\|u_{(0,0)}\|_C) \\ &+ p^*L+\frac{q^*L}{\Gamma(r_1)\Gamma(r_2)}\int_0^x\int_0^y (x-s)^{r_1-1}(y-t)^{r_2-1}\|u(s,t)\|dtds \\ &+ \frac{d^*}{\Gamma(r_1)\Gamma(r_2)}\int_0^x\int_0^y (x-s)^{r_1-1}(y-t)^{r_2-1}\|u_{(s,t)}\|_C dtds \\ &\leq \|\mu(x,y)\|+4L_1+4L_2\|u\|_E+p^*L \\ &+ \frac{q^*L}{\Gamma(r_1)\Gamma(r_2)}\int_0^x\int_0^y (x-s)^{r_1-1}(y-t)^{r_2-1}\|u\|_J dtds \\ &+ \frac{d^*}{\Gamma(r_1)\Gamma(r_2)}\int_0^x\int_0^y (x-s)^{r_1-1}(y-t)^{r_2-1}\|u\|_E dtds \\ &\leq \|\mu\|_J+4L_1+p^*L+\left(4L_2+q^*L^2+d^*L\right)\|u\|_E. \end{aligned}$$

It is obvious that

$$\|u\|_E = \max\{\|u\|_{\tilde{J}}, \|u\|_J\} \leq \max\{\|\phi\|_{\tilde{J}}, \|u\|_J\}.$$

As consequence, if $\|u\|_J \leq \|\phi\|_{\tilde{J}}$, then $\|u\|_E \leq \|\phi\|_{\tilde{J}}$.

On the contrary, when $\|u\|_J > \|\phi\|_{\tilde{J}}$, we have that $\|u\|_E = \|u\|_J$. So, from the previous inequalities, and condition (7.7), we arrive at

$$\|u\|_E \leq \frac{\|\mu\|_J+4L_1+p^*L}{1-4L_2-q^*L^2-d^*L} := M.$$

Thus,

$$\|u\|_E \leq \max\{\|\phi\|_{\tilde{J}}, M\} := M^*.$$

Set

$$U = \{u \in E : \|u\|_E < M^* + 1\}.$$

By our choice of U, there is no $u \in \partial U$ such that $u = \lambda N(u)$, for $\lambda \in (0,1)$.
As a consequence of steps 1 and 2 together with Theorem 1.56, we deduce that N has a fixed point u in $\overline{U}$ which is a solution to problem (7.1)-(7.3). □

7.2.3. The Phase Space $\mathcal{B}$

The notation of the phase space $\mathcal{B}$ plays an important role in the study of both qualitative and quantitative theory for functional differential equations. A usual choice is a semi-normed space satisfying suitable axioms, which was introduced by Hale and Kato (see [207]). For further applications see for instance the books [208, 226, 258] and their references. For any $(x,y) \in J$ denote $\mathcal{E}_{(x,y)} := ([0,x] \times \{0\}) \cup (\{0\} \times [0,y])$, furthermore in case $x = a,\ y = b$ we write simply $\mathcal{E}$. Consider the space $(\mathcal{B}, \|(\cdot,\cdot)\|_{\mathcal{B}})$ a semi-normed linear space of functions mapping $(-\infty,0] \times (-\infty,0]$ into $\mathbb{R}^n$, and satisfying the following fundamental axioms which were adapted from those introduced by Hale and Kato for ordinary differential functional equations:

(A_1) If $z : (-\infty,a] \times (-\infty,b] \to \mathbb{R}^n$ is a continuous function on J and $z_{(x,y)} \in \mathcal{B}$, for all $(x,y) \in \mathcal{E}$, then there are constants $H,K,M > 0$ such that for any $(x,y) \in J$ the following conditions hold:

(i) $z_{(x,y)}$ is in $\mathcal{B}$;

(ii) $\|z(x,y)\| \le H\|z_{(x,y)}\|_{\mathcal{B}}$,

(iii) $\|z_{(x,y)}\|_{\mathcal{B}} \le K \sup_{(s,t)\in[0,x]\times[0,y]} \|z(s,t)\| + M \sup_{(s,t)\in\mathcal{E}_{(x,y)}} \|z_{(s,t)}\|_{\mathcal{B}}$,

(A_2) For the function $z(\cdot,\cdot)$ in (A_1), $z_{(x,y)}$ is a $\mathcal{B}$-valued continuous function on J.

(A_3) The space $\mathcal{B}$ is complete.

Now, we present some examples of phase spaces [143, 144].

Example 7.3. *Let $\mathcal{B}$ be the set of all functions $\phi : (-\infty,0] \times (-\infty,0] \to \mathbb{R}^n$ which are continuous on $[-\alpha,0] \times [-\beta,0]$, $\alpha,\beta \ge 0$, with the semi-norm*

$$\|\phi\|_{\mathcal{B}} = \sup_{(s,t)\in[-\alpha,0]\times[-\beta,0]} \|\phi(s,t)\|.$$

Then we have $H = K = M = 1$. The quotient space $\widehat{\mathcal{B}} = \mathcal{B}/\|\cdot\|_{\mathcal{B}}$ is isometric to the space C of all continuous functions from $[-\alpha,0] \times [-\beta,0]$ into $\mathbb{R}^n$ with the supremum norm, this means that partial differential functional equations with finite delay are included in our axiomatic model.

Example 7.4. *Let $\gamma \in \mathbb{R}$ and let C_γ be the set of all continuous functions $\phi : (-\infty,0] \times (-\infty,0] \to \mathbb{R}^n$ for which a limit $\lim_{\|(s,t)\|\to\infty} e^{\gamma(s+t)}\phi(s,t)$ exists, with the norm*

$$\|\phi\|_{C_\gamma} = \sup_{(s,t)\in(-\infty,0]\times(-\infty,0]} e^{\gamma(s+t)}\|\phi(s,t)\|.$$

Then we have $H = 1$ *and* $K = M = \max\{e^{-\gamma(a+b)}, 1\}$.

Example 7.5. *Let* $\alpha, \beta, \gamma \geq 0$ *and let*

$$\|\phi\|_{CL_\gamma} = \sup_{(s,t)\in[-\alpha,0]\times[-\beta,0]} \|\phi(s,t)\| + \int_{-\infty}^{0}\int_{-\infty}^{0} e^{\gamma(s+t)}\|\phi(s,t)\|dtds.$$

be the seminorm for the space CL_γ *of all functions* $\phi : (-\infty,0]\times(-\infty,0] \to \mathbb{R}^n$ *which are continuous on* $[-\alpha,0]\times[-\beta,0]$ *measurable on* $((-\infty,-\alpha]\times(-\infty,0])\cup((-\infty,0]\times(-\infty,-\beta])$, *and such that* $\|\phi\|_{CL_\gamma} < \infty$. *Then*

$$H = 1,\ K = \int_{-\alpha}^{0}\int_{-\beta}^{0} e^{\gamma(s+t)}dtds,\ M = 2.$$

7.2.4. Existence Results with Infinite Delay

Set

$$\Omega := \{u : (-\infty,a]\times(-\infty,b] \to \mathbb{R}^n :\ u_{(x,y)} \in \mathcal{B} \text{ for } (x,y)\in\mathcal{E} \text{ and } u_{|_J} \in C(J)\}.$$

Let us start by defining what we mean by a solution of of the problem (7.4)-(7.6).

Definition 7.6. *A function* $u \in \Omega$ *is said to be a solution of (7.4)-(7.6) if* u *satisfies equations (7.4) and (7.6) on* J *and the condition (7.5) on* $\tilde{J}'$.

Now, we present conditions for the existence of a solution of problem (7.4)-(7.6).

(7.7.1) There exist nonnegative functions $\bar{p}, \bar{q}, \bar{d} \in C(J)$ such that

$$\|f(x,y,u,v)\| \leq \bar{p}(x,y) + \bar{q}(x,y)\|u\| + \bar{d}(x,y)\|v\|_{\mathcal{B}},$$

for all $(x,y) \in J,\ u \in \mathbb{R}^n$, and $v \in \mathcal{B}$.

(7.7.2) For any bounded set B in Ω, the set $\{(x,y) \to g(x,y,u_{(x,y)}) : u \in B\}$ is equicontinuous in Ω, and there exist constants $\bar{L}_1,\ \bar{L}_2 \geq 0$ such that

$$\|g(x,y,u)\| \leq \bar{L}_1 + \bar{L}_2\|u\|_{\mathcal{B}};\ (x,y)\in J \text{ and } u \in \mathcal{B}.$$

Set

$$\bar{p}^* = \|\bar{p}\|_J,\ \bar{q}^* = \|\bar{q}\|_J \text{ and } \bar{d}^* = \|\bar{d}\|_J.$$

Theorem 7.7. *Assume that the hypotheses* (7.7.1) *and* (7.7.2) *hold. If*

$$4\bar{L}_2 + \bar{q}^* L^2 + \bar{d}^* L < 1, \tag{7.10}$$

then problem (7.4)-(7.6) has at least one solution on $(-\infty,a]\times(-\infty,b]$.

Proof. Transform the problem (7.4)-(7.6) into a fixed point problem. Let $u \in \Omega$ and define the operator $\bar{N} : \Omega \to \Omega$ by,

$$(\bar{N}u)(x,y) = \begin{cases} \Phi(x,y); \ (x,y) \in \tilde{J}', \\ \mu(x,y) + g(x,y,u_{(x,y)}) + g(0,0,u_{(0,0)}) \\ -g(x,0,u_{(x,0)}) - g(0,y,u_{(0,y)}) \\ +I_\theta^r f(x,y,I_\theta^r u(x,y),u_{(x,y)}); \ (x,y) \in J. \end{cases} \tag{7.11}$$

As in Theorem 7.2, we can easily see that $\bar{N}$ maps Ω into itself.
Let $v(\cdot,\cdot) : (-\infty,a] \times (-\infty,b] \to \mathbb{R}^n$ be a function defined by,

$$v(x,y) = \begin{cases} \phi(x,y), & (x,y) \in \tilde{J}', \\ \mu(x,y), & (x,y) \in J. \end{cases}$$

Then $v_{(x,y)} = \phi$ for all $(x,y) \in \mathcal{E}$. For each $w \in C(J)$ with $w(x,y) = 0$ for each $(x,y) \in \mathcal{E}$ we denote by $\overline{w}$ the function defined by

$$\overline{w}(x,y) = \begin{cases} 0, & (x,y) \in \tilde{J}', \\ w(x,y) & (x,y) \in J. \end{cases}$$

If $u(\cdot,\cdot)$ satisfies the integral equation, $u(x,y) = (\bar{N}u)(x,y); \ (x,y) \in [-\alpha,a] \times [-\beta,b]$, we can decompose $u(\cdot,\cdot)$ as $u(x,y) = \overline{w}(x,y) + v(x,y); \ (x,y) \in [-\alpha,a] \times [-\beta,b]$, which implies $u_{(x,y)} = \overline{w}_{(x,y)} + v_{(x,y)}$, for every $(x,y) \in J$, and the function $w(\cdot,\cdot)$ satisfies

$$\begin{aligned} w(x,y) &= g(x,y,u_{(x,y)}) + g(0,0,\overline{w}_{(0,0)} + v_{(0,0)}) \\ &- g(x,0,\overline{w}_{(x,0)} + v_{(x,0)}) - g(0,y,\overline{w}_{(0,y)} + v_{(0,y)}) \\ &+ I_\theta^r f(x,y,I_\theta^r(\overline{w}(x,y) + v(x,y)),\overline{w}_{(x,y)} + v_{(x,y)}). \end{aligned}$$

Set

$$C_0 = \{w \in C(J) : \ w(x,y) = 0 \ \text{ for } (x,y) \in \mathcal{E}\},$$

and let $\|\cdot\|_{(a,b)}$ be the norm in C_0 defined by

$$\|w\|_{(a,b)} = \sup_{(x,y)\in\mathcal{E}} \|w_{(x,y)}\|_{\mathcal{B}} + \sup_{(x,y)\in J} \|w(x,y)\| = \sup_{(x,y)\in J} \|w(x,y)\|, \ w \in C_0.$$

C_0 is a Banach space with norm $\|\cdot\|_{(a,b)}$. Note that $u \in \Omega$ if and only if $w \in C_0$. Let the operator $P : \ C_0 \to C_0$ be defined by

$$\begin{aligned} (Pw)(x,y) &= g(x,y,u_{(x,y)}) + g(0,0,\overline{w}_{(0,0)} + v_{(0,0)}) \\ &- g(x,0,\overline{w}_{(x,0)} + v_{(x,0)}) - g(0,y,\overline{w}_{(0,y)} + v_{(0,y)}) \\ &+ \frac{1}{\Gamma(r_1)\Gamma(r_2)} \int_0^x \int_0^y (x-s)^{r_1-1}(y-t)^{r_2-1} \\ &\times f(s,t,I_\theta^r(\overline{w}(s,t) + v(s,t)),\overline{w}_{(s,t)} + v_{(s,t)}, g(s,t))dtds. \end{aligned} \tag{7.12}$$

Then the operator $\bar{N}$ has a fixed point in Ω if and only if P has a fixed point in C_0. As in the proof of the Theorem 7.2, we can show that the operator P satisfies all the conditions of Theorem 1.56. Indeed, to prove that P is continuous and completely continuous and by using (7.7.2) it suffices to show that the operator $\bar{P}: F \to F$ defined by,

$$\begin{aligned}(\bar{P}w)(x,y) &= \frac{1}{\Gamma(r_1)\Gamma(r_2)}\int_0^x\int_0^y (x-s)^{r_1-1}(y-t)^{r_2-1} \\ &\times f(s,t,I_\theta^r(\overline{w}(s,t)+v(s,t)),\overline{w}_{(s,t)}+v_{(s,t)},g(s,t))dtds. \end{aligned} \tag{7.13}$$

is continuous and completely continuous. Also, we can show that there exists an open set $U' \subseteq F$ with $u \neq \lambda P(u)$, for $\lambda \in (0,1)$ and $u \in \partial U'$. Consequently, by Theorem 1.56, we deduce that $\bar{N}$ has a fixed point u in $\overline{U'}$ which is a solution to problem (7.4)-(7.6). □

7.2.5. An Example

Consider the following neutral integro-differential equations of fractional order

$$\begin{aligned} &{}^cD_\theta^r\left(u(x,y) - \frac{1+2|u(x-1,y-2)|}{4e^{x+y+10}}\right) \\ &= \frac{3+(x+y)|u(x-1,y-2)|+xy|I_\theta^r u(x,y)|}{e^{10}(1+|u(x-1,y-2)|+|I_\theta^r u(x,y)|)};\ (x,y)\in[0,1]\times[0,1], \end{aligned} \tag{7.14}$$

$$u(x,y) = x+y^2;\ \text{if } (x,y)\in\tilde{J} := ([-1,1]\times[-2,1])\backslash((0,1]\times(0,1]), \tag{7.15}$$

$$\varphi(x) = x,\ \psi(y) = y^2;\ x,y\in[0,1]. \tag{7.16}$$

Set

$$f(x,y,u,v) = \frac{3+(x+y)|u|+xy|v|}{e^{10}(1+|u|+|v|)};\ (x,y)\in[0,1]\times[0,1],$$

and

$$g(x,y,u) = \frac{1+2|u|}{4e^{x+y+10}};\ (x,y)\in[0,1]\times[0,1].$$

We have $\mu(x,y) = x+y^2;\ (x,y)\in[0,1]\times[0,1]$. For each $u,\ v\in\mathbb{R}$ and $(x,y)\in[0,1]\times[0,1]$, we have

$$|f(x,y,u,v)| \le e^{-10}(3+(x+y)|u|+xy|v|),$$

and

$$|g(x,y,u)| \le \frac{1}{4}e^{-10}(1+2|u|).$$

Hence condition (7.1.1) is satisfied with $p^* = 3e^{-10}, q^* = 2e^{-10},\ d^* = e^{-10}$. Also, condition (7.1.2) is satisfied with $L_1 = \frac{1}{4}e^{-10}$ and $L_2 = \frac{1}{2}e^{-10}$. We shall show that the condition (7.10) holds for each $(r_1,r_2)\in(0,1]\times(0,1]$, with $a=b=1$. Indeed, $\Gamma(r_i) > 0.7,\ \Gamma(1+r_i) > 0.7;\ i=1,2$, and $L = \frac{1}{\Gamma(1+r_1)\Gamma(1+r_2)} < 2.1$. Then

$$4L_2 + q^*L^2 + d^*L = 2e^{-10} + e^{-10}(2L^2+L) < \frac{12.92}{e^{10}} < 1.$$

Consequently Theorem 7.2 implies that problem (7.14)-(7.16) has at least one solution defined on $[-1,1]\times[-2,1]$.

7.3. Integro-Differential Equations with Multiple Time Delay

7.3.1. Introduction

This section concerned with the existence and uniqueness of solutions for the following fractional order integro-differential equations for the system

$$^cD^r_\theta\left(u(x,y)-\sum_{i=1}^m g_i(x,y)u(x-\xi_i,y-\mu_i)\right)$$

$$= f(x,y,I^r_\theta u(x,y),u(x,y)); \text{ if } (x,y)\in J:=[0,a]\times[0,b], \quad (7.17)$$

$$u(x,y)=\Phi(x,y); \text{ if } (x,y)\in\tilde{J}:=[-\xi,a]\times[-\mu,b]\backslash(0,a]\times(0,b], \quad (7.18)$$

where $a,b>0$, $\theta=(0,0)$, $\xi_i,\mu_i\geq 0$; $i=1\dots,m$, $\xi=\max_{i=1\dots,m}\{\xi_i\}$, $\mu=\max_{i=1\dots,m}\{\mu_i\}$, I^r_θ is the left-sided mixed Riemann-Liouville integral of order $r=(r_1,r_2)\in(0,\infty)\times(0,\infty)$, $^cD^r_\theta$ is the fractional Caputo derivative of order r, $f:J\times\mathbb{R}^n\times\mathbb{R}^n\to\mathbb{R}^n$, $g_i:J\to\mathbb{R}$; $i=1\dots m$ are given continuous functions, and $\Phi:\tilde{J}\to\mathbb{R}^n$ is a given continuous function such that

$$\Phi(x,0)=\sum_{i=1}^m g_i(x,0)\Phi(x-\xi_i,-\mu_i);\ x\in[0,a],$$

and

$$\Phi(0,y)=\sum_{i=1}^m g_i(0,y)\Phi(-\xi_i,y-\mu_i);\ y\in[0,b].$$

We present three results for the problem (7.17)-(7.18), the first one is based on Schauder's fixed point theorem (Theorem 7.10), the second one is a uniqueness of the solution by using the Banach fixed point theorem (Theorem 7.11) and the last one on the nonlinear alternative of Leray-Schauder type (Theorem 7.12).

7.3.2. Main Results

$C:=C([-\xi,a]\times[-\mu,b])$ is a Banach space endowed with the norm

$$\|w\|_C=\sup_{(x,y)\in[-\xi,a]\times[-\mu,b]}\|w(x,y)\|.$$

Definition 7.8. *A function $u\in C$ is said to be a solution of (7.17)-(7.18) if u satisfies equation (7.17) on J and condition (7.18) on $\tilde{J}$.*

Lemma 7.9. *A function $u\in C$ is a solution of problem (7.17)-(7.18) if and only if $u(x,y)$ satisfies*

$$u(x,y)=\sum_{i=1}^m g_i(x,y)u(x-\xi_i,y-\mu_i)+I^r_\theta f(x,y,I^r_\theta u(x,y),u(x,y)),$$

on J and $u(x,y)=\Phi(x,y)$ on $\tilde{J}$.

Set

$$B = \max_{i=1\dots m}\{\sup_{(x,y)\in J}|g_i(x,y)|\}.$$

Theorem 7.10. *Assume*

(7.12.1) *There exists a positive function $h\in C(J)$ such that*

$$\|f(x,y,u,v)\|\le h(x,y),\ \textit{for all } (x,y)\in J \textit{ and } u,v\in\mathbb{R}^n.$$

If $mB<1$, then problem (7.17)-(7.18) has at least one solution u on $[-\xi,a]\times[-\mu,b]$.

Proof. Transform problem (7.17)-(7.18) into a fixed point problem. Consider the operator $N: C\to C$ defined by,

$$(Nu)(x,y)=\begin{cases}\Phi(x,y); & (x,y)\in\tilde{J},\\ \sum_{i=1}^{m} g_i(x,y)u(x-\xi_i,y-\mu_i) & \\ +I_\theta^r f(x,y,I_\theta^r u(x,y),u(x,y)); & (x,y)\in J.\end{cases} \tag{7.19}$$

From Lemma 7.9, the problem of finding the solutions of problem (7.17)-(7.18) is reduced to finding the solutions of the operator equation $N(u)=u$. Let

$$R\ge\frac{a^{r_1}b^{r_2}h^*}{(1-mB)\Gamma(1+r_1)\Gamma(1+r_2)},$$

$h^*=\|h\|_\infty$ and consider the set

$$B_R=\{u\in C:\|u\|_C\le R\}.$$

It is clear that B_R is a closed bounded and convex subset of C.
For every $u\in B_R$ and $(x,y)\in J$ we obtain by (7.12.1) that

$$\begin{aligned}\|(Nu)(x,y)\| &\le \sum_{i=1}^{m}\|g_i(x,y)\|\|u(x-\xi_i,y-\mu_i)\|\\ &+\frac{1}{\Gamma(r_1)\Gamma(r_2)}\int_0^x\int_0^y (x-s)^{r_1-1}(y-t)^{r_2-1}\|f(s,t,I_\theta^r u(s,t),u(s,t))\|dtds\\ &\le mB\|u\|_C+\frac{1}{\Gamma(r_1)\Gamma(r_2)}\int_0^x\int_0^y (x-s)^{r_1-1}(y-t)^{r_2-1}h(s,t)dtds\\ &\le mB\|u\|_C+h^*\frac{a^{r_1}b^{r_2}}{\Gamma(1+r_1)\Gamma(1+r_2)}\\ &\le mBR+(1-mB)R=R.\end{aligned}$$

On the other hand, for every $u\in B_R$ and $(x,y)\in\tilde{J}$, we obtain

$$\|(Nu)(x,y)\|=\|\Phi(x,y)\|\le R.$$

So we obtain that

$$\|N(u)\|_C\le R.$$

That is, $N(B_R)\subseteq B_R$. Since f is bounded on B_R, thus $N(B_R)$ is equicontinuous and the Schauder fixed point theorem shows that N has at least one fixed point $u^*\in B_R$ which is corresponding to the solution of (7.17)-(7.18). □

For the uniqueness we prove the following Theorem

Theorem 7.11. *Assume that following hypothesis hold*

$(7.14.1)$ *There exists a positive function* $l \in C(J)$ *such that*

$$\|f(x,y,u_1,u_2) - f(x,y,v_1,v_2)\| \le l_1(x,y)\|u_1 - v_1\| + l_2(x,y)\|u_2 - v_2\|,$$

for each $(x,y) \in J$ *and* $u_1,u_2,v_1,v_2 \in \mathbb{R}^n$.

If

$$mB + \frac{a^{r_1}b^{r_2}(a^{r_1}b^{r_2}l_1^* + r_1r_2l_2^*)}{r_1r_2\Gamma(1+r_1)\Gamma(1+r_2)} < 1, \tag{7.20}$$

where $l_k^* = \|l_k\|_\infty$; $k = 1,2$, *then problem (7.17)-(7.18) has a unique solution on* $[-\xi, a] \times [-\mu, b]$.

Proof. Consider the operator N defined in (7.19). Then by $(7.14.1)$, for every $u,v \in C$ and $(x,y) \in J$ we have

$$\begin{aligned}
\|(Nu)(x,y) - (Nv)(x,y)\| &\le \sum_{i=1}^m |g_i(x,y)| \|u(x-\xi_i, y-\mu_i) - v(x-\xi_i, y)\| \\
&+ \frac{1}{\Gamma(r_1)\Gamma(r_2)} \int_0^x \int_0^y (x-s)^{r_1-1}(y-t)^{r_2-1} \\
&\times \|f(s,t,I_\theta^r u(s,t), u(s,t)) - f(s,t,I_\theta^r v(s,t), v(s,t))\| dtds \\
&\le mB\|u-v\|_\infty \\
&+ \frac{1}{\Gamma(r_1)\Gamma(r_2)} \int_0^x \int_0^y (x-s)^{r_1-1}(y-t)^{r_2-1} \\
&\times (l_1(s,t)\|I_\theta^r(u-v)\|_C + l_2(s,t)\|u-v\|_C)\, dtds \\
&\le mB\|u-v\|_\infty + l_1^* \frac{a^{2r_1}b^{2r_2}}{r_1r_2\Gamma(1+r_1)\Gamma(1+r_2)} \|u-v\|_C \\
&+ l_2^* \frac{a^{r_1}b^{r_2}}{\Gamma(1+r_1)\Gamma(1+r_2)} \|u-v\|_C \\
&= \left(mB + \frac{a^{r_1}b^{r_2}(a^{r_1}b^{r_2}l_1^* + r_1r_2l_2^*)}{r_1r_2\Gamma(1+r_1)\Gamma(1+r_2)} \right) \|u-v\|_C.
\end{aligned}$$

Thus

$$\|N(u) - N(v)\|_C \le \left(mB + \frac{a^{r_1}b^{r_2}(a^{r_1}b^{r_2}l_1^* + r_1r_2l_2^*)}{r_1r_2\Gamma(1+r_1)\Gamma(1+r_2)} \right) \|u-v\|_C$$

Hence by (7.20), we have that N is a contraction mapping. Then in view of Banach fixed point Theorem, N has a unique fixed point which is corresponding to the solution of problem (7.17)-(7.18). □

Theorem 7.12. *Assume*

$(7.16.1)$ *There exist positive functions* $p,q,d \in C(J)$ *such that*

$$\|f(x,y,u,v)\| \le p(x,y) + q(x,y)\|u\| + d(x,y)\|v\|;$$

for all $(x,y) \in J$ *and* $u,v \in \mathbb{R}^n$.

If $mB < 1$, then problem (7.17)-(7.18) has at least one solution on $[-\xi, a] \times [-\mu, b]$.

Proof. Consider the operator N defined in (7.19). We shall show that the operator N is completely continuous. By the continuity of f and the Arzela-Ascoli theorem, we can easily obtain that N is completely continuous.

A priori bounds. We shall show there exists an open set $U \subseteq C$ with $u \neq \lambda N(u)$, for $\lambda \in (0,1)$ and $u \in \partial U$. Let $u \in C$ and $u = \lambda N(u)$ for some $0 < \lambda < 1$. Thus for each $(x,y) \in J$, we have

$$u(x,y) = \lambda \sum_{i=1}^{m} g_i(x,y)u(x-\xi_i, y-\mu_i) + \lambda I_\theta^r f(x,y,I_\theta^r u(x,y), u(x,y)).$$

This implies by (7.16.1) that, for each $(x,y) \in J$, we have

$$\begin{aligned}\|u(x,y)\| &\leq mB\|u(x,y)\| + \frac{p^* a^{r_1} b^{r_2}}{\Gamma(1+r_1)\Gamma(1+r_2)} \\ &+ \frac{q^* a^{r_1} b^{r_2}}{\Gamma(1+r_1)\Gamma(1+r_2)} \int_0^x \int_0^y (x-s)^{r_1-1}(y-t)^{r_2-1} u(s,t)\,dt\,ds \\ &+ \frac{d^*}{\Gamma(r_1)\Gamma(r_2)} \int_0^x \int_0^y (x-s)^{r_1-1}(y-t)^{r_2-1} u(s,t)\,dt\,ds\end{aligned}$$

where $p^* = \|p\|_\infty$, $q^* = \|q\|_\infty$ and $d^* = \|d\|_\infty$. Thus, for each $(x,y) \in J$, we get

$$\begin{aligned}\|u(x,y)\| &\leq \frac{p^* a^{r_1} b^{r_2}}{(1-mB)\Gamma(1+r_1)\Gamma(1+r_2)} \\ &+ \frac{q^* a^{r_1} b^{r_2}}{(1-mB)\Gamma(1+r_1)\Gamma(1+r_2)} \int_0^x \int_0^y (x-s)^{r_1-1}(y-t)^{r_2-1} u(s,t)\,dt\,ds \\ &+ \frac{d^*}{(1-mB)\Gamma(r_1)\Gamma(r_2)} \int_0^x \int_0^y (x-s)^{r_1-1}(y-t)^{r_2-1} u(s,t)\,dt\,ds \\ &\leq w + c\int_0^x \int_0^y (x-s)^{r_1-1}(y-t)^{r_2-1} u(s,t)\,dt\,ds,\end{aligned}$$

where

$$w := \frac{p^* a^{r_1} b^{r_2}}{(1-mB)\Gamma(1+r_1)\Gamma(1+r_2)}$$

and

$$c := \frac{q^* a^{r_1} b^{r_2}}{(1-mB)\Gamma(1+r_1)\Gamma(1+r_2)} + \frac{d^*}{(1-mB)\Gamma(r_1)\Gamma(r_2)}.$$

From Lemma 1.68, there exists $\delta := \delta(r_1, r_2) > 0$ such that, for each $(x,y) \in J$, we get

$$\begin{aligned}\|u\|_\infty &\leq w\left(1 + c\delta \int_0^x \int_0^y (x-s)^{r_1-1}(y-t)^{r_2-1}\,dt\,ds\right) \\ &\leq w\left(1 + \frac{c\delta a^{r_1} b^{r_2}}{r_1 r_2}\right) := M_0.\end{aligned}$$

Set $M := \max\{\|\Phi\|, M_0\}$ and

$$U = \{u \in C : \|u\|_C < M+1\}.$$

By our choice of U, there is no $u \in \partial U$ such that $u = \lambda N(u)$, for $\lambda \in (0,1)$. As a consequence of Theorem 1.56, we deduce that N has a fixed point u in $\overline{U}$ which is a solution to problem (7.17)-(7.18). □

7.3.3. Examples

As applications of our result, we give the following examples:

Example 1: Consider the following system of fractional integro-differential equations of the form

$$
\begin{aligned}
{}^{c}D_{\theta}^{r}\left[u(x,y) - \frac{x^3y}{8}u(x-\frac{3}{4},y-3) - \frac{x^4y^2}{12}u(x-2,y-\frac{1}{2}) - \frac{1}{4}u(x-1,y-\frac{3}{2})\right] \\
= f(x,y,I_{\theta}^{r},u);\ \text{if } (x,y) \in J := [0,1]\times[0,1], \qquad (7.21)\\
u(x,y) = 0;\ \text{if } (x,y) \in \tilde{J} := [-2,1]\times[-3,1]\backslash(0,1]\times(0,1], \qquad (7.22)
\end{aligned}
$$

where $m=3,\ r=(\frac{1}{2},\frac{1}{5})$ and

$$f(x,y,u) = e^{x+y}\frac{1}{1+|u|+|v|}.$$

Set

$$g_1(x,y) = \frac{x^3y}{8},\ g_2(x,y) = \frac{x^4y^2}{12},\ g_3(x,y) = \frac{1}{4}.$$

We have $B = \frac{1}{4}$ and

$$|f(x,y,u,v)| \le e^{x+y};\ \text{for all } (x,y) \in J \text{ and } u \in \mathbb{R}.$$

Then condition (7.12.1) is satisfied and $mB = \frac{3}{4} < 1$. In view of Theorem 7.10, problem (7.21)-(7.22) has a solution defined on $[-2,1]\times[-3,1]$.

Example 2: Consider now the system of fractional integro-differential equations of the form

$$
\begin{aligned}
{}^{c}D_{\theta}^{r}\left[u(x,y) - \frac{x^3y}{8}u(x-1,y-\frac{1}{2}) - \frac{x^4y^2}{12}u(x-\frac{2}{5},y-\frac{3}{4}) - \frac{1}{8}u(x-3,y-2)\right] \\
= f(x,y,I_{\theta}^{r}u,u);\ \text{if } (x,y) \in J := [0,1]\times[0,1], \qquad (7.23)\\
u(x,y) = \Phi(x,y);\ \text{if } (x,y) \in \tilde{J} := [-3,1]\times[-2,1]\backslash(0,1]\times(0,1], \qquad (7.24)
\end{aligned}
$$

where $m=3,\ r=(\frac{1}{2},\frac{1}{5}),\ f(x,y,u,v) = \frac{x+y}{20}\frac{2+|u|+|v|}{(1+|u|)(1+|v|)}$ and $\Phi : \tilde{J} \to \mathbb{R}$ is continuous with

$$\Phi(x,0) = \frac{1}{8}\Phi(x-3,-2),\ \Phi(0,y) = \frac{1}{8}\Phi(-3,y-2);\ x,y \in [0,1]. \qquad (7.25)$$

Notice that condition (7.25) is satisfied by $\Phi \equiv 0$.

Set

$$g_1(x,y) = \frac{x^3y}{8},\ g_2(x,y) = \frac{x^4y^2}{12},\ g_3(x,y) = \frac{1}{8}.$$

We have $B = \frac{1}{8}$. It is clear that f satisfies (7.14.1) with $l_1^* = l_2^* = \frac{1}{10}$. A simple computation show that condition (7.20) is satisfied. Hence by Theorem 7.11, problem (7.23)-(7.24) has a unique solution defined on $[-3,1]\times[-2,1]$.

7.4. Local Attractivity for Delay Partial Integro-Differential Equations

7.4.1. Introduction

In this section, we established sufficient conditions for the existence and the attractivity of solutions of the following system of delay integro-differential equations of fractional order of the form

$$^cD^r_\theta u(t,x) = f(t,x,I^{r_2}_{0,x}u(t,x),u(t-\tau_1,x-\xi_1),\dots,u(t-\tau_m,x-\xi_m));$$

$$\text{for } (t,x)\in J := \mathbb{R}_+\times[0,b], \tag{7.26}$$

$$u(t,x)=\Phi(t,x);\quad \text{for } (t,x)\in\tilde{J} := [-T,\infty)\times[-\xi,b]\backslash(0,\infty)\times(0,b], \tag{7.27}$$

$$\begin{cases} u(t,0)=\varphi(t); & t\in[0,\infty),\\ u(0,x)=\psi(x); & x\in[0,b], \end{cases} \tag{7.28}$$

where $b>0$, $\theta=(0,0)$, $\mathbb{R}_+=[0,\infty)$, $\tau_i,\xi_i\geq 0$; $i=1\dots,m$, $T=\max\limits_{i=1,\dots,m}\{\tau_i\}$, $\xi=\max\limits_{i=1,\dots,m}\{\xi_i\}$, $^cD^r_\theta$ is the Caputo fractional derivative of order $r=(r_1,r_2)\in(0,\infty)\times(0,\infty)$, $I^{r_2}_{0,x}$ is the partial Riemann-Liouville integral of order r_2 with respect to x, $f: J\times\mathbb{R}^m\to\mathbb{R}$ is a given continuous function, $\varphi:\mathbb{R}_+\to\mathbb{R}$, $\psi:[0,b]\to\mathbb{R}$ are absolutely continuous functions with $\lim_{t\to\infty}\varphi(t)=0$, and $\psi(x)=\varphi(0)$ for each $x\in[0,b]$, and $\Phi:\tilde{J}\to\mathbb{R}^n$ is continuous with $\varphi(t)=\Phi(t,0)$ for each $t\in\mathbb{R}_+$, and $\psi(x)=\Phi(0,x)$ for each $x\in[0,b]$.

7.4.2. Main Results

Definition 7.13. *A function $u\in BC$ is said to be a solution of (7.26)-(7.28) if u satisfies equation (7.26) on J, equation (7.27) on $\tilde{J}$ and condition (7.28) is satisfied.*

Now, we shall prove the following theorem concerning the existence and the attractivity of a solution of problem (7.26)-(7.28).

Theorem 7.14. *Assume that the function f satisfying the following hypothesis*

(7.19.1) *There exists continuous functions $p_i:\mathbb{R}_+\times[0,b]\to\mathbb{R}_+$ such that*

$$\left(1+\sum_{i=0}^m|u_i|\right)|f(t,x,u_0,u_1,u_2,\dots,u_m)|\leq\sum_{i=0}^m|u_i|p_i(t,x);$$

for $(t,x)\in\mathbb{R}_+\times[0,b]$ and for $u_i\in\mathbb{R}$; $i=0\dots m$. Moreover, assume that

$$\lim_{t\to\infty}I^r_\theta p_i(t,x)=0;\ x\in[0,b];\ i=0\dots m.$$

Then the problem (7.26)-(7.28) has at least one solution in the space BC. Moreover, solutions of problem (7.26)-(7.28) are locally asymptotically stable.

Proof. Set

$$\Phi^* := \sup_{(t,x)\in\tilde{J}} \Phi(t,x),\ \varphi^* := \sup_{t\in\mathbb{R}_+} \varphi(t) \textit{ and } p_i^* := \sup_{(t,x)\in\mathbb{R}_+\times[0,b]} I_\theta^r p_i(t,x);\ i=0\ldots m.$$

From (7.19.1), we infer that $p_i^*;\ i=0\ldots m$ are finite. Let us define the operator N such that, for any $u\in BC$,

$$(Nu)(t,x)=\begin{cases}\Phi(t,x); & (t,x)\in\tilde{J},\\ \varphi(t)+I_\theta^r f(t,x,I_{0,x}^{r_2}u(t,x), & \\ u(t-\tau_1,x-\xi_1),\ldots,u(t-\tau_m,x-\xi_m)); & (t,x)\in J.\end{cases} \tag{7.29}$$

From the assumption, we infer that $N(u)$ is continuous on $[-T,\infty)\times[-\xi,b]$. Now we prove that $N(u)\in BC$ for any $u\in BC$. For arbitrarily fixed $(t,x)\in J$ we have

$$\begin{aligned} &|(Nu)(t,x)|\le|\varphi(t)|\\ +\ & |I_\theta^r f(t,x,I_{0,x}^{r_2}u(t,x),u(t-\tau_1,x-\xi_1),\ldots,u(t-\tau_m,x-\xi_m))|\\ \le\ & |\varphi(t)|+\frac{1}{\Gamma(r_1)\Gamma(r_2)}\int_0^t\int_0^x (t-\tau)^{r_1-1}(x-s)^{r_1}\\ \times\ & \left(|I_{0,s}^{r_2}u(\tau,s)|p_0(\tau,s)+\sum_{i=1}^m|u(\tau-\tau_i,s-\xi_i)|p_i(\tau,s)\right)\\ \times\ & \left(1+|I_{0,s}^{r_2}u(\tau,s)|+\sum_{i=1}^m|u(\tau-\tau_i,s-\xi_i)|\right)^{-1} dsd\tau.\\ \le\ & \varphi^*+\sum_{i=0}^m p_i^*, \end{aligned}$$

and for $(t,x)\in\tilde{J}$, we have

$$|(Nu)(t,x)|=|\Phi(t,x)|\le\Phi^*.$$

Thus,

$$\|N(u)\|_{BC}\le\max\{\Phi^*,\varphi^*+\sum_{i=0}^m p_i^*\}:=\eta. \tag{7.30}$$

Hence, $N(u)\in BC$. This proves that the operator N maps BC into itself.
By Lemma 1.28, the problem of finding the solutions of the problem (7.26)-(7.28) is reduced to finding the solutions of the operator equation $N(u)=u$. Equation (7.30) yields that N transforms the ball $B_\eta:=B(0,\eta)$ into itself. We shall show that $N:B_\eta\to B_\eta$ satisfies the assumptions of Schauder's fixed point theorem [204]. The proof will be given in several steps.

Step 1: *N is continuous.*
Let $\{u_n\}_{n\in\mathbb{N}}$ be a sequence such that $u_n\to u$ in B_η. Then, for each $(t,x)\in[-T,\infty)\times[-\xi,b]$, we have

$$\begin{aligned} &|(Nu_n)(t,x)-(Nu)(t,x)|\\ \le\ & \tfrac{1}{\Gamma(r_1)\Gamma(r_2)}\int_0^t\int_0^x(t-\tau)^{r_1-1}(x-s)^{r_2-1}\\ \times\ & |f(\tau,s,I_{0,s}^{r_2}u_n(\tau,s),u_n(\tau-\tau_1,s-\xi_1),\ldots,u_n(\tau-\tau_m,s-\xi_m))\\ -\ & f(\tau,s,I_{0,s}^{r_2}u(\tau,s),u(\tau-\tau_1,s-\xi_1),\ldots,u(\tau-\tau_m,s-\xi_m)))|dsd\tau. \end{aligned} \tag{7.31}$$

Case 1. If $(t,x) \in \tilde{J} \cup ([0,T_0] \times [0,b])$, $T_0 > 0$, then, since $u_n \to u$ as $n \to \infty$ and $f, I^{r_2}_{0,\xi}$, are continuous, (7.31) gives

$$\|N(u_n) - N(u)\|_{BC} \to 0 \quad \text{as } n \to \infty.$$

Case 2. If $(t,x) \in (T_0,\infty) \times [0,b]$, $T_0 > 0$, then from (7.19.1) and (7.31), we get

$$\begin{aligned}
&|(Nu_n)(t,x) - (Nu)(t,x)| \\
\leq\ & \tfrac{2}{\Gamma(r_1)\Gamma(r_2)} \int_0^t \int_0^x (t-\tau)^{r_1-1}(x-s)^{r_2-1} \\
\times\ & \left(|I^{r_2}_{0,s}u(\tau,s)|p_0(\tau,s) + \sum_{i=1}^{m} |u(\tau-\tau_i, s-\xi_i)|p_i(\tau,s) \right) \\
\times\ & \left(1 + |I^{r_2}_{0,s}u(\tau,s)| + \sum_{i=1}^{m} |u_i(\tau-\tau_i, s-\xi_i)| \right)^{-1} dsd\tau \\
\leq\ & \sum_{i=1}^{m} \frac{2}{\Gamma(r_1)\Gamma(r_2)} \int_0^t \int_0^x (t-\tau)^{r_1-1}(x-s)^{r_2-1} p_i(\tau,s)dsd\tau \\
\leq\ & 2\sum_{i=1}^{m} I^r_\theta p_i(t,x).
\end{aligned} \tag{7.32}$$

Since $u_n \to u$ as $n \to \infty$ and $t \to \infty$, then (7.32) gives

$$\|N(u_n) - N(u)\|_{BC} \to 0 \quad \text{as } n \to \infty.$$

Step 2: *$N(B_\eta)$ is uniformly bounded.*
This is clear since $N(B_\eta) \subset B_\eta$ and B_η is bounded.

Step 3: *$N(B_\eta)$ is equicontinuous on every compact subset $[-T,a] \times [-\xi,b]$ of $[-T,a] \times [-\xi,\infty)$, $a > 0$.*
Let $(t_1,x_1),(t_2,x_2) \in [0,a] \times [0,b]$, $t_1 < t_2$, $x_1 < x_2$ and let $u \in B_\eta$. Thus we have

$$\begin{aligned}
&|(Nu)(t_2,x_2) - (Nu)(t_1,x_1)| \leq |\varphi(t_2) - \varphi(t_1)| \\
+\ & \frac{1}{\Gamma(r_1)\Gamma(r_2)} \int_0^{t_1} \int_0^{x_1} [(t_2-\tau)^{r_1-1}(x_2-s)^{r_2-1} \\
-\ & (t_1-\tau)^{r_1-1}(x_1-s)^{r_2-1}] \\
& \times |f(\tau,s,I^{r_2}_{0,s}u(\tau,s),u(\tau-\tau_1,s-\xi_1),\dots,u(\tau-\tau_m,s-\xi_m))|dsd\tau \\
+\ & \frac{1}{\Gamma(r_1)\Gamma(r_2)} \int_{t_1}^{t_2} \int_{x_1}^{x_2} (t_2-\tau)^{r_1-1}(x_2-s)^{r_2-1} \\
& \times |f(\tau,s,I^{r_2}_{0,s}u(\tau,s),u(\tau-\tau_1,s-\xi_1),\dots,u(\tau-\tau_m,s-\xi_m))|dsd\tau \\
+\ & \frac{1}{\Gamma(r_1)\Gamma(r_2)} \int_0^{t_1} \int_{x_1}^{x_2} (t_2-\tau)^{r_1-1}(x_2-s)^{r_2-1} \\
& \times |f(\tau,s,I^{r_2}_{0,s}u(\tau,s),u(\tau-\tau_1,s-\xi_1),\dots,u(\tau-\tau_m,s-\xi_m))|dsd\tau \\
+\ & \frac{1}{\Gamma(r_1)\Gamma(r_2)} \int_{t_1}^{t_2} \int_0^{x_1} (t_2-\tau)^{r_1-1}(x_2-s)^{r_2-1} \\
& \times |f(\tau,s,I^{r_2}_{0,s}u(\tau,s),u(\tau-\tau_1,s-\xi_1),\dots,u(\tau-\tau_m,s-\xi_m))|dsd\tau.
\end{aligned}$$

Thus

$$
\begin{aligned}
&|(Nu)(t_2,x_2)-(Nu)(t_1,x_1)| \leq |\varphi(t_2)-\varphi(t_1)| \\
+ & \frac{1}{\Gamma(r_1)\Gamma(r_2)}\int_0^{t_1}\int_0^{x_1}[(t_2-\tau)^{r_1-1}(x_2-s)^{r_2-1} \\
- & (t_1-\tau)^{r_1-1}(x_1-s)^{r_2-1}]\times\sum_{i=0}^{m}p_i(\tau,s)dsd\tau \\
+ & \frac{1}{\Gamma(r_1)\Gamma(r_2)}\int_{t_1}^{t_2}\int_{x_1}^{x_2}(t_2-\tau)^{r_1-1}(x_2-s)^{r_2-1}\sum_{i=0}^{m}p_i(\tau,s)dsd\tau \\
+ & \frac{1}{\Gamma(r_1)\Gamma(r_2)}\int_0^{t_1}\int_{x_1}^{x_2}(t_2-\tau)^{r_1-1}(x_2-s)^{r_2-1}\sum_{i=0}^{m}p_i(\tau,s)dsd\tau \\
+ & \frac{1}{\Gamma(r_1)\Gamma(r_2)}\int_{t_1}^{t_2}\int_0^{x_1}(t_2-\tau)^{r_1-1}(x_2-s)^{r_2-1}\sum_{i=0}^{m}p_i(\tau,s)dsd\tau.
\end{aligned}
$$

From continuity of $\varphi, p_i;\ i=0\ldots,m$ and as $t_1 \longrightarrow t_2$ and $x_1 \longrightarrow x_2$, the right-hand side of the above inequality tends to zero. The equicontinuity for the cases $t_1<t_2<0,\ x_1<x_2<0$ and $t_1\leq 0\leq t_2,\ x_1\leq 0\leq x_2$ is obvious.

Step 4: *$N(B_\eta)$ is equiconvergent.*
Let $(t,x)\in\mathbb{R}_+\times[0,b]$ and $u\in B_\eta$, then we have

$$
\begin{aligned}
&|(Nu)(t,x)| \leq |\varphi(t)| + |\frac{1}{\Gamma(r_1)\Gamma(r_2)}\int_0^t\int_0^x(t-\tau)^{r_1-1}(x-s)^{r_2-1} \\
&\times f(\tau,s,I_{0,s}^{r_2}u(\tau,s),u(\tau-\tau_1,s-\xi_1),\ldots,u(\tau-\tau_m,s-\xi_m))dsd\tau| \\
\leq\ & |\varphi(t)| + \frac{1}{\Gamma(r_1)\Gamma(r_2)}\int_0^t\int_0^x(t-\tau)^{r_1-1}(x-s)^{r_2-1}\sum_{i=0}^{m}p_i(\tau,s)dsd\tau \\
\leq\ & |\varphi(t)| + \sum_{i=0}^{m}I_\theta^r p_i(t,x).
\end{aligned}
$$

Thus, for each $x\in[0,b]$, we get

$$|(Nu)(t,x)| \longrightarrow 0,\ as\ t \longrightarrow +\infty.$$

Hence,

$$|(Nu)(t,x)-(Nu)(+\infty,x)| \longrightarrow 0,\ as\ t \longrightarrow +\infty.$$

As a consequence of Steps 1 to 4 together with the Lemma 1.69, we can conclude that $N: B_\eta \to B_\eta$ is continuous and compact. From an application of Schauder's theorem [204], we deduce that N has a fixed point u which is a solution of the problem (7.26)-(7.28).

Step 5: Now we investigate the uniform local attractivity for solutions of problem (7.26)-(7.28). Let us assume that u_0 is a solution of problem (7.26)-(7.28) with the conditions of this theorem. Consider the ball $B(u_0, \eta^*)$, where

$$\begin{aligned}\eta^* : &= \frac{1}{\Gamma(r_1)\Gamma(r_2)} \sup_{(t,x)\in J} \{ \int_0^t \int_0^x (t-\tau)^{r_1-1}(x-s)^{r_2-1} \\ &\times |f(\tau, s, I_{0,s}^{r_2}u(\tau,s), u(\tau-\tau_1, s-\xi_1), \dots, u(\tau-\tau_m, s-\xi_m)) \\ &- f(\tau, s, I_{0,s}^{r_2}u_0(\tau,s), u_0(\tau-\tau_1, s-\xi_1), \dots, \\ &u_0(\tau-\tau_m, s-\xi_m))|ds d\tau;\ u \in BC\}.\end{aligned}$$

Taking $u \in B(u_0, \eta^*)$, we have

$$\begin{aligned}&|(Nu)(t,x) - u_0(t,x)| = |(Nu)(t,x) - (Nu_0)(t,x)| \\ \leq\ & \frac{1}{\Gamma(r_1)\Gamma(r_2)} \int_0^t \int_0^x (t-\tau)^{r_1-1}(x-s)^{r_2-1} \\ &\times |f(\tau, s, I_{0,s}^{r_2}u(\tau,s), u(\tau-\tau_1, s-\xi_1), \dots, u(\tau-\tau_m, s-\xi_m)) \\ -\ & f(\tau, s, I_{0,s}^{r_2}u_0(\tau,s), u_0(\tau-\tau_1, s-\xi_1), \dots, u_0(\tau-\tau_m, s-\xi_m))|ds d\tau \\ \leq\ & \eta^*.\end{aligned}$$

Thus we observe that N is a continuous function such that $N(B(u_0, \eta^*)) \subset B(u_0, \eta^*)$. Moreover, if u is a solution of problem (7.26)-(7.28), then

$$\begin{aligned}&|u(t,x) - u_0(t,x)| = |(Nu)(t,x) - (Nu_0)(t,x)| \\ \leq\ & \tfrac{1}{\Gamma(r_1)\Gamma(r_2)} \int_0^t \int_0^x (t-\tau)^{r_1-1}(x-s)^{r_2-1} \\ &\times |f(\tau, s, I_{0,s}^{r_2}u(\tau,s), u(\tau-\tau_1, s-\xi_1), \dots, u(\tau-\tau_m, s-\xi_m)) \\ -\ & f(\tau, s, I_{0,s}^{r_2}u_0(\tau,s), u_0(\tau-\tau_1, s-\xi_1), \dots, u_0(\tau-\tau_m, s-\xi_m))|ds d\tau \\ \leq\ & \tfrac{2}{\Gamma(r_1)\Gamma(r_2)} \int_0^t \int_0^x (t-\tau)^{r_1-1}(x-s)^{r_2-1} \sum_{i=0}^m p_i(\tau,s) ds d\tau \\ \leq\ & 2\sum_{i=0}^m I_\theta^r p_i(t,x).\end{aligned} \tag{7.33}$$

By using (7.33) and the fact that $I_\theta^r p_i(t,x) \longrightarrow 0$ as $t \longrightarrow \infty$; $i = 0 \dots m$ we deduce that

$$\lim_{t\to\infty} |u(t,x) - u_0(t,x)| = 0.$$

Consequently, all solutions of problem (7.26)-(7.28) are locally asymptotically stable. □

7.4.3. An Example

As an application and to illustrate our results, we consider the following system of delay integro-differential equations of fractional order

$$^cD_\theta^r u(t,x) = f\left(t, x, I_{0,x}^{r_2}u(t,x), u\left(t-1, x-\frac{1}{4}\right), u\left(t-\frac{2}{3}, x-\frac{1}{5}\right)\right);$$

$$\text{for } (t,x) \in J := [0,\infty) \times [0,1], \tag{7.34}$$

$$u(t,x) = e^{-t}; \quad \text{for } (t,x) \in \tilde{J} := [-1,\infty) \times [-\frac{1}{4}, 1] \backslash (0,\infty) \times (0,1], \tag{7.35}$$

$$\begin{cases} u(t,0) = e^{-t}; & x \in [0,\infty), \\ u(0,x) = 1; & x \in [0,1], \end{cases} \tag{7.36}$$

where $r=(r_1,r_2)=(\frac{1}{4},\frac{1}{2})$ and

$$\begin{cases} f(t,x,u,v,w)=\dfrac{xt^{\frac{-3}{4}}\left(|v|\sin t+|w|e^{-\frac{1}{t}}\right)}{2+|u|+|v|+|w|};\\ \qquad (t,x)\in(0,\infty)\times[0,1]\ and\ u,v,w\in\mathbb{R},\\ f(t,x,u,v,w)=0; \quad (t,x)\in\{0\}\times[0,1]\ and\ u,v,w\in\mathbb{R}. \end{cases}$$

We have for each $x\in[0,\infty)$, $\mu(t)=e^{-t}\longrightarrow 0\ as\ t\to\infty$.
Let us notice that the function f satisfies assumption $(7.19.1)$, where

$$p_0(t,x)=0;\ (t,x)\in\mathbb{R}_+\times[0,1],$$

$$\begin{cases} p_1(t,x)=xt^{\frac{-3}{4}}|\sin t|;\ (t,x)\in(0,\infty)\times[0,1],\\ p_1(0,x)=0; \qquad x\in[0,1], \end{cases}$$

and

$$\begin{cases} p_2(t,x)=xt^{\frac{-3}{4}}e^{-\frac{1}{t}};\ (t,x)\in(0,\infty)\times[0,1],\\ p_2(0,x)=0; \qquad x\in[0,1]. \end{cases}$$

Also, for each $x\in[0,1]$, we get

$$I_\theta^r p_0(t,x)=0,$$

$$\begin{aligned} I_\theta^r p_1(t,x) &= \frac{1}{\Gamma(r_1)\Gamma(r_2)}\int_0^t\int_0^x(t-\tau)^{r_1-1}(x-s)^{r_2-1}p_1(\tau,s)dsd\tau\\ &= \frac{1}{\Gamma(\frac{1}{4})\Gamma(\frac{1}{2})}\int_0^t\int_0^x(t-\tau)^{\frac{-3}{4}}(x-s)^{\frac{-1}{2}}s\tau^{\frac{-3}{4}}|\sin\tau|dsd\tau\\ &\le \frac{1}{\Gamma(\frac{1}{4})\Gamma(\frac{1}{2})}\int_0^t\int_0^x(t-\tau)^{\frac{-3}{4}}(x-s)^{\frac{-1}{2}}s\tau^{\frac{-3}{4}}dsd\tau\\ &= \frac{\Gamma(\frac{1}{4})\Gamma(2)}{\Gamma(\frac{5}{4})\Gamma(\frac{5}{2})}t^{\frac{-1}{2}}x^{\frac{3}{2}}\longrightarrow 0\ as\ t\to\infty, \end{aligned}$$

and

$$\begin{aligned} I_\theta^r p_2(t,x) &= \frac{1}{\Gamma(r_1)\Gamma(r_2)}\int_0^t\int_0^x(t-\tau)^{r_1-1}(x-s)^{r_2-1}p_2(\tau,s)dsd\tau\\ &= \frac{1}{\Gamma(\frac{1}{4})\Gamma(\frac{1}{2})}\int_0^t\int_0^x(t-\tau)^{\frac{-3}{4}}(x-s)^{\frac{-1}{2}}s\tau^{\frac{-3}{4}}e^{-\frac{1}{\tau}}dsd\tau\\ &\le \frac{1}{\Gamma(\frac{1}{4})\Gamma(\frac{1}{2})}\int_0^t\int_0^x(t-\tau)^{\frac{-3}{4}}(x-s)^{\frac{-1}{2}}s\tau^{\frac{-3}{4}}dsd\tau\\ &= \frac{\Gamma(\frac{1}{4})\Gamma(2)}{\Gamma(\frac{5}{4})\Gamma(\frac{5}{2})}t^{\frac{-1}{2}}x^{\frac{3}{2}}\longrightarrow 0\ as\ t\to\infty. \end{aligned}$$

Thus

$$\lim_{t\longrightarrow\infty}I_\theta^r p_i(t,x)=0;\ x\in[0,1];\ i=0,1,2.$$

Hence by Theorem 7.14, the problem (7.34)-(7.36) has a solution defined on $[-1,\infty)\times[-\frac{1}{4},1]$ and all solutions are locally asymptotically stable on $[-1,\infty)\times[-\frac{1}{4},1]$.

7.5. Existence and Attractivity Results for Partial Integro-Differential Equations with Delay

7.5.1. Introduction

In this section, by means of integral inequalities and the fixed point approach, we initiate the question of global attractivity of the solution for the following system of partial integro-differential equations of fractional order of the form

$$^{c}D_{\theta}^{r}u(t,x)=f(t,x,u_{(t,x)},(Gu)(t,x));\quad \text{for } (t,x)\in J:=\mathbb{R}_{+}\times[0,b], \tag{7.37}$$

$$u(t,x)=\phi(t,x);\ \text{if } (t,x)\in\tilde{J}:=[-\alpha,\infty)\times[-\beta,b]\backslash(0,\infty)\times(0,b], \tag{7.38}$$

$$\begin{cases} u(t,0)=\varphi(t);\ t\in\mathbb{R}_{+},\\ u(0,x)=\psi(x);\ x\in[0,b], \end{cases} \tag{7.39}$$

where

$$(Gu)(t,x)=\frac{1}{\Gamma(r_1)\Gamma(r_2)}\int_0^t\int_0^x(t-s)^{r_1-1}(x-y)^{r_2-1}g(t,x,s,y,u_{(s,y)})dyds, \tag{7.40}$$

$\alpha,\beta,b>0$, $\theta=(0,0)$, $r=(r_1,r_2)\in(0,1]\times(0,1]$, $\mathbb{R}_{+}=[0,\infty)$, I_{θ}^{r} is the left-sided mixed Riemann-Liouville integral of order r, $^{c}D_{\theta}^{r}$ is the standard Caputo's fractional derivative of order r, $f:J\times C\to\mathbb{R}$, $g:J_1\times C\to\mathbb{R}$ are given continuous functions, $J_1:=\{(t,x,s,y):0\leq s\leq t<\infty,\ 0\leq y\leq x\leq b\}\}$, $\varphi:\mathbb{R}_{+}\to\mathbb{R}$, $\psi:[0,b]\to\mathbb{R}$ are absolutely continuous functions with $\lim_{t\to\infty}\varphi(t)=0$, and $\psi(x)=\varphi(0)$ for each $x\in[0,b]$, $\Phi:\tilde{J}\to\mathbb{R}$ is continuous with $\varphi(t)=\Phi(t,0)$ for each $t\in\mathbb{R}_{+}$, and $\psi(x)=\Phi(0,x)$ for each $x\in[0,b]$ and $C:=C([-\alpha,0]\times[-\beta,0])$ is the space of continuous functions on $[-\alpha,0]\times[-\beta,0]$ with the standard norm

$$\|u\|_{C}=\sup_{(t,x)\in[-\alpha,0]\times[-\beta,0]}|u(t,x)|.$$

We present our results for the problem (7.37)-(7.39) in the Banach space of continuous and bounded functions.

By $BC:=BC([-\alpha,\infty)\times[-\beta,b])$ we denote the Banach space of all bounded and continuous functions from $[-\alpha,\infty)\times[-\beta,b]$ into $\mathbb{R}$ equipped with the standard norm

$$\|u\|_{BC}=\sup_{(t,x)\in[-\alpha,\infty)\times[-\beta,b]}|u(t,x)|.$$

Definition 7.15. *A function $u\in BC$ with its mixed derivative D_{tx}^{2} exists and is integrable is said to be a solution of (7.37)-(7.39) if u satisfies equations (7.37) and (7.39) on J and the condition (7.38) on $\tilde{J}$.*

7.5.2. Existence and Uniqueness

Our first result is about the existence and uniqueness of the solution of problem (7.37)-(7.39).

Theorem 7.16. *Assume that following hypotheses hold*

(7.22.1) *The functions* φ *and* Φ *are bounded with*

$$\varphi^* = \sup_{t\in\mathbb{R}_+} |\varphi(t)| \text{ and } \Phi^* = \sup_{(t,x)\in\tilde{J}} |\Phi(t,x)|,$$

(7.22.2) *There exist positive functions* $p_1, p_2 \in BC(J)$ *such that*

$$|f(t,x,u_1,u_2) - f(t,x,v_1,v_2)| \leq p_1(t,x)\|u_1 - v_1\|_C + p_2(t,x)|u_2 - v_2|,$$

for each $(t,x) \in J$, $u_1, v_1 \in C$ *and* $u_2, v_2 \in \mathbb{R}$. *Moreover, assume that the function* $t \to \int_0^t\int_0^x (t-s)^{r_1-1}(x-y)^{r_2-1} f(s,y,0,(G0)(s,y))dyds$ *is bounded on J with*

$$f^* = \sup_{(t,x)\in J} \frac{1}{\Gamma(r_1)\Gamma(r_2)} \int_0^t \int_0^x (t-s)^{r_1-1}(x-y)^{r_2-1}$$

$$\times |f(s,y,0,(G0)(s,y))| dyds,$$

(7.22.3) *There exists a positive function* $q \in BC(J_1)$ *such that*

$$|g(t,x,s,y,u) - g(t,x,s,y,v)| \leq q(t,x,s,y)|u - v|,$$

for each $(t,x,s,y) \in J_1$ *and* $u, v \in \mathbb{R}$.

If

$$p_1^* + p_2^* q^* < 1, \tag{7.41}$$

where

$$p_i^* = \sup_{(t,x)\in J} [\frac{1}{\Gamma(r_1)\Gamma(r_2)} \int_0^t \int_0^x (t-s)^{r_1-1}(x-y)^{r_2-1} p_i(s,y)dyds];\ i = 1,2,$$

and

$$q^* = \sup_{(t,x)\in J} [\frac{1}{\Gamma(r_1)\Gamma(r_2)} \int_0^t \int_0^x (t-s)^{r_1-1}(x-y)^{r_2-1} q(t,x,s,y)dyds],$$

then problem (7.37)-(7.39) has a unique solution on $[-\alpha, \infty) \times [-\beta, b]$.

Proof. Let us define the operator $N : BC \to BC$ by

$$(Nu)(t,x) = \begin{cases} \Phi(t,x), & (t,x) \in \tilde{J}, \\ \varphi(t) + \frac{1}{\Gamma(r_1)\Gamma(r_2)} \int_0^t \int_0^x (t-s)^{r_1-1}(x-y)^{r_2-1} & \\ \times f(s,y,u_{(s,y)},(Gu)(s,y))dyds, & (t,x) \in J. \end{cases} \tag{7.42}$$

It is clear that the function $(t,x) \mapsto (Nu)(t,x)$ is continuous on $[-\alpha,\infty) \times [-\beta,b]$. Now we prove that $N(u) \in BC$ for any $u \in BC$. For each $(t,x) \in \tilde{J}$ we have

$$|\Phi(t,x)| \leq \sup_{(t,x)\in\tilde{J}} |\Phi(t,x)| = \Phi^*,$$

then $\Phi \in BC$. From (7.22.2), and for arbitrarily fixed $(t,x) \in J$ we have

$$
\begin{aligned}
|(Nu)(t,x)| = {} & |\varphi(t) + \frac{1}{\Gamma(r_1)\Gamma(r_2)} \int_0^t \int_0^x (t-s)^{r_1-1}(x-y)^{r_2-1} \\
& \times f(s,y,u_{(s,y)},(Gu)(s,y))dyds| \\
& \leq |\varphi(t)| + \frac{1}{\Gamma(r_1)\Gamma(r_2)} \int_0^t \int_0^x (t-s)^{r_1-1}(x-y)^{r_2-1} \\
& \times |f(s,y,u_{(s,y)},(Gu)(s,y)) - f(s,y,0,(G0)(s,y))|dyds \\
& + \frac{1}{\Gamma(r_1)\Gamma(r_2)} \int_0^t \int_0^x (t-s)^{r_1-1}(x-y)^{r_2-1}|f(s,y,0,(G0)(s,y))|dyds \\
& \leq |\varphi(t)| + \frac{1}{\Gamma(r_1)\Gamma(r_2)} \int_0^t \int_0^x (t-s)^{r_1-1}(x-y)^{r_2-1} \\
& \times \left(p_1(s,y)|u_{(s,y)}| + p_2(s,y)|(Gu)(s,y)|\right) dyds \\
& + \frac{1}{\Gamma(r_1)\Gamma(r_2)} \int_0^t \int_0^x (t-s)^{r_1-1}(x-y)^{r_2-1} \\
& \times |f(s,y,0,(G0)(s,y))|dyds \\
& \leq \varphi^* + f^* + p_1^*\|u\|_{BC} + \frac{1}{\Gamma(r_1)\Gamma(r_2)} \int_0^t \int_0^x (t-s)^{r_1-1}(x-y)^{r_2-1} \\
& \quad \times p_2(s,y)|(Gu)(s,y) - (G0)(s,y)|dyds. \qquad (7.43)
\end{aligned}
$$

But, (7.22.3) implies that

$$
\begin{aligned}
|(Gu)(t,x) - (G0)(t,x)| \leq {} & \frac{1}{\Gamma(r_1)\Gamma(r_2)} \int_0^t \int_0^x (t-s)^{r_1-1}(x-y)^{r_2-1} \\
& \times |g(t,x,s,y,u(s,y)) - g(t,x,s,y,0)|dyds \\
& \leq \frac{1}{\Gamma(r_1)\Gamma(r_2)} \int_0^t \int_0^x (t-s)^{r_1-1}(x-y)^{r_2-1} q(t,x,s,y)|u(s,y)|dyds \\
& \leq q^*\|u\|_{BC}.
\end{aligned}
$$

Thus, by (7.43) we get

$$
\begin{aligned}
|(Nu)(t,x)| \leq {} & \varphi^* + f^* + p_1^*\|u\|_{BC} \\
& + \frac{q^*\|u\|_{BC}}{\Gamma(r_1)\Gamma(r_2)} \int_0^t \int_0^x (t-s)^{r_1-1}(x-y)^{r_2-1} p_2(s,y)dyds \\
& \leq \varphi^* + f^* + p_1^*\|u\|_{BC} + p_2^*q^*\|u\|_{BC} \\
& \leq \varphi^* + f^* + (p_1^* + p_2^*q^*)\|u\|_{BC}.
\end{aligned}
$$

Hence $N(u) \in BC$. Let $u,\ v \in BC$. Using the hypotheses, for each $(t,x) \in J$, we have

$$
\begin{aligned}
|(Nu)(t,x) - (Nv)(t,x)| \leq {} & \frac{1}{\Gamma(r_1)\Gamma(r_2)} \int_0^t \int_0^x (t-s)^{r_1-1}(x-y)^{r_2-1} \\
& \times |f(s,y,u_{(s,y)},(Gu)(s,y)) - f(s,y,v_{(s,y)},(Gv)(s,y))|dyds \\
& \leq \frac{1}{\Gamma(r_1)\Gamma(r_2)} \int_0^t \int_0^x (t-s)^{r_1-1}(x-y)^{r_2-1}
\end{aligned}
$$

$$\times (p_1(s,y)\|u_{(s,y)} - v_{(s,y)}\|_C + p_2(s,y)|(Gu)(s,y) - (Gv)(s,y)|)dyds$$
$$\leq \frac{\|u-v\|_{BC}}{\Gamma(r_1)\Gamma(r_2)} \int_0^t \int_0^x (t-s)^{r_1-1}(x-y)^{r_2-1} p_1(s,y)dyds$$
$$+ \frac{\|u-v\|_{BC}}{\Gamma^2(r_1)\Gamma^2(r_2)} \int_0^t \int_0^x (t-s)^{r_1-1}(x-y)^{r_2-1}$$
$$\times p_2(s,y)\left(\int_0^s \int_0^y (s-\tau)^{r_1-1}(y-\xi)^{r_2-1} q(s,t,\tau,\xi)d\xi d\tau\right)dyds$$
$$\leq (p_1^* + p_2^* q^*)\|u-v\|_{BC}.$$

From (7.41), it follows from the Banach contraction principle that N has a unique fixed point in BC which is a solution of problem (7.37)-(7.39). □

7.5.3. Estimates on the Solutions

Denote by $D_1 := \frac{\partial}{\partial t}$, the partial derivative of a function defined on J (or J_1) with respect to the first variable, $D_2 := \frac{\partial}{\partial x}$, $D_2D_1 := \frac{\partial^2}{\partial t \partial x}$. In the sequel we will make use of the following Lemma due to Pachpatte.

Lemma 7.17. *[305] Let $u, e, p \in C(J)$, $k, D_1k, D_2k, D_2D_1k \in C(J_1)$ be positive functions. If $e(t,x)$ is nondecreasing in each variable $(t,x) \in J$ and*

$$u(t,x) \leq e(t,x) + \int_0^t \int_0^x p(s,y)$$
$$\times \left[u(s,y) + \int_0^s \int_0^y k(s,y,\tau,\xi)u(\tau,\xi)d\xi d\tau\right] dyds; \ (t,x) \in J, \tag{7.44}$$

then,

$$u(t,x) \leq e(t,x)[1$$
$$+ \int_0^t \int_0^x p(s,y)\exp\left(\int_0^s \int_0^y [p(\tau,\xi) + A(\tau,\xi)]d\xi d\tau\right) dyds\bigg]; \ (t,x) \in J, \tag{7.45}$$

where

$$A(t,x) = k(t,x,s,y) + \int_0^t D_1k(t,x,s,y)ds + \int_0^x D_2k(t,x,s,y)dy$$
$$+ \int_0^t \int_0^x D_2D_1k(t,x,s,y)dyds; \ (t,x) \in J. \tag{7.46}$$

Now, we shall prove the following theorem concerning the estimate on the solution of problem (7.37)-(7.39).

Theorem 7.18. *Set*

$$d = \varphi^* + f^*. \tag{7.47}$$

Assume that $(7.22.1)-(7.22.3)$ *and the following hypotheses hold*

(7.25.1) $p_1 = p_2$ *and there exists a positive function* $p \in BC(J)$ *such that,*

$$p_1(s,y) \le \Gamma(r_1)\Gamma(r_2)(t-s)^{1-r_1}(x-y)^{1-r_2}p(s,y);$$

for each $(t,x,s,y) \in J_1$,

(7.25.2) $k, D_1k, D_2k, D_2D_1k \in BC(J_1)$, *where*

$$k(t,x,s,y) = \frac{1}{\Gamma(r_1)\Gamma(r_2)}(t-s)^{r_1-1}(x-y)^{r_2-1}q(t,x,s,y).$$

If u is any solution of (7.37)-(7.39) on $[-\alpha,\infty)\times[-\beta,b]$, *then for each* $(t,x) \in J$,

$$|u(t,x)| \le d\left[1+\int_0^t\int_0^x p(s,y)\exp\left(\int_0^s\int_0^y [p(\tau,\xi)+A(\tau,\xi)]d\xi d\tau\right)dyds\right], \qquad (7.48)$$

where $A(t,x)$ *is defined by (7.46).*

Proof. Using the fact that u is a solution of (7.37)-(7.39) and from hypotheses, we have for each $(t,x) \in J$,

$$\begin{aligned}
|u(t,x)| &\le |\varphi(t)| \\
&+ \frac{1}{\Gamma(r_1)\Gamma(r_2)}\int_0^t\int_0^x (t-s)^{r_1-1}(x-y)^{r_2-1}|f(t,x,0,(G0)(t,x))|dyds \\
&+ \frac{1}{\Gamma(r_1)\Gamma(r_2)}\int_0^t\int_0^x (t-s)^{r_1-1}(x-y)^{r_2-1} \\
&\times |f(s,y,u_{(s,y)},(Gu)(s,y)) - f(s,y,0,(G0)(s,y))|dyds \\
&\le \varphi^* + f^* + \frac{1}{\Gamma(r_1)\Gamma(r_2)}\int_0^t\int_0^x (t-s)^{r_1-1}(x-y)^{r_2-1}p_1(s,y)[\|u_{(s,y)}\|_C \\
&+ \frac{1}{\Gamma(r_1)\Gamma(r_2)}\int_0^s\int_0^y (s-\tau)^{r_1-1}(y-\xi)^{r_2-1}q(s,y,\tau,\xi)|u(\tau,\xi)|d\xi d\tau\Bigg]dyds \\
&\le d + \int_0^t\int_0^x p(s,y)[\|u_{(s,y)}\|_C \\
&+ \frac{1}{\Gamma(r_1)\Gamma(r_2)}\int_0^s\int_0^y q(s,y,\tau,\xi)|u(\tau,\xi)|d\xi d\tau\Bigg]dyds \\
&\le d + \int_0^t\int_0^x p(s,y)\left[\|u_{(s,y)}\|_C + \int_0^s\int_0^y k(s,y,\tau,\xi)|u(\tau,\xi)|d\xi d\tau\right]dyds.
\end{aligned}$$

We consider the function w defined by

$$w(t,x) = \sup\{\|u(s,y)\| : -\alpha \le s \le t,\ -\beta \le y \le x\},\ 0 \le t < \infty,\ 0 \le x \le b.$$

Let $(t^*,x^*) \in [-\alpha,t]\times[-\beta,x]$ be such that $w(t,x) = |u(t^*,x^*)|$.
If $(t^*,x^*) \in \tilde{J}$, then $w(t,x) = \|\Phi\|_C$ and the previous inequality holds. If $(t^*,x^*) \in J$, then by the previous inequality, we have for $(t,x) \in J$,

$$w(t,x) \le d + \int_0^t\int_0^x p(s,y)\left[w(s,y) + \int_0^s\int_0^y k(s,y,\tau,\xi)w(\tau,\xi)d\xi d\tau\right]dyds.$$

From Lemma 7.17, we get

$$w(t,x) \le d[1+$$

$$\int_0^t \int_0^x p(s,y)\exp\left(\int_0^s \int_0^y [p(\tau,\xi)+A(\tau,\xi)]d\xi d\tau\right)dyds\Bigg];\ (t,x)\in J. \tag{7.49}$$

But, for every $(t,x)\in J$, $\|u_{(t,x)}\|_C \le w(t,x)$. Hence, from (7.49) yields (7.48). □

Theorem 7.19. *Set*

$$\overline{d} := f^* + \varphi^* p^*(1+q^*). \tag{7.50}$$

Assume that (7.22.1) – (7.22.3) *and* (7.25.1) – (7.25.2) *hold. If u is any solution of (7.37)-(7.39) on* $[-\alpha,\infty)\times[-\beta,b]$, *then*

$$|u(t,x)-\varphi(t)| \le \overline{d}[1+$$

$$\int_0^t \int_0^x p(s,y)\exp\left(\int_0^s \int_0^y [p(\tau,\xi)+A(\tau,\xi)]d\xi d\tau\right)dyds\Bigg],\ (t,x)\in J, \tag{7.51}$$

where A is given by (7.46).

Proof. Let $h(t,x) = |u(t,x)-\varphi(t)|$. Using the fact that u is a solution of (7.37)-(7.39) and hypotheses, for each $(t,x)\in J$, we have

$$\begin{aligned}
h(t,x) \le\ & \frac{1}{\Gamma(r_1)\Gamma(r_2)}\int_0^t\int_0^x (t-s)^{r_1-1}(x-y)^{r_2-1} \\
&\times |f(s,y,u_{(s,y)},(Gu)(s,y)) - f(s,y,\varphi(s),(G\varphi)(s))|dyds \\
&+\frac{1}{\Gamma(r_1)\Gamma(r_2)}\int_0^t\int_0^x (t-s)^{r_1-1}(x-y)^{r_2-1}|f(s,y,\varphi(s),(G\varphi)(s))|dyds \\
\le\ & \overline{d}+\frac{1}{\Gamma(r_1)\Gamma(r_2)}\int_0^t\int_0^x (t-s)^{r_1-1}(x-y)^{r_2-1} \\
&\times |f(s,y,u_{(s,y)},(Gu)(s,y)) - f(s,y,\varphi(s),(G\varphi)(s))|dyds \\
\le\ & \overline{d}+\int_0^t\int_0^x (t-s)^{r_1-1}(x-y)^{r_2-1}p(s,y) \\
&\times\left[h(s,y)+\int_0^s\int_0^y k(s,y,\tau,\xi)h(\tau,\xi)d\xi d\tau\right]dyds.
\end{aligned} \tag{7.52}$$

Now an application of Lemma 7.17, to (7.52) yields (7.51). □

7.5.4. Global Asymptotic Stability of Solutions

We next prove under more appropriate conditions on the functions involved in (7.37)-(7.39) that the solutions tends exponentially toward zero as $t\to\infty$.

Theorem 7.20. *Assume that* (7.25.1), (7.25.2) *ant the following hypotheses hold*

(7.29.1) *There exist constants* $\lambda > 0$ *and* $M \geq 0$ *such that*

$$|\varphi(t)| \leq Me^{-\lambda t}; \tag{7.53}$$

$$|f(t,x,u_1,u_2) - f(t,x,v_1,v_2)| \leq p_1(t,x)e^{-\lambda t}(\|u_1 - v_1\|_C + |u_2 - v_2|), \tag{7.54}$$

for each $(t,x) \in J,\ u_1, v_1 \in C,\ u_2, v_2 \in \mathbb{R}$,

$$|g(t,x,s,y,u) - g(t,x,s,y,v)| \leq q(t,x,s,y)|u - v|; \tag{7.55}$$

for each $(t,x,s,y) \in J_1,\ u, v \in \mathbb{R}$,
and $f(t,x,0,(G0)(t,x)) = 0$; *for each* $(t,x) \in J$ *and the functions* p, q *be as in Theorem 7.18,*

(7.29.2) $\int_0^\infty \int_0^x [p(s,y) + A(s,y)]dyds < \infty$, *where* A *is given by (7.46).*

If u *is any solution of (7.37)-(7.39) on* $[-\alpha,\infty) \times [-\beta,b]$, *then all solutions of problem (7.37)-(7.39) are uniformly globally attractive on* J.

Proof. From the hypotheses, for each $(t,x) \in J$, we have that

$$\begin{aligned}|u(t,x)| &\leq |\varphi(t)| + \frac{1}{\Gamma(r_1)\Gamma(r_2)} \int_0^t \int_0^x (t-s)^{r_1-1}(x-y)^{r_2-1} \\ &\quad \times |f(s,y,u_{(s,y)},(Gu)(s,y)) - g(s,y,0,(G0)(s,y))|dyds \\ &\quad + \frac{1}{\Gamma(r_1)\Gamma(r_2)} \int_0^t \int_0^x (t-s)^{r_1-1}(x-y)^{r_2-1}|f(s,y,0,(G0)(s,y))|dyds \\ &\leq Me^{-\lambda t} + \int_0^t \int_0^x p(s,y)e^{-\lambda t}[u_{(s,y)} + \frac{1}{\Gamma(r_1)\Gamma(r_2)} \\ &\quad \times \int_0^s \int_0^y (s-\tau)^{r_1-1}(y-\xi)^{r_2-1}q(s,y,\tau,\xi)|u(\tau,\xi)|d\xi d\tau\Big] dyds. \end{aligned} \tag{7.56}$$

From (7.56), we get

$$|u(t,x)|e^{\lambda t} \leq M + \int_0^t \int_0^x p(s,y)\left[u_{(s,y)} + k(s,y,\tau,\xi)|u(\tau,\xi)|d\xi d\tau\right] dyds. \tag{7.57}$$

Now an application of Lemma 7.17 to (7.57) yields

$$\begin{aligned}|u(t,x)|e^{\lambda t} &\leq M[1 \\ &+ \int_0^t \int_0^x p(s,y)\exp\left(\int_0^s \int_0^y [p(\tau,\xi) + A(\tau,\xi)]d\xi d\tau\right) dyds\Big];\ (t,x) \in J,\end{aligned} \tag{7.58}$$

Multiplying both sides of (7.58) by $e^{-\lambda t}$ and in view of (7.29.1), we get

$$\begin{aligned}|u(t,x)| &\leq M[e^{-\lambda t} \\ &+ \int_0^t \int_0^x p(s,y)\exp\left(-\lambda t + \int_0^s \int_0^y [p(\tau,\xi) + A(\tau,\xi)]d\xi d\tau\right) dyds\Big].\end{aligned}$$

Thus, for each $x \in [0,b]$, we get

$$\lim_{t \to \infty} |u(t,x)| = 0.$$

Hence, the solution u tends to zero as $t \to \infty$. Consequently, all solutions of problem (7.37)-(7.39) are uniformly globally attractive on $[-\alpha,\infty) \times [-\beta,b]$. □

7.5.5. An Example

To illustrate our results, we consider the following system of partial integro-differential equations of fractional order of the form

$$^{c}D_{\theta}^{r}u(t,x) = f(t,x,u_{(t,x)},(Gu)(t,x)); \quad \text{for } (t,x) \in J := \mathbb{R}_{+} \times [0,1], \tag{7.59}$$

$$u(t,x) = \frac{1}{1+t^2}; \text{ if } (t,x) \in \tilde{J} := [-1,\infty) \times [-2,1] \backslash (0,\infty) \times (0,1], \tag{7.60}$$

$$\begin{cases} u(t,0) = \frac{1}{1+t^2}; \ t \in \mathbb{R}_{+}, \\ u(0,x) = 1; \ x \in [0,1], \end{cases} \tag{7.61}$$

where

$$(Gu)(t,x) = \frac{1}{\Gamma(r_1)\Gamma(r_2)} \int_0^t \int_0^x (t-s)^{r_1-1}(x-y)^{r_2-1} g(t,x,s,y,u_{(s,y)}) dyds, \tag{7.62}$$

$r_1, r_2 \in (0,1]$,

$$\begin{cases} f(t,x,u,v) = \dfrac{x^2 t^{-r_1} \sin t}{2c(1+t^{-\frac{1}{2}})(1+|u(t+1,x+2)|+|v|)}; \\ \qquad for \ (t,x) \in J, \ t \neq 0 \ and \ u \in C, \ v \in \mathbb{R}, \\ f(0,x,u,v) = 0, \end{cases}$$

$$c := \frac{\Gamma(\frac{1}{2})}{\Gamma(\frac{1}{2}+r_1)} \left(1 + \frac{\Gamma(\frac{1}{2})e}{\Gamma(\frac{1}{2}+r_1)\Gamma(1+r_2)} \right),$$

$$\begin{cases} g(t,x,s,y,u) = \dfrac{t^{-r_1} s^{-\frac{1}{2}} e^{x-y-\frac{1}{s}-\frac{1}{t}}}{2c(1+t^{-\frac{1}{2}})(1+|u|)}; \\ \qquad for \ (t,x,s,y) \in J_1, \ st \neq 0 \ and \ u \in \mathbb{R}, \\ g(t,x,0,y,u) = g(0,x,s,y,u) = 0, \end{cases}$$

and

$$J_1 = \{(t,x,s,y) : 0 \le s \le t < \infty, \ 0 \le y \le x \le 1\}.$$

Set

$$\varphi(t) = \frac{1}{1+t^2}; \ t \in \mathbb{R}_{+}.$$

We can see that $(7.22.1)$ is satisfied because the function φ is continuous and bounded with $\varphi^* = 1$. For each $u_1, v_1 \in C$, $u_2, v_2 \in \mathbb{R}$ and $(t,x) \in J$, we have

$$|f(t,x,u_1,u_2) - f(t,x,s,v_1,v_2)| \le \frac{1}{2c(1+t^{-\frac{1}{2}})} \left(x^2 t^{-r_1} |\sin t|\right) (|u_1 - v_1| + |u_2 - v_2|),$$

and for each $u, v \in \mathbb{R}$ and $(t,x,s,y) \in J_1$, we have

$$|g(t,x,s,y,u) - g(t,x,s,y,v)| \le \frac{1}{2c(1+t^{-\frac{1}{2}})} \left(t^{-r_1} s^{-\frac{1}{2}} e^{x-y-t-\frac{1}{s}-\frac{1}{t}}\right) |u-v|.$$

Hence condition (7.22.2) is satisfied with

$$\begin{cases} p_1(t,x) = p_2(t,x) = \dfrac{x^2 t^{-r_1}|\sin t|}{2c(1+t^{-\frac{1}{2}})}; \ t \neq 0, \\ p_1(0,x) = p_2(0,x) = 0, \end{cases}$$

and condition (7.22.3) is satisfied with

$$\begin{cases} q(t,x,s,y) = \dfrac{1}{2c(1+t^{-\frac{1}{2}})}\left(t^{-r_1}s^{-\frac{1}{2}}e^{x-y-t-\frac{1}{s}-\frac{1}{t}}\right); \ st \neq 0, \\ q(t,x,0,y) = k(0,x,0,y) = 0. \end{cases}$$

We shall show that condition (7.41) holds with $b = 1$. Indeed

$$\begin{aligned} & \frac{1}{\Gamma(r_1)\Gamma(r_2)}\int_0^t\int_0^x (t-s)^{r_1-1}(x-y)^{r_2-1}p_1(s,y)dyds \\ \leq & \ \frac{1}{2c(1+t^{-\frac{1}{2}})\Gamma(r_1)\Gamma(r_2)}\int_0^t\int_0^1 (t-s)^{r_1-1}(1-y)^{r_2-1}x^2t^{-r_1}dyds \\ \leq & \ \frac{\Gamma(\frac{1}{2})et^{-\frac{1}{2}}}{2c(1+t^{-\frac{1}{2}})\Gamma(\frac{1}{2}+r_1)\Gamma(1+r_2)}, \end{aligned}$$

then

$$p_1^* = p_2^* \leq \frac{\Gamma(\frac{1}{2})}{2c\Gamma(\frac{1}{2}+r_1)}.$$

Also,

$$\begin{aligned} & \frac{1}{\Gamma(r_1)\Gamma(r_2)}\int_0^t\int_0^x (t-s)^{r_1-1}(x-y)^{r_2-1}q(t,x,s,y)dyds \\ \leq & \ \frac{1}{2c(1+t^{-\frac{1}{2}})\Gamma(r_1)\Gamma(r_2)}\int_0^t\int_0^1 (t-s)^{r_1-1}(1-y)^{r_2-1}t^{-r_1}s^{-\frac{1}{2}}e^x dyds \\ \leq & \ e^x t^{-r_1} t^{-\frac{1}{2}+r_1}\frac{\Gamma(\frac{1}{2})}{2c(1+t^{-\frac{1}{2}})\Gamma(\frac{1}{2}+r_1)\Gamma(1+r_2)} \\ \leq & \ \frac{\Gamma(\frac{1}{2})et^{-\frac{1}{2}}}{2c(1+t^{-\frac{1}{2}})\Gamma(\frac{1}{2}+r_1)\Gamma(1+r_2)}, \end{aligned}$$

then

$$q^* \leq \frac{e\Gamma(\frac{1}{2})}{2c\Gamma(\frac{1}{2}+r_1)\Gamma(1+r_2)}.$$

Thus,

$$p_1^* + p_2^* q^* \leq \frac{\Gamma(\frac{1}{2})}{2c\Gamma(\frac{1}{2}+r_1)}\left(1+\frac{\Gamma(\frac{1}{2})e}{\Gamma(\frac{1}{2}+r_1)\Gamma(1+r_2)}\right) = \frac{1}{2} < 1,$$

which is satisfied for each $r_1, r_2 \in (0,\infty)$. Consequently Theorem 7.16 implies that problem (7.59)-(7.61) has a unique solution defined on $[-1,\infty)\times[-2,1]$.

7.6. Notes and Remarks

The results of Chapter 7 are taken from Abbas *et al.* [1, 32, 33]. Other results may be found in [261, 347, 349, 350].

Chapter 8

Abstract Integral Equations

8.1. Introduction

This chapter is concerned with the existence and uniqueness of solutions for integral equations of Volterra type

$$y(t) = \int_0^t a(t-s)[Ay(s) + f(s, y(s))]ds; \; t \in J := [0, +\infty), \tag{8.1}$$

where $a : [0, \infty) \to \mathbb{R}$ is locally integrable, $A : D(A) \subset E \to E$ is the generator of an integral resolvent family defined on a complex Banach space $(E, \|\cdot\|)$, and $f : [0, +\infty) \times E \to E$ is a given function.
Let $B(E)$ be the Banach space of all linear bounded operators from E into E with norm

$$\|N\|_{B(E)} = \sup\{\|N(y)\| : \; \|y\| = 1\}.$$

A measurable function $y : J \to E$ is Bochner integrable if $\|y\|$ is Lebesgue integrable. For properties of the Bochner integral, see e.g. Yosida [371]. $L^1(J, E)$ denotes the Banach space of functions $y : J \to E$, which are Bochner integrable with norm

$$\|y\|_1 = \int_0^\infty \|y(t)\|dt.$$

The Laplace transformation of a function $f \in L^1_{loc}(\mathbb{R}_+, E)$ is defined by

$$\mathcal{L}(f)(\lambda) :=: \widehat{a}(\lambda) = \int_0^\infty e^{-\lambda t} f(t)dt, \quad Re(\lambda) > \omega,$$

if the integral is absolutely convergent for $Re(\lambda) > \omega$. In order to defined the mild solution of (8.1) we recall the following definition.

Definition 8.1. *Let A be a closed and linear operator with domain $D(A)$ defined on a Banach space E. We call A the generator of an integral resolvent if there exists $\omega > 0$ and a strongly continuous function $S : \mathbb{R}^+ \to B(E)$ such that*

$$\left(\frac{1}{\widehat{a}(\lambda)} I - A\right)^{-1} x = \int_0^\infty e^{-\lambda t} S(t)xdt, \; Re\lambda > \omega, x \in E.$$

In this case, $S(t)$ is called the integral resolvent family generated by A.

The following result is a direct consequence of ([266], Proposition 3.1 and Lemma 2.2).

Proposition 8.2. *Let $\{S(t)\}_{t\geq 0} \subset B(E)$ be an integral resolvent family with generator A. Then the following conditions are satisfied:*

a) $S(t)$ *is strongly continuous for* $t \geq 0$ *and* $S(0) = I$;

b) $S(t)D(A) \subset D(A)$ *and* $AS(t)x = S(t)Ax$ *for all* $x \in D(A)$, $t \geq 0$;

c) for every $x \in D(A)$ *and* $t \geq 0$,

$$S(t)x = a(t)x + \int_0^t a(t-s)AS(s)xds.$$

d) Let $x \in D(A)$. *Then* $\int_0^t a(t-s)S(s)xds \in D(A)$ *and*

$$S(t)x = a(t)x + A\int_0^t a(t-s)S(s)xds.$$

In particular, $S(0) = a(0)$.

Remark 8.3. *The uniqueness of resolvent is well-known (see Prüss [318]).*

If an operator A with domain $D(A)$ is the infinitesimal generator of an integral resolvent family $S(t)$ and $a(t)$ is a continuous, positive and nondecreasing function which satisfies $\lim_{t\to 0^+} \frac{\|S(t)\|_{B(E)}}{a(t)} < \infty$ then for all $x \in D(A)$ we have

$$Ax = \lim_{t\to 0^+} \frac{S(t)x - a(t)x}{(a*a)(t)},$$

see ([267], Theorem 2.1). For example, the case $a(t) \equiv 1$ corresponds to the generator of a C_0-semigroup and $a(t) = t$ actually corresponds to the generator of a sine family see [75]. A characterization of generators of integral resolvent families, analogous to the Hille-Yosida Theorem for C_0-semigroups, can be directly deduced from ([267], Theorem 3.4). More information on the C_0-semigroups and sine families can be found in [183, 184, 307].

Definition 8.4. A function $f : J \times E \to E$ is said to be an Carathéodory function if it satisfies:

(i) for each $t \in J$ the function $f(t,.) : E \to E$ is continuous;

(ii) for each $y \in E$ the function $f(.,y) : J \to E$ is measurable.

8.2. A Local Uniqueness Result for Abstract Integral Equations of Volterra Type in Banach Spaces

8.2.1. Introduction

We establish, in this Section, existence results for the equation (8.1) on a compact interval $J = [0,b]$.

8.2.2. Existence of Solutions

In order to define mild solutions for the equation (8.1), we need the following auxiliary lemma

Lemma 8.5. *Let $a \in L^1_{loc}(J,\mathbb{R})$. Assume that A generates an integral resolvent family $\{S(t)\}_{t\geq 0}$ on E, which is in addition integrable and $\overline{D(A)} = E$. Then the unique bounded solution of the problem*

$$y(t) = \int_0^t a(t-s)Ay(s)ds + \int_0^t a(t-s)f(s,y(s))ds, \tag{8.2}$$

is given by

$$y(t) = \int_0^t S(t-s)f(s,y(s))ds,\ t \in J. \tag{8.3}$$

This lemma leads us to the definition of a mild solution of equation (8.1).

Definition 8.6. *A function $y \in C(J,E)$ is said to be a mild solution of (8.1) if y satisfies the integral equation*

$$y(t) = \int_0^t S(t-s)f(s,y(s))ds,\ t \in J.$$

We assume that the following hypotheses are true:

(8.9.1) The operator solution $S(t)_{t\in J}$ is uniformly continuous, and there exists $M > 0$ such that

$$\|S(t)\| \leq M \text{ for every } t \in J.$$

(8.9.2) The function $f : J \times E \to E$ is Carathéodory;

(8.9.3) There exists $p \in L^1(J,\mathbb{R}_+)$ and a continuous and a continuous nondecreasing function $\psi : [0,\infty) \to [0,\infty)$ such that

$$\|F(t,x)\| \leq p(t)\psi(\|x\|) \text{ for a.e. } t \in J$$

(8.9.4) For each $t \in J$ and bounded $B \subset \mathcal{B}$ we have

$$\alpha(f(t,B)) \leq p(t)\alpha(B).$$

(8.9.5) There exists $r > 0$ such that

$$M\,\psi(r)\int_0^b p(s)ds \leq r,$$

Theorem 8.7. *Suppose that conditions $(8.9.1)-(8.9.5)$ hold. Then Equation (8.1) has has at least one solution.*

Proof. Consider the operator $N : C(J,E) \to C(J,E)$ defined by

$$Ny(t) := \int_0^t S(t-s)f(s,y(s))ds, \text{ for } t \in J.$$

Clearly, the fixed points of the operator N are solutions of Equation (8.1). Let y be a possible solution of Equation (8.1), that is for each $t \in J$

$$y(t) = \int_0^t S(t-s)f(s,y(s))ds.$$

Consider the set

$$D_r = \{y \in C(J,E) : \|y\|_\infty \leq r\},$$

where r is the constant defined in (8.9.5). Clearly, the subset D_r is closed, bounded, and convex. In order to apply Theorem 9.12, we give the proof in several steps.

Step 1: N is continuous
Let y_n be a sequence such that $y_n \longrightarrow y$ in $C(J,E)$. Then

$$\begin{aligned} |N(y_n)(t) - N(y)(t)| &= |\int_0^t S(t-s)\, [f(s,y_n(s)) - f(s,y(s))]\, ds| \\ &\leq M \int_0^b |f(s,y_n(s)) - f(s,y(s))|\, ds. \end{aligned}$$

Since f is of Carathéodory type, then by the Lebesgue dominated convergence theorem we have

$$\|N(y_n)(t) - N(y)(t)\| \to 0 \text{ as } n \to +\infty.$$

Step 2: N maps D_r into itself.
For each $y \in D_r$, by (8.9.3) and (8.9.5), we have for each $t \in J$

$$\begin{aligned} |Ny(t)| &\leq M \int_0^t |f(s,y(s))|\, ds \\ &\leq M \int_0^t p(s)\, \psi(\|y(s)\|)\, ds \\ &\leq M \int_0^b p(s)\, \psi(r)\, ds \\ &\leq r. \end{aligned}$$

Step 3: $N(D_r)$ is bounded and equicontinuous.
By Step 2, it is obvious that $N(D_r) \subset D_r$ is bounded. For the equicontinuity of $N(D_r)$. Let $\tau_1, \tau_2 \in J$ with $\tau_2 > \tau_1$, and $y \in D_r$,

$$\begin{aligned} |N(y)(\tau_2) - N(y)(\tau_1)| &\leq |\int_0^{\tau_1} [S(\tau_2 - s) - S(\tau_1 - s)]f(s,y(s)ds| \\ &\quad + |\int_{\tau_1}^{\tau_2} S(\tau_2 - s)f(s,y(s)|ds \\ &\leq \psi(r) + \int_0^{\tau_1} |S(\tau_2 - s) - S(\tau_1 - s)|p(s)ds \\ &\quad + M\psi(r) \int_{\tau_1}^{\tau_2} p(s)ds. \end{aligned}$$

The right-hand of the above inequality tends to zero as $\tau_2 - \tau_1 \to 0$, since $S(t)$ is uniformly continuous operator.
Now let V be a subset of D_r such that $V \subset \overline{conv}(N(V) \cup \{0\})$. V is bounded and equicontinuous and therefore the function $t \to v(t) = \alpha(V(t))$ is continuous on J. By (8.9.4), and the properties of the measure α we have for each $t \in J$

$$\begin{aligned} v(t) &\leq \alpha(N(V)(t) \cup \{0\}) \\ &\leq \alpha(N(V)(t)) \\ &\leq \alpha\left(\int_0^t S(t-s)\, f(s,y(s))\, ds\right) \\ &\leq M\int_0^t p(s)\alpha(y(s) : y \in V)ds \\ &\leq M\int_0^t p(s)\alpha(V(s))ds \\ &\leq M\int_0^t p(s)v(s)ds. \end{aligned}$$

By Gronwall Lemma it follows that $v(t) = 0$ for each for each $t \in J$, and then $V(t)$ is relatively compact in E. In view of the Arzela-Ascoli theorem, V is relatively compact in D_r. Applying now Theorem 9.12 we conclude that N has a fixed point $y \in D_r$. Hence N has a fixed point which is solution of (8.1). □

8.3. A Global Uniqueness Result for Abstract Integral Equations of Volterra Type in Banach Spaces

8.3.1. Introduction

In this section, we study the existence of mild solutions, defined on the positive real line for the abstract integral equations (8.1).

8.3.2. Existence of Solutions

In order to define mild solutions for (8.1), we need the following auxiliary lemma.

Lemma 8.8. *Let $a \in L^1_{loc}(J, \mathbb{R})$. Assume that A generates an integral resolvent family $\{S(t)\}_{t\geq 0}$ on E, which is in addition integrable and $\overline{D(A)} = E$. Then the unique bounded solution of the problem*

$$y(t) = \int_0^t a(t-s)Ay(s)ds + \int_0^t a(t-s)f(s,y(s))ds, \tag{8.4}$$

is given by

$$y(t) = \int_0^t S(t-s)f(s,y(s))ds,\ t \in J. \tag{8.5}$$

This lemma leads us to the definition of a mild solution of equation (8.1).

Definition 8.9. *A function $y \in C(J,E)$ is said to be a mild solution of (8.1) if y satisfies the integral equation*

$$y(t) = \int_0^t S(t-s)f(s,y(s))ds, \; t \in J.$$

We assume that the following hypotheses are true:

(8.13.1) The function $f : J \times E \to E$ is Carathéodory;

(8.13.2) There exists $M \geq 1$ such that

$$\|S(t)\|_{B(E)} \leq M \text{ for every } t \geq 0,$$

(8.13.3) There exists $l \in L^\infty_{loc}(J, \mathbb{R}_+)$ such that

$$\|f(t,u) - f(t,\overline{u})\| \leq l(t)\|u - \overline{u}\|, \text{ for each } t \in J, \; u, \; \overline{u} \in E.$$

Set

$$l_n^* = \|l\|_{L^\infty([0,n])}, \; f_n = \int_0^n f(s,0)ds.$$

For every $n \in \mathbb{N}$, we define seminorms in $C(J,E)$ according to the formula

$$\|y\|_n = \sup\{e^{-\tau M l_n^* t}\|y(t)\| : t \leq n\},$$

where $\tau > 1$. Then $C([0,\infty),E)$ is a Fréchet space with the family of semi-norms $\|.\|_n$.

Theorem 8.10. *Suppose that conditions* $(8.13.1) - (8.13.3)$ *hold. Then the equation (8.1) has a unique solution.*

Proof. Consider the operator $N : C(J,E) \to C(J,E)$ defined by

$$Ny(t) := \int_0^t S(t-s)f(s,y(s))ds, \text{ for } t \in J.$$

Clearly, the fixed points of the operator N are solutions of Equation (8.1). Let y be a possible solution of Equation (8.1), that is for each $t \in J$

$$y(t) = \int_0^t S(t-s)f(s,y(s))ds.$$

Thus for $t \in [0,n]$,

$$\begin{aligned} \|y(t)\| &\leq \int_0^t \|S(t-s)\|_{B(E)}\|f(s,y(s))\|ds \\ &\leq M\int_0^t l(s)\|y(s)\|ds + M\int_0^t l(s)\|f(s,0)\|ds \\ &\leq Ml_n^* f_n + Ml_n^* \int_0^t \|y(s)\|ds. \end{aligned}$$

Using Gronwall's inequality [150], for each $t \in [0,n]$, we get

$$\|y(t)\| \leq Ml_n^* f_n e^{Ml_n^* n} := b_n.$$

Set

$$U := \{y \in C(J,E) : \|y\|_n \leq b_n + 1, \text{ for each } n \in \mathbb{N}\}.$$

Clearly, U is a closed subset of $C(J,E)$. Let us show that $N : U \to C(J,E)$ is a contraction operator. Indeed, consider $y, \bar{y} \in U$. Then, for every $t \in [0,n]$ and $n \in \mathbb{N}$, we have

$$\begin{aligned}
\|Ny(t) - N\bar{y}(t))\| &= \left\| \int_0^t S(t-s)f(s,y(s))ds - \int_0^t S(t-s)f(s,\bar{y}(s)) \right\| ds \\
&\leq M \int_0^t l_n(s)\|y(s) - \bar{y}(s)\|ds \\
&\leq Ml_n^* \int_0^t \|y(s) - \bar{y}(s)\|ds \\
&\leq Ml_n^* \int_0^t e^{\tau l_n^* Ms} e^{-\tau l_n^* Ms} \|y(s) - \bar{y}(s)\|ds \\
&\leq \frac{1}{\tau} e^{\tau l_n^* Mt} \|y - \bar{y}\|_n.
\end{aligned}$$

Therefore,

$$\|N(y) - N(\bar{y})\|_n \leq \frac{1}{\tau}\|y - \bar{y}\|_n.$$

Hence, the operator N is a contraction for all $n \in \mathbb{N}$. It follows from the choice of U that there is no $y \in \partial U^n$ such that $y = \lambda N(y)$ for some $\lambda \in (0,1)$. As a consequence of the nonlinear alternative , we conclude that N has a unique fixed point, which is a mild solution of (8.1). □

8.4. Notes and Remarks

The results of Chapter 8 are taken from Benchohra and Litimein [100].

Chapter 9

Abstract Integro-Differential Equations

9.1. Introduction

In this chapter, we prove sufficient conditions for the existence and uniqueness of mild solutions for semilinear integro-differential equations of fractional order with infinite delay

$$y'(t) - \int_0^t \frac{(t-s)^{\alpha-2}}{\Gamma(\alpha-1)} Ay(s)ds = f(t, y_t), \quad \text{a.e. } t \in J \tag{9.1}$$

$$y_0 = \phi \in \mathcal{B}, \tag{9.2}$$

where $1 < \alpha < 2$ and $A : D(A) \subset E \to E$ is the generator of an integral resolvent family defined on a complex Banach space $(E, |\cdot|)$, the convolution integral in the equation is known as the Riemann-Liouville fractional integral and $f : [0, +\infty) \times E \to E$ is a given function. For any continuous function y defined on $(-\infty, +\infty)$ and any $t \geq 0$, we denote by y_t the element of $\mathcal{B}$ defined by $y_t(\theta) = y(t+\theta)$ for $\theta \in (-\infty, 0]$. Here $y_t(\cdot)$ represents the history of the state from each time $\theta \in (-\infty, 0]$ up to the present time t. We assume that the histories y_t belongs to some abstract phase space $\mathcal{B}$. We will employ an axiomatic definition of the phase space $\mathcal{B}$ introduced by Hale and Kato in [207] and follow the terminology used in [226]. Thus, $(\mathcal{B}, \|\cdot\|_{\mathcal{B}})$ will be a seminormed linear space of functions mapping $(-\infty, 0]$ into E, and satisfying the following axioms :

(A_1) If $y : (-\infty, b) \to E, b > 0$, is continuous on $[0, b]$ and $y_0 \in \mathcal{B}$, then for every $t \in [0, b)$ the following conditions hold :
(i) $y_t \in \mathcal{B}$;
(ii) There exists a positive constant H such that $|y(t)| \leq H \|y_t\|_{\mathcal{B}}$;
(iii) There exist two functions $K(\cdot), M(\cdot) : \mathbb{R}_+ \to \mathbb{R}_+$ independent of y with K continuous and M locally bounded such that :

$$\|y_t\|_{\mathcal{B}} \leq K(t) \sup\{\, |y(s)| : 0 \leq s \leq t\} + M(t) \|y_0\|_{\mathcal{B}}.$$

(A_2) For the function y in (A_1), y_t is a $\mathcal{B}-$valued continuous function on $[0, b]$.

(A_3) The space $\mathcal{B}$ is complete.

Denote $K_b = \sup\{K(t) : t \in [0,b]\}$ and $M_b = \sup\{M(t) : t \in [0,b]\}$.

Remark 9.1. 1. *(ii) is equivalent to* $|\phi(0)| \leq H\|\phi\|_{\mathcal{B}}$ *for every* $\phi \in \mathcal{B}$.

2. *Since* $\|\cdot\|_{\mathcal{B}}$ *is a seminorm, two elements* $\phi, \psi \in \mathcal{B}$ *can verify* $\|\phi - \psi\|_{\mathcal{B}} = 0$ *without necessarily* $\phi(\theta) = \psi(\theta)$ *for all* $\theta \leq 0$.

3. *From the equivalence of in the first remark, we can see that for all* $\phi, \psi \in \mathcal{B}$ *such that* $\|\phi - \psi\|_{\mathcal{B}} = 0$ *: We necessarily have that* $\phi(0) = \psi(0)$.

We now indicate some examples of phase spaces . For other details we refer, for instance to the book by Hino *et al* [226].

Example 9.2. *Let:*

BC *the space of bounded continuous functions defined from* $(-\infty, 0]$ *to* E;

BUC *the space of bounded uniformly continuous functions defined from* $(-\infty, 0]$ *to* E;

$$C^{\infty} := \left\{\phi \in BC : \lim_{\theta \to -\infty} \phi(\theta) \text{ exist in } E\right\};$$

$$C^{0} := \left\{\phi \in BC : \lim_{\theta \to -\infty} \phi(\theta) = 0\right\}, \text{ endowed with the uniform norm}$$

$$\|\phi\| = \sup\{|\phi(\theta)| : \theta \leq 0\}.$$

We have that the spaces BUC, C^{∞} *and* C^0 *satisfy conditions* $(A_1) - (A_3)$. *However,* BC *satisfies* $(A_1), (A_3)$ *but* (A_2) *is not satisfied.*

Example 9.3. *The spaces* C_g, UC_g, C_g^{∞} *and* C_g^0. *Let* g *be a positive continuous function on* $(-\infty, 0]$. *We define:*

$$C_g := \left\{\phi \in C((-\infty,0],E) : \frac{\phi(\theta)}{g(\theta)} \text{ is bounded on } (-\infty,0]\right\};$$

$$C_g^0 := \left\{\phi \in C_g : \lim_{\theta \to -\infty} \frac{\phi(\theta)}{g(\theta)} = 0\right\}, \text{ endowed with the uniform norm}$$

$$\|\phi\| = \sup\left\{\frac{|\phi(\theta)|}{g(\theta)} : \theta \leq 0\right\}.$$

Then we have that the spaces C_g *and* C_g^0 *satisfy conditions* (A_3). *We consider the following condition on the function* g.

(g_1) *For all* $a > 0$, $\displaystyle\sup_{0 \leq t \leq a} \sup\left\{\frac{g(t+\theta)}{g(\theta)} : -\infty < \theta \leq -t\right\} < \infty.$

They satisfy conditions (A_1) *and* (A_2) *if* (g_1) *holds.*

Example 9.4. *The space C_γ: For any real positive constant γ, we define the functional space C_γ by*

$$C_\gamma := \left\{ \phi \in C((-\infty,0],E) : \lim_{\theta \to -\infty} e^{\gamma\theta}\phi(\theta) \text{ exists in } E \right\}$$

endowed with the following norm

$$\|\phi\| = \sup\{e^{\gamma\theta}|\phi(\theta)| : \theta \leq 0\}.$$

Then in the space C_γ the axioms $(A_1)-(A_3)$ are satisfied.

9.2. A local Uniqueness Result for Functional Integro-Differential Equations in Banach Spaces

9.2.1. Introduction

In this Section, we are concerned with the existence and uniqueness of solutions defined on a compact real interval $J = [0,b]$ for the problem (9.1)-(9.2).

9.2.2. Existence of Solutions

Before starting and proving this result, we give the definition of its mild solution.

Definition 9.5. *We say that the function $y : (-\infty,b] \to E$ is a mild solution of (9.1)-(9.2) if $y(t) = \phi(t)$ for all $t \leq 0$ and y satisfies the following integral equation*

$$y(t) = S(t)\phi(0) + \int_0^t S(t-s)\, f(s,y_s)\, ds \quad \text{for each } t \in [0,b]. \tag{9.3}$$

Theorem 9.6. *Assume the conditions,*

(9.6.1) *The operator solution $S(t)_{t\in J}$ is compact for $t > 0$ and there exists a constant $M \geq 1$ such that*

$$\|S(t)\|_{B(E)} \leq M \quad \text{for every } t \geq 0.$$

(9.6.2) *The function $f : J \times \mathcal{B} \to E$ is Carathéodory.*

(9.6.3) *There exists a function $p \in L^1(J;\mathbb{R}_+)$ and a continuous nondecreasing function $\psi : \mathbb{R}_+ \to (0,\infty)$ and such that :*

$$|f(t,u)| \leq p(t)\, \psi(\|u\|_{\mathcal{B}}) \text{ for a.e. } t \in J \text{ and each } u \in \mathcal{B}.$$

And

$$\int_{c_b}^{+\infty} \frac{ds}{\psi(s)} > K_b M \int_0^b p(s)\, ds, \tag{9.4}$$

with

$$c_b = (M_b + L^\phi + K_b M H)\|\phi\|_{\mathcal{B}}.$$

Then the problem (9.1)-(9.2) has at least one solution on $(-\infty,b]$.

Proof. Let

$$B=\left\{y:(-\infty,b]\to E: y|_{[0,b]} \text{ is continuous and } y_0\in\mathcal{B}\right\},$$

where $y|_{[0,b]}$ is the restriction of y to the real compact interval $[0,b]$. We transform the problem (9.1)-(9.2) into a fixed-point problem. Consider the operator $N:B\to B$ defined by:

$$N(y)(t)=\begin{cases}\phi(t), & \text{if } t\le 0;\\ S(t)\,\phi(0)+\int_0^t S(t-s)\,f(s,y_s)\,ds, & \text{if } t\in[0,b].\end{cases} \tag{9.5}$$

Clearly, fixed points of the operator N are mild solutions of the problem (e231)-(e232).

For $\phi\in\mathcal{B}$, we will define the function $x(.):(-\infty,b]\to E$ by

$$x(t)=\begin{cases}\phi(t), & \text{if } t\le 0;\\ S(t)\,\phi(0), & \text{if } t\in[0,b].\end{cases}$$

Then $x_0=\phi$. For each function $z\in B$ with $z_0=0$, we denote by $\bar{z}$ the function defined by

$$\bar{z}(t)=\begin{cases}0, & \text{if } t\le 0;\\ z(t), & \text{if } t\in[0,b].\end{cases}$$

If $y(\cdot)$ satisfies (9.3), we can decompose it as $y(t)=\bar{z}(t)+x(t)$, $t\ge 0$, which implies $y_t=\bar{z}_t+x_t$, for every $t\in J$ and the function $z(\cdot)$ satisfies

$$z(t)=\int_0^t S(t-s)\,f(s,\bar{z}_s+x_s)\,ds \quad \text{for } t\in[0,b].$$

Let

$$B_b=\{z\in B: z_0=0\in\mathcal{B}\}.$$

For any $z\in B_b$ we have

$$\|z\|_b=\|z_0\|_{\mathcal{B}}+\sup\{|z(s)|:0\le s<b\}=\sup\{|z(s)|:0\le s<b\}.$$

Thus $(B_b,\|\cdot\|_b)$ is a Banach space. We define the operator $F:B_b\to B_b$ by :

$$F(z)(t)=\int_0^t S(t-s)\,f(s,\bar{z}_s+x_s)\,ds \quad \text{for } t\in[0,b]. \tag{9.6}$$

Obviously the operator N has a fixed point is equivalent to F has one, so it turns to prove that F has a fixed point. We shall show that the operators $F:B_b\to B_b$ is continuous and completely continuous.

Step 1: F is continuous.
Let z_n be a sequence such that $z_n\longrightarrow z$ in B_b. Then

$$\begin{aligned}|F(z_n)(t)-F(z)(t)| &= \left|\int_0^t S(t-s)\,[f(s,\bar{z}_{n_s}+x_s)-f(s,\bar{z}_s+x_s)]\,ds\right|\\ &\le M\int_0^t |f(s,\bar{z}_{n_s}+x_s)-f(s,\bar{z}_s+x_s)|\,ds.\end{aligned}$$

Since $f(s,.)$ is continuous, we have by the Lebesgue dominated convergence theorem

$$\|f(.,\bar{z}_{n.}+x_.)-f(.,\bar{z}_.+x_.)\|_{L^1}\to 0 \text{ as } n\to+\infty.$$

Thus F is continuous.

Step 2: F maps bounded sets into bounded sets in B_b.
Indeed, it is enough show that, for any $d>0$, there exists a positive constant ℓ such that for each $z\in B_d=\{z\in B_b:\|z\|_b\le d\}$ one has $\|F(z)\|_b\le\ell$. Let $z\in B_d$. By the hypotheses (9.6.1) and (9.6.2), we have for each $t\in[0,b]$

$$\begin{aligned}|Fz(t)| &\le M\int_0^t|f(s,\bar{z}_s+x_s)|\,ds\\ &\le M\int_0^t p(s)\,\psi(\|\bar{z}_s+x_s\|_{\mathcal{B}})\,ds.\end{aligned}$$

Using the assumption (A1), we get

$$\begin{aligned}\|\bar{z}_s+x_s\|_{\mathcal{B}} &\le \|\bar{z}_s\|_{\mathcal{B}}+\|x_s\|_{\mathcal{B}}\\ &\le K(s)|z(s)|+M(s)\|z_0\|_{\mathcal{B}}+K(s)|x(s)|+M(s)\|x_0\|_{\mathcal{B}}\\ &\le K_b|z(s)|+K_bM|\phi(0)|+M_b\|\phi\|_{\mathcal{B}}\\ &\le K_b|z(s)|+K_bMH\|\phi\|_{\mathcal{B}}+M_b\|\phi\|_{\mathcal{B}}\\ &\le K_b|z(s)|+(K_bMH+M_b)\|\phi\|_{\mathcal{B}}.\end{aligned}$$

Set

$$c_b:=(M_b+K_bMH)\|\phi\|_{\mathcal{B}},\quad D_b:=K_bd+c_b.$$

Then, we have

$$|F(z)(t)|\le M\psi(D_b)\int_0^b p(s)\,ds.$$

Thus

$$\|F(z)\|_b\le M\psi(D_b)\|p\|_{L^1}:=\ell.$$

Step 3: F maps bounded sets into equicontinuous sets in B_b.
Let $\tau_1,\tau_2\in[0,b]$ with $\tau_2>\tau_1$, B_d be a bounded set as in Step 2, and $z\in B_d$. Then

$$\begin{aligned}|F(z)(\tau_2)-F(z)(\tau_1)| &\le \left|\int_0^{\tau_1}[S(\tau_2-s)-S(\tau_1-s)]f(s,\bar{z}_s+x_s)\,ds\right|\\ &\quad+\left|\int_{\tau_1}^{\tau_2}S(\tau_2-s)f(s,\bar{z}_s+x_s)\,ds\right|\\ &\le \int_0^{\tau_1}|S(\tau_2-s)-S(\tau_1-s)||f(s,\bar{z}_s+x_s)|\,ds\\ &\quad+\int_{\tau_1}^{\tau_2}|S(\tau_2-s)||f(s,\bar{z}_s+x_s)|\,ds\\ &\le \psi(D_b)\int_0^{\tau_1}|S(\tau_2-s)-S(\tau_1-s)|p(s)ds\\ &\quad+M\psi(D_b)\int_{\tau_1}^{\tau_2}p(s)ds.\end{aligned}$$

The right-hand of the above inequality tends to zero as $\tau_2 - \tau_1 \to 0$, since $S(t)$ is uniformly continuous. As a consequence of Steps 1 to 3 together with the Arzela-Ascoli theorem it suffices to show that the operator F maps B_d into a precompact set in E. Let $t \in [0,b]$ be fixed and let ε be a real number satisfying $0 < \varepsilon < t$. For $z \in B_d$ we define

$$F_\varepsilon(z)(t) = S(\varepsilon)\int_0^{t-\varepsilon} S(t-s-\varepsilon)f(s,\bar{z}_s+x_s)\,ds.$$

Since $S(t)$ is a compact operator, the set $Z_\varepsilon(t) = \{F_\varepsilon(z)(t) : z \in B_d\}$ is pre-compact in E for every ε, $0 < \varepsilon < t$. Moreover

$$\begin{aligned}|F(z)(t) - F_\varepsilon(z)(t)| &\leq \int_{t-\varepsilon}^{t} \|S(t-s)\||f(s,\bar{z}_s+x_s)|ds\\ &\leq M\psi(D_b)\int_{t-\varepsilon}^{t} p(s)ds.\end{aligned}$$

Therefore the set $Z(t) = \{F(z)(t) : z \in B_d\}$ is precompact in E. Hence the operator F is completely continuous.

Step 4: (A priori bounds on solutions) We now show there exists an open set $U \subseteq Z$ with $z \neq \lambda F(z)$ for $\lambda \in (0,1)$ and $z \in \partial U$ Let $z \in C_0$ be such that $z = \lambda F(z)$ for some $\lambda \in [0,1)$. By the hypotheses (9.6.1), (9.6.2), we have for each $t \in [0,b]$

$$\begin{aligned}|z(t)| &\leq \int_0^t \|S(t-s)\|_{B(E)}\ \||f(s,\bar{z}_s+x_s)|\ ds\\ &\leq M\int_0^t p(s)\ \psi(\|\bar{z}_s+x_s\|_{\mathcal{B}})\ ds\\ &\leq M\int_0^t p(s)\ \psi(K_b|z(s)|+(M_b+\phi+K_bMH)\|\phi\|_{\mathcal{B}})\ ds\\ &\leq M\int_0^t p(s)\ \psi(K_b|z(s)|+c_b)\ ds.\end{aligned}$$

Then

$$K_b|z(t)|+c_b \leq c_b+K_bM\int_0^t p(s)\ \psi(K_b|z(s)|+c_b)\ ds.$$

We consider the function μ defined by

$$\mu(t) := \sup\{\ K_b|z(s)|+c_b\ :\ 0\leq s\leq t\ \},\quad 0\leq t<b.$$

Let $t^\star \in [0,t]$ be such that $\mu(t) = K_b|z(t^\star)|+c_b\|\phi\|_{\mathcal{B}}$. By the previous inequality, we have

$$\mu(t) \leq c_b+K_bM\int_0^t p(s)\ \psi(\mu(s))\ ds\quad \text{for}\quad t\in[0,b].$$

Let us take the right-hand side of the above inequality as $v(t)$. Then, we have

$$\mu(t) \leq v(t)\ \text{ for all } t\in[0,b].$$

From the definition of v, we have

$$v(0) = c_b \quad \text{and} \quad v'(t) = K_b M p(t)\, \psi(\mu(t)) \ \text{ a.e. } t \in [0,b].$$

Using the nondecreasing character of ψ; we get

$$v'(t) \leq K_b M\, p(t)\, \psi(v(t)) \ \text{ a.e. } t \in [0,b].$$

Using the condition (9.4), this implies that for each $t \in [0,b]$ we have

$$\int_{c_b}^{v(t)} \frac{ds}{\psi(s)} \leq K_b M \int_0^t p(s)\, ds \leq K_b M \int_0^b p(s)\, ds < \int_{c_b}^{+\infty} \frac{ds}{\psi(s)}.$$

Thus, for every $t \in [0,b]$, there exists a constant Λ such that $v(t) \leq \Lambda$ and hence $\mu(t) \leq \Lambda$. Since $\|z\|_b \leq \mu(t)$, we have $\|z\|_b \leq \Lambda$. Set

$$U = \{z \in Z : \|z\|_\infty < \Lambda + 1\}.$$

Then $F : \overline{U} \to Z$ is continuous and completely continuous. From the choice of U, there is no $z \in \partial U$ such that $z = \lambda F(z)$, for $\lambda \in (0,1)$. As a consequence of the nonlinear alternative of Leray-Schauder type [204], we deduce that F has a fixed point z in U, then the operator N has one, which is a solution of problem (9.1)-(9.2). □

We next introduce some additional conditions that lead to uniqueness of the solution of (9.1)-(9.2).

Theorem 9.7. *Assume* (9.6.1). *Moreover,*

(9.8.1) *There exists* $l > 0$ *such that :*

$$|f(t,u) - f(t,v)| \leq l\ \|u - v\|_{\mathcal{B}} \quad \text{for all } u, v \in \mathcal{B}.$$

If

$$M\, l\, k\, b < 1, \tag{9.7}$$

then there exists a unique solution for the problem (9.1)-(9.2).

Proof. We prove that the problem (9.1)-(9.2) has unique solution. Then we prove only that the operator F defined in Theorem 9.6 has a unique fixed point. We shall show that F is a contraction operator. Indeed, consider $z, z^* \in B_b$; thus for each $t \in [0,b]$

$$\begin{aligned} |F(z)(t) - F(z^*)(t)| &\leq \int_0^t \|S(t-s)\|_{B(E)}\ |f(s, \overline{z}_s + x_s) \\ &\qquad - f(s, \overline{z^*}_s + x_s)|\ ds \\ &\leq \int_0^t M\, l \|\overline{z}_s - \overline{z^*}\|_{\mathcal{B}}\, ds. \end{aligned}$$

Using (A1), we obtain

$$\begin{aligned}|F(z)(t)-F(z^*)(t)| &\leq \int_0^t M\, l\, K\, \left|\overline{z}(s)-\overline{z^*}(s)\right|\, ds\\ &\leq M\, l\, K \int_0^t \left|\overline{z}(s)-\overline{z^*}(s)\right|\, ds\\ &\leq M\, l\, K\, b\|\overline{z}-\overline{z^*}\|_{B_b}.\end{aligned}$$

Therefore,

$$\|F(z)-F(z^*)\|_{B_b} \leq L\|\overline{z}-\overline{z^*}\|_{B_b},$$

where

$$L := M\, K\, l\, b.$$

As a consequence of the Banach fixed point theorem, we deduce that F has a unique fixed point which is a solution to (9.1)-(9.2). □

9.3. A Global Existence and Uniqueness Result for Functional Integro-Differential Equations in Fréchet Spaces

9.3.1. Introduction

We provide in this section, sufficient conditions for the existence of the unique mild solution on the semi-infinite interval for the problem (9.1)-(9.2) by using a recent nonlinear alternative of Leray-Schauder type due to Frigon and Granas for contractions maps in Fréchet spaces .

9.3.2. Existence of Solutions

Definition 9.8. *We say that the function* $y : \mathbb{R} \to E$ *is a mild solution of* $(9.1)-(9.2)$ *if* $y(t) = \phi(t)$ *for all* $t \leq 0$ *and* y *satisfies the following integral equation*

$$y(t) = S(t)\phi(0) + \int_0^t S(t-s)\, f(s, y_s)\, ds \quad \textit{for each } t \geq 0. \tag{9.8}$$

We introduce the following hypotheses:

(9.11.1) There exists a constant $\widehat{M} \geq 1$ such that

$$\|S(t)\|_{B(E)} \leq \widehat{M} \quad \text{for every } t \geq 0.$$

(9.11.2) There exists a function $p \in L^1_{loc}(J; \mathbb{R}_+)$ and a continuous nondecreasing function $\psi : \mathbb{R}_+ \to (0, \infty)$ and such that :

$$|f(t,u)| \leq p(t)\ \psi(\|u\|_{\mathcal{B}}) \text{ for a.e. } t \in J \text{ and each } u \in \mathcal{B}.$$

(9.11.3) For all $R > 0$, there exists $l_R \in L^1_{loc}(J;\mathbb{R}_+)$ such that :

$$|f(t,u) - f(t,v)| \leq l_R(t)\ \|u - v\|_{\mathcal{B}}$$

for all $u, v \in \mathcal{B}$ with $\|u\|_{\mathcal{B}} \leq R$ and $\|v\|_{\mathcal{B}} \leq R$.

Consider the following space

$$B_{+\infty} = \left\{ y : \mathbb{R} \to E : y|_{[0,T]} \text{ continuous for } T > 0 \text{ and } y_0 \in \mathcal{B} \right\},$$

where $y|_{[0,T]}$ is the restriction of y to the real compact interval $[0,T]$.

Let us fix $\tau > 1$. For every $n \in \mathbb{N}$, we define in $B_{+\infty}$ the semi-norms by :

$$\|y\|_n := \sup\ \{ \ e^{-\tau\ L_n^*(t)}\ |y(t)| : t \in [0,n]\ \}$$

where $L_n^*(t) = \int_0^t \bar{l}_n(s)\ ds$, $\bar{l}_n(t) = K_n \widehat{M} l_n(t)$ and l_n is the function from (9.11.3). Then $B_{+\infty}$ is a Fréchet space with those family of semi-norms $\|\cdot\|_{n\in\mathbb{N}}$.

Theorem 9.9. *Assume that* (9.11.1) – (9.11.3) *hold, suppose that*

$$\int_0^{+\infty} \frac{ds}{\psi(s)} > K_n \widehat{M} \int_0^n p(s)\ ds, \quad n \in \mathbb{N}. \tag{9.9}$$

Then the problem (9.1) – (9.2) *has a unique mild solution on* $(-\infty, +\infty)$.

Proof. We transform the problem (9.1)-(9.2) into a fixed-point problem. Consider the operator $N : B_{+\infty} \to B_{+\infty}$ defined by :

$$N(y)(t) = \begin{cases} \phi(t), & \text{if } t \leq 0; \\ S(t)\ \phi(0) + \displaystyle\int_0^t S(t-s)\ f(s, y_s)\ ds, & \text{if } t \in J. \end{cases} \tag{9.10}$$

Clearly, fixed points of the operator N are mild solutions of the problem (9.1) – (9.2).

For $\phi \in \mathcal{B}$, we will define the function $x(.) : \mathbb{R} \to E$ by

$$x(t) = \begin{cases} \phi(t), & \text{if } t \leq 0; \\ S(t)\ \phi(0), & \text{if } t \in J. \end{cases}$$

Then $x_0 = \phi$. For each function $z \in B_{+\infty}$ with $z_0 = 0$, we denote by $\bar{z}$ the function defined by

$$\bar{z}(t) = \begin{cases} 0, & \text{if } t \leq 0; \\ z(t), & \text{if } t \in J. \end{cases}$$

If $y(\cdot)$ satisfies (9.8), we can decompose it as $y(t) = z(t) + x(t)$, $t \geq 0$, which implies $y_t = z_t + x_t$, for every $t \in J$ and the function $z(\cdot)$ satisfies

$$z(t) = \int_0^t S(t-s)\ f(s, z_s + x_s)\ ds \quad \text{for } t \in J.$$

Let

$$B^0_{+\infty} = \{z \in B_{+\infty} : z_0 = 0 \in \mathcal{B}\}.$$

We define the operator $F : B^0_{+\infty} \to B^0_{+\infty}$ by :

$$F(z)(t) = \int_0^t S(t-s)\, f(s, z_s + x_s)\, ds \quad \text{for } t \in J. \tag{9.11}$$

Obviously the operator N has a fixed point is equivalent to F has one, so it turns to prove that F has a fixed point. Let $z \in B^0_{+\infty}$ be such that $z = \lambda F(z)$ for some $\lambda \in [0,1)$. By the hypotheses (9.11.1), (9.11.2), we have for each $t \in [0,n]$

$$\begin{aligned} |z(t)| &\leq \int_0^t \|S(t-s)\|_{B(E)}\, |f(s, z_s + x_s)|\, ds \\ &\leq \widehat{M} \int_0^t p(s)\, \psi(\|z_s + x_s\|_{\mathcal{B}})\, ds \\ &\leq \widehat{M} \int_0^t p(s)\, \psi\Big(K_n |z(s)| + (M_n + K_n \widehat{M} H)\|\phi\|_{\mathcal{B}}\Big)\, ds \end{aligned}$$

Set $c_n := (M_n + K_n \widehat{M} H)\|\phi\|_{\mathcal{B}}$. Then, we have

$$|z(t)| \leq \widehat{M} \int_0^t p(s)\, \psi(K_n |z(s)| + c_n)\, ds$$

Then

$$K_n |z(t)| + c_n \leq c_n + K_n \widehat{M} \int_0^t p(s)\, \psi(K_n |z(s)| + c_n)\, ds.$$

We consider the function μ defined by

$$\mu(t) := \sup\{\, K_n |z(s)| + c_n \ :\ 0 \leq s \leq t \,\}, \quad 0 \leq t < +\infty.$$

Let $t^\star \in [0,t]$ be such that $\mu(t) = K_n |z(t^\star)| + c_n \|\phi\|_{\mathcal{B}}$. By the previous inequality, we have

$$\mu(t) \leq c_n + K_n \widehat{M} \int_0^t p(s)\, \psi(\mu(s))\, ds \quad \text{for} \quad t \in [0,n].$$

Let us take the right-hand side of the above inequality as $v(t)$. Then, we have

$$\mu(t) \leq v(t) \quad \text{for all } t \in [0,n].$$

From the definition of v, we have

$$v(0) = c_n \quad \text{and} \quad v'(t) = \widehat{M} K_n p(t)\, \psi(\mu(t)) \quad \text{a.e. } t \in [0,n].$$

Using the nondecreasing character of ψ, we get

$$v'(t) \leq \widehat{M}\, K_n\, p(t)\, \psi(v(t)) \ \text{ a.e. } t \in [0,n].$$

Using the condition (9.9), this implies that for each $t \in [0,n]$ we have

$$\int_{c_n}^{v(t)} \frac{ds}{\psi(s)} \leq \widehat{M}\, K_n \int_0^t p(s)\, ds \leq \widehat{M}\, K_n \int_0^n p(s)\, ds < \int_{c_n}^{+\infty} \frac{ds}{\psi(s)}.$$

Thus, for every $t \in [0,n]$, there exists a constant Λ_n such that $v(t) \leq \Lambda_n$ and hence $\mu(t) \leq \Lambda_n$. Since $\|z\|_n \leq \mu(t)$, we have $\|z\|_n \leq \Lambda_n$.

Set

$$Z = \{\, z \in B^0_{+\infty} \ :\ \sup_{0 \leq t \leq n} |z(t)| \leq \Lambda_n + 1, \quad \forall\ n \in \mathbb{N}\}.$$

Clearly, Z is a closed subset of $B^0_{+\infty}$.

We shall show that $F : Z \to B^0_{+\infty}$ is a contraction operator. Indeed, consider $z, \overline{z} \in Z$, thus using (9.11.1) and (9.11.3) for each $t \in [0,n]$ and $n \in \mathbb{N}$

$$\begin{aligned} |F(z)(t) - F(\overline{z})(t)| &\leq \int_0^t \|S(t-s)\|_{B(E)}\ |f(s, z_s + x_s) \\ &\qquad - f(s, \overline{z}_s + x_s)|\ ds \\ &\leq \int_0^t \widehat{M}\ l_n(s)\ \|z_s - \overline{z}_s\|_{\mathcal{B}}\ ds. \end{aligned}$$

Using (A_1), we obtain

$$\begin{aligned} |F(z)(t) - F(\overline{z})(t)| &\leq \int_0^t \widehat{M}\ l_n(s)(K(s)\ |z(s) - \overline{z}(s)| + M(s)\|z_0 - \overline{z}_0\|_{\mathcal{B}})\ ds \\ &\leq \int_0^t \widehat{M}\ l_n(s)\ K_n\ |z(s) - \overline{z}(s)|\ ds \\ &\leq \int_0^t \left[\overline{l}_n(s)\ e^{\tau\ L_n^*(s)}\right] \left[e^{-\tau\ L_n^*(s)}\ |z(s) - \overline{z}(s)|\right]\ ds \\ &\leq \int_0^t \left[\frac{e^{\tau\ L_n^*(s)}}{\tau}\right]'\ ds\ \|z - \overline{z}\|_n \\ &\leq \frac{1}{\tau}\ e^{\tau\ L_n^*(t)}\ \|z - \overline{z}\|_n. \end{aligned}$$

Therefore,

$$\|F(z) - F(\overline{z})\|_n \leq \frac{1}{\tau}\ \|z - \overline{z}\|_n.$$

So, the operator F is a contraction for all $n \in \mathbb{N}$. From the choice of Z there is no $z \in \partial Z^n$ such that $z = \lambda\, F(z)$, $\lambda \in (0,1)$. The Theorem 1.66 shows that $(S1)$ holds. Thus, we deduce that the operator F has a unique fixed-point $z^\star$. Then $y^\star(t) = z^\star(t) + x(t)$, $t \in (-\infty, +\infty)$ is a fixed point of the operator N, which is the unique mild solution of the problem (9.1) – (9.2). □

9.4. Abstract Fractional Integro-Differential Equations with State-Dependent Delay

9.4.1. Introduction

This section is concerned to the existence of solutions defined on a compact real interval for semilinear integro-differential equations of fractional order of the form

$$y'(t)-\int_0^t \frac{(t-s)^{\alpha-2}}{\Gamma(\alpha-1)}Ay(s)ds = f(t,y_{\rho(t,y_t)}), \quad \text{a.e. } t\in J:=[0,b] \tag{9.12}$$

$$y_0=\phi\in\mathcal{B}, \tag{9.13}$$

where $1<\alpha<2$ and $A: D(A)\subset E\to E$ is a closed linear operator, and $(E,|\cdot|)$ is a Banach space. The convolution integral in the equation is known as the Riemann-Liouville fractional integral, $f: J\times\mathcal{B}\to E$ and $\rho: J\times\mathcal{B}\to(-\infty,b]$ are suitable functions. For any continuous function y defined on $(-\infty,b]$ and any $t\geq 0$, we denote by y_t the element of $\mathcal{B}$ defined by $y_t(\theta)=y(t+\theta)$ for $\theta\in(-\infty,0]$. Here $y_t(\cdot)$ represents the history of the state from each time $\theta\in(-\infty,0]$ up to the present time t. We assume that the histories y_t belongs to some abstract phase space $\mathcal{B}$.

Now let us recall some fundamental facts of the notion of Kuratowski measure of noncompactness.

Definition 9.10. *([86]) Let E be a Banach space and Ω_E the family of bounded subsets of E. The Kuratowski measure of noncompactness is the map $\alpha:\Omega_E\to[0,\infty)$ defined by*

$$\alpha(B)=\inf\{\varepsilon>0: B\subseteq\cup_{i=1}^n B_i \text{ and } diam(B_i)\leq\varepsilon\}.$$

Properties 9.11. *The Kuratowski measure of noncompactness satisfies the following properties (for more details see [86]).*

(a) $\alpha(B)=0\Longleftrightarrow\overline{B}$ *is compact (B is relatively compact).*

(b) $\alpha(B)=\alpha(\overline{B})$.

(c) $A\subset B\Longrightarrow\alpha(A)\leq\alpha(B)$.

(d) $\alpha(A+B)\leq\alpha(A)+\alpha(B)$.

(e) $\alpha(cB)=|c|\alpha(B)$; $c\in\mathbb{R}$.

(f) $\alpha(convB)=\alpha(B)$.

Theorem 9.12. *([59, 282]) Let D be a bounded, closed and convex subset of a Banach space such that $0\in D$, and let N be a continuous mapping of D into itself. If the implication*

$$V=\overline{conv}N(V) \quad or \quad V=N(V)\cup 0\Longrightarrow\alpha(V)=0$$

holds for every subset V of D, then N has a fixed point.

Lemma 9.13. *([206]) If $H \subset C(J,E)$ is a bounded and equicontinuous set, then*

(ii) $\alpha(H) = \sup_{0 \le t \le b} \alpha(H(t)).$

(ii) $\alpha\left(\int_0^b x(s)ds : x \in H\right) \le \int_0^b \alpha(H(s))ds$ *for* $t \in J,$

where $H(s) = \{x(s) : x \in H\},\ s \in J.$

Lemma 9.14. *([206]) Let D be a bounded, closed and convex subset of Banach space X. If the operator $N : D \to D$ is a strict set contraction, i.e there is a constant $0 \le \lambda < 1$ such that $\alpha(N(S)) \le \lambda\alpha(S)$ for any set $S \subset D$ then N has a fixed point in D.*

9.4.2. Existence of Solutions

Let us start by defining what we mean by a solution of the problem $(9.12)-(9.13)$.

Definition 9.15. *A function $y : (-\infty, b] \to E$ is said to be a mild solution of* $(9.12)-(9.13)$ *if $y_0 = \phi, y_{\rho(s,y_s)} \in \mathcal{B}$ for every $s \in J$ and*

$$y(t) = S(t)\phi(0) + \int_0^t S(t-s)\ f(s, y_{\rho(s,y_s)})\ ds \quad for\ each\ t \in J. \tag{9.14}$$

Set

$$\mathcal{R}(\rho^-) = \{\rho(s,\phi) : (s,\phi) \in J \times \mathcal{B}, \rho(s,\phi) \le 0\}.$$

We always assume that $\rho : J \times \mathcal{B} \to (-\infty, b]$ is continuous. Additionally, we introduce the following hypothesis:

(H_ϕ) The function $t \to \phi_t$ is continuous from $\mathcal{R}(\rho^-)$ into $\mathcal{B}$ and there exists a continuous and bounded function $L^\phi : \mathcal{R}(\rho^-) \to (0, \infty)$ such that

$$\|\phi_t\|_{\mathcal{B}} \le L^\phi(t)\|\phi\|_{\mathcal{B}} \quad \text{for every } t \in \mathcal{R}(\rho^-).$$

Remark 9.16. *The condition (H_ϕ), is frequently verified by functions continuous and bounded. For more details, see for instance [226].*

Lemma 9.17. *([220], Lemma 2.4) If $y : (-\infty, b] \to E$ is a function such that $y_0 = \phi$, then*

$$\|y_s\|_{\mathcal{B}} \le (M_b + L^\phi)\|\phi\|_{\mathcal{B}} + K_b \sup\{|y(\theta)|; \theta \in [0, max\{0,s\}]\},\ s \in \mathcal{R}(\rho^-) \cup J,$$

where $L^\phi = \sup_{t \in \mathcal{R}(\rho^-)} L^\phi(t).$

Our first result is based on the Mönch fixed point theorem.

Theorem 9.18. *Assume (H_ϕ). Moreover,*

$(9.23.1)$ *The operator solution $S(t)_{t \in J}$ is uniformly continuous, and there exists $M > 0$ such that*

$$\|S(t)\| \le M\ for\ every\ t \in J.$$

(9.23.2) *The function $f : J \times \mathcal{B} \to E$ is Carathéodory.*

(9.23.3) *There exists a function $p \in L^1(J;\mathbb{R}_+)$ and a continuous nondecreasing function $\psi : \mathbb{R}_+ \to (0,\infty)$ such that*

$$|f(t,u)| \leq p(t)\ \psi(\|u\|_{\mathcal{B}}) \text{ for a.e. } t \in J \text{ and each } u \in \mathcal{B},$$

(9.23.4) *For each $t \in J$ and bounded $B \subset \mathcal{B}$ we have*

$$\alpha(f(t,B)) \leq p(t)\alpha(B).$$

(9.23.5) *There exists $r > 0$ such that*

$$M\|\phi\|_{\mathcal{B}} + M\ \psi\left((M_b + L^{\phi})\|\phi\|_{\mathcal{B}} + K_b r\right) \int_0^b p(s)ds \leq r,$$

Then the problem (9.12)-(9.13) has at least one solution on $(-\infty, b]$.

Proof. Let $Y = \{u \in C(J,E) : u(0) = \phi(0) = 0\}$ endowed with the uniform convergence topology and $N : Y \to Y$ be the operator defined by

$$N(y)(t) = S(t)\ \phi(0)$$

$$+ \int_0^t S(t-s)\ f(s, \bar{y}_{\rho(s,\bar{y}_s)})\ ds, \quad t \in J,$$

where $\bar{y} : (-\infty, b] \to E$ is such that $\bar{y}_0 = \phi$ and $\bar{y} = y$ on J. Let $\bar{\phi} : (-\infty, b] \to E$ be the extension of ϕ to $(-\infty, b]$ such that $\bar{\phi}(\theta) = \phi(0) = 0$ on J. Let the set

$$B_r = \{y \in Y : \|y\|_\infty \leq r\},$$

where r is the constant defined in (9.23.5). Clearly, the subset B_r is closed, bounded, and convex. In order to apply Theorem 9.12, we give the proof in several steps.

Step 1: N is continuous on B_r.
Let y^n be a sequence such that $y^n \longrightarrow y$ in Y. Then

$$\begin{aligned} |N(y^n)(t) - N(y)(t)| &= \left| S(t)\phi(0) + \int_0^t S(t-s)\ [f(s, \overline{y^n}_{\rho(s,\overline{y^n}_s)}) - f(s, \bar{y}_{\rho(s,\bar{y}_s)})]\ ds \right| \\ &\leq M \int_0^t |f(s, \overline{y^n}_{\rho(s,\overline{y^n}_s)}) - f(s, \bar{y}_{\rho(s,\bar{y}_s)})|\ ds. \end{aligned}$$

Since $f(s,.)$ is continuous, we have by the Lebesgue dominated convergence theorem

$$\|N(y^n)(t) - N(y)(t)\| \to 0 \text{ as } n \to +\infty.$$

Thus N is continuous.

Step 2: N maps B_r into itself.
If $y \in B_r$, from Lemma 9.17 follows that

$$\|\bar{y}_{\rho(t,\bar{y}_t)}\|_{\mathcal{B}} \leq (M_b + L^{\phi})\|\phi\|_{\mathcal{B}} + K_b r$$

and For each $y \in B_r$, by (9.23.3) and (9.23.5), we have for each $t \in J$

$$\begin{aligned} |N(y)(t)| &\leq M\|\phi\|_{\mathcal{B}} + M\int_0^t |f(s,\bar{y}_{\rho(s,\bar{y}_s)}|\, ds \\ &\leq M\|\phi\|_{\mathcal{B}} + M\int_0^t p(s)\, \psi\left(\|\bar{y}_{\rho(s,\bar{y}_s)}\|_{\mathcal{B}}\right)\, ds \\ &\leq M\|\phi\|_{\mathcal{B}} + M\int_0^t p(s)\, \psi\left((M_b + L^{\phi})\|\phi\|_{\mathcal{B}} + K_b r\right)\, ds \\ &\leq M\|\phi\|_{\mathcal{B}} + M\, \psi\left((M_b + L^{\phi})\|\phi\|_{\mathcal{B}} + K_b r\right)\int_0^b p(s)ds \\ &\leq r. \end{aligned}$$

Step 3: $N(B_r)$ is bounded and equicontinuous.
By Step 2, it is obvious that $N(B_r) \subset B_r$ is bounded. For the equicontinuity of $N(B_r)$. Let $\tau_1, \tau_2 \in J$ with $\tau_2 > \tau_1$, and $y \in B_r$,

$$\begin{aligned} |N(y)(\tau_2) - N(y)(\tau_1)| &\leq |S(\tau_2) - S(\tau_1)|\|\phi\|_{\mathcal{B}} + |\int_0^{\tau_1} [S(\tau_2 - s) - S(\tau_1 - s)]f(s,\bar{y}_{\rho(s,\bar{y}_s)})ds| \\ &\quad + |\int_{\tau_1}^{\tau_2} S(\tau_2 - s)f(s,\bar{y}_{\rho(s,\bar{y}_s)})|ds \\ &\leq |S(\tau_2) - S(\tau_1)|\|\phi\|_{\mathcal{B}} + \int_0^{\tau_1} |S(\tau_2 - s) - S(\tau_1 - s)||f(s,\bar{y}_{\rho(s,\bar{y}_s)})|\, ds \\ &\quad + \int_{\tau_1}^{\tau_2} |S(\tau_2 - s)f(s,\bar{y}_{\rho(s,\bar{y}_s)})|\, ds \\ &\leq |S(\tau_2) - S(\tau_1)|\|\phi\|_{\mathcal{B}} + \psi\left((M_b + L^{\phi})\|\phi\|_{\mathcal{B}} + K_b r\right) \\ &\quad \int_0^{\tau_1} |S(\tau_2 - s) - S(\tau_1 - s)|p(s)ds \\ &\quad + M\psi\left((M_b + L^{\phi})\|\phi\|_{\mathcal{B}} + K_b r\right)\int_{\tau_1}^{\tau_2} p(s)ds. \end{aligned}$$

The right-hand of the above inequality tends to zero as $\tau_2 - \tau_1 \to 0$, since $S(t)$ is uniformly continuous operator.
Now let V be a subset of B_r such that $V \subset \overline{conv}(N(V) \cup \{0\})$. V is bounded and equicontinuous and therefore the function $t \to v(t) = \alpha(V(t))$ is continuous on J. By (9.21.4), Theorem 2.1 in [341] and the properties of the measure α we have for each $t \in J$

$$\begin{aligned} v(t) &\leq \alpha(N(V)(t) \cup \{0\}) \\ &\leq \alpha(N(V)(t)) \\ &\leq \alpha\left(S(t)\, \phi(0) + \int_0^t S(t-s)\, f(s,\bar{y}_{\rho(s,\bar{y}_s)})\, ds\right) \end{aligned}$$

$$\begin{aligned}
&\leq \alpha(S(t)\,\phi(0)) + \alpha\left(\int_0^t S(t-s)\, f(s,\overline{y}_{\rho(s,\overline{y}_s)})\, ds\right)\\
&\leq M\int_0^t p(s)\alpha(\overline{y}_{\rho(s,y_s)} : \overline{y} \in V)ds\\
&\leq M\int_0^t p(s)K(s) \sup_{0\leq\tau\leq s} \alpha(V(\tau))ds\\
&\leq M\int_0^t p(s)K(s)\alpha(V(s))ds\\
&= M\int_0^t p(s)K(s)v(s)ds.
\end{aligned}$$

Gronwall's Lemma implies that $v(t) = 0$ for each $t \in J$, and then $V(t)$ is relatively compact in E. In view of the Arzela-Ascoli theorem, V is relatively compact in B_r. Applying now Theorem 9.12 we conclude that N has a fixed point $y \in B_r$. Hence N has a fixed point which is solution to the problem (9.12)-(9.13). □

Proposition 9.19. Assume that $(H_\phi), (9.23.1), (9.23.2), (9.23.5)$ are satisfied, then a slight modification of the proof (i.e. use the Darbo's fixed point theorem) guarantees that $(9.23.4)$ could be replaced by

$(9.25.1)$ There exists a nonnegative function $l \in L^1(J, \mathbb{R}^+)$ such that

$$\alpha(f(t,B)) \leq l(t)\alpha(B),\ \ t \in J,$$

and consider the Kuratowski measure of noncompactness α_C defined on the family of bounded subsets of the space $C(J,E)$ by

$$\alpha_C(H) = \sup_{t\in J} e^{-\tau L(t)}\alpha(H(t)),$$

where $L(t) = \displaystyle\int_0^t \widetilde{l}(s)ds,\ \ \widetilde{l}(t) = Ml(t)K(t),\ \tau > 1.$

Proof. We show that the operator $N : B_r \longrightarrow B_r$ is a strict set contraction. We know that $N : B_r \longrightarrow B_r$ is bounded and continuous, we need to prove that there exists a constant $0 \leq \lambda < 1$ such that $\alpha_C(NH) \leq \lambda\alpha_C(H)$ for $H \subset B_r$. For each $t \in J$ we have

$$\alpha((NH)(t)) \leq M\int_0^t \alpha(f(s,\overline{y}_{\rho(s,\overline{y}_s)})) : \overline{y} \in H)ds.$$

This implies by $(9.25.1)$ and Theorem 2.1 in [341]

$$\begin{aligned}
\alpha((NH)(t)) &\leq \int_0^t M\, l(s)\alpha(\overline{y}_{\rho(s,\overline{y}_s)} : \overline{y} \in H)ds\\
&\leq \int_0^t Ml(s)K(s) \sup_{0\leq\tau\leq s} \alpha(H(\tau))ds\\
&\leq \int_0^t Ml(s)K(s)\alpha(H(s))ds\\
&= \int_0^t \widetilde{l}(s)\alpha(H(s))ds
\end{aligned}$$

$$
\begin{aligned}
&= \int_0^t e^{\tau L(s)} e^{-\tau L(s)} \widetilde{l}(s) \alpha(H(s)) ds \\
&\leq \int_0^t \widetilde{l}(s)\, e^{\tau L(s)} \sup_{s\in[0,t]} e^{-\tau L(s)} \alpha(H(s)) ds \\
&\leq \sup_{t\in[0,b]} e^{-\tau L(t)} \alpha(H(t)) \int_0^t \widetilde{l}(s) e^{\tau L(s)} ds \\
&= \alpha_C(H) \int_0^t \left(\frac{e^{\tau L(s)}}{\tau} \right)' ds \\
&\leq \alpha_C(H) \frac{1}{\tau} e^{\tau L(t)}.
\end{aligned}
$$

Therefore,

$$\alpha_C(NH) \leq \frac{1}{\tau} \alpha_C(H).$$

So, the operator N is a set contraction. As a consequence of Theorem 9.14, we deduce that N has a fixed point which is a mild solution to the problem (9.12)-(9.13). □

9.4.3. An Example

To apply our abstract results, we consider the fractional differential equation with state-dependent delay of the form

$$
\begin{cases}
\dfrac{\partial u}{\partial t}(t,\xi) - \dfrac{1}{\Gamma(\mu-1)} \displaystyle\int_t^0 (t-s)^{\mu-2} L_\xi u(s,\xi) ds \\
\quad = \dfrac{e^{-\gamma t + t} |u(t-\sigma(u(t,0)),\xi)|}{3(e^{-t}+e^t)(1+|u(t-\sigma(u(t,0),\xi))|)}, \qquad t \in [0,b],\ \xi \in [0,\pi], \\
u(t,0) = u(t,\pi) = 0, \qquad t \in [0,\infty), \\
u(\theta,\xi) = u_0(\theta,\xi), \quad \theta \in (-\infty,0],\ \xi \in [0,\pi],
\end{cases}
\tag{9.15}
$$

where $1 < \mu < 2$, $\sigma \in C(\mathbb{R},[0,\infty))$, $\gamma > 0$, L_ξ stands for the operator with respect to the spatial variable ξ which is given by:

$$L_\xi = \frac{\partial^2}{\partial \xi^2} - r, \quad \text{with} \quad r > 0.$$

Take $E = L^2([0,\pi],\mathbb{R})$ and the operator $A := L_\xi : D(A) \subset E \to E$ with domain

$$D(A) := \{\, u \in E \ : u'' \in E,\ u(0) = u(\pi) = 0 \,\}.$$

Clearly A is densely defined in E and is sectorial. Hence A is a generator of a solution operator on E. For the phase space, we choose $\mathcal{B} = \mathcal{B}_\gamma$ defined by

$$\mathcal{B}_\gamma = \{\phi \in C((-\infty,0],\mathbb{R}) : \lim_{\theta \to -\infty} e^{\gamma\theta}\phi(\theta) \text{ exists}\}$$

with the norm

$$\|\phi\|_\gamma = \sup_{\theta\in(-\infty,0]} e^{\gamma\theta}|\phi(\theta)|.$$

Notice that the phase space $\mathcal{B}_\gamma$ satisfies axioms (9.23.1), (9.23.2), and (9.23.3) (see [226] for more details). Set

$$y(t)(\xi) = u(t,\xi), \quad t\in[0,b],\ \xi\in[0,\pi].$$

$$\phi(\theta)(\xi) = u_0(\theta,\xi), \quad t\in[0,b],\ \theta\leq 0.$$

$$f(t,\phi)(\xi) = \frac{e^{-\gamma t+t}|\phi(0,\xi)|}{3(e^{-t}+e^{t})(1+|\phi(0,\xi)|)}, \quad t\in[0,\infty),\ \xi\in[0,\pi].$$

$$\rho(t,\phi) = t-\sigma(\phi(0,0)).$$

Let $\phi\in\mathcal{B}_\gamma$ be such that (H_ϕ) holds, and let $t\to\phi_t$ be continuous on $\mathcal{R}(\rho^-)$. Then by Theorem 9.18 there exists at least one mild solution of (9.15).

9.5. Integro-Differential Equations with State-Dependent Delay on an Unbounded Domain

9.5.1. Introduction

This section deals with the existence of mild solutions, defined on the positive semi-infinite real interval $J := [0,+\infty)$, for semilinear integro-differential equations of fractional order

$$y'(t) - \int_0^t \frac{(t-s)^{\alpha-2}}{\Gamma(\alpha-1)}Ay(s)ds = f(t, y_{\rho(t,y_t)}), \quad \text{a.e. } t\in J, \tag{9.16}$$

$$y_0 = \phi\in\mathcal{B}, \tag{9.17}$$

where $1<\alpha<2$ and $A: D(A)\subset E\to E$ is the generator of an integral resolvent family defined on a complex Banach space $(E,|\cdot|)$, the convolution integral in the equation is known as the Riemann-Liouville fractional integral, $f:[0,+\infty)\times\mathcal{B}\to E$ and $\rho:[0,\infty)\times\mathcal{B}\to\mathbb{R}$ are appropriated functions. For any continuous function y defined on $(-\infty,+\infty)$ and any $t\geq 0$, we denote by y_t the element of $\mathcal{B}$ defined by $y_t(\theta) = y(t+\theta)$ for $\theta\in(-\infty,0]$. Here $y_t(\cdot)$ represents the history of the state from each time $\theta\in(-\infty,0]$ up to the present time t. We assume that the histories y_t belongs to some abstract phase space $\mathcal{B}$.

9.5.2. Existence of Solutions

Now we define the mild solution for the initial value problem (9.16)-(9.17).

Definition 9.20. *We say that the function $y:(-\infty,+\infty)\to E$ is a mild solution of (9.16) – (9.17) if $y(t)=\phi(t)$ for all $t\leq 0$ and y satisfies the following integral equation*

$$y(t) = S(t)\phi(0) + \int_0^t S(t-s)\, f(s, y_{\rho(s,y_s)})\, ds \quad \textit{for each } t\in J. \tag{9.18}$$

Set

$$\mathcal{R}(\rho^-) = \{\rho(s,\varphi) : (s,\varphi) \in J \times \mathcal{B}, \rho(s,\varphi) \leq 0\}.$$

We always assume that $\rho : J \times \mathcal{B} \to \mathbb{R}$ is continuous. Additionally, we introduce following hypothesis:

(H_φ) The function $t \to \varphi_t$ is continuous from $\mathcal{R}(\rho^-)$ into $\mathcal{B}$ and there exists a continuous and bounded function $L^\phi : \mathcal{R}(\rho^-) \to (0,\infty)$ such that

$$\|\phi_t\|_{\mathcal{B}} \leq L^\phi(t)\|\phi\|_{\mathcal{B}} \quad \text{for every } t \in \mathcal{R}(\rho^-).$$

Remark 9.21. *The condition (H_φ), is frequently verified by functions continuous and bounded. For more details, see for instance [226].*

Theorem 9.22. *Assume that*

(9.30.1) *The operator solution $S(t)_{t\in J}$ is compact for $t > 0$,*

(9.30.2) *The function $f : J \times \mathcal{B} \to E$ is Carathéodory,*

(9.30.3) *There exists a function $p \in L^1(J;\mathbb{R}_+)$ and a continuous nondecreasing function $\psi : \mathbb{R}_+ \to (0,\infty)$ such that*

$$|f(t,u)| \leq p(t)\ \psi(\|u\|_{\mathcal{B}}) \text{ for a.e. } t \in J \text{ and each } u \in \mathcal{B},$$

(9.30.4) *For each $n \in \mathbb{N}$, there exists $r_n > 0$ such that*

$$r_n \geq M\ \psi(K_n r_n + c_n)\int_0^n p(s)ds$$

where

$$c_n := (M_n + L^\phi + K_n MH)\|\phi\|_{\mathcal{B}},$$

(9.30.5) *For each $t \in J$ and bounded set $B \subset \mathcal{B}$,, the set $\{f(t, y_{\rho(t,y_t)}), y \in B\}$is relatively compact in E.*

Then the problem (9.16) – (9.17) *has a mild solution on* $(-\infty,+\infty)$.

Proof. The proof will be given in two parts. Fix $n \in \mathbb{N}$ and consider the problem

$$y'(t) - \int_0^t \frac{(t-s)^{\alpha-2}}{\Gamma(\alpha-1)}Ay(s)ds = f(t, y_{\rho(t,y_t)}), \quad \text{a.e. } t \in J_n := [0,n], \tag{9.19}$$

$$y_0 = \phi \in \mathcal{B}. \tag{9.20}$$

Let

$$B_n = \left\{y : (-\infty,n] \to E : y|_{[0,n]} \text{ continuous and } y_0 \in \mathcal{B}\right\},$$

where $y|_{[0,n]}$ is the restriction of y to the real compact interval $[0,n]$.
Part I: We begin by showing that the problem (9.19) – (9.20) has a solution $y_n \in B_n$. Consider the operator $N : B_n \to B_n$ defined by :

$$N(y)(t) = \begin{cases} \phi(t), & \text{if } t \le 0; \\ S(t)\,\phi(0) + \displaystyle\int_0^t S(t-s)\, f(s, y_{\rho(s,y_s)})\, ds, & \text{if } t \in J_n. \end{cases} \tag{9.21}$$

Clearly, fixed points of the operator N are mild solutions of the problem (9.19) – (9.20).

For $\phi \in \mathcal{B}$, we will define the function $x(.) : \mathbb{R} \to E$ by

$$x(t) = \begin{cases} \phi(t), & \text{if } t \le 0; \\ S(t)\,\phi(0), & \text{if } t \in J_n. \end{cases}$$

Then $x_0 = \phi$. For each function $z \in B_n$ with $z_0 = 0$, we denote by $\bar{z}$ the function defined by

$$\bar{z}(t) = \begin{cases} 0, & \text{if } t \le 0; \\ z(t), & \text{if } t \in J_n. \end{cases}$$

If $y(\cdot)$ satisfies (9.18), we can decompose it as $y(t) = \bar{z}(t) + x(t)$, $t \ge 0$, which implies $y_t = z_t + x_t$, for every $t \in J_n$ and the function $z(\cdot)$ satisfies

$$z(t) = \int_0^t S(t-s)\, f(s, \bar{z}_{\rho(s,\bar{z}_s+x_s)} + x_{\rho(s,\bar{z}_s+x_s)})\, ds \quad \text{for } t \in J_n.$$

Let

$$B_n^0 = \{z \in B_n : z_0 = 0 \in \mathcal{B}\}.$$

For any $z \in B_n^0$ we have

$$\|z\|_n = \|z_0\|_{\mathcal{B}} + \sup\{|z(s)| : 0 \le s \le n\} = \sup\{|z(s)| : 0 \le s \le n\}.$$

Thus $(B_n^0, \|\cdot\|_n)$ is a Banach space . We define the operator $F : B_n^0 \to B_n^0$ by:

$$F(z)(t) = \int_0^t S(t-s)\, f(s, \bar{z}_{\rho(s,\bar{z}_s+x_s)} + x_{\rho(s,\bar{z}_s+x_s)})\, ds \quad \text{for } t \in J_n. \tag{9.22}$$

Obviously the operator N has a fixed point is equivalent to F has one, so it turns to prove that F has a fixed point. We shall show that the operators F satisfies the assumptions of Schauder's fixed point theorem. The proof will be given in several steps.
Let

$$C_n = \{z \in B_n^0, \|z\|_n \le r_n\},$$

where r_n is the constant from (9.28.4). It is clear that C_n is a closed, convex subset of B_n^0.

Step 1: F is continuous. Let z_q be a sequence such that $z_q \longrightarrow z$ in B_n^0.
Then for each $t \in J_n$

$$\begin{aligned} |F(z_q)(t) - F(z)(t)| &= \Big|\int_0^t S(t-s)\,[f(s,\overline{z}_{q\rho(s,\overline{z}_{qs}+x_s)} + x_{\rho(s,\overline{z}_{qs}+x_s)}) \\ &- f(s,\overline{z}_{\rho(s,\overline{z}_s+x_s)} + x_{\rho(s,\overline{z}_s+x_s)})]\,ds\Big| \\ &\leq M\int_0^t |f(s,\overline{z}_{q\rho(s,\overline{z}_{qs}+x_s)} + x_{\rho(s,\overline{z}_{qs}+x_s)}) \\ &- f(s,\overline{z}_{\rho(s,\overline{z}_s+x_s)} + x_{\rho(s,\overline{z}_s+x_s)})|\,ds. \end{aligned}$$

Since $f(s,.)$ is continuous, we have by the Lebesgue dominated convergence theorem

$$\|F(z_q) - F(z)\|_n \to 0 \text{ as } q \to +\infty.$$

Thus F is continuous.
Step 2: $F(C_n) \subset C_n$. Let $z \in C_n$, we show that $F(z) \in C_n$. For each $t \in J_n$ we have

$$\begin{aligned} |Fz(t)| &\leq M\int_0^t |f(s,\overline{z}_{\rho(s,\overline{z}_s+x_s)} + x_{\rho(s,\overline{z}_s+x_s)})|\,ds \\ &\leq M\int_0^t p(s)\,\psi\left(\|\overline{z}_{\rho(s,\overline{z}_s+x_s)} + x_{\rho(s,\overline{z}_s+x_s)}\|_{\mathcal{B}}\right)ds \\ &\leq M\int_0^t p(s)\,\psi\left(K_n r_n + (M_n + L^{\phi} + K_n\widehat{M}H)\|\phi\|_{\mathcal{B}}\right)ds. \end{aligned}$$

Set

$$c_n := (M_n + L^{\phi} + K_n MH)\|\phi\|_{\mathcal{B}}$$

Then, we have

$$\|Fz\|_n \leq M\,\psi(K_n r_n + c_n)\int_0^n p(s)ds.$$

By (9.30.4), we have

$$\|Fz\|_n \leq r_n.$$

Step 3: $F(C_n)$ is an equicontinuous set. Let $\tau_1, \tau_2 \in J_n$ with $\tau_2 > \tau_1$, and $z \in C_n$. Then

$$\begin{aligned} &|F(z)(\tau_2) - F(z)(\tau_1)| \\ \leq\ & \Big|\int_0^{\tau_1} [S(\tau_2 - s) - S(\tau_1 - s)]f(s,\overline{z}_{\rho(s,\overline{z}_s+x_s)} + x_{\rho(s,\overline{z}_s+x_s)})\,ds\Big| \\ &+ \Big|\int_{\tau_1}^{\tau_2} S(\tau_2 - s)|f(s,\overline{z}_{\rho(s,\overline{z}_s+x_s)} + x_{\rho(s,\overline{z}_s+x_s)})|\,ds\Big| \\ \leq\ & \int_0^{\tau_1} |S(\tau_2 - s) - S(\tau_1 - s)||f(s,\overline{z}_{\rho(s,\overline{z}_s+x_s)} + x_{\rho(s,\overline{z}_s+x_s)})|\,ds \\ &+ \int_{\tau_1}^{\tau_2} |S(\tau_2 - s)||f(s,\overline{z}_{\rho(s,\overline{z}_s+x_s)} + x_{\rho(s,\overline{z}_s+x_s)})|\,ds \\ \leq\ & \psi(r_n K_n + c_n)\int_0^{\tau_1} |S(\tau_2 - s) - S(\tau_1 - s)|p(s)ds \end{aligned}$$

$$+M\ \psi(r_nK_n+c_n)\int_{\tau_1}^{\tau_2}p(s)ds.$$

The right-hand of the above inequality tends to zero as $\tau_2-\tau_1\to 0$, since $S(t)$ is uniformly continuous. As a consequence of Steps 1 to 3, (9.30.5) together with the Arzela-Ascoli theorem, the operator F is completely continuous. Therefore, we deduce from Schauder's fixed point theorem that F has a fixed point $z_n\in C_n$ which is a solution of problem (9.19)-(9.20).

Part II: The diagonalization process

We now use the following diagonalization process. For $k\in\mathbb{N}$, let

$$u_k(t)=\begin{cases} z_k(t), & t\in[0,n_k];\\ z_k(n_k), & t\in[n_k,\infty).\end{cases} \tag{9.23}$$

Here $\{n_k\}_k\in\mathbb{N}^*$ is a sequence of numbers satisfying

$$0<n_1<n_2<...<n_k<\ldots\uparrow\infty.$$

Let $S=\{u_k\}_{k=1}^{\infty}$. For $k\in\mathbb{N}$ and $t\in[0,n_1]$ we have

$$u_{n_k}(t)=\int_0^t S(t-s)f(s,(\overline{u}_{n_k})_{\rho(s,(\overline{u}_{n_k})_s+x_s)}+x_{\rho(s,(\overline{u}_{n_k})_s+x_s)})ds.$$

Thus, for $k\in\mathbb{N}$ and $t,h\in[0,n_1]$ we have

$$u_{n_k}(t)-u_{n_k}(h)=\int_0^t[S(t-s)-S(h-s)]f(s,(\overline{u}_{n_k})_{\rho(s,(\overline{u}_{n_k})_s+x_s)}+x_{\rho(s,(\overline{u}_{n_k})_s+x_s)})ds$$

and by (9.30.3), we have

$$|u_{n_k}(t)-u_{n_k}(h)|\le\psi(r_{n_1}K_{n_1}+c_{n_1})\int_0^t|S(t-s)-S(h-s)|p(s)ds.$$

The Arzela-Ascoli Theorem guarantees that there is a subsequence F_1^* of $\mathbb{N}$ and a function $v_1\in B_{n_1}^0$ with $u_{n_k}\to v_1$ in $B_{n_1}^0$ as $k\to\infty$ through F_1^*. Let $N_1=N_1^*\setminus\{1\}$. Also for $k\in\mathbb{N}$ and $t,h\in[0,n_2]$ we have

$$|u_{n_k}(t)-u_{n_k}(x)|\le\psi(r_{n_2}K_{n_2}+c_{n_2})\int_0^t|S(t-s)-S(h-s)|p(s)ds.$$

The Arzela-Ascoli Theorem guarantees that there is a subsequence F_2^* of F_1 and a function $v_2\in B_{n_2}^0$ with $u_{n_k}\to v_2$ in $B_{n_2}^0$ as $k\to\infty$ through F_2^*. Note that $v_1=v_2$ on $[0,n_1]$ since $F_2^*\subseteq F_1$. Let $F_2=F_2^*\setminus\{2\}$. Proceed inductively to obtain for $m\in\{3,4,\ldots\}$ a subsequence F_m^* of F_{m-1} and a function $v_m\in B_{n_m}^0$ with $u_{n_k}\to v_m$ in $B_{n_m}^0$ as $k\to\infty$ through F_m^*. Let $F_m=F_m^*\setminus\{m\}$.

Define a function z as follows. Fix $t\in(0,\infty)$ and let $m\in\mathbb{N}$ with $s\le n_m$. Then define $z(t)=v_m(t)$. Then $z\in B_\infty$ and $z_0=0$.

Again fix $t\in[0,\infty)$ and let $m\in\mathbb{N}$ with $s\le n_m$. Then for $n\in F_m$ we have

$$u_{n_k}(t)=\int_0^t S(t-s)f(s,(\overline{u}_{n_k})_{\rho(s,(\overline{u}_{n_k})_s+x_s)}+x_{\rho(s,(\overline{u}_{n_k})_s+x_s)})ds.$$

Let $n_k \to \infty$ through F_m to obtain

$$v_m(t) = \int_0^t S(t-s) f(s, \overline{v}_{m_{\rho(s,\overline{v}_{m_s}+x_s)}} + x_{\rho(s,\overline{v}_{m_s}+x_s)}) ds.$$

i.e

$$z(t) = \int_0^t S(t-s) f(s, \overline{z}_{\rho(s,\overline{z}_s+x_s)} + x_{\rho(s,\overline{z}_s+x_s)}) ds.$$

We can use this method for each $h \in [0, n_m]$, and for each $m \in \mathbb{N}$. Thus the constructed function y is a mild solution of (9.16) – (9.17). This completes the proof of the theorem. □

9.5.3. An Example

To apply our abstract results, we consider the fractional differential equation with state dependent delay of the form

$$\begin{cases} \dfrac{\partial u}{\partial t}(t,\xi) - \dfrac{1}{\Gamma(\mu-1)} + \displaystyle\int_t^0 (t-s)^{\mu-2} L u_\xi u(s,\xi) ds \\ \quad = \dfrac{e^{-\gamma t + t}|u(t-\sigma(u(t,0)),\xi)|}{3(e^{-t}+e^t)(1+|u(t-\sigma(u(t,0),\xi))|)}, & t \in [0,\infty),\ \xi \in [0,\pi], \\ u(t,0) = u(t,\pi) = 0, & t \in [0,\infty), \\ u(\theta,\xi) = u_0(\theta,\xi), & \theta \in (-\infty,0],\ \xi \in [0,\pi], \end{cases} \tag{9.24}$$

Where $1 < \mu < 2$, $\sigma \in C(\mathbb{R}, [0,\infty))$, $\gamma > 0$, L_ξ stands for the operator with respect to the spatial variable ξ which is given by:

$$L_\xi = \frac{\partial^2}{\partial \xi^2} - r, \quad \text{with} \quad r > 0.$$

Take $E = L^2([0,\pi], \mathbb{R})$ and the operator $A := L_\xi : D(A) \subset E \to E$ with domain

$$D(A) := \{ u \in E : u'' \in E,\ u(0) = u(\pi) = 0 \}.$$

Clearly A is densely defined in E and is sectorial. Hence A is a generator of a solution operator on E. For the phase space, we choose $\mathcal{B} = \mathcal{B}_\gamma$ defined by

$$\mathcal{B}_\gamma = \{\phi \in C((-\infty,0], \mathbb{R}) : \lim_{\theta \to -\infty} e^{\gamma\theta}\phi(\theta) \text{ exists}\}$$

with the norm

$$\|\phi\|_\gamma = \sup_{\theta \in (-\infty,0]} e^{\gamma\theta}|y(\theta)|.$$

Notice that the phase space $\mathcal{B}_\gamma$ satisfies axioms (A_1), (A_2) (see[226] for more details). Set

$$y(t)(\xi) = u(t,\xi),\ \ t \in [0,\infty),\ \xi \in [0,\pi].$$

$$\phi(\theta)(\xi) = u_0(\theta,\xi),\ \ t \in [0,\infty),\ \theta \le 0.$$

$$f(t,\varphi)(\xi) = \frac{e^{-\gamma t + t}\varphi(0,\xi)}{3(e^{-t}+e^t)(1+\varphi(0,\xi))},\ \ t \in [0,\infty),\ \xi \in [0,\pi].$$

$$\rho(t,\varphi) = t - \sigma(\varphi(0,0))$$

Theorem 9.23. *Let $\varphi \in \mathcal{B}_\gamma$ be such that (H_φ) holds,and let $t \to \varphi_t$ be continuous on $\mathcal{R}(\rho^-)$. Then there exists a mild solution of* (9.24).

9.6. Notes and Remarks

The results of Chapter 9 are taken from Benchohra *et al.* [101, 102, 104]. Other results may be found in [55].

Chapter 10

Abstract Integro-Differential Inclusions

10.1. Introduction

In this chapter, we provide sufficient conditions for the existence of mild solutions of a class of semilinear functional integro-differential inclusions with state dependent delay.

10.2. Abstract Integro-Differential Inclusions with State-Dependent Delay

10.2.1. Introduction

This section studied the existence of mild solutions for semilinear functional integro-differential inclusions of fractional order of the form

$$y'(t) - \int_0^t \frac{(t-s)^{\alpha-2}}{\Gamma(\alpha-1)} Ay(s)ds \in F(t, y_{\rho(t,y_t)}), \quad \text{a.e. } t \in J = [0,b] \tag{10.1}$$

$$y_0 = \phi \in \mathcal{B}, \tag{10.2}$$

where $1 < \alpha < 2$ and $A : D(A) \subset E \to E$ is the generator of a solution operator defined on a complex Banach space $(E, |\cdot|)$, the convolution integral in the equation is known as the Riemann-Liouville fractional integral and $F : [0,b] \times \mathcal{B} \to \mathcal{P}(E)$ is a multi-valued map ($\mathcal{P}(E)$ is the family of nonempty subsets of E). For any continuous function y defined on $(-\infty, b]$ and any $t \geq 0$, we denote by y_t the element of $\mathcal{B}$ defined by $y_t(\theta) = y(t+\theta)$ for $\theta \in (-\infty, 0]$. Here $y_t(\cdot)$ represents the history of the state from each time $\theta \in (-\infty, 0]$ up to the present time t. We assume that the histories y_t belongs to some abstract phase space $\mathcal{B}$. The Laplace transformation of a function $f \in L^1_{loc}(\mathbb{R}_+, E)$ is defined by

$$\mathcal{L}(f)(\lambda) :=: \widehat{f}(\lambda) = \int_0^\infty e^{-\lambda t} f(t)dt, \quad Re(\lambda) > \omega,$$

if the integral is absolutely convergent for $Re(\lambda) > \omega$. In order to defined the mild solution of the problems $(10.1)-(10.2)$ we recall the following definition.

Definition 10.1. *Let A be a closed and linear operator with domain $D(A)$ defined on a Banach space E. We call A the generator of a solution operator if there exists $\omega > 0$ and a strongly continuous function $S : \mathbb{R}^+ \to B(E)$ such that*

$$\{\lambda^\alpha : Re(\lambda) > \omega\} \subset \rho(A),$$

and

$$\lambda^{\alpha-1}(\lambda^\alpha - A)^{-1}x = \int_0^\infty e^{-\lambda t} S(t)xdt,\ Re\lambda > \omega, x \in E.$$

In this case, $S(t)$ is called the solution operator generated by A.

The following result is a direct consequence of ([266], Proposition 3.1 and Lemma 2.2).

Proposition 10.2. *Let $\{S(t)\}_{t\geq 0} \subset B(E)$ be the solution operator with generator A. Then the following conditions are satisfied:*

a) $S(t)$ is strongly continuous for $t \geq 0$ and $S(0) = I$;

b) $S(t)D(A) \subset D(A)$ and $AS(t)x = S(t)Ax$ for all $x \in D(A)$, $t \geq 0$;

c) for every $x \in D(A)$ and $t \geq 0$,

$$S(t)x = x + \int_0^t \frac{(t-s)^{\alpha-1}}{\Gamma(\alpha)} AS(s)xds.$$

d) Let $x \in D(A)$. Then $\int_0^t \frac{(t-s)^{\alpha-1}}{\Gamma(\alpha)} S(s)xds \in D(A)$ and

$$S(t)x = x + A\int_0^t \frac{(t-s)^{\alpha-1}}{\Gamma(\alpha)} S(s)xds.$$

Remark 10.3. *The concept of a solution operator, as defined above, is closely related to the concept of a resolvent family (see Prüss [318]). Because of the uniqueness of the Laplace transform, in the border case $\alpha = 1$, the family $S(t)$ corresponds to a C_0 semigroup (see [89, 183]), whereas in the case $\alpha = 2$ a solution operator corresponds to the concept of cosine family (see [75]).*

More information on the C_0-semigroups and sine families can be found in [183, 184, 307].

Definition 10.4. *A solution operator $\{S(t)\}_{t>0}$ is called uniformly continuous if*

$$\lim_{t\to s} \|S(t) - S(s)\|_{B(E)} = 0.$$

10.2.2. Main Results

In this section, we are concerned with the existence of solutions for the problem (10.1)-(10.2).

Definition 10.5. *We say that the function $y : (-\infty, b] \to E$ is a mild solution of* $(10.1)-(10.2)$ *if $y(t) = \phi(t)$ for all $t \leq 0$, the restriction of $y(\cdot)$ to the interval $[0, b]$ is continuous and there exists $v(\cdot) \in L^1(J, E)$, such that $v(t) \in f(t, y_{\rho(t,y_t)})$ a.e $t \in [0, b]$, and y satisfies the following integral equation*

$$y(t) = S(t)\phi(0) + \int_0^t S(t-s)\, v(s)\, ds \quad \textit{for each } t \in [0, b]. \tag{10.3}$$

Set

$$\mathcal{R}(\rho^-) = \{\rho(s, \varphi) : (s, \varphi) \in J \times \mathcal{B}, \rho(s, \varphi) \leq 0\}.$$

We always assume that $\rho : J \times \mathcal{B} \to (-\infty, b]$ is continuous. Additionally, we state following hypothesis:

(H_ϕ) The function $t \to \phi_t$ is continuous from $\mathcal{R}(\rho^-)$ into $\mathcal{B}$ and there exists a continuous and bounded function $L^\phi : \mathcal{R}(\rho^-) \to (0, \infty)$ such that

$$\|\phi_t\|_{\mathcal{B}} \leq L^\phi(t)\|\phi\|_{\mathcal{B}} \quad \text{for every } t \in \mathcal{R}(\rho^-).$$

Remark 10.6. *The condition (H_ϕ), is frequently verified by continuous and bounded functions. For more details, see for instance [226].*

The Convex Case

In this section, we are concerned with the existence of solutions for the problem $(10.1)-(10.2)$ when the right hand side has convex values. Initially, we assume that F is a compact and convex valued multivalued map. Our first result is based on the Bohnenblust-Karlin fixed point theorem [114]. We assume the following hypotheses:

(10.8.1) The operator solution $S(t)_{t \in J}$ is compact for $t > 0$, and there is $M > 0$ such that

$$\|S(t)\|_{B(E)} \leq M, \text{ for each } t \in J.$$

(10.8.2) The multivalued map $F : J \times \mathcal{B} \to P_{cp,cv}(E)$ is Carathéodory.

(10.8.3) For each $r > 0$, there exists a positive function $h_r \in L^1(J; \mathbb{R}_+)$ such that

$$\|F(t, z)\| := sup\{\|v\| : v \in F(t, z)\} \leq h_r(t) \text{ a.e. } t \in J,$$

and for $z \in \mathcal{B}$ with $\|z\|_{\mathcal{B}} \leq r$; and

$$\liminf_{r \to \infty} \frac{1}{r} \int_0^b h_r(s) ds \leq \frac{1}{M}.$$

(10.8.4) For each $t \in J$ and $\rho > 0$ the set $\{F(t, \phi) : \|\phi\|_{\mathcal{B}} \leq \rho\}$ is relatively compact in E.

Theorem 10.7. *Assume that* $(H_\phi), (10.8.1)-(10.8.4)$ *hold. Then the problem* $(10.1)-(10.2)$ *has a mild solution on* $(-\infty, b]$.

Proof. We will transform the problem $(10.1)-(10.2)$ into a fixed point problem. To this end we set

$$B = \{y : (-\infty, b] \to E : y|_J \text{ is continuous and } y_0 \in \mathcal{B}\},$$

where $y|_J$ is the restriction of y to the real compact interval J. We transform the problem $(10.1)-(10.2)$ into a fixed-point problem. Consider the multivalued operator $N : B \to \mathcal{P}(B)$ defined by $N(h) = \{h \in B\}$ with

$$h(t) = \begin{cases} \phi(t), & \text{if } t \leq 0; \\ S(t)\,\phi(0) + \int_0^t S(t-s)\,v(s)\,ds, \quad v \in S_{F,y_{\rho(s,y_s)}} & \text{if } t \in J. \end{cases} \tag{10.4}$$

For $\phi \in \mathcal{B}$, we will define the function $x(.) : (-\infty, b] \to E$ by

$$x(t) = \begin{cases} \phi(t), & \text{if } t \leq 0; \\ S(t)\,\phi(0), & \text{if } t \in J. \end{cases}$$

Then $x_0 = \phi$. For each function $z \in B$ with $z_0 = 0$, we denote by $\bar{z}$ the function defined by

$$\bar{z}(t) = \begin{cases} 0, & \text{if } t \leq 0; \\ z(t), & \text{if } t \in J. \end{cases}$$

If $y(\cdot)$ satisfies (10.3), we can decompose it as $y(t) = \bar{z}(t) + x(t)$, which implies $y_t = z_t + x_t$, for every $t \in J$, and the function $z(\cdot)$ satisfies

$$z(t) = \int_0^t S(t-s)\,v(s)\,ds \quad t \in J,$$

where $v(s) \in S_{F,z_{\rho(s,z_s+x_s)}+x_{\rho(s,z_s+x_s)}}$.
Let

$$B_b^0 = \{z \in B : z_0 = 0 \in \mathcal{B}\}.$$

For any $z \in B_b^0$ we have

$$\|z\|_b = \|z_0\|_{\mathcal{B}} + \sup\{|z(s)| : 0 \leq s < b\} = \sup\{|z(s)| : 0 \leq s < b\}.$$

Thus $(B_b^0, \|\cdot\|_b)$ is a Banach space. We define the operator $P : B_b^0 \to \mathcal{P}(B_b^0)$ by : $P(z) := \{h \in B_b^0\}$ with

$$h(t) = \int_0^t S(t-s)\,v(s)ds, \quad v(s) \in S_{F,\bar{z}_{\rho(s,\bar{z}_s+x_s)}+x_{\rho(s,\bar{z}_s+x_s)}}, \quad t \in J. \tag{10.5}$$

Obviously the operator N has a fixed point is equivalent to P has one, so it turns to prove that P has a fixed point. We shall show that the operators P satisfy all conditions of Bohnenblust-Karlin Theorem. For better readability, we break the proof into a sequence of steps.
Let

$$K_r = \{z \in B_b^0 : z(0) = 0, \|z\|_b \leq r\},$$

where r is the constant given by (10.8.3). It is clear that K_r is a closed, convex, bounded set in B_b^0.

Step 1: *$P(z)$ is convex for each $z \in B_b^0$.*

Indeed, if h_1 and h_2 belong to $P(z)$, then there exist $v_1,\ v_2 \in S_{F,\bar{z}_{\rho(s,\bar{z}_s+x_s)}+x_{\rho(s,\bar{z}_s+x_s)}}$ such that, for $t \in J$, we have

$$h_i(t) = \int_0^t S(t-s)v_i(s)ds, \quad i = 1,2.$$

Let $0 \leq d \leq 1$. Then, for each $t \in J$, we have

$$(dh_1 + (1-d)h_2)(t) = \int_0^t S(t-s)[dv_1(s) + (1-d)v_2(s)]ds.$$

Since $S_{F,\bar{z}_{\rho(s,\bar{z}_s+x_s)}+x_{\rho(s,\bar{z}_s+x_s)}}$ is convex (because F has convex values), we have

$$dh_1 + (1-d)h_2 \in P(z).$$

Step 2: $P(K_r) \subset K_r$.

If it is not true, then for each positive number r, there exists a function $z_r \in K_r$ and $h \in P(z_r)$ such that $\|h(t)\| > r$ for some $t \in J$. Then we have

$$\begin{aligned} r < |h(t)| &\leq \int_0^t |S(t-s)|_{B(E)}|v(s)|ds \\ &\leq M\int_0^b h_r(s)ds. \end{aligned}$$

Dividing both sides by r and taking the lower limit and from (10.8.3), we get

$$\begin{aligned} 1 &< \lim_{r\to\infty} M \inf \frac{1}{r}\int_0^b h_r(s)ds \\ &\leq 1. \end{aligned}$$

which yields to a contradiction. Hence there exists a positive number r' such that $P(K_{r'}) \subset K_{r'}$.

Step 3: *$P(K_{r'})$ is relatively compact.*

Since $K_{r'}$ is bounded and $P(K_{r'}) \subset K_{r'}$, it is clear that $P(K_{r'})$ is bounded. It remains to show that $P(K_{r'})$ is equicontinuous. Let $\tau_1, \tau_2 \in J$ with $\tau_2 > \tau_1$ and $z \in P(K_{r'})$. Then

$$\begin{aligned} |h(\tau_2) - h(\tau_1)| &\leq |\int_0^{\tau_1} [S(\tau_2 - s) - S(\tau_1 - s)]v(s)ds| \\ &\quad + |\int_{\tau_1}^{\tau_2} S(\tau_2 - s)|v(s)|ds| \\ &\leq \int_0^{\tau_1} |S(\tau_2 - s) - S(\tau_1 - s)|h_{r'}(s)ds \\ &\quad + \int_{\tau_1}^{\tau_2} |S(\tau_2 - s)|h_{r'}(s)ds. \end{aligned}$$

The right-hand side of the above inequality tends to zero independently of $z \in K_{r'}$ as $\tau_2 - \tau_1 \to 0$, since $S(t)$ is uniformly continuous. As a consequence of Steps 1 to 3, (10.8.4) together with the Arzela-Ascoli theorem the operator P is completely continuous.

Step 4: *P has a closed graph.*

Let $z_n \to z_*$, $h_n \in P(z_n)$, and $h_n \to h_*$. We need to show that $h_* \in P(z_*)$. Now, $h_n \in P(z_n)$ implies there exists $v_n \in S_{F,\bar{z}_{n\rho(s,\bar{z}_{ns}+x_s)}+x_{\rho(s,\bar{z}_{ns}+x_s)}}$ such that, for each $t \in J$,

$$h_n(t) = \int_0^t S(t-s)v_n(s)ds.$$

We must show that there exists $v_* \in S_{F,\bar{z}_{*\rho(s,\bar{z}_{*s}+x_s)}+x_{\rho(s,\bar{z}_{*s}+x_s)}}$ such that for each $t \in J$,

$$h_*(t) = \int_0^t S(t-s)v_*(s)ds.$$

Consider the linear continuous operator $\Gamma : L^1(J,E) \longrightarrow B_b^0$ defined by

$$(\Gamma v)(t) = \int_0^t S(t-s)v(s)ds.$$

From the definition of Γ, we know that

$$h_n(t) \in \Gamma(S_{F,z_{n\rho(s,z_{ns}+x_s)}+x_{\rho(s,z_{ns}+x_s)}}).$$

Since $z_n \longrightarrow z_*$ and $\Gamma \circ S_P$ is a closed graph operator (see [259]), there exists $v_* \in S_{F,z_{*\rho(s,z_{*s}+x_s)}+x_{\rho(s,z_{*s}+x_s)}}$ such that

$$h_*(t) = \int_0^t S(t-s)v_*(s)ds.$$

Hence $h_* \in P(z_*)$, As a consequence of Bohnenblust-Karlin's fixed point theorem, we deduce that P has a fixed point z on the interval $(-\infty, b]$, so $y = \bar{z} + x$ is a fixed point of the operator N which is the mild solution of problem $(10.1)-(10.2)$. □

Theorem 10.8. Assume that $(10.8.1), (10.8.2), (10.8.4)$ are satisfied then a slight modification of the proof (i.e. use the usual Leray-Schauder alternative) guarantees that $(10.8.3)$ could be replaced by

$(10.10.1)$ There exists a function $p \in L^1(J;\mathbb{R}_+)$ and a continuous nondecreasing function $\psi_F : \mathbb{R}_+ \to (0,\infty)$ such that

$$|F(t,u)| \leq p(t)\ \psi(\|u\|_{\mathcal{B}}) \text{ for a.e. } t \in J \text{ and each } u \in \mathcal{B},$$

with

$$\int_{c_b}^{+\infty} \frac{ds}{\psi(s)} > K_b M \int_0^b p(s)\, ds, \tag{10.6}$$

where

$$c_b = (M_b + L^{\phi} + K_b M H)\|\phi\|_{\mathcal{B}}.$$

Proof. Let z be solutions of the inclusion $z \in \lambda P(z)$, for some $\lambda \in (0,1)$, then there exists $v \in S_{F,\overline{z}_{\rho(s,\overline{z}_s+x_s)}+x_{\rho(s,\overline{z}_s+x_s)}}$ such that

$$\begin{aligned}
|z(t)| &\le \int_0^t \|S(t-s)\|_{B(E)} \, \|F(s,\overline{z}_{\rho(s,\overline{z}_s+x_s)}+x_{\rho(s,\overline{z}_s+x_s)})| \, ds \\
&\le M\int_0^t p(s)\,\psi\left(\|\overline{z}_{\rho(s,\overline{z}_s+x_s)}+x_{\rho(s,\overline{z}_s+x_s)}\|_{\mathcal{B}}\right) ds \\
&\le M\int_0^t p(s)\,\psi\left(K_b|z(s)|+(M_b+L^{\phi}+K_bMH)\|\phi\|_{\mathcal{B}}\right) ds.
\end{aligned}$$

Set $c_b := (M_b+L^{\phi}+K_bMH)\|\phi\|_{\mathcal{B}}$. Then, we have

$$|z(t)| \le M\int_0^t p(s)\psi(K_b|z(s)|+c_b)\,ds.$$

Then

$$K_b|z(t)|+c_b \le c_b+K_bM\int_0^t p(s)\psi(K_b|z(s)|+c_b)\,ds.$$

We consider the function μ defined by

$$\mu(t) := \sup\{\, K_b|z(s)|+c_b \ : \ 0\le s\le t \,\}, \quad t\in J.$$

Let $t^{\star} \in [0,t]$ be such that $\mu(t) = K_b|z(t^{\star})|+c_b\|\phi\|_{\mathcal{B}}$. By the previous inequality, we have

$$\mu(t) \le c_b+K_bM\int_0^t p(s)\,\psi(\mu(s))\,ds \quad \text{for} \quad t\in J.$$

Let us take the right-hand side of the above inequality as $v(t)$. Then, we have

$$\mu(t)\le v(t) \ \text{ for all } t\in J.$$

From the definition of v, we have

$$v(0)=c_b \quad \text{and} \quad v'(t)=K_bMp(t)\,\psi(\mu(t)) \ \text{ a.e. } t\in J.$$

Using the nondecreasing character of ψ, we get

$$v'(t)\le K_bM\,p(t)\,\psi(v(t)) \ \text{ a.e. } t\in J.$$

Using the condition (10.6), this implies that for each $t\in J$ we have

$$\int_{c_b}^{v(t)} \frac{ds}{\psi(s)} \le K_bM\int_0^t p(s)ds \le K_bM\int_0^b p(s)\,ds < \int_{c_b}^{+\infty}\frac{ds}{\psi(s)}.$$

Thus, for every $t\in J$, there exists a constant Λ such that $v(t)\le\Lambda$ and hence $\mu(t)\le\Lambda$. Since $\|z\|_b\le\mu(t)$, we have $\|z\|_b\le\Lambda$. Set

$$U=\{z\in B_b^0 : \|z\|_\infty<\Lambda+1\}.$$

As in Theorem 10.7, the operator $P:\overline{U}\to P(Z)$ is completely continuous. From the choice of U, there is no $z\in\partial U$ such that $z=\lambda P(z)$, for $\lambda\in(0,1)$. As a consequence of the nonlinear alternative of Leray-Schauder type [204], we deduce that F has a fixed point z in U, then the problem (10.1)-(10.2) has at least one mild solution on $(-\infty,b]$. □

The Nonconvex Case

This section is devoted to proving the existence of solutions for $(10.1)-(10.2)$ with a nonconvex valued right hand side. Our result is based on the fixed point theorem for contraction multivalued maps given by Covitz and Nadler.

Theorem 10.9. *Assume that the following hypotheses hold:*

(10.12.1 $F : J \times \mathcal{B} \to P_{cp}(E)$ *has the property that* $F(\cdot, u) : J \to P_{cp}(E)$ *is measurable, for each* $u \in \mathcal{B}$,

(10.12.2) there exists $l \in L^1(J, \mathbb{R}^+)$ *such that*

$$H_d(F(t,u), F(t,\overline{u})) \leq l(t)\|u - \overline{u}\| \quad \textit{for every } u,\ \overline{u} \in \mathcal{B},$$

and

$$d(0, F(t,0)) \leq l(t) \quad a.e.\ t \in J.$$

Then $(10.1)-(10.2)$ *has at least one mild solution on* $(-\infty, b]$.

Remark 10.10. For each $z \in B_b^0$, the set $S_{F,z}$ is nonempty since, by (10.12.1), F has a measurable selection (see [134, Theorem III.6]).

Proof. Let $P : B_b^0 \to \mathcal{P}(B_b^0)$ where P is defined in Theorem 10.7 are solutions of the problem (10.1)-(10.2). We shall show that P satisfies the assumptions of Lemma 1.63. The proof will be given in two steps.

Step 1:$P(z) \in \mathcal{P}_{cl}(B_b^0)$ *for all* $z \in B_b^0$.
Let $(z_n)_{n\geq 0} \in P(z)$ be such that $z_n \to \tilde{z} \in B_b^0$. Then there exists $v_n \in S_{F,\overline{z}_{\rho(s,\overline{z}_s+x_s)}+x_{\rho(s,\overline{z}_s+x_s)}}$ such that, for each $t \in J$,

$$z_n(t) = \int_0^t S(t-s)v_n(s)ds.$$

Using the fact that F has compact values, we may pass to a subsequence if necessary to obtain that v_n converges to v in $L^1(J,E)$ and hence $v \in S_{F,\overline{z}_{\rho(s,\overline{z}_s+x_s)}+x_{\rho(s,\overline{z}_s+x_s)}}$. Thus, for each $t \in J$,

$$z_n(t) \to \tilde{z}(t) = \int_0^t S(t-s)v(s)ds,$$

so $\tilde{z} \in P(z)$.

Step 2: *There exists* $\gamma < 1$ *such that*

$$H_d(F(z), F(z_*)) \leq \gamma\|z - z_*\|_\infty \text{for all} z,\ z_* \in B_b^0.$$

Let $z,\ z_* \in B_b^0$ and $h \in P(z)$. Then, there exists $v(t) \in F(t, \overline{z}_{\rho(s,\overline{z}_s+x_s)} + x_{\rho(s,\overline{z}_s+x_s)})$ such that, for each $t \in J$,

$$h(t) = \int_0^t S(t-s)v(s)ds.$$

From (10.12.2) it follows that

$$H_d(F(t,\overline{z}_{\rho(s,\overline{z}_s+x_s)}+x_{\rho(s,\overline{z}_s+x_s)}),F(t,\overline{z_*}_{\rho(s,\overline{z_*}_s+x_s)}+x_{\rho(s,\overline{z_*}_s+x_s)})\leq l(t)|z(t)-z_*(t)|.$$

Hence, there exists $w\in F(t,\overline{z}_{\rho(s,\overline{z}_s+x_s)}+x_{\rho(s,\overline{z}_s+x_s)})$ such that

$$|v(t)-w|\leq l(t)|z(t)-z_*(t)|,\quad t\in J.$$

Consider $U:J\to\mathcal{P}(E)$ given by

$$U(t)=\{w\in E:|v(t)-w|\leq l(t)|z(t)-z_*(t)|\}.$$

Since the multivalued operator $V(t)=U(t)\cap F(t,\overline{z_*}_{\rho(s,\overline{z_*}_s+x_s)}+x_{\rho(s,\overline{z_*}_s+x_s)}))$ is measurable (see Proposition [134, III.4]), there exists a function $v_2(t)$ which is a measurable selection for V. Thus, $\overline{v}(t)\in F(t,\overline{z_*}_{\rho(s,\overline{z_*}_s+x_s)}+x_{\rho(s,\overline{z_*}_s+x_s)}))$, and for each $t\in J$,

$$|v(t)-\overline{v}(t)|\leq l(t)|z(t)-z_*(t)|.$$

For each $t\in J$, define

$$\overline{h}(t)=\int_0^t S(t-s)\overline{v}(s)ds.$$

Then, for $t\in J$,

$$\begin{aligned}|h(t)-\overline{h}(t)| &\leq \int_0^t ||S(s-t)||_{B(E)}|v(s)-\overline{v}(s)|ds\\ &\leq M\int_0^t l(s)||z-z_*||\\ &\leq \int_0^t \overline{l}(s)||z-z_*||ds\\ &\leq \frac{1}{\tau}e^{\tau L(t)}||z-z_*||_{\overline{B}},\end{aligned}$$

where $\tau>1$, $L(t)=\int_0^t Ml(s)ds$ and $||.||_{\overline{B}}$ is the Bielecki-type norm on B_b^0 defined by

$$||z||_{\overline{B}}=\sup\{e^{-\tau L(t)}||z(t)||:t\in J\}.$$

Therefore,

$$||h-\overline{h}||_{\overline{B}}\leq\frac{1}{\tau}||z-\overline{z}||_{\overline{B}}.$$

By an analogous relation, obtained by interchanging the roles of z and z_*, it follows that

$$H_d(P(z),P(z_*))\leq\frac{1}{\tau}||z-z_*||_{\overline{B}}.$$

So, P is a contraction, and thus, by Lemma 1.63 it has a fixed point z, which is a mild solution to (10.1)-(10.2). □

10.2.3. An Example

To apply our abstract results, we consider the fractional differential equation with state dependent delay of the form

$$\begin{cases} \frac{\partial u}{\partial t}(t,\xi) - \frac{1}{\Gamma(\mu-1)} \int_t^0 (t-s)^{\mu-2} L_\xi u(s,\xi)ds \\ \in [f_1(t,u(t-\sigma(u(t,0)),\xi)), f_2(t,u(t-\sigma(u(t,0)),\xi))], \quad t \in J := [0,b], \ \xi \in [0,\pi], \\ u(t,0) = u(t,\pi) = 0, \qquad t \in J, \\ u(\theta,\xi) = u_0(\theta,\xi), \qquad \theta \in (-\infty,0], \ \xi \in [0,\pi], \end{cases} \tag{10.7}$$

where $1 < \mu < 2$, $\sigma \in C(\mathbb{R},[0,\infty))$, $\gamma > 0$, L_ξ stands for the operator with respect to the spatial variable ξ which is given by:

$$L_\xi = \frac{\partial^2}{\partial \xi^2} - r, \quad \text{with} \quad r > 0,$$

$f_1, f_2 : J \times \mathcal{B} \to \mathbb{R}$ are measurable in t and continuous in y. We assume that for each $t \in J$, $f_1(t,\cdot)$ is lower semi-continuous (i.e., the set $\{y \in \mathcal{B} : f_1(t,y) > \nu\}$ is open for all $\nu \in \mathbb{R}$), and assume that for each $t \in J$, $f_2(t,\cdot)$ is upper semi-continuous (i.e., the set $\{y \in \mathcal{B} : f_2(t,y) < \nu\}$ is open for each $\nu \in \mathbb{R}$). Take $E = L^2([0,\pi],\mathbb{R})$ and the operator $A := L_\xi : D(A) \subset E \to E$ with domain

$$D(A) := \{\, u \in E \ : u'' \in E, \ u(0) = u(\pi) = 0 \,\}.$$

Clearly A is densely defined in E and is sectorial. Hence A is a generator of a solution operator on E. For the phase space , we choose $\mathcal{B} = \mathcal{B}_\gamma$ defined by

$$\mathcal{B}_\gamma = \{\phi \in C((-\infty,0],\mathbb{R}) : \lim_{\theta \to -\infty} e^{\gamma\theta}\phi(\theta) \text{ exists}\}$$

with the norm

$$\|\phi\|_\gamma = \sup_{\theta \in (-\infty,0]} e^{\gamma\theta}|\phi(\theta)|.$$

Notice that the phase space $\mathcal{B}_\gamma$ satisfies axioms (A_1), (A_2) and (A_3) (see [226] for more details). Set

$$\begin{aligned} y(t)(\xi) &= u(t,\xi), \ \ t \in [0,b], \ \xi \in [0,\pi], \\ \phi(\theta)(\xi) &= u_0(\theta,\xi), \ \ t \in [0,b], \ \theta \le 0, \\ F(t,\varphi)(\xi) &= [f_1(t,\varphi(0,\xi)), f_2(t,\varphi(0,\xi))], \ t \in [0,b], \ \xi \in [0,\pi], \\ \rho(t,\varphi) &= t - \sigma(\varphi(0,0)). \end{aligned}$$

The mutivalued map F is u.s.c. with compact convex values ([153]). Hence (10.10.1) and (10.10.2) are satisfied. An application of Theorem 10.8 yields the following result.

Theorem 10.11. *Let $\phi \in \mathcal{B}_\gamma$ be such that (H_ϕ) holds, and let $t \to \phi_t$ be continuous on $\mathcal{R}(\rho^-)$. Moreover we assume that* (10.12.1) *is satisfied. Then there exists at least one mild solution of* (10.7).

Corollary 4. *Let $\phi \in \mathcal{B}_\gamma$ be continuous and bounded. Assume that* (10.12.1) *holds. Then there exists at least one mild solution of* (10.7) *on $(-\infty,b]$.*

10.3. Abstract Integro-Differential Inclusions with State-Dependent Delay in Fréchet Spaces

10.3.1. Introduction

In this section,a recent nonlinear alternative of Leray-Schauder type due to Frigon for contractive multivalued maps in Fréchet spaces [186] is used to investigate the existence and uniqueness solutions for the following fractional integro-differential inclusions in Fréchet spaces.

$$y'(t)-\int_0^t \frac{(t-s)^{\alpha-2}}{\Gamma(\alpha-1)}Ay(s)ds \in F(t,y_{\rho(t,y_t)}),\quad \text{a.e. } t\in J=[0,\infty), \tag{10.8}$$

$$y_0=\phi\in \mathcal{B}. \tag{10.9}$$

10.3.2. Main Results

Definition 10.12. *We say that the function $y:(-\infty,+\infty)\to E$ is a mild solution of (10.8)-(10.9) if $y(t)=\phi(t)$ for all $t\leq 0$, the restriction of $y(.)$ to the interval $[0,\infty)$ is continuous and there exists $v(.)\in L^1(J,E)$.,such that $v(t)\in f(t,y_{\rho(t,y_t)})$ a.e $t\in[0,\infty)$, and y satisfies the following integral equation*

$$y(t)=S(t)\phi(0)+\int_0^t S(t-s)\,v(s)\,ds \quad for\ each\ t\in[0,+\infty). \tag{10.10}$$

Set

$$\mathcal{R}(\rho^-)=\{\rho(s,\varphi):(s,\varphi)\in J\times\mathcal{B},\rho(s,\varphi)\leq 0\}.$$

We always assume that $\rho:J\times\mathcal{B}\to(-\infty,+\infty)$ is continuous. Additionally, we introduce following hypothesis:

(10.18.1) The operator solution $S(t)_{t\in J}$ is compact for $t>0$.

(10.18.2) The multifunction $F:J\times\mathcal{B}\to\mathcal{P}(E)$ is L^1_{loc}-Carathéodory with compact and convex values for each $u\in\mathcal{B}$ and there exist a function $p\in L^1_{loc}(J,\mathbb{R}^+)$ and a continuous nondecreasing function $\psi:J\to(0,\infty)$ such that:

$$\|F(t,u)\|_{\mathcal{P}(E)}\leq p(t)\psi(\|u\|_{\mathcal{B}})\ \text{ for a.e } t\in J \text{ and each } u\in\mathcal{B}.$$

(10.18.3) For all $R>0$, there exists $l_R\in L^1_{loc}(J,\mathbb{R}_+)$ such that:

$$H_d(F(t,u),F(t,v))\leq l_R(t)\|u-v\|_{\mathcal{B}}$$

for each $t\in J$ and for all $u,v\in\mathcal{B}$ with $\|u\|_{\mathcal{B}}\leq R$ and $\|v\|_{\mathcal{B}}\leq R$ and

$$d(0,F(t,0))\leq l_R(t)\ \text{ a.e. } t\in J.$$

For every $n\in\mathbb{N}$, we define in $B_{+\infty}$ the family of semi-norms by:

$$\|y\|_n:=\sup\{\,e^{-\tau\,L_n^*(t)}\,|y(t)|:t\in[0,n]\,\}$$

where $L_n^*(t) = \int_0^t \bar{l}_n(s)\, ds$, $\bar{l}_n(t) = K_n\, Ml_n(t)$ and l_n is the function from (10.18.3). Then $B_{+\infty}$ is a Fréchet space with the family of semi-norms $\|.\|_{n\in\mathbb{N}}$. In what follows we will choose $\tau > 1$.

Theorem 10.13. *Suppose that hypotheses* (10.18.1) – (10.18.3) *are satisfied and moreover*

$$\int_0^{+\infty} \frac{ds}{\psi(s)} > K_n M \int_0^n p(s)\, ds \quad \textit{for each } n \in \mathbb{N} \tag{10.11}$$

Then the problem (6.17)-(6.18) has a unique mild solution on $(-\infty, +\infty)$.

Proof. Transform the problem (6.17)-(6.18) into a fixed point problem. Consider the multivalued operator $N : B_{+\infty} \to \mathcal{P}(B_{+\infty})$ defined by: $N(h) = \{h \in B_{+\infty}\}$, with

$$h(t) = \begin{cases} \phi(t), & \text{if } t \leq 0; \\ S(t)\, \phi(0) + \displaystyle\int_0^t S(t-s)\, v(s)\, ds, \quad v \in S_{F,y_{\rho(s,y_s)}} & \text{if } t \in J. \end{cases} \tag{10.12}$$

For $\phi \in \mathcal{B}$, we will define the function $x(.) : \mathbb{R} \to E$ by

$$x(t) = \begin{cases} \phi(t), & \text{if } t \leq 0; \\ S(t)\, \phi(0), & \text{if } t \in J. \end{cases}$$

Then $x_0 = \phi$. For each function $z \in B_{+\infty}$ with $z_0 = 0$, we denote by $\bar{z}$ the function defined by

$$\bar{z}(t) = \begin{cases} 0, & \text{if } t \leq 0; \\ z(t), & \text{if } t \in J. \end{cases}$$

If $y(\cdot)$ satisfies (10.10), we can decompose it as $y(t) = \bar{z}(t) + x(t)$, $t \geq 0$, which implies $y_t = z_t + x_t$, for every $t \in J$ and the function $z(\cdot)$ satisfies

$$z(t) = \int_0^t S(t-s)\, v(s)\, ds \quad t \in J,$$

where $v(s) \in S_{F,z_{\rho(s,z_s+x_s)}+x_{\rho(s,z_s+x_s)}}$.
Let

$$B_{+\infty}^0 = \{z \in B_{+\infty} : z_0 = 0 \in \mathcal{B}\}.$$

We define the operator $P : B_{+\infty}^0 \to \mathcal{P}(B_{+\infty}^0)$ by:
$P(z) := \{h \in B_{+\infty}^0\}$ with

$$h(t) = \int_0^t S(t-s)\, v(s) ds, \quad v(s) \in S_{F,\bar{z}_{\rho(s,\bar{z}_s+x_s)}+x_{\rho(s,\bar{z}_s+x_s)}}, \quad t \in J. \tag{10.13}$$

Obviously the operator N has a fixed point is equivalent to P has one, so it turns to prove that P has a fixed point. Let $z \in B_{+\infty}^0$ be a possible fixed point of the operator P. Given

$n \in \mathbb{N}$, then z should be solution of the inclusion $z \in \lambda P(z)$ for some $\lambda \in (0,1)$ and there exists $v \in S_{F,z}$ such that, for each $t \in [0,n]$, we have

$$\begin{aligned} |z(t)| &\leq \int_0^t \|S(t-s)\|_{B(E)}\, |v(s)|\, ds \\ &\leq M \int_0^t p(s)\, \psi\left(\|z_{\rho(s,z_s+x_s)} + x_{\rho(s,z_s+x_s)}\|_{\mathcal{B}}\right) ds \\ &\leq M \int_0^t p(s)\, \psi\left(K_n|z(s)| + (M_n + L^{\phi} + K_n MH)\|\phi\|_{\mathcal{B}}\right) ds \end{aligned}$$

Set $c_n := (M_n + L^{\phi} + K_n MH)\|\phi\|_{\mathcal{B}}$. Then, we have

$$|z(t)| \leq M \int_0^t p(s)\, \psi(K_n|z(s)| + c_n)\, ds$$

Then

$$K_n|z(t)| + c_n \leq c_n + K_n M \int_0^t p(s)\, \psi(K_n|z(s)| + c_n)\, ds.$$

We consider the function μ defined by

$$\mu(t) := \sup\{\, K_n|z(s)| + c_n \,:\, 0 \leq s \leq t\,\}, \quad 0 \leq t < +\infty.$$

Let $t^{\star} \in [0,t]$ be such that

$$\mu(t) = K_n|z(t^{\star})| + c_n\|\phi\|_{\mathcal{B}}.$$

By the previous inequality, we have

$$\mu(t) \leq c_n + K_n\, M \int_0^t p(s)\, \psi(\mu(s))\, ds \quad \text{for} \quad t \in [0,n].$$

Let us take the right-hand side of the above inequality as $v(t)$. Then, we have

$$\mu(t) \leq v(t) \ \text{ for all } t \in [0,n].$$

From the definition of v, we have

$$v(0) = c_n \quad \text{and} \quad v'(t) = K_n M p(t)\, \psi(\mu(t)) \quad \text{a.e. } t \in [0,n].$$

Using the nondecreasing character of ψ, we get

$$v'(t) \leq K_n M\, p(t)\, \psi(v(t)) \ \text{ a.e. } t \in [0,n].$$

Using the condition (10.11), this implies that for each $t \in [0,n]$ we have

$$\int_{c_n}^{v(t)} \frac{ds}{\psi(s)} \leq K_n M \int_0^t p(s)\, ds \leq K_n M \int_0^n p(s)\, ds < \int_{c_n}^{+\infty} \frac{ds}{\psi(s)}.$$

Thus, for every $t \in [0,n]$, there exists a constant Λ_n such that $v(t) \leq \Lambda_n$ and hence $\mu(t) \leq \Lambda_n$. Since $\|z\|_n \leq \mu(t)$, we have $\|z\|_n \leq \Lambda_n$.
Set

$$U = \{\, z \in B^0_{+\infty} \,:\, \sup_{0 \leq t \leq n} |z(t)| < \Lambda_n + 1, \ \text{ for all } n \in \mathbb{N}\}.$$

Clearly, U is an open subset of $B^0_{+\infty}$.

We shall show that $P:\overline{U}\to \mathcal{P}(B^0_{+\infty})$ is a contraction and an admissible operator.

First, we prove that P is a contraction. Let $z,\overline{z}\in B^0_{+\infty}$ and $h\in P(z)$. Then there exists $v(t)\in F(t,z_{\rho(s,z_s+x_s)}+x_{\rho(s,z_s+x_s)})$ such that for each $t\in[0,n]$

$$h(t)=\int_0^t S(t-s)v(s)ds.$$

From (10.18.3) it follows that

$$H_d(F(t,z_{\rho(s,z_s+x_s)}+x_{\rho(s,z_s+x_s)}),F(t,\overline{z}_{\rho(s,z_s+x_s)}+x_{\rho(s,z_s+x_s)}))$$
$$\leq l_n(t)\|z_{\rho(s,z_s+x_s)}-\overline{z}_{\rho(s,z_s+x_s)}\|_{\mathcal{B}}.$$

Hence, there is $w\in F(t,\overline{z}_{\rho(s,z_s+x_s)}+x_{\rho(s,z_s+x_s)})$ such that

$$|v(t)-w|\leq l_n(t)\|z_{\rho(s,z_s+x_s)}-\overline{z}_{\rho(s,z_s+x_s)}\|_{\mathcal{B}}\ \ t\in[0,n].$$

Consider $U_*:[0,n]\to\mathcal{P}(E)$, given by

$$U_*=\{w\in E:|v(t)-w|\leq l_n(t)\|z_{\rho(s,z_s+x_s)}-\overline{z}_{\rho(s,z_s+x_s)}\|_{\mathcal{B}}.\}$$

Since the multivalued operator $V_*(t)=U_*(t)\cap F(t,\overline{z}_{\rho(s,z_s+x_s)}+x_{\rho(s,z_s+x_s)})$ is measurable (in [134], see Proposition III.4), there exists a function $\overline{v}(t)$, which is a measurable selection for V_*. So, $\overline{v}(t)\in F(t,\overline{z}_{\rho(s,z_s+x_s)}+x_{\rho(s,z_s+x_s)})$ and using $(A1)$, we obtain for each $t\in[0,n]$

$$\begin{aligned}|v(t)-\overline{v}(t)| &\leq l_n(t)\|z_{\rho(s,z_s+x_s)}-\overline{z}_{\rho(s,z_s+x_s)}\|_{\mathcal{B}}\\ &\leq l_n(t)[K(t)|z(t)-\overline{z}(t)|+M(t)\|z_0-\overline{z}_0\|_{\mathcal{B}}]\\ &\leq l_n(t)K_n|z(t)-\overline{z}(t)|\end{aligned}$$

Let us define, for each $t\in[0,n]$

$$\overline{h}(t)=\int_0^t S(t-s)\overline{v}(s)ds.$$

Then we have

$$\begin{aligned}|h(t)-\overline{h}(t)| &\leq \int_0^t\|S(t-s)\||v(s)-\overline{v}(s)|ds\\ &\leq \int_0^t MK_nl_n(s)|z(s)-\overline{z}(s)|ds\\ &\leq \int_0^t[\overline{l}_n(s)e^{\tau L_n^*(s)}][e^{-\tau L_n^*(s)}|z(s)-\overline{z}(s)|]ds\\ &\leq \int_0^t\left[\frac{e^{\tau L_n^*(s)}}{\tau}\right]' ds\ \|z-\overline{z}\|_n\\ &\leq \frac{1}{\tau}e^{\tau L_n^*(t)}\|z-\overline{z}\|_n.\end{aligned}$$

Therefore,

$$\|h-\overline{h}\|_n \leq \frac{1}{\tau}\ \|z-\overline{z}\|_n.$$

By an analogous relation, obtained by interchanging the roles of z and $\overline{z}$, it follows that

$$H_d(P(z),P(\overline{z})) \leq \frac{1}{\tau}\ \|z-\overline{z}\|_n$$

So, P is a contraction for all $n \in \mathbb{N}$.

Now we shall show that P is an admissible operator. Let $z \in B^0_{+\infty}$. Set, for every $n \in \mathbb{N}$, the space

$$B^0_n = \left\{y : (-\infty, n] \to E : y|_{[0,n]} \in C([0,n],E), y_0 \in \mathcal{B}\right\},$$

and let us consider the multivalued operator

$$P(z) = \{h \in B^0_n : h(t) = \int_0^t S(t-s)v(s)ds,\ t \in [0,n]\}$$

where $v \in S^n_{F,y} = \{v \in L^1([0,n],E) : v(t) \in F(t, y_{\rho(t,y_t)})$ for a.e. $t \in [0,n]\}$.

From $(10.18.1)-(10.18.3)$ and since P is a multivalued map with compact values, we can prove that for every $z \in B^0_n$, $P(z) \in \mathcal{P}_{cp}(B^0_n)$ and there exist $z_* \in B^0_n$ such that $z_* \in P(z_*)$. Let $h \in B^0_n$, $\overline{z} \in \overline{U}$ and $\varepsilon > 0$. Assume that $z_* \in P(\overline{z})$, then we have

$$\begin{aligned}|\overline{z}(t)-z_*(t)| &\leq |\overline{z}(t)-h(t)|+|z_*(t)-h(t)| \\ &\leq e^{\tau\, L^*_n(t)}\|\overline{z}-P(\overline{z})\|_n+\|z_*-h\|.\end{aligned}$$

Since h is arbitrary, we may suppose that $h \in B(z_*,\varepsilon) = \{h \in B^0_n : \|h-z_*\|_n \leq \varepsilon\}$. Therefore,

$$\|\overline{z}-z_*\|_n \leq \|\overline{z}-P(\overline{z})\|_n+\varepsilon.$$

If z is not in $P(\overline{z})$, then $\|z_*-P(\overline{z})\| \neq 0$. Since $P(\overline{z})$ is compact, there exists $x \in P(\overline{z})$ such that $\|z_*-P(\overline{z})\| = \|z_*-x\|$. Then we have

$$\begin{aligned}|\overline{z}(t)-x(t)| &\leq |\overline{z}(t)-h(t)|+|x(t)-h(t)| \\ &\leq e^{\tau\, L^*_n(t)}\|\overline{z}-P(\overline{z})\|_n+\|x(t)-h(t)\|.\end{aligned}$$

Therefore,

$$\|\overline{z}-x\|_n \leq \|\overline{z}-P(\overline{z})\|_n+\varepsilon.$$

So, P is an admissible operator contraction. From the choice of U there is no $z \in \partial U$ such that $z = \lambda P(z)$ for some $\lambda \in (0,1)$. As a consequence of Theorem 1.67, we deduce that the operator P has a fixed point z^*. Then $y^*(t) = z^*(t)+x(t)$, $t \in (-\infty,+\infty)$ is a fixed point of the operator N, which is a mild solution of the problem (6.17)-(6.18). □

10.3.3. Application to Control Theory

This section is devoted to an application of the argument used in the previous section to the controllability of fractional integro-differential inclusions with state-dependent delay . More precisely we will consider the following problem:

$$\begin{cases} y'(t) - \int_0^t \frac{(t-s)^{\alpha-2}}{\Gamma(\alpha-1)} Ay(s)ds \in F(t, y_{\rho(t,y_t)}) \\ \quad +(Bu)(t), \quad \text{a.e. } t \in J = [0,\infty) \\ y_0 = \phi \in \mathcal{B}, \end{cases} \tag{10.14}$$

where α, F, A, ϕ are as Section 5.1, B is a bounded linear operator from U into E and the control parameter $u(.)$ belongs to $L^2(J,U)$, a space of admissible controls, and U is a Banach space .

Definition 10.14. *We say that the function $y : (-\infty, +\infty) \to E$ is a mild solution of (10.14) if $y(t) = \phi(t)$ for all $t \leq 0$, the restriction of $y(.)$ to the interval $[0,\infty)$ is continuous and there exists $v(.) \in L^1(J,E)$, such that $v(t) \in F(t, y_{\rho(t,y_t)})$ a.e $t \in [0,\infty)$, and y satisfies the following integral equation*

$$y(t) = S(t)\phi(0) + \int_0^t S(t-s)\, v(s)\, ds + \int_0^t (Bu)(s)\, ds \quad \textit{for each } t \in [0,+\infty). \tag{10.15}$$

Definition 10.15. *The system (10.14) is said to be controllable if for any continuous function $\phi \in \mathcal{B}$ and any $x_1 \in E$ and for each $n \in \mathbb{N}$ there exists a control $u \in L^2([0,n],E)$ such that the mild solution $y(.)$ of (10.14) satisfies $y(n) = x_1$.*

Let us introduce the following hypothesis:

(10.23.1) For each $n > 0$ the linear operator $W : L^2([0,n],U) \to E$ defined by

$$Wu = \int_0^n S(n-s)Bu(s)ds,$$

has an invertible operator W^{-1} which takes values in $L^2([0,n],U)/KerW$ and there exist positive constants M_1, M_2 such that $\|B\| \leq M_1$ and $\|W^{-1}\| \leq M_2$,

Remark 10.16. *The question of the existence of the operator W and its inverse is discussed in the paper by Quinn and Carmichael [322].*

Theorem 10.17. *Assume that hypotheses* $(10.18.1)-(10.18.3)$ *and* $(10.23.1)$ *hold. Then the problem (10.14) is controllable.*

Proof. Let us fix $\tau > 1$. For every $n \in \mathbb{N}$, we define in $B_{+\infty}$ the semi-norms by:

$$\|y\|_n := \sup \{ e^{-\tau \overline{L}_n(t)}\, |y(t)| : t \in [0,n] \};$$

where $\overline{L}_n(t) = \int_0^t \overline{l}_n(s)\, ds$, $\overline{l}_n(t) = max(MK_n l_n(t), K_n M^2 M_1 M_2 l_n(t))$. Then $B_{+\infty}$ is a Fréchet space with those family of semi-norms $\|\cdot\|_{n\in\mathbb{N}}$. Transform (10.14)into a fixed point problem. Consider the operator $N_1 : B_{+\infty} \to \mathcal{B}_{+\infty}$ defined by: $N_1(h) = \{h \in B_{+\infty}\}$, with

$$h(t) = \begin{cases} \phi(t), & \text{if } t \leq 0; \\ S(t)\,\phi(0) + \int_0^t S(t-s)\, v(s)\, ds & \\ \quad + \int_0^t S(t-s)Bu_y^n(s)ds, \quad v \in S_{F,y_{\rho(s,y_s)}} & \text{if } t \in J. \end{cases} \tag{10.16}$$

Using the assumption (10.23.1), for arbitrary function $y(.)$ define the control

$$u_y^n(t) = W^{-1}[y_1 - S(t)\,\phi(0) - \int_0^t S(n-s)\,v(s)\,ds](t).$$

Next we will prove that $\widetilde{N}$ has a fixed point. Let $x(.) : \mathbb{R} \to E$ be the function defined by

$$x(t) = \begin{cases} \phi(t), & \text{if } t \leq 0; \\ S(t)\,\phi(0), & \text{if } t \in J. \end{cases}$$

Then $x_0 = \phi$. For each function $z \in B_{+\infty}$ with $z_0 = 0$, we denote by $\bar{z}$ the function defined by

$$\bar{z}(t) = \begin{cases} 0, & \text{if } t \leq 0; \\ z(t), & \text{if } t \in J. \end{cases}$$

If $y(\cdot)$ satisfies (10.15), we can decompose it as $y(t) = \bar{z}(t) + x(t)$, $t \geq 0$, which implies $y_t = z_t + x_t$, for every $t \in J$ and the function $z(\cdot)$ satisfies

$$z(t) = \int_0^t S(t-s)\,v(s)\,ds + \int_0^t Bu_{z+x}^n(s)\,ds \quad t \in J,$$

where $v(s) \in S_{F,z_{\rho(s,z_s+x_s)}+x_{\rho(s,z_s+x_s)}}$.
Let

$$B_{+\infty}^0 = \{z \in B_{+\infty} : z_0 = 0 \in \mathcal{B}\}.$$

We define the operator $P_1 : B_{+\infty}^0 \to \mathcal{P}(B_{+\infty}^0)$ by : $P_1(z) := \{h \in B_{+\infty}^0\}$ with

$$h(t) = \int_0^t S(t-s)\,v(s)ds$$

$$+\int_0^t S(t-s)\,Bu_{z+x}(s)\,ds, \quad v(s) \in S_{F,\bar{z}_{\rho(s,\bar{z}_s+x_s)}+x_{\rho(s,\bar{z}_s+x_s)}}, \quad t \in J. \tag{10.17}$$

Obviously the operator N_1 has a fixed point is equivalent to P_1 has one, so it turns to prove that P_1 has a fixed point . Let $z \in B_{+\infty}^0$ be a possible fixed point of the operator P_1. Given $n \in \mathbb{N}$, then z should be solution of the inclusion $z \in \lambda P_1(z)$ for some $\lambda \in (0,1)$ and there exists $v \in S_{F,z}$ such that, for each $t \in [0,n]$, we have

$$\begin{aligned}
|z(t)| &\leq \int_0^t \|S(t-s)\|_{B(E)}\,|v(s)|\,ds + \int_0^t \|S(t-s)\|_{B(E)}\,\|B\|\,\|u_{z+x}^n(s)\|\,ds \\
&\leq M\int_0^t p(s)\,\psi(\|z_{\rho(s,z_s+x_s)} + x_{\rho(s,z_s+x_s)}\|_{\mathcal{B}})\,ds + M\,M_1\,M_2 \\
&\quad \int_0^t \left[\|y_1\| + M\|\phi\| + M\int_0^n p(\tau)\psi(\|z_{\rho(\tau,z_\tau+x_\tau)} + x_{\rho(\tau,z_\tau+x_\tau)}\|)\,d\tau\right] ds \\
&\leq M\int_0^t p(s)\,\psi\left(K_n|z(s)| + (M_n + L^\phi + K_nMH)\|\phi\|_{\mathcal{B}}\right)\,ds \\
&\quad + nMM_1M_2[\|y_1\| + M\|\phi\|] + nM^2\,M_1\,M_2 \\
&\quad \int_0^n p(s)\,\psi\left(K_n|z(s)| + (M_n + L^\phi + K_nMH)\|\phi\|_{\mathcal{B}}\right)\,ds
\end{aligned}$$

Set $c_n := (M_n + L^\phi + K_n MH)\|\phi\|_{\mathcal{B}}$. Then, we have

$$\begin{aligned} |z(t)| &\leq M\int_0^t p(s)\,\psi(K_n|z(s)|+c_n)\,ds + nM\,M_1\,M_2[\|y_1\|+M\|\phi\|] \\ &\quad + nM^2\,M_1\,M_2\int_0^n p(s)\,\psi(K_n|z(s)|+c_n)\,ds \end{aligned}$$

Then

$$\begin{aligned} K_n|z(t)|+c_n &\leq c_n + K_nM\int_0^t p(s)\,\psi(K_n|z(s)|+c_n)\,ds \\ &\quad + K_n nM\,M_1\,M_2[\|y_1\|+M\|\phi\|] \\ &\quad + K_n\,M^2\,M_1\,M_2\,n\int_0^n p(s)\,\psi(K_n|z(s)|+c_n)\,ds \end{aligned}$$

Set

$$\beta_n = K_n nM\,M_1\,M_2[\|y_1\|+M\|\phi\|]+c_n.$$

So

$$\begin{aligned} K_n|z(t)|+c_n &\leq \beta_n + K_nM\int_0^t p(s)\,\psi(K_n|z(s)|+c_n)\,ds \\ &\quad + K_n\,M^2\,M_1\,M_2\,n\int_0^n p(s)\,\psi(K_n|z(s)|+c_n)\,ds \end{aligned}$$

We consider the function μ defined by

$$\mu(t) := \sup\{\,K_n|z(s)|+c_n\;:\;0\leq s\leq t\,\},\quad 0\leq t<+\infty.$$

By the previous inequality, we have

$$\begin{aligned} \mu(t) &\leq \beta_n + K_nM\int_0^t p(s)\,\psi(\mu(s))\,ds + K_n\,M^2\,M_1\,M_2\,n \\ &\quad \int_0^n p(s)\,\psi(\mu(s))ds \;\text{ for }\; t\in[0,n]. \end{aligned}$$

Let us take the right-hand side of the above inequality as $v(t)$. Then, we have

$$\mu(t)\leq v(t)\;\text{ for all } t\in[0,n].$$

From the definition of v, we have

$$v(0) = \beta_n + K_n\,M^2\,M_1\,M_2\,n\int_0^n p(s)\,\psi(\mu(s))ds,$$

and

$$v'(t) = K_nMp(t)\,\psi(\mu(t))\quad \text{a.e. } t\in[0,n].$$

Using the nondecreasing character of ψ, we get

$$v'(t)\leq K_nM\,p(t)\,\psi(v(t))\;\text{ a.e. } t\in[0,n].$$

This implies that for each $t \in [0,n]$ we have

$$\int_{c_n}^{v(t)} \frac{ds}{\psi(s)} \leq K_n M \int_0^t p(s)\, ds \leq K_n M \int_0^n p(s)\, ds < \int_{c_n}^{+\infty} \frac{ds}{\psi(s)}.$$

Thus, for every $t \in [0,n]$, there exists a constant Λ_n such that $v(t) \leq \Lambda_n$ and hence $\mu(t) \leq \Lambda_n$. Since $\|z\|_n \leq \mu(t)$, we have $\|z\|_n \leq \Lambda_n$.
Set

$$U_1 = \{\, z \in B^0_{+\infty} \ : \ \sup_{0 \leq t \leq n} |z(t)| < \Lambda_n + 1, \ \text{ for all } \ n \in \mathbb{N}\}.$$

Clearly, U is an open subset of $B^0_{+\infty}$.
We can easily prove that P_1 is a contraction and an admissible multivalued map. From the choice of U_1 there is no $z \in \partial U_1$ such that $z = \lambda P_1(z)$for some $\lambda \in (0,1)$. As a consequence of Theorem 1.67 we deduce that the operator P_1 has a fixed point z^*. Then $y^*(t) = z^*(t) + x(t)$, $t \in (-\infty, +\infty)$ is a fixed point of the operator N_1, which is a mild solution of the problem (10.14). □

10.4. Notes and Remarks

The results of Chapter 10 are taken from Benchohra *et al.* [92, 103].

Chapter 11

Weak Solutions for Nonlinear Fractional Differential Equations

11.1. Introduction

In this chapter, we consider the existence of weak solutions for some classes of fractional order nonlinear differential equations. The main tool is based of the measure of weak noncompactness together with appropriates fixed point theorems.

11.2. Fractional Calculus in Banach Spaces with Weak Topologies

In this section, we shall be concerned with fractional calculus in Banach spaces with weak topologies.

11.2.1. Pettis Fractional Integrals

In this section we give the definitions of the Pettis fractional integrals on a finite interval on the real line and present some of their properties. More detailed information may be found in the book by Samko *et al.* ([337], section 2).

Let $\Omega = [0,T]$ such that $(T > 0)$ be a finite interval on the real axis $\mathbb{R}$. According to the custom $L^p(\Omega)$, $1 \leq p \leq \infty$ will be denote the Banach space of real valued measurable functions x defined over Ω and having $|x|^p$ Lebesgue integrable function on the interval Ω, and $L^\infty(\Omega)$ denote the Banach space of real valued essentially and measurable function defined over Ω. E is a Banach space with norm $\|\cdot\|$ and dual E^*.

We will let $C(\Omega,E)$ denote the Banach space of strong continuous functions $f:\Omega \to E$ while $P[\Omega,E]$ denotes the space of all E-valued Pettis integral functions in the interval Ω. We introduce the following definitions from [312].

Definition 11.1. *For $1 \leq p \leq \infty$, we define the class $\mathcal{H}^p(E)$ to be the class of all functions $x : \Omega \to E$ having $\varphi x \in L^p(\Omega)$ for every $\varphi \in E^*$. The class $\mathcal{H}_0^p(E)$ is defined by*

$$\mathcal{H}_0^p(E) = \{x \in P[\Omega, E] : \varphi x \in L^p(\Omega), \quad \forall \varphi \in E^*\}.$$

Remark 11.2. *The space $\mathcal{H}_0^p(E)$ coincide with the space $P[\Omega, E]$. This is due to the fact that " if the function $x : \Omega \to E$ is Pettis integrable, $\varphi x \in L^1(\Omega)$ for every $\varphi \in E^*$." Moreover, it is not difficult to see that*

$$\mathcal{H}_0^1(E) \supset \mathcal{H}_0^2(E) \supset \cdots \supset \mathcal{H}_0^\infty(E).$$

The following result is due to J. Pettis (see [312] Theorem 3.4 and corrollary 3.41).

Proposition 11.3. *Let p and q be a constants of conjugate exponents. In order that x be in $\mathcal{H}_0^p(E)$, it is necessary and sufficient that $x(\cdot)u(\cdot)$ be Pettis integrable for every $u(\cdot) \in L^q(\Omega)$.*

Proposition 11.4. *If $x(\cdot)$ is Pettis integrable and $u(\cdot)$ is measurable and essentially bounded real-valued function, then $x(\cdot)u(\cdot)$ is Pettis integrable.*

Proposition 11.5. *if strongly measurable $x(\cdot)$ is in $\mathcal{H}^p(E)$, $p > 1$, then $x(\cdot)$ is in $\mathcal{H}_0^p(E)$. If E is weakly complete, this is also true for $p = 1$. In particular, if $p \in [1, \infty]$, $q \in [1, \infty)$, then $\mathcal{H}^p(L^q(\Omega)) = \mathcal{H}_0^p(L^q(\Omega))$.*

Definition 11.6. *Let $x : \Omega \to E$. The left-sided fractional Pettis integral (shortly LS-FPI) of x of order $\alpha > 0$ is defined by*

$$I_+^\alpha x(t) = \int_0^t \frac{(t-s)^{\alpha-1}}{\Gamma(\alpha)} f(s)ds \tag{11.1}$$

where the sign " $\int$ " denotes the Pettis integral and Γ is the gamma function. We can white $I_+^\alpha x(t) = x(t) * \phi_\alpha(t)$, where $\phi_\alpha(t) = \frac{t^{\alpha-1}}{\Gamma(\alpha}$ for $t > 0$ and $\phi_\alpha(t) = 0$ for $t \leq 0$, and $\phi_\alpha \to \delta(t)$ as $\alpha \to 0$, where δ is the delta function.

Similary, we have

Definition 11.7. *Let $x : \Omega \to E$. The right-sided fractional Pettis integral (shortly RS-FPI) of x of order $\alpha > 0$ is defined by and*

$$I_-^\alpha x(t) = \int_t^T \frac{(s-t)^{\alpha-1}}{\Gamma(\alpha)} x(s)dt. \tag{11.2}$$

Lemma 11.8. *Let $\alpha > 0$ and $p > \max\{1, 1/\alpha\}$. Assume that the strongly measurable $x : \Omega \to E$ having $\varphi x \in L^p(\Omega)$ for every $\varphi \in E^*$. Then the LS-FPI and RS-FPI of x exists for almost every $t \in \Omega$ as a weakly continuous function from Ω to E. Moreover,*

$$\varphi((I_+^\alpha x)(t)) = I_+^\alpha \varphi(x(t)), \qquad \text{for all } \varphi \in E^*$$

and

$$\varphi((I_-^\alpha x)(t)) = I_-^\alpha \varphi(x(t)), \qquad \text{for all } \varphi \in E^*$$

For the proof of this Lemma see[332].

Since weak continuity implies strong measurability See([225], p.73), we have the following lemma which is an immediate consequence of the above result.

Lemma 11.9. *Let $\alpha > 0$. Assume $x : \Omega \to E$ be a weakly continuous function on Ω. Then the LS-FPI $I^{\alpha}_{+}x(t)$ and the RS-FPI $I^{\alpha}_{-}x(t)$ of x exists for almost every $t \in \Omega$ as a weakly continuous function from Ω to E. Moreover,*

$$\varphi((I^{\alpha}_{+})x(t)) = I^{\alpha}_{+}\varphi(x(t)), \qquad \varphi((I^{\alpha}_{-}x)(t)) = I^{\alpha}_{-}\varphi(x(t)) \qquad \text{for all } \varphi \in E^*.$$

Lemma 11.10. *Let $\alpha > 0$. Assume that $x : \Omega \to E$ is a Pettis-integrable function such that $\varphi(x(\cdot)) \in L^{\infty}(\Omega)$ for each $\varphi \in E^*$. If $\alpha > 0$ then the LS-FPI and the RS-FPI exists as a functions from Ω into E and*

$$\varphi(I^{\alpha}_{+}x(t)) = I^{\alpha}_{+}\varphi(x(t)), \qquad \varphi((I^{\alpha}_{-}x)(t)) = I^{\alpha}_{-}\varphi(x(t)) \text{for all } \varphi \in E^*.$$

Proof. Only the proof in case of the LS-FPI is given since the case of the RS-FPI is very similar.

Set $h(t) = \frac{(t-s)^{\alpha-1}}{\Gamma(\alpha)}, t \in \Omega$ and $s < t$. Since x is Pettis-integrable function and $\varphi(x(\cdot)) \in L^{\infty}(\Omega)$, then according to Proposition 11.35, the function $x(\cdot)h(\cdot)$ is Pettis-integrable, hence from the definition of Pettis-integral there exists a function, denoted by $I^{\alpha}_{+}x$ from Ω into E so that

$$\varphi(I^{\alpha}_{+}x(t)) = \int_0^t \frac{(t-s)^{\alpha-1}}{\Gamma(\alpha)}\varphi(x(s))ds = I^{\alpha}_{+}\varphi(x(t)) \qquad \text{for all } \varphi \in E^*.$$

By definition

$$I^{\alpha}_{+}x(t) = \int_0^t \frac{(t-s)^{\alpha-1}}{\Gamma(\alpha)}x(s)ds, \qquad t \in \Omega.$$

□

Remark 11.11. *When $\alpha = n \in \mathbb{N}$, the definitions (11.1) and (11.2) coincide with the nth integrals of the form*

$$I^{\alpha}_{+}x(t) = \int_0^t \frac{(t-s)^{\alpha-1}}{(n-1)!}x(s)ds \tag{11.3}$$

and

$$I^{\alpha}_{-}x(t) = \int_t^T \frac{(t-s)^{\alpha-1}}{(n-1)!}x(s)ds. \tag{11.4}$$

It can be directly verified that the Pettis fractional integration of the power functions $t^{\beta-1}$ and $(T-t)^{\beta-1}$ yield power functions of the same form.

Property 11.12. *If $\alpha \geq 0$ and $\beta \in \mathbb{R}$, then*

$$I^{\alpha}_{+}(t^{\beta-1}) = \frac{\Gamma(\beta)}{\Gamma(\beta+\alpha)}t^{\beta+\alpha-1}, \tag{11.5}$$

and

$$I^{\alpha}_{-}((T-t)^{\beta-1}) = \frac{\Gamma(\beta)}{\Gamma(\beta+\alpha)}(T-t)^{\beta+\alpha-1}, \tag{11.6}$$

In the following lemma, we prove the boundedness of the Pettis fractional integration operators I_+^α and I_-^α.

Lemma 11.13. *(a) The fractional integration operators I_+^α, I_-^α with $\alpha > 0$ are bounded in $L^p(0,T)$ $(1 \le p \le \infty)$;*

$$\|I_+^\alpha x\|_p \le K\|x\|_p, \qquad \|I_-^\alpha x\|_p \le K\|x\|_p \qquad \left(K = \frac{T^\alpha}{\Gamma(\alpha+1)}\right). \tag{11.7}$$

(b) If $0 < \alpha < 1$ and $1 < p < \frac{1}{\alpha}$, then the operators I_+^α and I_-^α are bounded from $L^p(0,T)$ into $L^q(0,T)$, where $q = \frac{p}{1-\alpha p}$.

Some properties of the fractional integration operators I_+^α and I_-^α are given by the following result (see [337]).

Lemma 11.14. *Let $\alpha, \beta > 0$ and $p > \max\{1, 1/\alpha, 1/\beta\}$. Assume that the strongly measurable $x : \Omega \to E$ having $\varphi x \in L^p(\Omega)$ for every $\varphi \in E^*$. then,*

(a) $(I_+^\alpha I_+^\beta x)(t) = (I_+^{\alpha+\beta} x)(t)$.

(b) $\lim\limits_{\alpha \to n} I^\alpha x(t) = I^n x(t), n = 1, 2, \cdots$ weakly uniformly,

(c) $\lim\limits_{\alpha \to 0} I^\alpha x(t) = x(t)$, weakly,

(d) If $\varphi(x(t))$ is bounded for each $\varphi \in E^*$ then $\lim\limits_{t \to 0} I_+^\alpha x(t) = 0$.

Proof. From the properties of the fractional calculus, we have

(a) By Lemma 11.8, $I_+^\alpha x$ and $I_+^{\alpha+\beta} x$ exist and they are weakly continuous as functions from Ω to E. By Lemma 11.9, $I_+^\alpha I_+^\beta x$ exists also and again it is weakly continuous on Ω. Therefore, for any $\varphi \in E^*$ wa have

$$\begin{aligned}\varphi(I_+^\alpha I_+^\beta x(t)) &= I_+^\alpha \varphi(I_+^\beta x(t)) = I_+^\alpha I_+^\beta \varphi(x(t))\\ &= I_+^{\alpha+\beta} \varphi(x(t)) = \varphi(I_+^{\alpha+\beta} x(t)),\end{aligned}$$

which implies $\varphi(I_+^\alpha I_+^\beta x(t) - I_+^{\alpha+\beta} x(t)) = 0$, hence we obtain $I_+^\alpha I_+^\beta x(t) = I_+^{\alpha+\beta} x(t)$.

(b) For any $\varphi \in E^*$ we have

$$|\varphi(I_+^\alpha x(t)) - \varphi(I_+^n x(t))| = |I_+^\alpha \varphi(x(t)) - I_+^n \varphi(x(t))|,$$

since $\varphi x(\cdot) \in L^p(\Omega)$, we deduce from the properties of the fractional order in the space $L^p(\Omega)$ that $I_+^\alpha \varphi x(t)) \to \varphi(I_+^n \varphi(x(t))$ uniformly on Ω, hence $\varphi(I_+^\alpha x(t)) \to \varphi(I_+^n x(t))$ uniformly on Ω, that is $I^\alpha x(t) \to I_+^n x(t)$ weakly uniformly on Ω.

(c) we have

$$\lim_{\alpha \to 0} \varphi(I_+^\alpha x(t)) = \lim_{\alpha \to 0} I_+^\alpha(\varphi x(t)) = \lim_{\alpha \to 0} \varphi(x(t)) * \phi_\alpha(t) = \varphi(x(t)) * \delta(t) = \varphi(x(t)).$$

(d) The proof of this part follows immediately from Proposition 11.38 and the inequality

$$|\varphi(I_+^\alpha x(t))| = |I_+^\alpha \varphi(x(t))| \le K\frac{t^\alpha}{\Gamma(1+\alpha)}, \qquad \text{where } K = \|\varphi x\|_p.$$

□

Remark 11.15. *(a) In the case $E = \mathbb{R}$, it is well known consequence of an inequality of Young that the linear fractional integral operators I_+^α and I_-^α sends $L^q(\Omega)$ continuously into $L^p(\Omega)$ if $p \in [1,\infty]$ satisfies $q > (1/(\alpha+(1/p))($ see [201]).*

(b) In particular, $I_+^\alpha : L^p(\Omega) \to L^p(\Omega)$ is compact for each $p \in [1,\infty]$. Moreover, for , the map $I_+^\alpha : L^p(\Omega) \to C(\Omega)$ is compact (see [336, 337]).

An example. Let $I = [0,1]$ and $x : I \to L^\infty(I)$ be defined by $x(t) = \chi_{[0,t]}$. This function is weakly measurable and for each $\varphi \in L^{\infty *}$, we have $\varphi x \in L^1(I)$ (each φx is a function of bounded variation) (cf. [172, 173, 195]). Thus, according to Lemma 11.8, $I^\alpha x$ exists. To calculate the fractional Pettis integral of x, let $\rho \in L^1(I)$ and let φ be the element in $L^{\infty *}$ corresponding to ρ. For any $t > 0$, we have

$$\begin{aligned}
&\int_0^t \varphi\left(\frac{(t-s)^{\alpha-1}}{\Gamma(\alpha)}x(s)\right)ds \\
&= \int_0^t \frac{(t-s)^{\alpha-1}}{\Gamma(\alpha)}\varphi(x(s))ds = \int_0^t \frac{(t-s)^{\alpha-1}}{\Gamma(\alpha)}\int_0^1 \rho(\theta)\chi_{[0,s]}d\theta ds \\
&= \int_0^t \frac{(t-s)^{\alpha-1}}{\Gamma(\alpha)}\int_0^s \rho(\theta)d\theta ds = \int_0^t\int_0^s \frac{(t-s)^{\alpha-1}}{\Gamma(\alpha)}\rho(\theta)d\theta ds \\
&= \int_0^t\int_0^t \frac{(t-s)^{\alpha-1}}{\Gamma(\alpha)}\rho(\theta)dsd\theta = \int_0^t \frac{(t-\theta)^\alpha}{\Gamma(1+\alpha)}\rho(\theta)d\theta \\
&= \int_0^1 \frac{(t-\theta)^\alpha}{\Gamma(1+\alpha)}\rho(\theta)\chi_{[0,t]}(\theta)d\theta = \varphi\left(\frac{(t-s)^\alpha}{\Gamma(1+\alpha)}\chi_{[0,t]}(s)\right).
\end{aligned}$$

Hence

$$I^\alpha x(t) = \int_0^t \frac{(t-s)^{\alpha-1}}{\Gamma(\alpha)}x(s)ds = \frac{(t-\cdot)^\alpha}{\Gamma(1+\alpha)}\chi_{[0,t]}(\cdot) \in L^\infty.$$

Remark 11.16. *As $\alpha \to 1$, we get the same result du to R. Geitz (see [195]).*

Now, we will prove a result which is an immediate consequence of the Lebesgue Dominated Convergence theorem for the Pettis-integral [[194], Corollary 4].

Theorem 11.17. *Let $x : \Omega \to E$. Suppose there is a sequence (x_n) of weakly continuous functions from Ω into E such that $\lim_{n\to\infty} \varphi(x_n) = \varphi(x)$ a.e. for $\varphi \in E^*$. If there is a scalar function $\psi \in L^1(\Omega)$ with $\|x_n(\cdot)\| < \psi(\cdot)$ a.e for all n, then x is Pettis integrable and $\varphi(I^\alpha x_n(t)) \to \varphi(I^\alpha x(t))$ for each $\varphi \in E^*$ and $t \in \Omega$.*

Proof. Since (x_n) is weakly continuous, $I^\alpha x_n(t)$ exists for almost every $t \in \Omega$. Moreover, for $t \in \Omega$ and a.e. $s \in (0,t)$ we have

$$\|\frac{(t-s)^{\alpha-1}}{\Gamma(\alpha)}x_n(s)\| \le \frac{(t-s)^{\alpha-1}}{\Gamma(\alpha)}\psi(s) \quad \text{for all } n$$

and

$$\begin{aligned}\lim_{n\to\infty}\varphi\left(\frac{(t-s)^{\alpha-1}}{\Gamma(\alpha)}x_n(s)\right) &= \lim_{n\to\infty}\frac{(t-s)^{\alpha-1}}{\Gamma(\alpha)}\varphi(x_n(s))\\ &= \frac{(t-s)^{\alpha-1}}{\Gamma(\alpha)}\varphi(x(s))\\ &= \varphi\left(\frac{(t-s)^{\alpha-1}}{\Gamma(\alpha)}x(s)\right) \text{ a.e. for all } \varphi\in E^*.\end{aligned}$$

By applying the Lebesgue dominated convergence theorem for the Pettis integral we complete the proof. □

11.2.2. Fractional Weak Derivative

In this section we give the definitions of fractional weak derivatives on a finite interval on the real line and present some of their properties. More detailed information may be found in Salem *et al.* [335].

Definition 11.18. *The left-sided fractional weak derivative and the right-sided fractional weak derivative (Shortly LS-Fractional weak derivative and RS-Fractional weak derivative)* $D^\alpha_+ x$ *and* $D^\alpha_- x$ *of order* $\alpha \geq 0$ *are defined respectively by*

$$D^\alpha_+ x(t) = \left(\frac{d}{dt}\right)^n (I_+^{n-\alpha}x)(t) = \frac{1}{\Gamma(n-\alpha)}\left(\frac{d}{dt}\right)^n \int_0^t (t-s)^{n-\alpha-1}x(s)ds \tag{11.8}$$

and

$$D^\alpha_- x(t) = \left(-\frac{d}{dt}\right)^n (I_-^{n-\alpha}x)(t) = \frac{1}{\Gamma(n-\alpha)}\left(-\frac{d}{dt}\right)^n \int_t^T (s-t)^{n-\alpha-1}x(s)ds \tag{11.9}$$

respectively, where $n = [\alpha]+1$, *and* $0 < t < T$

Lemma 11.19. *Let* $x : [0,T] \to E$ *be a weakly continuous function on* $[0,T]$ *such that the real valued function* $I_+^{n-\alpha}\varphi x$ *is n-times differentiable for each* $\varphi \in E^*$. *Then the fractional weak derivatives 11.8 11.9 of x of order* $\alpha \in (n-1,n]$, *exists.*

Proof. According to Lemma 11.9, the LS-FPI of x of order $n-\alpha$ exists as a weakly continuous function from $[0,T]$ to E. Moreover,

$$\varphi(I_+^{n-\alpha}x(t)) = I_+^{n-\alpha}\varphi(x(t)), \quad \text{for all } \varphi \in E^*.$$

Then

$$\left(\frac{d}{dt}\right)^n (I_+^{n-\alpha}\varphi x)(t) = \left(\frac{d}{dt}\right)^n \varphi(I_+^{n-\alpha}x)(t).$$

Then, x has a LS-fractional derivative of order $\alpha \in (n-1,n]$. The Proof for the RS-fractional derivative is similar.

In particular, when $\alpha = n \in \mathbb{N}$, then

$$(D^0_+x)(t) = D^0_- x(t) = x(t); \qquad D^n_+ x(t) = x^{(n)}(t),$$

$$\text{and } D^n_- x(t) = (-1)^n x^{(n)}(t), \tag{11.10}$$

where $x^{(n)}(t)$ is the usual derivative of $x(t)$ of order $n \in \mathbb{N}$.
If $0 < \alpha < 1$, then

$$D^\alpha_+ x(t) = \frac{1}{\Gamma(1-\alpha)} \left(\frac{d}{dt}\right) \int_0^t (t-s)^{-\alpha} x(s) ds, \tag{11.11}$$

$$D^\alpha_- x(t) = \frac{1}{\Gamma(1-\alpha)} \left(\frac{d}{dt}\right) \int_t^T (s-t)^{-\alpha} x(s) ds. \tag{11.12}$$

Property 11.20. *If $\alpha \geq 0$ and $\beta > 0$, then*

$$D^\alpha_+ (t^{\beta-1})(s) = \frac{\Gamma(\beta)}{\Gamma(\beta-\alpha)} s^{\beta-\alpha-1}, \tag{11.13}$$

and

$$D^\alpha_- ((T-t)^{\beta-1})(s) = \frac{\Gamma(\beta)}{\Gamma(\beta-\alpha)} (T-s)^{\beta-\alpha-1}. \tag{11.14}$$

Remark 11.21. *In particular, if $\beta = 1$ and $\alpha \geq 0$, then the weak fractional derivatives of a constant are, in general not equal to zero:*

$$(D^\alpha_+ 1)(t) = \frac{t^{-\alpha}}{\Gamma(1-\alpha)}, \qquad (D^\alpha_- 1)(t) = \frac{(T-t)^{-\alpha}}{\Gamma(1-\alpha)} \qquad (0 < \alpha < 1). \tag{11.15}$$

On the other hand, for $j = 1, 2, \cdots, [\alpha]+1$,

$$D^\alpha_+ (t^{\alpha-j})(s) = 0, \qquad D^\alpha_- ((t-T)^{\alpha-j})(s) = 0. \tag{11.16}$$

From (11.16) we derive the following result.

Corollary 5. *Let $\alpha > 0$ and $n = [\alpha]+1$.*

(a) The equality $(D^\alpha_+ y)(x) = 0$ is valid, if and only if,

$$y(x) = \sum_{j=1}^n c_j x^{\alpha-j},$$

where $c_j \in \mathbb{R} (j = 1, \cdots, n)$ are arbitrary constants.
In particular, when $0 < \alpha \leq 1$, the relation $(D^\alpha_+ y)(x) = 0$ holds if, and only if, $y(x) = cx^{\alpha-1}$ with any $c \in \mathbb{R}$.

(b) The equality $(D^\alpha_- y)(x) = 0$ is valid, if and only if,

$$y(x) = \sum_{j=1}^n d_j (T-x)^{\alpha-j},$$

where $d_j \in \mathbb{R} (j = 1, \cdots, n)$ are arbitrary constants.
In particular, when $0 < \alpha \leq 1$, the relation $(D^\alpha_- y)(x) = 0$ holds if, and only if, $y(x) = d(T-x)^{\alpha-1}$ with any $d \in \mathbb{R}$.

The next result characterizes the conditions for the existence of fractional derivatives D_+^α and D_-^α in the space $AC^n[0,T]$.

Lemma 11.22. *Let $\alpha \geq 0$ and $n = [\alpha]+1$. If $y(x) \in AC^n[0,T]$, then the fractional derivatives D_+^α and D_-^α exists almost everywhere on $[0,T]$ and can be represented in the forms*

$$(D_+^\alpha y)(x) = \sum_{k=0}^{n-1} \frac{y^{(k)}(0)}{\Gamma(1+k-\alpha)} x^{k-\alpha} + \frac{1}{\Gamma(n-\alpha)} \int_0^x \frac{y^{(n)}(t)dt}{(x-t)^{\alpha-n+1}} \tag{11.17}$$

and

$$(D_-^\alpha y)(x) = \sum_{k=0}^{n-1} \frac{(-1)^k y^{(k)}(T)}{\Gamma(1+k-\alpha)} (T-x)^{k-\alpha} + \frac{(-1)^n}{\Gamma(n-\alpha)} \int_x^T \frac{y^{(n)}(t)dt}{(t-x)^{\alpha-n+1}} \tag{11.18}$$

respectively.

Corollary 6. *If $0 < \alpha < 1$ and $y(x) \in AC[0,T]$, then*

$$(D_+^\alpha y)(x) = \frac{1}{\Gamma(1-\alpha)}[\frac{y(0)}{x^\alpha} + \int_0^x \frac{y'(t)dt}{(x-t)^\alpha}] \tag{11.19}$$

and

$$(D_-^\alpha y)(x) = \frac{1}{\Gamma(1-\alpha)}[\frac{y(T)}{(T-x)^\alpha} - \int_x^T \frac{y'(t)dt}{(t-x)^\alpha}] \tag{11.20}$$

□

The following assertion shows that the fractional differentiation is an operation inverse to the fractional integration from the left.

Lemma 11.23. *If $\alpha > 0$ and $x(t) \in L^p(0,T), (1 \leq p \leq \infty)$, then the following equalities*

$$(D_+^\alpha I_+^\alpha x)(t) = x(t), \text{and } (D_-^\alpha I_-^\alpha x)(t) = x(t) \tag{11.21}$$

hold almost everywhere on $[0,T]$.

From Lemmas 11.22 and 11.23, we drive the following composition relations between fractional differentiation and fractional integration operators.

Property 11.24. *If $\alpha > \beta > 0$, then, for $x(t) \in L^p(0,T), (1 \leq p \leq \infty)$, the relations*

$$(D_+^\beta I_+^\alpha x)(t) = (I_+^{\alpha-\beta} x)(t), \text{and } (D_-^\beta I_-^\alpha x)(t) = (I_-^{\alpha-\beta} x)(t) \tag{11.22}$$

hold almost everywhere on $[0,T]$.

11.2.3. Caputo Fractional Weak Derivatives

In this section we present the definitions and some properties of the Caputo fractional weak derivatives. Let $[0,T]$ be a finite interval of the real line $\mathbb{R}$, and let $(D^{\alpha}_{+}y)(x)$ and $(D^{\alpha}_{-}y)(x)$ be the Riemann-Liouville weak fractional derivatives of order $\alpha > 0$ defined by (11.8) and (11.9) respectively.

Definition 11.25. *Let $x : \Omega \to E$. We define the left-sided and the the right-sided Caputo fractional derivative of x of order $\alpha > 0$ on Ω by*

$$({}^{c}D^{\alpha}_{+}x)(t) = \left(D^{\alpha}_{+}[x(s) - \sum_{k=0}^{n-1} \frac{x^{(k)}(0)}{k!} s^{k}] \right)(t) \tag{11.23}$$

$$({}^{c}D^{\alpha}_{-}x)(t) = \left(D^{\alpha}_{-}[x(s) - \sum_{k=0}^{n-1} \frac{x^{(k)}(T)}{k!} (T-s)^{k}] \right)(t) \tag{11.24}$$

respectively,where $n = [\alpha]+1$ for $\alpha \notin \mathbb{N}$; $n = \alpha$ for $\alpha \in \mathbb{N}$.

In particular, when $0 < \alpha < 1$, the relations (11.23) and (11.24) take the following form

$$({}^{c}D^{\alpha}_{+}x)(t) = (D^{\alpha}_{+}[x(s) - x(0)])(t).$$

$$({}^{c}D^{\alpha}_{-}x)(t) = (D^{\alpha}_{-}[x(s) - x(T)])(t).$$

If $\alpha \notin \mathbb{N}$ and $x(t)$ is a function for which the Caputo fractional derivatives $({}^{c}D^{\alpha}_{+}x()t)$ and $({}^{c}D^{\alpha}_{-}x()t)$ of order α exists together with the Riemann-Liouville fractional derivatives $(D^{\alpha}_{+}x()t)$ and $(D^{\alpha}_{-}x)(t)$, then they are connected with each other by the following relations

$$({}^{c}D^{\alpha}_{+}x)(t) = (D^{\alpha}_{+})x(t) - \sum_{k=0}^{n-1} \frac{x^{(k)}(0)}{\Gamma(k-\alpha+1)} t^{k-\alpha} \tag{11.25}$$

$$({}^{c}D^{\alpha}_{-}x)(t) = (D^{\alpha}_{+})x(t) - \sum_{k=0}^{n-1} \frac{x^{(k)}(T)}{\Gamma(k-\alpha+1)} (T-t)^{k-\alpha} \tag{11.26}$$

In particular, when $0 < \alpha < 1$ we have

$$({}^{c}D^{\alpha}_{+}x)(t) = (D^{\alpha}_{+})x(t) - \frac{x(0)}{\Gamma(1-\alpha)} t^{-\alpha}$$

$$({}^{c}D^{\alpha}_{-}x)(t) = (D^{\alpha}_{+})x(t) - \frac{x(T)}{\Gamma(1-\alpha)} (T-t)^{-\alpha}$$

Remark 11.26. *The Caputo fractional derivatives (11.23) and (11.24) coincide with the Riemann-Liouville weak fractional derivative (11.8) and (11.9) in the following cases*

$$({}^{c}D^{\alpha}_{+}x)(t) = (D^{\alpha}_{+}x)(t)$$

if $x(0) = x'(0) = \cdots = x^{(n-1)}(0) = 0 \quad (n = [\alpha]+1),$

$$({}^{c}D^{\alpha}_{-}x)(t) = (D^{\alpha}_{-}x(t)$$

if $x(T) = x'(T) = \cdots = x^{(n-1)}(T) = 0 \quad (n = [\alpha]+1)$. *In particular, when* $0 < \alpha < 1$, *we have*

$$({}^cD^\alpha_+ x)(t) = (D^\alpha_+ x)(t) \qquad \text{when} \quad x(0) = 0,$$

$$({}^cD^\alpha_- x)(t) = (D^\alpha_- x)(t) \qquad \text{when} \quad x(T) = 0.$$

If $\alpha = n$ *and the usual derivative* $x^{(n)}$ *of order n exists, then* $({}^cD^\alpha_+ x)(t)$ *and* $({}^cD^\alpha_- x)(t)$ *coincide with* $x^{(n)}$.

$$({}^cD^\alpha_+ x)(t) = x^{(n)}(t) \quad \text{and} \quad ({}^cD^\alpha_- x(t)) = (-1)^n x^{(n)}(t).$$

The Caputo fractional derivatives are defined for functions for which the Riemann-Liouville fractional derivatives of the right-hand sides exist. In particular, they are defined for $x(t)$ belonging to the space $AC^n[0,T]$ of absolutely continuous functions. Thus the following statement holds.

Theorem 11.27. *Let* $\alpha \geq 0$ *and* $n = [\alpha]+1$. *If* $x \in AC^n[0,T]$, *then the Caputo fractional derivatives* $({}^cD^\alpha_+ x)(t)$ *and* $({}^cD^\alpha_- x)(t)$ *exists almost every where on* $[0,T]$.

(a) if $\alpha \notin \mathbb{N}$ *then* $({}^cD^\alpha_+ x)(t)$ *and* $({}^cD^\alpha_- x)(t)$ *are represented by*

$$({}^cD^\alpha_+ x)(t) = \frac{1}{\Gamma(n-\alpha)} \int_0^t \frac{x^{(n)}(s)ds}{(t-s)^{\alpha-n+1}} = (I^{n-\alpha}_+ D^n_+ x)(t)$$

$$({}^cD^\alpha_- x)(t) = \frac{(-1)^n}{\Gamma(n-\alpha)} \int_t^T \frac{x^{(n)}(s)ds}{(s-t)^{\alpha-n+1}} = (-1)^n (I^{n-\alpha}_- D^n_- x)(t)$$

respectively, where $D = d/dt$.

The Caputo derivatives $({}^cD^\alpha_+ x)(t)$ and $({}^cD^\alpha_- x)(t)$ have properties similar to those of the fractional weak derivative $(D^\alpha_+ x)(t)$ and $(D^\alpha_- x)(t)$ given in (11.13) and (11.14) but different from those in (11.15).

Property 11.28. *Let* $\alpha \geq 0$ *and* $n = [\alpha]+1$. *Also let* $\beta > 0$. *Then the following relations hold*

$${}^cD^\alpha_+ (t^{\beta-1})(s) = \frac{\Gamma(\beta)}{\Gamma(\beta-\alpha)} s^{\beta-\alpha-1}, \tag{11.27}$$

and

$${}^cD^\alpha_- ((T-t)^{\beta-1})(s) = \frac{\Gamma(\beta)}{\Gamma(\beta-\alpha)} (T-s)^{\beta-\alpha-1}. \tag{11.28}$$

In particular, the Caputo fractional weak derivatives of a constant still zero, i.e.

$$({}^cD^\alpha_+ 1)(t) = 0, \qquad ({}^cD^\alpha_- 1)(t) = 0. \tag{11.29}$$

From (11.15) and (11.29) we thus recognize the following statements about functions, which for $t > 0$ admit the same fractional derivative of order α, with $n-1 < \alpha < n$, $n \in \mathbb{N}$

$$D^\alpha_+ x(t) = D^\alpha_+ y(t) \Leftrightarrow x(t) = y(t) + \sum_{j=1}^n c_j t^{\alpha-j},$$

and

$$^{c}D^{\alpha}_{+}x(t) = D^{\alpha}_{+}y(t) \Leftrightarrow x(t) = y(t) + \sum_{j=1}^{n} c_j t^{n-j},$$

where c_j are arbitrary constants. For proving all main results in this thesis, we present the following auxiliary lemmas.

Lemma 11.29. *[244] Let $\alpha > 0$ and $x \in L^{\infty}(0,T)$ or $C[0,T]$, then*

$$^{c}D^{\alpha}_{+}I^{\alpha}_{+}x(t) = x(t).$$

Lemma 11.30. *[244] Let $\alpha > 0$ and $n = [\alpha]+1$. If $x \in AC^{n}[0,T]$ or $C^{n}[0,T]$ then*

$$(I^{\alpha c}_{+}D^{\alpha}_{+}x)(t) = x(t) - \sum_{k=0}^{n-1} \frac{x^{(k)}(0)}{k!}t^{k}.$$

Lemma 11.31. *[244] Let $\alpha > 0$ then*

$$(I^{\alpha c}_{+}D^{\alpha}_{+}x)(t) = x(t) + c_0 + c_1 t + c_2 t^2 + \cdots + c_{n-1}t^{n-1},$$

for arbitrary $c_j \in \mathbb{R}$, $i = 0,1,2,\cdots,n-1, \quad n = [\alpha]+1$.

For further readings and details, we refer to the books by Kilbas *et al* [244], Podlubny [314] and Samko *et al* [337].

11.3. Weak Solutions For Nonlinear Differential Equations on Reflexive Banach Spaces

11.3.1. Introduction

In this section we investigate the existence of solutions for the boundary value problem with fractional order differential equations and nonlinear integral conditions of the form

$$^{c}D^{\alpha}x(t) = f(t,x(t)), \quad \text{for each } t \in I = [0,T], \tag{11.30}$$

$$x(0) - x'(0) = \int_0^T g(s,x(s))ds, \tag{11.31}$$

$$x(T) + x'(T) = \int_0^T h(s,x(s))ds, \tag{11.32}$$

where $^{c}D^{\alpha}, 1 < \alpha \leq 2$ is the Caputo fractional derivative, f, g and $h : I \times E \to E$ are given functions satisfying some assumptions that will be specified later, and E is a reflexive Banach space with norm $\|\cdot\|$.

Definition 11.32. *A function $h : E \to E$ is said to be weakly sequentially continuous if h takes each weakly convergent sequence in E to weakly convergent sequence in E (ie for any $(x_n)_n$ in E with $x_n(t) \to x(t)$ in (E,w) for each $t \in I$ then $h(x_n(t)) \to h((x(t)))$ in (E,w) for each $t \in I$).*

Note that if E is weakly complete and $\varphi(x(\cdot))$ is differentiable for every $\varphi \in E^*$, then x is weakly differentiable. If x is weakly continuous on $[0,T]$, then x is strongly measurable, hence weakly measurable. Also it can be easily proved, that weak differentiability implies weak continuity.

Definition 11.33. *[312] The function $x : J \to E$ is said to be Pettis integrable on J if and only if there is an element $x_I \in E$ corresponding to each $I \subset J$ such that $\varphi(x) = \int_I \varphi(x(s))ds$ for all $\varphi \in E^*$ where the integral on the right is supposed to exist in the sense of Lebesgue. By definition, $x_I = \int_I x(s)ds$. We denote the space of Pettis integrable by $P[J,E]$.*

The immediate consequence of the definition of the Pettis integral is: if $x : J \to E$ is Pettis integrable on J, then $\varphi(x(\cdot))$ is Lebesgue integrable on J for every $\varphi \in E^*$. We point out, that a bounded weakly measurable function $x : J \to E$ need not to be Pettis integrable even E is reflexive.

Lemma 11.34. *In reflexive Banach space, the weakly measurable function $x : J \to E$ is Pettis integrable if, and only if $\varphi(x(\cdot))$ is Lebesgue integrable on J for every $\varphi \in E^*$.*

Proposition 11.35. *[163, 312] If $x(\cdot)$ is Pettis integrable and $h(\cdot)$ is a measurable and essentially bounded real-valued function, then $x(\cdot)h(\cdot)$ is Pettis integrable.*

Remark 11.36. *The Pettis integral is also a generalisation of the Bochner integral ([163, 312]). This notion is strictly relative to weak topologies in Banach spaces.*

Definition 11.37. *[150] Let E be a Banach space , Ω_E the bounded subsets of E and B_1 the unit ball of E. The De Blasi measure of weak noncompactness is the map $\beta : \Omega_E \to [0,\infty]$ defined by*
$\beta(X) = \inf\{\varepsilon > 0 :$ *there exists a weakly compact subset* Ω *of* $E : X \subset \varepsilon B_1 + \Omega\}$.

De Blasi measure of noncompactness satisfies some properties

(a) $A \subset B \Rightarrow \beta(A) \leq \beta(B)$,

(b) $\beta(A) = 0 \Leftrightarrow A$ is relatively compact,

(c) $\beta(A \cup B) = \max\{\beta(A), \beta(B)\}$,

(d) $\beta(\overline{A}^{\omega}) = \beta(A)$, ($\overline{A}^{\omega}$ denotes the weak closure of A),

(e) $\beta(A+B) \leq \beta(A) + \beta(B)$,

(f) $\beta(\lambda A) = |\lambda|\beta(A)$,

(g) $\beta(conv(A)) = \beta(A)$,

(h) $\beta(\cup_{|\lambda| \leq h} \lambda A = h\beta(A)$.

The following result follows directly from the Hahn-Banach theorem.

Proposition 11.38. *Let E be a normed space with $x_0 \neq 0$ then there exists $\varphi \in E^*$ with $\|\varphi\| = 1$ and $\varphi(x_0) = \|x_0\|$.*

For completeness we recall the definitions of Pettis-integral and Caputo derivative of fractional order.

Definition 11.39. *[335] Let $h : I \to E$ a function. The fractional Pettis integral of the function h of order $\alpha \in \mathbb{R}_+$ is defined by*

$$I^{\alpha}h(t) = \int_0^t \frac{(t-s)^{\alpha-1}}{\Gamma(\alpha)} h(s)ds$$

where the sign "$\int$" denotes the Pettis integral.

Definition 11.40. *[244] For a function $h : I \to E$, the Caputo fractional-order derivative of h, is defined by*

$$^cD^{\alpha}h(t) = \frac{1}{\Gamma(n-\alpha)} \int_0^t \frac{h^{(n)}(s)ds}{(t-s)^{1-n+\alpha}}$$

here $n = [\alpha]+1$ and $[\alpha]$ denotes the integer part of α.

Theorem 11.41. *[297] Let Q be a closed convex and equicontinuous subset of a metrizable locally convex vector space $C(I,E)$ such that $0 \in Q$. Assume that $T : Q \to Q$ is wk-sequentially continuous. If the implication*

$$\overline{V} = \overline{conv}(\{0\} \cup T(V)) \Rightarrow V \text{ is relatively weakly compact,} \tag{11.33}$$

holds for every subset $V \subset Q$, then T has a fixed point.

11.3.2. Existence of Solutions

Let us start by defining what we mean by a solution of the problem (11.30)-(11.32).

Definition 11.42. *A function $x \in AC^1(I,E_w)$ is said to be solution of (11.30)-(11.32) if x satisfies (11.30)-(11.32).*

Let $\sigma, \sigma_1, \sigma_2 : I \to E$ be continuous functions and consider the linear boundary value problem

$$^cD^{\alpha}x(t) = \sigma(t), \quad t \in I, \tag{11.34}$$

$$x(0) - x'(0) = \int_0^T \sigma_1(s)ds \tag{11.35}$$

$$x(T) + x'(T) = \int_0^T \sigma_2(s)ds. \tag{11.36}$$

Lemma 11.43. *[93] Let $1 < \alpha \leq 2$ and let $\sigma, \sigma_1, \sigma_2 : I \to E$ be continuous. A function x is a solution of the fractional integral equation*

$$x(t) = P(t) + \int_0^T G(t,s)\sigma(s)ds \tag{11.37}$$

with

$$P(t) = \frac{(T+1-t)}{T+2} \int_0^T \sigma_1(s)ds + \frac{(t+1)}{T+2} \int_0^T \sigma_2(s)ds \tag{11.38}$$

and

$$G(t,s)=\begin{cases} \frac{(t-s)^{\alpha-1}}{\Gamma(\alpha)}-\frac{(1+t)(T-s)^{\alpha-1}}{(T+2)\Gamma(\alpha)}-\frac{(1+t)(T-s)^{\alpha-2}}{(T+2)\Gamma(\alpha-1)}, & 0\le s\le t,\\ -\frac{(1+t)(T-s)^{\alpha-1}}{(T+2)\Gamma(\alpha)}-\frac{(1+t)(T-s)^{\alpha-2}}{(T+2)\Gamma(\alpha-1)}, & t\le s<T, \end{cases} \tag{11.39}$$

if and only if x is a solution of the fractional boundary value problem (11.34)-(11.36).

Let

$$\tilde{G}=\sup\left\{\int_0^T |G(t,s)|ds,\ t\in I\right\}.$$

To establish our main result concerning existence of solutions of (11.30)-(11.32), we impose suitable conditions on the functions involved in that problem namely we assume that

(11.50.1) For each $t\in I$, $f(t,.)$, $g(t,.)$ and $h(t,.)$ are weakly sequentially continuous,

(11.50.2) For each $x\in C(I,E), f(\cdot,x(\cdot)), g(\cdot,x(\cdot))$ and $h(\cdot,x(\cdot))$ are Pettis integrable on I,

(11.50.3) There exist $p_g,p_h\in L^1(I,\mathbb{R}^+)$, $p_f\in L^\infty(I,\mathbb{R}^+)$, such that:

$$\|f(t,x)\|\le p_f(t)\|x\|, \text{ for a.e. } t\in I \text{ and each } x\in E,$$

$$\|g(t,x)\|\le p_g(t)\|x\|, \text{ for a.e. } t\in I \text{ and each } x\in E,$$

$$\|h(t,x)\|\le p_h(t)\|x\|, \text{ for a.e. } t\in I \text{ and each } x\in E.$$

Theorem 11.44. *Let E be a reflexive Banach space, and assume the assumptions* $(11.50.1)-(11.50.3)$ *be satisfied. If*

$$\frac{T+1}{T+2}\int_0^T [p_g(s)+p_h(s)]ds+\tilde{G}\|p_f\|_{L^\infty}<1, \tag{11.40}$$

then the boundary value problem (11.30)-(11.32) has at least one solution.

Proof. We shall reduce the existence of solutions of the boundary value problem (11.30)-(11.32) to fixed point problem. To this end we consider the operator $T: C(I,E)\to C(I,E)$ defined by

$$(Tx)(t)=P_x(t)+\int_0^T G(t,s)f(s,x(s))ds \tag{11.41}$$

with

$$P_x(t)=\frac{(T+1-t)}{T+2}\int_0^T g(s,x(s))ds+\frac{(t+1)}{T+2}\int_0^T h(s,x(s))ds$$

and $G(\cdot,\cdot)$ is the Green's function defined by (11.39). Clearly the fixed points of the operator T are solutions of the problem (11.30)-(11.32).
First notice that, for $x\in C(I,E), f(\cdot,x(\cdot))\in P(I,E)$ (assumption (11.50.2)). Since, $s\mapsto G(t,s)\in L^\infty(I)$ then $G(t,\cdot)f(\cdot,x(\cdot)), \forall t\in I$ is Pettis integrable and thus, the operator T

makes sense.
Let $R \in \mathbb{R}_+^*$, and consider the set

$$\begin{aligned} Q &= \{x \in C(I,E) : \|x\|_\infty \leq R \\ &\quad \text{and } \|x(t_1) - x(t_2)\| \leq \frac{|t_1 - t_2|R}{T+2} \int_0^T (p_h(s) + p_g(s))ds \\ &\quad + R\|p_f\|_{L^\infty} \int_0^T \|G(t_2,s) - G(t_1,s)\|ds \text{ for } t_1, t_2 \in I\}. \end{aligned}$$

Clearly, the subset Q is closed, convex and equicontinuous . We shall show that T satisfies the assumptions of Theorem 11.41.

Step 1: T maps Q into itself.

Takes $x \in Q, t \in [0,T]$ and assume that $Tx(t) \neq 0$. Then there exists $\varphi \in E^*$ such that $\|Tx(t)\| = \varphi(Tx(t))$. Thus

$$\begin{aligned} \|Tx(t)\| &= \varphi(Tx(t)) \\ &= \varphi(P_x(t) + \int_0^T G(t,s)f(s,x(s))ds) \\ &\leq \varphi(P_x(t)) + \varphi(\int_0^T G(t,s)f(s,x(s))ds) \\ &\leq \|P_x(t)\| + \int_0^T \|G(t,s)\|\varphi(f(s,x(s)))ds \\ &\leq \frac{T+1}{T+2}R\int_0^T [p_g(s) + p_h(s)]ds + \tilde{G}R\|p_f\|_{L^\infty} \\ &\leq R. \end{aligned}$$

Let $t_1, t_2 \in I, t_1 < t_2, x \in Q$, so $Tx(t_2) - Tx(t_1) \neq 0$ then there exist $\varphi \in E^*$ such that $\|Tx(t_2) - Tx(t_1)\| = \varphi(Tx(t_2) - Tx(t_1))$

$$\begin{aligned} &\|Tx(t_2) - Tx(t_1)\| \\ &= \varphi(P_x(t_2) - P_x(t_1) + \int_0^T (G(t_2,s) - G(t_1,s))f(s,x(s))ds) \\ &= \varphi(P_x(t_2) - P_x(t_1)) + \varphi(\int_0^T (G(t_2,s) - G(t_1,s))f(s,x(s))ds) \\ &\leq \|P_x(t_2) - P_x(t_1)\| + \int_0^T \|G(t_2,s) - G(t_1,s)\|\|f(s,x(s)\|ds \\ &\leq \frac{(t_2 - t_1)R}{T+2}\int_0^T (p_h(s) + p_g(s))ds \\ &+ R\|p_f\|_{L^\infty}\int_0^T \|G(t_2,s) - G(t_1,s)\|ds. \end{aligned}$$

Thus $T(Q) \subset Q$.

Step 2: T is wk-sequentially continuous.

Let (x_n) be a sequence in Q and let $(x_n(t)) \to x(t)$ in (E,w) for each $t \in I$. Fix $t \in I$. Since f, g and h satisfies assumptions (11.50.1), we have $f(t,x_n(t))$, $g(t,x_n(t))$ and $h(t,x_n(t))$ converging weakly uniformly to $f(t,x(t))$, $g(t,x(t))$ and $h(t,x(t))$ respectively. Hence the Lebesgue Dominated Convergence theorem for Pettis integral implies $Tx_n(t)$ converging weakly uniformly to $Tx(t)$ in E_w. We do it for each $t \in I$ so $Tx_n \to Tx$. Then $T : Q \to Q$ is weakly sequentially continuous.

Now let V be a subset of Q such that $V = \overline{conv}(T(V) \cup \{0\})$. Obviously $V(t) \subset \overline{conv}(T(V) \cup \{0\}), \forall t \in I$. $TV(t) \subset TQ(t), t \in I$ is bounded in E. So $TV(t)$ is weakly relatively compact since a subset of a reflexive Banach space is weakly relatively compact if and only if it is bounded in the norm topology. So

$$\begin{aligned}\beta(V(t)) &\leq \beta(\overline{conv}(T(V)) \cup \{0\}) \\ &\leq \beta(TV(t)) \\ &= 0.\end{aligned}$$

Thus, V is relatively weakly compact.
Applying now Theorem 11.41, we conclude that T has a fixed point which is an solution of the problem (11.30)-(11.32). □

11.3.3. An Example

In this section we give an example to illustrate the usefulness of our main results. Let us consider the following fractional boundary value problem,

$$^cD^r y_n(t) = \frac{2}{19+e^t}|y_n(t)|, \ t \in J := [0,1], \ 1 < r \leq 2, \tag{11.42}$$

$$y_n(0) - y_n'(0) = \int_0^1 \frac{1}{5+e^{5s}}|y_n(s)|ds,, \tag{11.43}$$

$$y_n(1) + y_n'(1) = \int_0^1 \frac{1}{3+e^{3s}}|y_n(s)|ds. \tag{11.44}$$

Let

$$E = l^1 = \{u = (u_1, u_2, \ldots, u_n, \ldots) : \sum_{n=1}^{\infty} |u_n| < \infty\}$$

with the norm

$$\|u\|_E = \sum_{n=1}^{\infty} |u_n|.$$

Set

$$u = (u_1, u_2, \ldots, u_n, \ldots) \text{ and } f = (f_1, f_2, \ldots, f_n, \ldots).$$

$$f_n(t, x_n) = \frac{2}{19+e^t}x_n, \ (t, x_n) \in J \times [0, \infty),$$

$$g_n(t, x_n) = \frac{1}{5+e^{5t}}x_n, \ (t, x_n) \in [0,1] \times [0, \infty),$$

$$h_n(t,x_n) = \frac{1}{3+e^{3t}}x_n,\ (t,x_n) \in [0,1]\times[0,\infty).$$

Clearly, conditions $(11.13.1),(11.13.2)$ hold with

$$p_f(t) = \frac{2}{19+e^t},\ p_g(t) = \frac{1}{5+e^{5t}},\ p_h(t) = \frac{1}{3+e^{3t}}.$$

From (11.39) the function G is given by

$$G(t,s) = \begin{cases} \frac{(t-s)^{r-1}}{\Gamma(r)} - \frac{(1+t)(1-s)^{r-1}}{3\Gamma(r)} - \frac{(1+t)(1-s)^{r-2}}{3\Gamma(r-1)}, & 0 \le s \le t \\ -\frac{(1+t)(1-s)^{r-1}}{3\Gamma(r)} - \frac{(1+t)(1-s)^{r-2}}{3\Gamma(r-1)}, & t \le s < 1. \end{cases} \tag{11.45}$$

From (11.45), we have

$$\begin{aligned} \int_0^1 G(t,s)ds &= \int_0^t G(t,s)ds + \int_t^1 G(t,s)ds \\ &= \frac{t^r}{\Gamma(r+1)} + \frac{(1+t)(1-t)^r}{3\Gamma(r+1)} - \frac{(1+t)}{3\Gamma(r+1)} \\ &\quad + \frac{(1+t)(1-t)^{r-1}}{3\Gamma(r)} - \frac{(1+t)}{3\Gamma(r)} \\ &\quad - \frac{(1+t)(1-t)^r}{3\Gamma(r+1)} - \frac{(1+t)(1-t)^{r-1}}{3\Gamma(r)}. \end{aligned}$$

A simple computation gives

$$\tilde{G} < \frac{3}{\Gamma(r+1)} + \frac{2}{\Gamma(r)}.$$

Condition (11.40) is satisfied with $T=1$. Indeed

$$\begin{aligned} \tfrac{T+1}{T+2}[\|p_g\|_{L^\infty} + \|p_h\|_{L^\infty}] + \tilde{G}\|p_f\|_{L^\infty} &< \tfrac{2}{3}[\tfrac{1}{6}+\tfrac{1}{4}] + \tfrac{3}{10\Gamma(r+1)} + \tfrac{2}{10\Gamma(r)} \\ &= \tfrac{5}{18} + \tfrac{3}{10\Gamma(r+1)} + \tfrac{1}{5\Gamma(r)} < 1. \end{aligned}$$

which is satisfied for each $r \in (1,2]$. Then by Theorem 11.44 the problem (11.42)-(11.44) has a solution on $[0,1]$.

11.3.4. Concluding Remarks

In this section we have presented an existence result of weak solutions for the boundary value problem (6.17)-(6.19) in the case when the Banach space E is reflexive. However, in the nonreflexive case, hypotheses (11.50.1)-(11.50.3) are not sufficient for the application of Theorem 11.41, in particular condition (11.33).

Let us introduce the following conditions

(C1) For each bounded set $Q \subset E$ and each $t \in I$, the sets $f(t,Q)$, $g(t,Q)$, $h(t,Q)$ are weakly relatively compact in E,

(C2) For each bounded set $Q \subset E$ and each $t \in I$,

$$\beta(f(t,Q)) \leq p_f(t)\beta(Q),$$

$$\beta(g(t,Q)) \leq p_g(t)\beta(Q),$$

$$\beta(h(t,Q)) \leq p_h(t)\beta(Q).$$

We then have the following results.

Theorem 11.45. *Let E be a Banach space , and assume the assumptions* $(11.50.1)-(11.50.3),(C1)$ *be satisfied. If (11.40) is satisfied, then the boundary value problem (11.30)-(11.32) has at least one solution.*

Theorem 11.46. *Let E be a Banach space, and assume the assumptions* $(11.50.1)-(11.50.3),(C2)$ *be satisfied. If If (11.40) is satisfied, then the boundary value problem (11.30)-(11.32) has at least one solution.*

11.4. Weak Solutions of Boundary Value Problem on Unbounded Domains in Banach Spaces

11.4.1. Introduction

In this section we investigate the existence of weak solutions for the boundary value problem with fractional order differential equation on the half-line in a Banach space of the form

$$D^{\alpha}x(t) = f(t,x(t),D^{\alpha-1}x(t)), \quad \text{for each } t \in J = [0,+\infty), \tag{11.46}$$

$$x(0) = 0, \; D^{\alpha-1}x(+\infty) = x_{\infty}, \tag{11.47}$$

where D^{α} and $D^{\alpha-1}, 1 < \alpha \leq 2$, are the standard Riemann-Liouville fractional derivatives, $D^{\alpha-1}x(+\infty) = \lim_{t\to+\infty} D^{\alpha-1}x(t)$, $f : J \times E \times E \to E$ is a given function satisfying some assumptions that will be specified later, and E is a real Banach space with norm $\|\cdot\|$. Let E^* the dual of E. and $E_w = (E,\sigma(E,E^*))$ denotes the space E with its weak topology. $C(J,E_w)$ denotes the space of continuous functions $x : J \to E_w$.

11.4.2. Existence of Solutions

Let us start by defining what we mean by a weak solution of the problem (11.46)-(11.47).

Definition 11.47. *A function $x \in C(J,E_{\omega})$ is said to be a weak solution of the problem (11.46)-(11.47) if x satisfies (11.46)-(11.47).*

The following result provides the integral formulation for the boundary value problem (11.46)-(11.47).

Lemma 11.48. *Let $1 < \alpha \leq 2$ and let $h \in C(J,E)$ be a given function. Then the boundary value problem*

$$D^{\alpha}x(t) = h(t),\ t \in J, \tag{11.48}$$

$$x(0) = 0,\ D^{\alpha-1}x(+\infty) = x_{\infty} \tag{11.49}$$

has a unique solution given by

$$x(t) = \frac{t^{\alpha-1}}{\Gamma(\alpha)}\left[x_{\infty} - \int_0^{+\infty} h(s)ds\right] + \frac{1}{\Gamma(\alpha)}\int_0^t (t-s)^{\alpha-1}h(s)ds. \tag{11.50}$$

Proof. Assume that x satisfies the equation (11.48), then

$$x(t) = \frac{1}{\Gamma(\alpha)}\int_0^t (t-s)^{\alpha-1}h(s)ds + c_1 t^{\alpha-1} + c_2 t^{\alpha-2}.$$

From (11.49), a simple calculation yields $c_2 = 0$ and

$$c_1 = \frac{1}{\Gamma(\alpha)}D^{\alpha-1}x(+\infty) - \frac{1}{\Gamma(\alpha)}\int_0^{\infty} h(s)ds.$$

Hence we get equation (11.50). Conversely, it is clear that if x satisfies equation (11.50), then equations (11.48)-(11.49) hold, which completes the proof. □

To establish our main result concerning the existence of weak solutions of (11.46)-(11.47), we list the following hypotheses:

(11.61.1) For any compact interval $I \subset J$ and $t \in I$, the function $f(t,\cdot,\cdot)$ is weakly sequentially continuous.

(11.61.2) For each, $x \in Y$, $f(\cdot,x(\cdot),D^{\alpha-1}x(\cdot))$ is Pettis integrable on J.

(11.61.3) There exists nonnegative functions $a,b,c \in L^1(J,\mathbb{R}^+)$ such that

$$\|f(t,x,y)\| \leq a(t)\|x\| + b(t)\|y\| + c(t),$$

and

$$\int_0^{+\infty} [(1+t^{\alpha-1})a(t) + b(t)]dt < \frac{\Gamma(\alpha)}{2}.$$

(11.61.4) For bounded sets $B_1, B_2 \subset E$ and for each $t \in J$ there exists a nonnegative function $l \in L^1(J,\mathbb{R}^+)$ such that

$$\beta(f(t,B_1,B_2)) \leq l(t)\max(\beta(B_1),\beta(B_2)).$$

Remark 11.49. *We can derive from (11.61.3) that for $x \in Y$ we have*

$$\begin{aligned}
&\int_0^{+\infty} \|f(s,x(s),D^{\alpha-1}x(s))\|ds \\
\leq\ & \int_0^{+\infty} [(1+s^{\alpha-1})a(s)\|x\|_X + b(s)\|D^{\alpha-1}x(s)\| + c(s)]ds \\
\leq\ & \|x\|_Y \int_0^{+\infty} [(1+s^{\alpha-1})a(s) + b(s)]ds + \int_0^{+\infty} c(s)ds \\
<\ & +\infty.
\end{aligned}$$

Let us consider the operator $T : Y \to Y$ defined by

$$\begin{aligned}(Tx)(t) &= \frac{t^{\alpha-1}}{\Gamma(\alpha)}[x_\infty - \int_0^{+\infty} f(s,x(s),D^{\alpha-1}x(s))ds] \\ &+ \frac{1}{\Gamma(\alpha)}\int_0^t (t-s)^{\alpha-1} f(s,x(s),D^{\alpha-1}x(s))ds.\end{aligned}$$

Clearly the fixed points of the operator T are weak solutions of the problem (11.46)-(11.47). First notice that for $x \in Y$, $f(\cdot,x(\cdot),D^{\alpha-1}(\cdot)) \in P(J,E)$. For $t \in J$, define the function $L : J \to \mathbb{R}$ by

$$L(s) = \begin{cases} (t-s)^{\alpha-1}, & \text{if } s \in [0,t], \\ 0 \quad , & \text{if } s \notin [0,t]. \end{cases}$$

It is clear that $L \in L^\infty(J,E)$ so, from Proposition 11.35, the function $L(\cdot)f(\cdot,x(\cdot),D^{\alpha-1}(\cdot))$, is Pettis integrable, and thus the operator T makes sense.
The following two lemmas present some properties of the operator T, that are necessary for the proof of our main result.

Lemma 11.50. *Let condition (11.61.3) be satisfied, and B be a bounded subset of Y. Then*

(a) $\{\frac{TB(t)}{1+t^{\alpha-1}}\}$ *and* $\{D^{\alpha-1}TB(t)\}$ *are equicontinuous on any compact interval of J.*

(b) For given $\varepsilon > 0$, *there exists* $N > 0$ *such that*

$$\left\| \frac{Tx(t_1)}{1+t_1^{\alpha-1}} - \frac{Tx(t_2)}{1+t_2^{\alpha-1}} \right\| < \varepsilon$$

and

$$\|D^{\alpha-1}Tx(t_1) - D^{\alpha-1}Tx(t_2)\| < \varepsilon \textit{ for any } t_1,t_2 \geq N \textit{ and } x \in B.$$

Proof. Take $x \in B$, and $t \in J$ such that $\|x\|_Y \leq R$, $R > 0$, and let $I \subset J$ be a compact interval and $t_1,t_2 \in I$ with $t_1 < t_2$ such that

$$\frac{Tx(t_1)}{1+t_1^{\alpha-1}} - \frac{Tx(t_2)}{1+t_2^{\alpha-1}} \neq 0$$

and $D^{\alpha-1}Tx(t_2) - D^{\alpha-1}Tx(t_1) \neq 0$. Then there exists $\varphi \in E^*$ such that

$$\left\| \frac{Tx(t_1)}{1+t_1^{\alpha-1}} - \frac{Tx(t_2)}{1+t_2^{\alpha-1}} \right\| = \varphi(\frac{Tx(t_1)}{1+t_1^{\alpha-1}} - \frac{Tx(t_2)}{1+t_2^{\alpha-1}})$$

and

$$\|D^{\alpha-1}Tx(t_2) - D^{\alpha-1}Tx(t_1)\| = \varphi(D^{\alpha-1}Tx(t_2) - D^{\alpha-1}Tx(t_1)).$$

Thus

$$\|\frac{Tx(t_1)}{1+t_1^{\alpha-1}} - \frac{Tx(t_2)}{1+t_2^{\alpha-1}}\| = \varphi\left(\frac{Tx(t_1)}{1+t_1^{\alpha-1}} - \frac{Tx(t_2)}{1+t_2^{\alpha-1}}\right)$$

$$
\begin{aligned}
&= \varphi\left(\frac{x_\infty - \int_0^{+\infty} f(s,x(s),D^{\alpha-1}x(s))ds}{\Gamma(\alpha)}\left(\frac{t_2^{\alpha-1}}{1+t_2^{\alpha-1}} - \frac{t_1^{\alpha-1}}{1+t_1^{\alpha-1}}\right)\right. \\
&+ \frac{1}{\Gamma(\alpha)}\int_0^{t_2}\frac{(t_2-s)^{\alpha-1}}{1+t_2^{\alpha-1}}f(s,x(s),D^{\alpha-1}x(s))ds \\
&- \left.\frac{1}{\Gamma(\alpha)}\int_0^{t_1}\frac{(t_1-s)^{\alpha-1}}{1+t_1^{\alpha-1}}f(s,x(s),D^{\alpha-1}x(s))ds\right) \\
&= \varphi\left(\frac{x_\infty - \int_0^{+\infty} f(s,x(s),D^{\alpha-1}x(s))ds}{\Gamma(\alpha)}(\frac{t_2^{\alpha-1}}{1+t_2^{\alpha-1}} - \frac{t_1^{\alpha-1}}{1+t_1^{\alpha-1}})\right. \\
&+ \frac{1}{\Gamma(\alpha)}\int_0^{t_1}[\frac{(t_2-s)^{\alpha-1}}{1+t_2^{\alpha-1}} - \frac{(t_1-s)^{\alpha-1}}{1+t_1^{\alpha-1}}]f(s,x(s),D^{\alpha-1}x(s))ds \\
&+ \left.\frac{1}{\Gamma(\alpha)}\int_{t_1}^{t_2}\frac{(t_2-s)^{\alpha-1}}{1+t_2^{\alpha-1}}f(s,x(s),D^{\alpha-1}x(s))ds\right) \\
&\leq \frac{\|x_\infty - \int_0^{+\infty} f(s,x(s),D^{\alpha-1}x(s))ds\|}{\Gamma(\alpha)}|\frac{t_2^{\alpha-1}}{1+t_2^{\alpha-1}} - \frac{t_1^{\alpha-1}}{1+t_1^{\alpha-1}}| \\
&+ \frac{1}{\Gamma(\alpha)}\int_0^{t_1}|\frac{(t_2-s)^{\alpha-1}}{1+t_2^{\alpha-1}} - \frac{(t_1-s)^{\alpha-1}}{1+t_1^{\alpha-1}}|\|f(s,x(s),D^{\alpha-1}x(s))\|ds \\
&+ \frac{1}{\Gamma(\alpha)}\int_{t_1}^{t_2}|\frac{(t_2-s)^{\alpha-1}}{1+t_2^{\alpha-1}}|\|f(s,x(s),D^{\alpha-1}x(s))\|ds \\
&\leq \frac{\|x_\infty\| + I_*R\|a\|_{L^1} + R\|b\|_{L^1} + \|c\|_{L^1}}{\Gamma(\alpha)}|\frac{t_2^{\alpha-1}}{1+t_2^{\alpha-1}} - \frac{t_1^{\alpha-1}}{1+t_1^{\alpha-1}}| \\
&+ \frac{1}{\Gamma(\alpha)}\int_0^{t_1}|\frac{(t_2-s)^{\alpha-1}}{1+t_2^{\alpha-1}} - \frac{(t_1-s)^{\alpha-1}}{1+t_1^{\alpha-1}}|(R(I_*a(s)+b(s))+c(s))ds \\
&+ \frac{1}{\Gamma(\alpha)}\int_{t_1}^{t_2}|\frac{(t_2-s)^{\alpha-1}}{1+t_2^{\alpha-1}}|(R(I_*a(s)+b(s))+c(s))ds,
\end{aligned}
$$

where

$$I_* = 1 + (\sup\{r : r \in I\})^{\alpha-1},$$

and,

$$
\begin{aligned}
\|D^{\alpha-1}Tx(t_2) - D^{\alpha-1}Tx(t_1)\| &= \varphi(D^{\alpha-1}Tx(t_2) - D^{\alpha-1}Tx(t_1)) \\
&\leq \int_{t_1}^{t_2}(R(I_*a(s)+b(s))+c(s))ds.
\end{aligned}
$$

As $t_1 \to t_2$, the right-hand side of the two above inequalities tends to zero. Then $\{\frac{TB(t)}{1+t^{\alpha-1}}\}$ and $\{D^{\alpha-1}TB(t)\}$ are equicontinuous on I. In view of condition (11.24.3) and the boundedness of B, there exists $M > 0$ such that

$$\|x\|_Y \leq M \text{ for each } x \in B$$

and

$$\begin{aligned}
&\int_0^{+\infty}\|f(s,x(s),D^{\alpha-1}x(s)\|ds\\
<&\int_0^{+\infty}(a(s)\|x(s)\|+b(s)\|D^{\alpha-1}x(s)\|+c(s))ds\\
<&\int_0^{+\infty}((1+s^{\alpha-1})a(s)\frac{\|x(s)\|}{1+s^{\alpha-1}}+b(s)\|D^{\alpha-1}x(s)\|)ds\\
&+\int_0^{+\infty}c(s)ds\\
<&\int_0^{+\infty}M((1+s^{\alpha-1})a(s)+b(s))ds+\int_0^{+\infty}c(s)ds.
\end{aligned}$$

Thus for given $\varepsilon>0$, there exists $L>0$ such that

$$\int_L^{+\infty}M((1+s^{\alpha-1})a(s)+b(s))ds+\int_L^{+\infty}c(s)ds<\varepsilon. \tag{11.51}$$

On the other hand, since $\lim_{t\to+\infty}\frac{t^{\alpha-1}}{1+t^{\alpha-1}}=1$, there exists $T_1>0$ such that for, $t_1,t_2>T_1$

$$\left|\frac{t_1^{\alpha-1}}{1+t_1^{\alpha-1}}-\frac{t_2^{\alpha-1}}{1+t_2^{\alpha-1}}\right|\le\left|\frac{t_1^{\alpha-1}}{1+t_1^{\alpha-1}}-1\right|+\left|\frac{t_2^{\alpha-1}}{1+t_2^{\alpha-1}}-1\right|<\varepsilon. \tag{11.52}$$

Note also that $\lim_{t\to+\infty}\frac{(t-s)^{\alpha-1}}{1+t^{\alpha-1}}=1$, hence there exists $T_2>0$ such that for, $t_1,t_2>T_2$ and for each $s\in[0,L]$

$$|\frac{(t_1-s)^{\alpha-1}}{1+t_1^{\alpha-1}}-\frac{(t_2-s)^{\alpha-1}}{1+t_2^{\alpha-1}}|\le|\frac{(t_1-s)^{\alpha-1}}{1+t_1^{\alpha-1}}-1|+|\frac{(t_2-s)^{\alpha-1}}{1+t_2^{\alpha-1}}-1|<\varepsilon. \tag{11.53}$$

Choose $T>\max(T_1,T_2,L)$ and let $t_1,t_2>T$, with $t_1<t_2$. Then by (11.51)-(11.53) we obtain

$$\begin{aligned}
&\left\|\frac{(Tx)(t_1)}{1+t_1^{\alpha-1}}-\frac{(Tx)(t_2)}{1+t_2^{\alpha-1}}\right\|\\
=&\ \varphi\left(\frac{(Tx)(t_1)}{1+t_1^{\alpha-1}}-\frac{(Tx)(t_2)}{1+t_2^{\alpha-1}}\right)\\
\le&\ \frac{\|x_\infty-\int_0^{+\infty}f(s,x(s),D^{\alpha-1}x(s))ds\|}{\Gamma(\alpha)}\left|\frac{t_2^{\alpha-1}}{1+t_2^{\alpha-1}}-\frac{t_1^{\alpha-1}}{1+t_1^{\alpha-1}}\right|\\
+&\ \varphi\left(\frac{1}{\Gamma(\alpha)}\int_0^{t_2}\frac{(t_2-s)^{\alpha-1}}{1+t_2^{\alpha-1}}f(s,x(s),D^{\alpha-1}x(s))ds\right.\\
-&\left.\frac{1}{\Gamma(\alpha)}\int_0^{t_1}\frac{(t_1-s)^{\alpha-1}}{1+t_1^{\alpha-1}}f(s,x(s),D^{\alpha-1}x(s))ds\right)\\
\le&\ \frac{\|x_\infty-\int_0^{+\infty}f(s,x(s),D^{\alpha-1}x(s))ds\|}{\Gamma(\alpha)}|\frac{t_2^{\alpha-1}}{1+t_2^{\alpha-1}}-\frac{t_1^{\alpha-1}}{1+t_1^{\alpha-1}}|
\end{aligned}$$

$$
\begin{aligned}
&+ \frac{1}{\Gamma(\alpha)}\int_0^L |\frac{(t_1-s)^{\alpha-1}}{1+t_1^{\alpha-1}} - \frac{(t_2-s)^{\alpha-1}}{1+t_2^{\alpha-1}}|\|f(s,x(s),D^{\alpha-1}x(s))\|ds\\
&+ \frac{1}{\Gamma(\alpha)}\int_L^{t_1} |\frac{(t_1-s)^{\alpha-1}}{1+t_1^{\alpha-1}}|\|f(s,x(s),D^{\alpha-1}x(s))\|ds\\
&+ \frac{1}{\Gamma(\alpha)}\int_L^{t_2} |\frac{(t_2-s)^{\alpha-1}}{1+t_2^{\alpha-1}}|\|f(s,x(s),D^{\alpha-1}x(s))\|ds\\
&\leq \frac{\|x_\infty\| + \int_0^{+\infty}\|f(s,x(s),D^{\alpha-1}x(s))\|ds}{\Gamma(\alpha)}\left|\frac{t_2^{\alpha-1}}{1+t_2^{\alpha-1}} - \frac{t_1^{\alpha-1}}{1+t_1^{\alpha-1}}\right|\\
&+ \frac{1}{\Gamma(\alpha)}\int_0^L \left|\frac{(t_1-s)^{\alpha-1}}{1+t_1^{\alpha-1}} - \frac{(t_2-s)^{\alpha-1}}{1+t_2^{\alpha-1}}\right|(a(s)\|x(s)\| + b(s)\|D^{\alpha-1}x(s)\| + c(s))ds\\
&+ \frac{1}{\Gamma(\alpha)}\int_L^{t_1} |\frac{(t_1-s)^{\alpha-1}}{1+t_1^{\alpha-1}}|(a(s)\|x(s)\|\\
&+ b(s)\|D^{\alpha-1}x(s)\| + c(s))ds\\
&+ \frac{1}{\Gamma(\alpha)}\int_L^{t_2} |\frac{(t_2-s)^{\alpha-1}}{1+t_2^{\alpha-1}}|(a(s)\|x(s)\| + b(s)\|D^{\alpha-1}x(s)\| + c(s))ds.
\end{aligned}
$$

Then

$$
\begin{aligned}
&\left\|\frac{(Tx)(t_1)}{1+t_1^{\alpha-1}} - \frac{(Tx)(t_2)}{1+t_2^{\alpha-1}}\right\|\\
&\leq \frac{\|x_\infty\| + M_*}{\Gamma(\alpha)}|\frac{t_2^{\alpha-1}}{1+t_2^{\alpha-1}} - \frac{t_1^{\alpha-1}}{1+t_1^{\alpha-1}}|\\
&+ \frac{1}{\Gamma(\alpha)}\int_0^L |\frac{(t_1-s)^{\alpha-1}}{1+t_1^{\alpha-1}} - \frac{(t_2-s)^{\alpha-1}}{1+t_2^{\alpha-1}}|(M((1+L^{\alpha-1})a(s)+b(s))+c(s))ds\\
&+ \frac{1}{\Gamma(\alpha)}\int_L^{t_1} (\frac{(t_1-s)^{\alpha-1}}{1+t_1^{\alpha-1}} M((1+s^{\alpha-1})a(s)+b(s))+c(s))ds\\
&+ \frac{1}{\Gamma(\alpha)}\int_L^{t_2} (\frac{(t_2-s)^{\alpha-1}}{1+t_2^{\alpha-1}}|M((1+s^{\alpha-1})a(s)+b(s))+c(s))ds\\
&\leq \frac{\|x_\infty\| + M_*}{\Gamma(\alpha)}|\frac{t_2^{\alpha-1}}{1+t_2^{\alpha-1}} - \frac{t_1^{\alpha-1}}{1+t_1^{\alpha-1}}|\\
&+ \frac{1}{\Gamma(\alpha)}\int_0^L |\frac{(t_1-s)^{\alpha-1}}{1+t_1^{\alpha-1}} - \frac{(t_2-s)^{\alpha-1}}{1+t_2^{\alpha-1}}|(M((1+L^{\alpha-1})a(s)+b(s))+c(s))ds\\
&+ \frac{1}{\Gamma(\alpha)}\int_L^{t_1} (M((1+s^{\alpha-1})a(s)+b(s))+c(s))ds\\
&+ \frac{1}{\Gamma(\alpha)}\int_L^{t_2} (M((1+s^{\alpha-1})a(s)+b(s))+c(s))ds\\
&\leq \frac{\|x_\infty\| + M_*}{\Gamma(\alpha)}\varepsilon + \frac{1}{\Gamma(\alpha)}\int_0^L (M((1+s^{\alpha-1})a(s)+b(s))+c(s))ds\varepsilon + \frac{2}{\Gamma(\alpha)}\varepsilon.
\end{aligned}
$$

where

$$M_* = \int_0^{+\infty} (M((1+s^{\alpha-1})a(s)+b(s))+c(s))ds,$$

and,

$$\begin{aligned}\|D^{\alpha-1}Tx(t_2)-D^{\alpha-1}Tx(t_1)\| &= \varphi(D^{\alpha-1}Tx(t_2)-D^{\alpha-1}Tx(t_1))\\ &\le \int_{t_1}^{t_2} \|f(s,x(s),D^{\alpha-1}x(s)\|ds\\ &\le \int_L^{+\infty} \|f(s,x(s),D^{\alpha-1}x(s)\|ds\\ &\le \int_L^{+\infty} M((1+s^{\alpha-1})a(s)+b(s))ds+\int_L^{+\infty} c(s)ds\\ &\le \varepsilon.\end{aligned}$$

This completes the proof of Lemma 11.50. □

The proof of the following auxiliary result follows from the ideas in [342], so we omit it.

Lemma 11.51. *Suppose that condition (11.61.3) holds and B is a bounded subset of Y. Then*

$$\beta_Y(TB) = \max\left(\sup_{t\in J}\beta(\frac{TB(t)}{1+t^{\alpha-1}}), \sup_{t\in J}\beta(D^{\alpha-1}TB(t))\right).$$

Now we are in the position to give the main result of this work.

Theorem 11.52. *Let E be a Banach space , and assume that assumptions* $(11.61.1)-(11.61.4)$ *be satisfied. If*

$$\frac{1}{\Gamma(\alpha)}\int_0^{+\infty}(1+t^{\alpha-1})l(t)dt<1, \tag{11.54}$$

then there exists at least one weak solution $x\in Y$ of (11.46)-(11.47).

Proof. Let R be a constant such that

$$R \ge \frac{\|x_\infty\|+2\int_0^{+\infty}c(t)dt}{\Gamma(\alpha)-2\int_0^{+\infty}[(1+t^{\alpha-1})a(t)+b(t)]dt},$$

and consider the set:

$$B=\{x\in Y : \|x\|_Y\le R\}.$$

Then we can derive that $T : B\to B$. Indeed, if $x\in B$, $t\in J$, $(Tx)(t)\neq 0$ and $D^{\alpha-1}(Tx)(t)\neq 0$, then there exists $\varphi\in E^*$ such that $\|(Tx)(t)\|=\varphi((Tx)(t))$ and $\|D^{\alpha-1}Tx(t)\|=\varphi(D^{\alpha-1}(Tx)(t))$. Thus by condition (11.24.3), we get

$$\begin{aligned}\frac{\|(Tx)(t)\|}{1+t^{\alpha-1}} &= \frac{\varphi((Tx)(t))}{1+t^{\alpha-1}}\\ &= \frac{1}{1+t^{\alpha-1}}\varphi\left[\frac{t^{\alpha-1}}{\Gamma(\alpha)}[x_\infty-\int_0^{+\infty} f(s,x(s),D^{\alpha-1}x(s))ds]\right.\end{aligned}$$

$$
\begin{aligned}
&+ \frac{1}{\Gamma(\alpha)}\int_0^t (t-s)^{\alpha-1} f(s,x(s),D^{\alpha-1}x(s))ds\Big] \\
&\leq \frac{t^{\alpha-1}}{\Gamma(\alpha)(1+t^{\alpha-1})}\left[\|x_\infty\| + \int_0^{+\infty} \varphi(f(s,x(s),D^{\alpha-1}x(s)))ds\right] \\
&+ \frac{1}{\Gamma(\alpha)}\int_0^t \frac{(t-s)^{\alpha-1}}{1+t^{\alpha-1}}\varphi(f(s,x(s),D^{\alpha-1}x(s)))ds \\
&\leq \frac{1}{\Gamma(\alpha)}\Big[\|x_\infty\| + \int_0^{+\infty} \|f(s,x(s),D^{\alpha-1}x(s))\|ds \\
&\quad + \int_0^{+\infty} \|f(s,x(s),D^{\alpha-1}x(s))\|ds\Big] \\
&\leq \frac{\|x_\infty\|}{\Gamma(\alpha)} + \frac{2}{\Gamma(\alpha)}\int_0^{+\infty} \big(\|x\|_Y[(1+s^{\alpha-1})a(s)+b(s)]+c(s)\big)\,ds \\
&\leq \frac{\|x_\infty\|}{\Gamma(\alpha)} + \frac{2}{\Gamma(\alpha)}\left[R\int_0^{+\infty}[(1+s^{\alpha-1})a(s)+b(s)]ds + \int_0^{+\infty} c(s)ds\right] \\
&\leq R.
\end{aligned}
$$

We remark that

$$
(D^{\alpha-1}Tx)(t) = x_\infty - \int_0^{+\infty} f(s,x(s),D^{\alpha-1}x(s))ds + \int_0^t f(s,x(s),D^{\alpha-1}x(s))ds.
$$

Then

$$
\begin{aligned}
& \|(D^{\alpha-1}Tx)(t)\| \\
&= \varphi(D^{\alpha-1}Tx(t)) \\
&= \varphi(x_\infty - \int_0^{+\infty} f(s,x(s),D^{\alpha-1}x(s))ds + \int_0^t f(s,x(s),D^{\alpha-1}x(s))ds) \\
&\leq \|x_\infty\| + 2\left[\|x\|_Y\int_0^{+\infty}[(1+s^{\alpha-1})a(s)+b(s)]ds + \int_0^{+\infty} c(s)ds\right] \\
&\leq R.
\end{aligned}
$$

Hence $\|Tx\|_Y \leq R$ and we conclude that T maps B into itself. Choose the set $Q = \bar{co}(TB)$, clearly Q is a bounded, convex and closed subset of B. We shall show that $T : Q \to Q$ satisfies the assumptions of Theorem 1.64. T maps Q into itself , in fact $TQ \subset TB \subset Q$. In what follows, we divide the proof into several steps.

Step 1: T is weakly-sequentially continuous.
Let $(x_n)_n$ be a sequence in Q and let $(x_n(t))_n \to x(t)$ in E_ω for each $t \in J$. Then $\|x_n\|_Y \leq R$ for $n \geq 1$ and $\|x\|_Y \leq R$. Now, by (11.24.3), for any $\varepsilon > 0$, there exists $L > 0$ such that

$$
\int_L^{+\infty} [(1+t^{\alpha-1})a(t)+b(t)]dt < \frac{\Gamma(\alpha)}{6R}\varepsilon, \int_L^{+\infty} c(t)dt < \frac{\Gamma(\alpha)}{6}\varepsilon. \tag{11.55}
$$

Condition (11.61.1) yields that there exists $N > 0$ such that for $n > N$ and $t \in J$,

$$
\|f(t,x_n(t),D^{\alpha-1}x_n(t)) - f(t,x(t),D^{\alpha-1}x(t))\| < \frac{\Gamma(\alpha)}{3L}\varepsilon \tag{11.56}
$$

Therefore, for $t \in [0,L]$ and $n > N$, we can obtain that

$$\begin{aligned}
\frac{\|(Tx_n)(t)-(Tx)(t)\|}{1+t^{\alpha-1}} &= \frac{\varphi((Tx_n)(t)-(Tx)(t))}{1+t^{\alpha-1}} \\
&\leq \frac{1}{\Gamma(\alpha)}\int_0^t \frac{t^{\alpha-1}-(t-s)^{\alpha-1}}{1+t^{\alpha-1}} \\
&\times \|f(s,x_n(s),D^{\alpha-1}x_n(s))-f(s,x(s),D^{\alpha-1}x(s))\|ds \\
&+ \frac{1}{\Gamma(\alpha)}\int_t^{+\infty}\frac{t^{\alpha-1}}{1+t^{\alpha-1}}\|f(s,x_n(s),D^{\alpha-1}x_n(s))-f(s,x(s),D^{\alpha-1}x(s))\|ds \\
&\leq \frac{1}{\Gamma(\alpha)}\int_0^t \|f(s,x_n(s),D^{\alpha-1}x_n(s))-f(s,x(s),D^{\alpha-1}x(s))\|ds \\
&+ \frac{1}{\Gamma(\alpha)}\int_t^L \|f(s,x_n(s),D^{\alpha-1}x_n(s))-f(s,x(s),D^{\alpha-1}x(s))\|ds \\
&+ \frac{1}{\Gamma(\alpha)}\int_L^{+\infty} \|f(s,x_n(s),D^{\alpha-1}x_n(s))-f(s,x(s),D^{\alpha-1}x(s))\|ds \\
&\leq \frac{1}{\Gamma(\alpha)}\int_0^L \|f(s,x_n(s),D^{\alpha-1}x_n(s))-f(s,x(s),D^{\alpha-1}x(s))\|ds \\
&+ \frac{1}{\Gamma(\alpha)}\int_L^{+\infty} \|f(s,x_n(s),D^{\alpha-1}x_n(s))-f(s,x(s),D^{\alpha-1}x(s))\|ds \\
&\leq \frac{1}{3L}\varepsilon Ł + \frac{2}{\Gamma(\alpha)}\int_L^{+\infty} c(s)ds \\
&+ \frac{2}{\Gamma(\alpha)}\int_L^{+\infty}[(1+s^{\alpha-1})a(s)+b(s)]\|x_n\|ds \\
&\leq \frac{\varepsilon}{3}+\frac{2R}{\Gamma(\alpha)}\frac{\Gamma(\alpha)}{6R}\varepsilon+\frac{2}{\Gamma(\alpha)}\frac{\Gamma(\alpha)}{6}\varepsilon < \varepsilon.
\end{aligned}$$

And,

$$\begin{aligned}
& \|(D^{\alpha-1}Tx_n)(t)-(D^{\alpha-1}Tx)(t)\| \\
&= \varphi((D^{\alpha-1}Tx_n)(t)-(D^{\alpha-1}Tx)(t)) \\
&= \varphi(-\int_t^{+\infty} f(s,x_n(s),D^{\alpha-1}x_n(s))ds+\int_t^{+\infty} f(s,x(s),D^{\alpha-1}x(s))ds) \\
&\leq \int_t^L \|f(s,x_n(s),D^{\alpha-1}x_n(s))-f(s,x(s),D^{\alpha-1}x(s))\|ds
\end{aligned}$$

$$\begin{aligned}
&+ \int_L^{+\infty} \|f(s,x_n(s),D^{\alpha-1}x_n(s))-f(s,x(s),D^{\alpha-1}x(s))\|ds \\
&\leq \frac{\Gamma(\alpha)}{3L}\varepsilon L+2\int_L^{+\infty}\|f(s,x(s),D^{\alpha-1}x(s))\|ds \\
&\leq \frac{\Gamma(\alpha)}{3}\varepsilon+2[\|x\|\int_L^{+\infty}[(1+s^{\alpha-1})a(s)+b(s)]ds+\int_L^{+\infty}c(s)ds] \\
&\leq \frac{\Gamma(\alpha)}{3}\varepsilon+2R\frac{\Gamma(\alpha)}{6R}\varepsilon+2\frac{\Gamma(\alpha)}{6}\varepsilon \\
&\leq \Gamma(\alpha)\varepsilon < \varepsilon.
\end{aligned}$$

And for $t > L$ and $n > N$, again from (11.55)-(11.56) we derive that

$$\begin{aligned}
&\frac{\|(Tx_n)(t)-(Tx)(t)\|}{1+t^{\alpha-1}}\\
= &\frac{\varphi((Tx_n)(t)-(Tx)(t))}{1+t^{\alpha-1}}\\
\leq &\frac{1}{\Gamma(\alpha)}\int_0^t \frac{t^{\alpha-1}-(t-s)^{\alpha-1}}{1+t^{\alpha-1}}\|f(s,x_n(s),D^{\alpha-1}x_n(s))-f(s,x(s),D^{\alpha-1}x(s))\|ds\\
+ &\frac{1}{\Gamma(\alpha)}\int_t^{+\infty} \frac{t^{\alpha-1}}{1+t^{\alpha-1}}\|f(s,x_n(s),D^{\alpha-1}x_n(s))-f(s,x(s),D^{\alpha-1}x(s))\|ds\\
\leq &\frac{1}{\Gamma(\alpha)}\int_0^L \|f(s,x_n(s),D^{\alpha-1}x_n(s))-f(s,x(s),D^{\alpha-1}x(s))\|ds\\
+ &\frac{1}{\Gamma(\alpha)}\int_L^t \|f(s,x_n(s),D^{\alpha-1}x_n(s))-f(s,x(s),D^{\alpha-1}x(s))\|ds\\
+ &\frac{1}{\Gamma(\alpha)}\int_t^{+\infty} \|f(s,x_n(s),D^{\alpha-1}x_n(s))-f(s,x(s),D^{\alpha-1}x(s))\|ds\\
\leq &\frac{1}{\Gamma(\alpha)}\int_0^L \|f(s,x_n(s),D^{\alpha-1}x_n(s))-f(s,x(s),D^{\alpha-1}x(s))\|ds\\
+ &\frac{1}{\Gamma(\alpha)}\int_L^{+\infty} \|f(s,x_n(s),D^{\alpha-1}x_n(s))-f(s,x(s),D^{\alpha-1}x(s))\|ds\\
\leq &\frac{1}{\Gamma(\alpha)}\int_0^L \|f(s,x_n(s),D^{\alpha-1}x_n(s))-f(s,x(s),D^{\alpha-1}x(s))\|ds\\
+ &\frac{1}{\Gamma(\alpha)}\int_L^{+\infty} \|f(s,x_n(s),D^{\alpha-1}x_n(s))-f(s,x(s),D^{\alpha-1}x(s))\|ds\\
\leq &\frac{1}{\Gamma(\alpha)}\int_0^L \|f(s,x_n(s),D^{\alpha-1}x_n(s))-f(s,x(s),D^{\alpha-1}x(s))\|ds\\
+ &\frac{2R}{\Gamma(\alpha)}\int_L^{+\infty} [(1+s^{\alpha-1})a(s)+b(s)]ds+\frac{2}{\Gamma(\alpha)}\int_L^{+\infty} c(s)ds\\
\leq &\frac{\varepsilon}{3}+\frac{\varepsilon}{3}+\frac{\varepsilon}{3}=\varepsilon.
\end{aligned}$$

On the other hand,

$$\begin{aligned}
&\|(D^{\alpha-1}Tx_n)(t)-(D^{\alpha-1}Tx)(t)\|\\
= &\varphi(D^{\alpha-1}Tx_n(t)-D^{\alpha-1}Tx(t))\\
= &\varphi(-\int_t^{+\infty} f(s,x_n(s),D^{\alpha-1}x_n(s))ds+\int_t^{+\infty} f(s,x(s),D^{\alpha-1}x(s))ds))\\
\leq &\int_t^{+\infty} \|f(s,x_n(s),D^{\alpha-1}x_n(s))-f(s,x(s),D^{\alpha-1}x(s))\|ds\\
\leq &2\int_t^{+\infty} \|f(s,x_n(s),D^{\alpha-1}x_n(s))-f(s,x(s),D^{\alpha-1}x(s))\|ds\\
\leq &2R\int_t^{+\infty} [(1+s^{\alpha-1})a(s)+b(s)]ds+2\int_t^{+\infty} c(s)ds
\end{aligned}$$

$$\leq \frac{2R\Gamma(\alpha)}{6R}\varepsilon + 2\frac{\Gamma(\alpha)}{6}\varepsilon$$
$$\leq \frac{2}{3}\varepsilon < \varepsilon.$$

Thus, we conclude that $\|Tx_n - Tx\|_Y \leq \varepsilon$ as $n > N$, namely, T is weakly-sequentially continuous.

Step 2: The implication (1.6) holds. Now let V be a subset of Q such that $V = \overline{conv}(T(V) \cup \{0\})$. Obviously $V(t) \subset \overline{conv}(T(V) \cup \{0\}), \forall t \in I$. $(TV)(t) \subset (TQ)(t), t \in J$ is bounded in E. We need to prove that $\beta_Y(V) = \max\left(\sup_{t\in J}\beta(\frac{V(t)}{1+t^{\alpha-1}}), \sup_{t\in J}\beta(D^{\alpha-1}V(t))\right) = 0$, i.e., for any $t \in J$, $\beta(\frac{V(t)}{1+t^{\alpha-1}}) = 0$ and $\beta(D^{\alpha-1}V(t)) = 0$. For this, let us define

$$\begin{aligned} T_n x(t) &= \frac{x_\infty}{\Gamma(\alpha)}t^{\alpha-1} - \frac{1}{\Gamma(\alpha)}\int_0^t [t^{\alpha-1} - (t-s)^{\alpha-1}]f(s,x(s),D^{\alpha-1}x(s))ds \\ &- \frac{1}{\Gamma(\alpha)}\int_t^n t^{\alpha-1}f(s,x(s),D^{\alpha-1}x(s))ds. \end{aligned}$$

Then,

$$\begin{aligned} D^{\alpha-1}(T_n x)(t) &= x_\infty - \int_0^n f(s,x(s),D^{\alpha-1}x(s))ds + \int_0^t f(s,x(s),D^{\alpha-1}x(s))ds \\ &= x_\infty - \int_t^n f(s,x(s),D^{\alpha-1}x(s))ds. \end{aligned}$$

Using (11.61.3), it is easy to show that

$$\left\|\frac{(T_n x)(t)}{1+t^{\alpha-1}} - \frac{(Tx)(t)}{1+t^{\alpha-1}}\right\| \leq \frac{1}{\Gamma(\alpha)}[R\int_n^{+\infty}((1+s^{\alpha-1})a(s)+b(s))ds$$
$$+\int_n^{+\infty} c(s)ds]$$

and

$$\|D^{\alpha-1}(T_n x)(t) - D^{\alpha-1}(Tx)(t)\| \leq R\int_n^{+\infty}((1+s^{\alpha-1})a(s)+b(s))ds + \int_n^{+\infty} c(s)ds.$$

This shows that

$$d\left(\frac{(T_n x)(t)}{1+t^{\alpha-1}}, \frac{(Tx)(t)}{1+t^{\alpha-1}}\right) \to 0 \text{ as } n \to \infty$$

and

$$d(D^{\alpha-1}(T_n x)(t), D^{\alpha-1}(Tx)(t)) \to 0 \text{ as } n \to +\infty,$$

for each $t \in J$ where $d(\cdot)$ denotes the Hausdorff metric in the space E. Thus by the property of the measure of weak noncompactness β, we get

$$\lim_{n\to\infty}\beta\left(\frac{(T_n x)(t)}{1+t^{\alpha-1}}\right) = \beta\left(\frac{(Tx)(t)}{1+t^{\alpha-1}}\right)$$

and

$$\lim_{n\to\infty}\beta(D^{\alpha-1}(T_nx)(t))=\beta(D^{\alpha-1}(Tx)(t)).$$

Taking into account that $\{\frac{Q(t)}{1+t^{\alpha-1}}\}$ and $\{D^{\alpha-1}Q(t)\}$ are equicontinuous on any compact interval of J, we obtain that $\{\frac{V(t)}{1+t^{\alpha-1}}\}$ and $\{D^{\alpha-1}V(t)\}$ are equicontinuous on any compact interval of J. By (11.61.1), and (11.61.3) it is easy to see that the set $\{f(t,x(t),D^{\alpha-1}x(t)),x\in V\}$ is bounded and equicontinuous on $[0,n]$. By (11.61.4), Lemma 11.51 and the properties of β, for $t\in J$ we have

$$\begin{aligned}
&\beta\left(\frac{(T_nV)(t)}{1+t^{\alpha-1}}\right)=\beta(\{\frac{(T_nx)(t)}{1+t^{\alpha-1}},x\in V\})\\
\leq\ & \frac{1}{\Gamma(\alpha)}\int_0^t\beta(\{f(t,x(t),D^{\alpha-1}x(t)):x\in V\})\\
+\ & \frac{1}{\Gamma(\alpha)}\int_t^n\beta(\{f(t,x(t),D^{\alpha-1}x(t)):x\in V\})\\
\leq\ & \frac{1}{\Gamma(\alpha)}\int_0^n\beta(\{f(t,x(t),D^{\alpha-1}x(t)):x\in V\})\\
\leq\ & \frac{1}{\Gamma(\alpha)}\int_0^n l(t)\max(\beta(V(t)),\beta(D^{\alpha-1}V(t)))dt\\
\leq\ & \frac{1}{\Gamma(\alpha)}\int_0^n(1+t^{\alpha-1})l(t)\max(\beta(\frac{V(t)}{1+t^{\alpha-1}}),\beta(\frac{D^{\alpha-1}V(t)}{1+t^{\alpha-1}}))dt\\
\leq\ & \frac{1}{\Gamma(\alpha)}\int_0^n(1+t^{\alpha-1})l(t)\max(\beta(\frac{V(t)}{1+t^{\alpha-1}}),\beta(D^{\alpha-1}V(t)))dt\\
\leq\ & \frac{1}{\Gamma(\alpha)}\int_0^n(1+t^{\alpha-1})l(t)\beta_Y(V)dt\\
\leq\ & \beta_Y(V)\frac{1}{\Gamma(\alpha)}\int_0^n(1+t^{\alpha-1})l(t)dt.
\end{aligned}$$

On the other hand,

$$\begin{aligned}
\beta(D^{\alpha-1}T_nV(t)) &= \beta(\{D^{\alpha-1}T_nx(t):x\in V\}\\
&\leq \int_t^n\beta(\{f(t,x(t),D^{\alpha-1}x(t)):x\in V\})
\end{aligned}$$

$$\begin{aligned}
\leq\ & \int_0^n l(t)\max(\beta(V(t)),\beta(D^{\alpha-1}V(t)))dt\\
\leq\ & \int_0^n(1+t^{\alpha-1})l(t)\max(\beta(\frac{V(t)}{1+t^{\alpha-1}}),\beta(\frac{D^{\alpha-1}V(t)}{1+t^{\alpha-1}}))dt\\
\leq\ & \int_0^n(1+t^{\alpha-1})l(t)\max(\beta(\frac{V(t)}{1+t^{\alpha-1}}),\beta(D^{\alpha-1}V(t)))dt\\
\leq\ & \int_0^n(1+t^{\alpha-1})l(t)\beta_C(V)dt\\
\leq\ & \beta_Y(V)\frac{1}{\Gamma(\alpha)}\int_0^n(1+t^{\alpha-1})l(t)dt.
\end{aligned}$$

Then for any $t \in J$,

$$\max(\beta(\frac{TV(t)}{1+t^{\alpha-1}}), \beta(D^{\alpha-1}TV(t))) \le \beta_Y(V)\int_0^{+\infty}(1+t^{\alpha-1})l(t)dt.$$

thus,

$$\beta_Y(V) \le \beta_Y(TV) \le \beta_Y(V)\frac{1}{\Gamma(\alpha)}\int_0^{+\infty}(1+t^{\alpha-1})l(t)dt.$$

By (11.54), it follows that $\beta_Y(V) = 0$ and then for each $t \in J$, $V(t)$ is relatively weakly compact in E. Consequently, Theorem 1.64 implies that problem (11.46)-(11.47) has a weak solution defined on J. □

11.5. Weak Solutions for Nonlinear Differential Equations with Integral Boundary Conditions in Banach Spaces

11.5.1. Introduction

In this section we investigate the existence of weak solutions, for the boundary value problem, for fractional differential equations with mixed boundary conditions of the form

$$^cD^\alpha x(t) = f(t, x(t)), \quad \text{for each } t \in I = [0, T], \tag{11.57}$$

$$x(0) + \mu\int_0^T x(s)ds = x(T), \tag{11.58}$$

where $^cD^\alpha, 0 < \alpha \le 1$ is the Caputo fractional derivative, $f : I \times E \to E$ is given function satisfying some assumptions that will be specified later, E is a Banach space with norm $\|\cdot\|$ and $\mu \in \mathbb{R}^*$.

11.5.2. Existence of Solutions

Let us start by defining what we mean by a solution of the problem (11.57)-(11.58).

Definition 11.53. *A function $x \in C(I, E_\omega)$ is said to be a solution of (11.57)-(11.58) if x satisfies the equation $^cD^\alpha x(t) = f(t, x(t))$ on I, and the condition (7.18).*

Lemma 11.54. *Let $0 < \alpha \le 1$ and let $h \in C(I, E)$ be a given function, then the boundary value problem*

$$^cD^\alpha x(t) = h(t),\ t \in I, \tag{11.59}$$

$$x(0) + \mu\int_0^T x(s)ds = x(T),\ \mu \in \mathbb{R}^* \tag{11.60}$$

has a unique solution given by

$$x(t) = \int_0^T G(t, s)h(s)ds, \tag{11.61}$$

where $G(t,s)$ is the Green's function defined by the formula

$$G(t,s)=\begin{cases}\frac{-(T-s)^{\alpha}+\alpha T(t-s)^{\alpha-1}}{T\Gamma(\alpha+1)}+\frac{(T-s)^{\alpha-1}}{T\mu\Gamma(\alpha)}, & \text{if } 0\le s<t,\\ \frac{-(T-s)^{\alpha}}{T\Gamma(\alpha+1)}+\frac{(T-s)^{\alpha-1}}{T\mu\Gamma(\alpha)}, & \text{if } t\le s<T.\end{cases} \quad (11.62)$$

Proof. By Lemma 11.54, we can reduce the problem (11.59)-(11.60) to an equivalent integral equation

$$x(t)=I^{\alpha}h(t)-c_0=\int_0^t\frac{(t-s)^{\alpha-1}}{\Gamma(\alpha)}h(s)ds-c_0,$$

for some constant $c_0\in\mathbb{R}$. We have by integration (using Fubini's integral theorem)

$$\begin{aligned}\int_0^T x(s)ds &= \int_0^T\left(\int_0^s\frac{(s-\tau)^{\alpha-1}}{\Gamma(\alpha)}h(\tau)d\tau-c_0\right)ds\\ &= \int_0^T\left(\int_\tau^T\frac{(s-\tau)^{\alpha-1}}{\Gamma(\alpha)}ds\right)h(\tau)d\tau-c_0T\\ &= \int_0^T\frac{(T-\tau)^{\alpha}}{\alpha\Gamma(\alpha)}h(\tau)d\tau-c_0T.\end{aligned}$$

Applying the boundary condition (11.60), we get

$$x(0)=-c_0$$

$$x(T)=\int_0^T\frac{(T-s)^{\alpha-1}}{\Gamma(\alpha)}h(s)ds-c_0,$$

that is

$$c_0=\frac{1}{T}\int_0^T\left(-\frac{(T-s)^{\alpha-1}}{\mu\Gamma(\alpha)}+\frac{(T-s)^{\alpha}}{\Gamma(\alpha+1)}\right)h(s)ds.$$

Therefore, the unique solution of (11.59)-(11.60) has the form

$$\begin{aligned}x(t) &= \int_0^t\frac{(t-s)^{\alpha-1}}{\Gamma(\alpha)}h(s)ds+\frac{1}{T}\int_0^T\left(\frac{(T-s)^{\alpha-1}}{\mu\Gamma(\alpha)}-\frac{(T-s)^{\alpha}}{\Gamma(\alpha+1)}\right)h(s)ds,\\ &= \int_0^t\left(\frac{-(T-s)^{\alpha}+\alpha T(t-s)^{\alpha-1}}{T\Gamma(\alpha+1)}+\frac{(T-s)^{\alpha-1}}{T\mu\Gamma(\alpha)}\right)h(s)ds\\ &+ \int_t^T\left(-\frac{(T-s)^{\alpha}}{T\Gamma(\alpha+1)}+\frac{(T-s)^{\alpha-1}}{T\mu\Gamma(\alpha)}\right)h(s)ds\\ &= \int_0^T G(t,s)h(s)ds,\end{aligned}$$

which completes the proof. □

Remark 11.55. *The function $t\mapsto\int_0^T|G(t,s)|ds$ is continuous on I, and hence is bounded. Let*

$$\tilde{G}=sup\left\{\int_0^T|G(t,s)|ds,\ t\in I\right\}.$$

To establish our main result concerning existence of solutions (11.57)-(11.58), we list the following hypotheses:

(11.67.1) For each $t \in I$, the function $f(t,.)$ is weakly sequentially continuous.

(11.67.2) For each $x \in C(I,E)$, the function $f(\cdot,x(\cdot))$ is Pettis integrable on I.

(11.67.3) There exist $p \in L^{\infty}(I,E)$ and a continuous nondecreasing function $\psi : [0,\infty) \to (0,\infty)$ such that

$$\|f(t,x)\| \leq p(t)\psi(\|x\|).$$

(11.67.4) There exist a constant $R > 0$ such that

$$\frac{R}{\|p\|_{L^{\infty}}\psi(R)\tilde{G}} > 1.$$

(11.67.5) For each bounded set $Q \subset E$, and each $t \in I$, the following inequality holds

$$\beta(f(t,Q)) \leq p(t)\beta(Q).$$

Theorem 11.56. *Let E be a Banach space , and assume assumptions* $(11.67.1)-(11.67.5)$ *be satisfied. If*

$$\|p\|_{L^{\infty}}\tilde{G} < 1, \tag{11.63}$$

then the boundary value problem (11.57)-(11.58) has at least one solution.

Proof. We shall reduce the existence of solutions of the boundary value problem (11.57)-(11.58) to fixed point problem. To this end we consider the operator $T : C(I,E) \to C(I,E)$ defined by

$$(Tx)(t) = \int_0^T G(t,s)f(s,x(s))ds \tag{11.64}$$

where $G(\cdot,\cdot)$ is the Green's function defined by (11.62). Clearly the fixed points of the operator T are solutions of the problem (11.57)-(11.58). First notice that, for $x \in C(I,E)$, we have $f(\cdot,x(\cdot)) \in P(I,E)$ (assumption (11.67.2)). Since, $s \mapsto G(t,s) \in L^{\infty}(I)$ then $G(t,\cdot)f(\cdot,x(\cdot)), \forall t \in I$ is Pettis integrable, and thus, the operator T makes sense. Let $R \in \mathbb{R}_+^*$, and consider the set

$$\begin{aligned} Q &= \{x \in C(I,E) : \|x\|_{\infty} \leq R \text{ and} \\ &\quad \|x(t_1) - x(t_2)\| \leq \|p\|_{L^{\infty}}\psi(R)\int_0^T |G(t_2,s) - G(t_1,s)|ds \text{ for } t_1,t_2 \in I\}. \end{aligned}$$

Clearly, the subset Q is closed, convex and equicontinuous. We shall show that T satisfies the assumptions of Theorem 1.64.

Step 1: T maps Q into itself.

Take $x \in Q, t \in [0,T]$ and assume that $Tx(t) \neq 0$. Then there exists $\varphi \in E^*$ such that $\|Tx(t)\| = \varphi(Tx(t))$. Thus

$$\begin{aligned}
\|Tx(t)\| &= \varphi(Tx(t)) \\
&= \varphi\left(\int_0^T G(t,s)f(s,x(s))ds\right) \\
&\leq \int_0^T |G(t,s)|\varphi(f(s,x(s)))ds \\
&\leq \int_0^T |G(t,s)|p(s)\psi(\|x\|)ds \\
&\leq \|p\|_{L^\infty}\psi(R)\tilde{G} \\
&\leq R.
\end{aligned}$$

Let $t_1, t_2 \in I, t_1 < t_2, x \in Q$, so $Tx(t_2) - Tx(t_1) \neq 0$. Then there exist $\varphi \in E^*$ such that

$$\|Tx(t_2) - Tx(t_1)\| = \varphi(Tx(t_2) - Tx(t_1)).$$

Thus

$$\begin{aligned}
\|Tx(t_2) - Tx(t_1)\| &= \varphi\left(\int_0^T G(t_2,s)f(s,x(s))ds - \int_0^T G(t_1,s)f(s,x(s))ds\right) \\
&\leq \int_0^T |G(t_2,s) - G(t_1,s)|\|f(s,x(s)\|ds \\
&\leq \|p\|_{L^\infty}\psi(R)\int_0^T |G(t_2,s) - G(t_1,s)|ds.
\end{aligned}$$

Hence $T(Q) \subset Q$.

Step 2: T is weakly sequentially continuous.

Let (x_n) be a sequence in Q and let $(x_n(t)) \to x(t)$ in (E,w) for each $t \in I$. Fix $t \in I$. Since f satisfies assumptions (11.67.1), we have $f(t,x_n(t))$, converging weakly uniformly to $f(t,x(t))$. Hence the Lebesgue Dominated Convergence theorem for Pettis integral implies $Tx_n(t)$ converging weakly uniformly to $Tx(t)$ in E_w. We do it for each $t \in I$ so $Tx_n \to Tx$. Then $T : Q \to Q$ is weakly sequentially continuous.

Step 3: The implication (1.6) holds.
Now let V be a subset of Q such that $V = \overline{conv}(T(V) \cup \{0\})$. Obviously $V(t) \subset \overline{conv}(T(V) \cup \{0\}), \forall t \in I$. $TV(t) \subset TQ(t), t \in J$ is bounded in E. By Assumption (11.35.5), and the properties of the measure β we have for each $t \in I$

$$\begin{aligned}
\upsilon(t) &\leq \beta(T(V(t)) \cup \{0\}) \\
&\leq \beta(T(V(t))) \\
&\leq \int_0^T |G(t,s)|p(s)\beta(V(s))ds \\
&\leq \|p\|_{L^\infty}\int_0^T |G(t,s)|\upsilon(s)ds \\
&\leq \|p\|_{L^\infty}\|\upsilon\|_\infty\tilde{G}.
\end{aligned}$$

This means that

$$\|\upsilon\|_\infty(1-\|p\|_{L^\infty}\tilde{G})\leq 0.$$

By (11.63) it follows that $\|\upsilon\|_\infty=0$, that is $\upsilon(t)=0$ for each $t\in I$, and then $V(t)$ is relatively weakly compact in E. Applying now Theorem 1.64 we conclude that T has a fixed point which is a solution of the problem (11.57)-(11.58). □

11.5.3. An Example

As an application of our results we consider the following partial hyperbolic fractional differential equations of the form

$$({}^cD_0^\alpha x_n)(t)=\frac{1}{3e^{t+4}}(1+|x_n(t)|),\ \text{ if } t\in I=[0,1], \tag{11.65}$$

$$x_n(0)+\int_0^1 x_n(s)ds=x_n(1). \tag{11.66}$$

Let

$$E=l^1=\{x=(x_1,x_2,\ldots,x_n,\ldots):\sum_{n=1}^{\infty}|x_n|<\infty\}$$

with the norm

$$\|x\|_E=\sum_{n=1}^{\infty}|x_n|.$$

Set

$$x=(x_1,x_2,\ldots,x_n,\ldots)\text{ and }f=(f_1,f_2,\ldots,f_n,\ldots).$$

$$f_n(t,x_n)=\frac{1}{3e^{t+4}}(1+|x_n|),\ t\in[0,1].$$

For each $x_n\in\mathbb{R}$ and $t\in[0,1]$ we have

$$|f_n(t,x_n)|\leq\frac{1}{3e^{t+4}}(1+|x_n|). \tag{11.67}$$

Hence conditions (11.35.1), (11.35.2) and (11.35.3) are satisfied with

$$p(t)=\frac{1}{3e^{t+4}},\ t\in[0,1],$$

and

$$\psi(u)=1+u,\ u\in[0,\infty).$$

By (11.67), for any bounded set $B\subset l^1$, we have

$$\beta(f(t,B))\leq\frac{1}{3e^{t+4}}\beta(B),\ \text{for each } t\in[0,1].$$

Hence (11.35.5) is satisfied. From (11.62) we have

$$G(t,s)=\begin{cases}\frac{-(1-s)^\alpha+\alpha(t-s)^{\alpha-1}}{\Gamma(\alpha+1)}+\frac{(1-s)^{\alpha-1}}{\Gamma(\alpha)}, & \text{if } 0\leq s<t,\\ \frac{-(1-s)^\alpha}{\Gamma(\alpha+1)}+\frac{(1-s)^{\alpha-1}}{\Gamma(\alpha)}, & \text{if } t\leq s<1.\end{cases} \tag{11.68}$$

Using (11.68) and the fact that $\int_0^1 G(t,s)ds=\int_0^t G(t,s)ds+\int_t^1 G(t,s)ds$ we get $\tilde{G}<7$ and (11.35.4) is satisfied for $R>\frac{7}{3e^4-7}$. It is also clear to see that (11.63) is satisfied. Consequently Theorem 11.56 implies that problem (11.65)-(11.66) has a solution defined on $[0,1]$.

11.6. Weak Solutions for Hyperbolic Partial Differential Equations in Banach Spaces

11.6.1. Introduction

In this section, we will investigate the existence of solutions of an initial value problem (IVP for short) for the hyperbolic fractional differential equation of the form

$$({}^cD_0^\alpha u)(x,y) = f(x,y,u(x,y)), \quad (x,y) \in I := [0,a] \times [0,b], \tag{11.69}$$

$$u(x,0) = \phi(x), \quad u(0,y) = \psi(y),\ x \in [0,a],\ y \in [0,b], \tag{11.70}$$

where ${}^cD_0^\alpha$ is the Caputo fractional derivative of order $\alpha = (\alpha_1, \alpha_2) \in (0,1] \times (0,1]$, $f : I \times E \to E$ is a given function satisfying some assumptions that will be specified later, $\phi : [0,a] \to E$ and $\psi : [0,b] \to E$ are absolutely continuous functions with $\phi(0) = \psi(0)$, a, $b > 0$, and E is a Banach space with norm $\|.\|$.
Next we investigate the following nonlocal hyperbolic fractional differential equation

$$({}^cD_0^\alpha u)(x,y) = f(x,y,u(x,y)), \quad (x,y) \in I, \tag{11.71}$$

$$u(x,0) + Q(u) = \phi(x),\ u(0,y) + K(u) = \psi(y),\ x \in [0,a],\ y \in [0,b], \tag{11.72}$$

where f, ϕ, ψ are as in problem (11.69)-(11.70) are $Q, K : C(I,E) \to E$ are continuous functions.

11.6.2. Existence of Solutions

Definition 11.57. *A function $u \in AC(I,E)$ such that $\frac{\partial^2 u}{\partial x \partial y}$ exists and it is integrable on I is said to be a solution of (11.69)-(11.70) if u satisfies equation (11.69) a.e. on I and condition (11.70).*

Let $h \in L^1(I,E)$ and consider the following linear problem

$$\begin{cases} ({}^cD_0^r u)(x,y) = h(x,y),\ (x,y) \in I \\ u(x,0) = \phi(x),\ u(0,y) = \psi(y). \end{cases} \tag{11.73}$$

Set

$$\tilde{h}(x,y) = \phi(x) + \psi(y) - \phi(0),\ x \in [0,a],\ y \in [0,b]. \tag{11.74}$$

To establish our main result concerning existence of solutions of (11.69)-(11.70) we impose suitable conditions on the functions involved in that problem namely we assume that

(11.70.1) For a.e. $(x,y) \in I$, $f(x,y,\cdot)$ is weakly sequentially continuous.

(11.70.2) For each $u \in C(I,E)$, $f(\cdot,\cdot,u(\cdot,\cdot))$ is Pettis integrable a.e. on I.

(11.70.3) There exists $p \in L^\infty(I, \mathbb{R}_+)$ such that

$$\|f(x,y,u)\| \le p(x,y), \text{ for a.e. } (x,y) \in I, \text{ and each } u \in E.$$

(11.70.4) For each bounded set $B \subset E$ and for each $(x,y) \in J$ we have

$$\beta(f(x,y,B)) \leq p(x,y)\beta(B).$$

Let $p^* = \|p\|_{L^\infty}$. Now we are in the position to formulate and prove the following existence result.

Theorem 11.58. *Assume that assumptions* $(11.70.1)-(11.70.4)$ *are satisfied. If*

$$\frac{2p^* a^{\alpha_1} b^{\alpha_2}}{\Gamma(\alpha_1+1)\Gamma(\alpha_2+1)} < 1 \tag{11.75}$$

then the initial value problem (11.69)-(11.70) has at least one solution.

Proof. We shall reduce the existence of solutions of the problem (7.26)-(7.27) to fixed point problem. To this end we consider the operator $T : C(I,E) \to C(I,E)$ defined by

$$\begin{aligned}(Tu)(x,y) &= \frac{1}{\Gamma(\alpha_1)\Gamma(\alpha_2)}\int_0^x\int_0^y (x-s)^{\alpha_1-1}(y-t)^{\alpha_2-1}f(s,t,u(s,t))dtds \\ &\quad +\tilde{h}(x,y).\end{aligned}$$

where $\tilde{h}(\cdot,\cdot)$ is the function defined by (11.74). Clearly the fixed points of the operator T are solutions of the problem (11.69)-(11.70). First notice that, for all $u \in C(I,E)$, $f(\cdot,\cdot,u(\cdot,\cdot)) \in P(I,E)$ (11.70.2). From (11.70.3) $(x-\cdot)^{\alpha_1-1}(y-\cdot)^{\alpha_2-1}f(\cdot,\cdot,u(\cdot,\cdot)), \forall (x,y) \in I$ is Pettis integrable and thus, the operator T makes sense. Let $R > 0$ be such that

$$R \geq \|\tilde{h}\|_\infty + \frac{p^* a^{\alpha_1} b^{\alpha_2}}{\Gamma(\alpha_1+1)\Gamma(\alpha_2+1)}, \tag{11.76}$$

and consider the set

$$\begin{aligned}Q &= \{u \in C(I,E) : \|u\|_\infty \leq R \\ &\quad \text{and } \|u(x_1,y_1)-u(x_2,y_2)\| \leq |\phi(x_1)-\phi(x_2)| + |\psi(x_1)-\psi(x_2)| \\ &\quad + \frac{p^*}{\Gamma(\alpha_1+1)\Gamma(\alpha_2+1)}[2(x_2-x_1)^{\alpha_1}(y_2-y_1)^{\alpha_2} + x_1^{\alpha_1}y_1^{\alpha_2} - x_2^{\alpha_1}y_2^{\alpha_2}] \\ &\quad \text{for } (x_1,y_1),(x_2,y_2) \in I\}.\end{aligned}$$

Clearly, the subset Q is closed, convex and equicontinuous. We shall show that T satisfies the assumptions of Theorem 1.64

Step 1: T maps Q into itself.
Takes $u \in Q, (x,y) \in I$ and assume that $Tu(x,y) \neq 0$. Then there exists $\varphi \in E^*$ such that $\|Tu(x,y)\| = \varphi(Tu(x,y))$. Thus

$$\begin{aligned}\|Tu(x,y)\| &= \varphi(Tu(x,y)) \\ &= \varphi(\tilde{h}(x,y) \\ &\quad + \frac{1}{\Gamma(\alpha_1)\Gamma(\alpha_2)}\int_0^x\int_0^y (x-s)^{\alpha_1-1}(y-t)^{\alpha_2-1}f(s,t,u)dtds\Big)\end{aligned}$$

$$
\begin{aligned}
&= \varphi(\tilde{h}(x,y)) + \varphi\left(\frac{1}{\Gamma(\alpha_1)\Gamma(\alpha_2)}\int_0^x\int_0^y (x-s)^{\alpha_1-1}(y-t)^{\alpha_2-1}f(s,t,u)dtds\right)\\
&\leq |\tilde{h}(x,y)|\\
&\quad +\frac{1}{\Gamma(\alpha_1)\Gamma(\alpha_2)}\int_0^x\int_0^y (x-s)^{\alpha_1-1}(y-t)^{\alpha_2-1}p(s,t)dtds\\
&\leq \|\tilde{h}\|_\infty + \frac{p^*}{\Gamma(\alpha_1)\Gamma(\alpha_2)}\int_0^x\int_0^y (x-s)^{\alpha_1-1}(y-t)^{\alpha_2-1}dtds\\
&\leq \|\tilde{h}\|_\infty + \frac{p^*}{\Gamma(\alpha_1+1)\Gamma(\alpha_2+1)}a^{\alpha_1}b^{\alpha_2}\\
&\leq R.
\end{aligned}
$$

Let $(x_1,y_1),(x_2,y_2)\in I, x_1<x_2, y_1<y_2$ and $u\in Q$ with

$$Tu(x_1,y_1)-Tu(x_2,y_2)\neq 0.$$

Then there exist $\varphi\in E^*$ with $\|\varphi\|=1$ such that

$$
\begin{aligned}
&\|Tu(x_1,y_1)-Tu(x_2,y_2)\|\\
=\;& \varphi(Tu(x_1,y_1)-Tu(x_2,y_2))\\
=\;& \varphi(\phi(x_1)-\phi(x_2)+\psi(y_1)-\psi(y_2)\\
&+\frac{1}{\Gamma(\alpha_1)\Gamma(\alpha_2)}\int_{x_1}^{x_2}\int_{y_1}^{y_2}(x_2-s)^{\alpha_1-1}(y_2-t)^{\alpha_2-1}f(s,t,u)dtds\\
&+\frac{1}{\Gamma(\alpha_1)\Gamma(\alpha_2)}\int_0^{x_1}\int_0^{y_1}[(x_2-s)^{\alpha_1-1}(y_2-t)^{\alpha_2-1}\\
-\;& (x_1-s)^{\alpha_1-1}(y_1-t)^{\alpha_2-1}]f(s,t,u)dtds)\\
\leq\;& |\phi(x_1)-\phi(x_2)+\psi(y_1)-\psi(y_2)|\\
&+\frac{p^*}{\Gamma(\alpha_1)\Gamma(\alpha_2)}\int_{x_1}^{x_2}\int_{y_1}^{y_2}(x_2-s)^{\alpha_1-1}(y_2-t)^{\alpha_2-1}dtds\\
&+\frac{p^*}{\Gamma(\alpha_1)\Gamma(\alpha_2)}\int_0^{x_1}\int_0^{y_1}[(x_1-s)^{\alpha_1-1}(y_1-t)^{\alpha_2-1}\\
-\;& (x_2-s)^{\alpha_1-1}(y_2-t)^{\alpha_2-1}]dtds\\
\leq\;& |\phi(x_1)-\phi(x_2)|+|\psi(y_1)-\psi(y_2)|\\
&+\frac{p^*}{\Gamma(\alpha_1+1)\Gamma(\alpha_2+1)}(x_2-x_1)^{\alpha_1}(y_2-y_1)^{\alpha_2}\\
&+\frac{p^*}{\Gamma(\alpha_1+1)\Gamma(\alpha_2+1)}[(x_2-x_1)^{\alpha_1}(y_2-y_1)^{\alpha_2}+{x_1}^{\alpha_1}{y_1}^{\alpha_2}-{x_2}^{\alpha_1}{y_2}^{\alpha_2}]\\
\leq\;& |\phi(x_1)-\phi(x_2)|+|\psi(y_1)-\psi(y_2)|\\
&+\frac{p^*}{\Gamma(\alpha_1+1)\Gamma(\alpha_2+1)}[2(x_2-x_1)^{\alpha_1}(y_2-y_1)^{\alpha_2}+{x_1}^{\alpha_1}{y_1}^{\alpha_2}-{x_2}^{\alpha_1}{y_2}^{\alpha_2}].
\end{aligned}
$$

Thus $T(Q)\subset Q$.

Step 2: T is *wk*-sequentially continuous.
Let (u_n) be a sequence in Q and let $(u_n(x,y))\to u(x,y)$ in (E,ω) for each $(x,y)\in I$.

Fix $(x,y) \in I$, since f satisfies Assumption (11.70.1), we have $f(x,y,u_n(x,y))$ converging weakly uniformly to $f(x,y,u(x,y))$. Hence the Lebesgue Dominated Convergence Theorem for Pettis integral implies $Tu_n(x,y)$ converging weakly uniformly to $Tu(x,y)$ in (E,ω). We do it for each $(x,y) \in I$ so $Tu_n \to Tu$. Then $T : Q \to Q$ is *wk*-sequentially continuous.

Step 3: The implication (1.6) holds.
Let V be a subset of Q such that $\overline{V} = \overline{conv}(T(V) \cup \{0\})$. Obviously $V(x,y) \subset \overline{conv}(T(V(x,y)) \cup \{0\}), \forall (x,y) \in I$. Further, as V is bounded and equicontinuous , by Ambrosetti Lemma (cf.[118], Lemma 3) the function $(x,y) \to \upsilon(x,y) = \beta(V(x,y))$ is continuous on I. Since functions ϕ and ψ is continuous on I, the set $\overline{\{\phi(x)+\psi(y)-\phi(0),(x,y) \in I\}} \subset E$ is compact. By (11.70.3) and the properties of the measure β, for any $(x,y) \in I$ we have

$$\begin{aligned}
\upsilon(x,y) &\leq \beta(T(V)(x,y) \cup \{0\}) \\
&\leq \beta(T(V)(x,y)) \\
&\leq \frac{2}{\Gamma(\alpha_1)\Gamma(\alpha_2)} \int_0^x \int_0^y (x-s)^{\alpha_1-1}(y-t)^{\alpha_2-1} p(s,t)\beta(V(s,t))dtds \\
&\leq \frac{2}{\Gamma(\alpha_1)\Gamma(\alpha_2)} \int_0^x \int_0^y (x-s)^{\alpha_1-1}(y-t)^{\alpha_2-1} p(s,t)\upsilon(s,t)dtds \\
&\leq \|\upsilon\|_\infty \frac{2}{\Gamma(\alpha_1)\Gamma(\alpha_2)} \int_0^x \int_0^y (x-s)^{\alpha_1-1}(y-t)^{\alpha_2-1} p(s,t)ftds \\
&\leq \|\upsilon\|_\infty \frac{2p^* a^{\alpha_1} b^{\alpha_2}}{\Gamma(\alpha_1+1)\Gamma(\alpha_2+1)}.
\end{aligned}$$

This means that

$$\|\upsilon\|_\infty \leq \|\upsilon\|_\infty \frac{2p^* a^{\alpha_1} b^{\alpha_2}}{\Gamma(\alpha_1+1)\Gamma(\alpha_2+1)}.$$

By (11.75) it follows that $\|\upsilon\|_\infty = 0$, that is $\upsilon(x,y) = \beta(V(x,y)) = 0$ for each $(x,y) \in I$ and then by Theorem 2 in [277] V is weakly relatively compact in $C(I,E)$. Applying now Theorem 1.64 we conclude that T has a fixed point which is a solution of the problem (11.69)-(11.70). □

Now we present an existence result for the nonlocal problem (11.71)-(11.72).

Definition 11.59. *A function $u \in AC(I,E)$ is said to be a solution of (11.71)-(11.72) if u satisfies equations (11.71) and (11.72) on I.*

Theorem 11.60. *Assume (11.70.1)-(11.70.4) and the following conditions*

(11.73.1) There exists $\tilde{k} > 0$ such that

$$\|Q(u)\| \leq \tilde{k}\|u\|_\infty, \text{ for each } u \in C(I,E)$$

(11.73.2) There exists $k^* > 0$ such that

$$\|K(u)\| \leq k^*\|u\|_\infty, \text{ for each } u \in C(I,E)$$

(11.73.3) For any bounded set $B \subset C(I,E)$

$$\alpha(Q(B)) \leq \tilde{k}\alpha(B);$$

(11.73.4) For any bounded set $B \subset C(I,E)$

$$\alpha(K(B)) \leq k^*\alpha(B),$$

hold. If

$$\tilde{k} + k^* + \frac{2p^* a^{\alpha_1} b^{\alpha_2}}{\Gamma(\alpha_1+1)\Gamma(\alpha_2+1)} < 1,$$

then there exists a solution for problem (11.71)-(11.72) on I.

Proof. Consider the operator $\widetilde{T} : C(I,E) \to C(I,E)$ defined by

$$\begin{aligned}(\widetilde{T}u)(x,y) &= \frac{1}{\Gamma(\alpha_1)\Gamma(\alpha_2)} \int_0^x \int_0^y (x-s)^{\alpha_1-1}(y-t)^{\alpha_2-1} f(s,t,u(s,t))dtds \\ &\quad + \tilde{h}(x,y) - Q(u) - K(u).\end{aligned}$$

where $\tilde{h}(\cdot,\cdot)$ is the function defined by (11.74). Clearly the fixed points of the operator $\widetilde{T}$ are solutions of the problem (11.71)-(11.72). Let $\tilde{R} > 0$ be such that

$$\tilde{R} \geq \|\tilde{h}\|_\infty + \|Q\|_\infty + \|K\|_\infty + \frac{p^* a^{\alpha_1} b^{\alpha_2}}{\Gamma(\alpha_1+1)\Gamma(\alpha_2+1)} \tag{11.77}$$

and consider the set

$$\begin{aligned}\tilde{Q} &= \{u \in C(I,E) : \|u\|_\infty \leq \tilde{R} \\ &\quad \text{and } \|u(x_1,y_1) - u(x_2,y_2)\| \leq |\phi(x_1) - \phi(x_2)| + |\psi(x_1) - \psi(x_2)| \\ &\quad + \frac{p^*}{\Gamma(\alpha_1+1)\Gamma(\alpha_2+1)}[2(x_2-x_1)^{\alpha_1}(y_2-y_1)^{\alpha_2} + x_1^{\alpha_1}y_1^{\alpha_2} - x_2^{\alpha_1}y_2^{\alpha_2}] \\ &\quad \text{for } (x_1,y_1),(x_2,y_2) \in I\}.\end{aligned}$$

Clearly, the subset $\tilde{Q}$ is closed, convex and equicontinuous. We can easily show (as in Theorem 11.58) that $\tilde{T}$ satisfies the assumptions of Theorem 1.64. □

11.6.3. An Example

As an application of our results we consider the following partial hyperbolic fractional differential equations of the form

$$({}^cD_0^\alpha u_n)(x,y) = \frac{1}{3e^{x+y+4}} \frac{u_n(x,y)}{1+|u_n(x,y)|}, \text{ if } (x,y) \in J = [0,1] \times [0,1], \tag{11.78}$$

$$u_n(x,0) = \frac{1}{2^n}x,\ u_n(0,y) = \frac{1}{2^n}y^2,\ x \in [0,1],\ y \in [0,1],\ n = 1,2,\ldots. \tag{11.79}$$

Let

$$E = l^1 = \{u = (u_1,u_2,\ldots,u_n,\ldots) : \sum_{n=1}^\infty |u_n| < \infty\}$$

with the norm

$$\|u\|_E = \sum_{n=1}^\infty |u_n|.$$

Set

$$u=(u_1,u_2,\ldots,u_n,\ldots) \text{ and } f=(f_1,f_2,\ldots,f_n,\ldots).$$

$$f_n(x,y,u_n)=\frac{1}{3e^{x+y+4}}\frac{u_n}{1+|u_n|},\ (x,y)\in[0,1]\times[0,1].$$

A simple computations show that conditions (11.38.1)-(11.38.4) are satisfied. Theorem 1.64 implies that problem (11.78)-(11.79) has a solution defined on $[0,1]\times[0,1]$.

11.7. Pettis Integral Equations with Multiple Time Delay in Banach Spaces

11.7.1. Introduction

In this section, we will investigate the existence of solutions for the following fractional integral equation

$$u(x,y)=\sum_{i=1}^{m} g_i(x,y)u(x-\xi_i,y-\mu_i)+I_\theta^\alpha f(x,y,u(x,y));\quad (x,y)\in J_a\times J_b. \tag{11.80}$$

$$u(x,y)=\Psi(x,y);\quad (x,y)\in\tilde{J}=[-\xi,a]\times[-\mu,b]\backslash(0,a]\times(0,b], \tag{11.81}$$

where $J_a=[0,a]$, $J_b=[0,b]$ for $a,b>0$, $\theta=(0,0)$, $\xi=\max_{i=1\ldots m}\{\xi_i\},\mu=\max_{i=1\ldots m}\{\mu_i\}$, I_θ^α is the left sided mixed Pettis integral of order α, $\alpha=(\alpha_1,\alpha_2)\in(0,\infty)\times(0,\infty)$, $f:J_a\times J_b\times E\to E$ is a given function satisfying some assumptions that will be specified later, $g_i:J_a\times J_b\to\mathbb{R}$, $i=1,\ldots,m$ are given continuous functions, and $\Psi:\tilde{J}\to E$ is a given continuous function such that

$$\Psi(0,y)=\sum_{i=1}^{m} g_i(0,y)\Psi(-\xi_i,y-\mu_i);\ y\in[0,b],$$

and

$$\Psi(x,0)=\sum_{i=1}^{m} g_i(x,0)\Psi(x-\xi_i,-\mu_i);\ x\in[0,a],$$

E is a Banach space with norm $\|.\|$.

11.7.2. Existence of Solutions

Definition 11.61. *A function $u\in C(J,E)$ is said to be a solution of (11.80)-(11.81) if u satisfies equation (11.80) on $J_a\times J_b$ and condition (11.81) on $\tilde{J}$*

Set

$$G=\max_{i=1\ldots m}\{\sup_{(x,y)\in J_a\times J_b}|g_i(x,y)|\}.$$

We are now in the position to state and prove our existence result for the problem (11.80)-(11.81). We first list the following hypotheses.

(11.76.1) For each $(x,y)\in J_a\times J_b$, $f(x,y,\cdot)$ is weakly sequentially continuous.

(11.76.2) For each $u \in C(J_a \times J_b, E), f(\cdot,\cdot,u(\cdot,\cdot))$ is Pettis integrable on $J_a \times J_b$.

(11.76.3) There exists $p \in L^\infty(J_a \times J_b, \mathbb{R}_+)$ such that

$$\|f(x,y,u)\| \leq p(x,y), \text{ for } (x,y) \in J_a \times J_b \text{ and each } u \in E.$$

(11.76.4) Let $r_0 > 0$ be arbitrary (but fixed). For any $\varepsilon > 0$ and for any subset $X \subset B_{r_0}$, there exists a closed subset $I_\varepsilon \subset J_a \times J_b$ such that $\mu(J_a \times J_b \backslash I_\varepsilon) < \varepsilon$ and

$$\beta(f(T \times X)) \leq \sup_{(x,y)\in T} p(x,y)\beta(X),$$

for each closed subset T of I_ε, where μ denotes the Lebesgue measure in $\mathbb{R}^2$.

The main result in this section reads as follows.

Theorem 11.62. *Assume that assumptions* (11.76.1) – (11.76.4) *hold. If*

$$mG + \frac{p^* a^{\alpha_1} b^{\alpha_2}}{\Gamma(\alpha_1+1)\Gamma(\alpha_2+1)} < 1, \tag{11.82}$$

where $p^* = \|p\|_\infty$, then problem (11.80)-(11.81) has at least one solution on J.

Proof. To transform the problem (11.80)-(11.81) into a fixed point problem, we define the operator $T : C(J,E) \to C(J,E)$ as

$$T(u)(x,y) = \begin{cases} \Psi(x,y); & (x,y) \in \tilde{J}, \\ \sum_{i=1}^{m} g_i(x,y)u(x-\xi_i, y-\mu_i) & \\ +I_\theta^\alpha f(x,y,u(x,y)); & (x,y) \in J_a \times J_b. \end{cases} \tag{11.83}$$

Now we prove that T satisfies all the assumptions of Theorem 1.64 and thus T has a fixed point which is a solution of problem (11.80)-(11.81). First notice that, for all $u \in C(J_a \times J_b, E)$, $f(x,y,u(x,y))$ is Pettis integrable for a.e. $(x,y) \in J_a \times J_b$ (Assumption (11.76.2)) then $\varphi(f(x,y,u(x,y))) \in L^1(J_a \times J_b)$ for any $\varphi \in E^*$. From the definition of the integral of fractional order we have

$$\begin{aligned} I^\alpha \varphi(f(x,y,u(x,y))) &= \int_0^x \int_0^y \frac{(x-s)^{\alpha_1-1}(y-t)^{\alpha_2-1}}{\Gamma(\alpha_1)\Gamma(\alpha_2)} \varphi(f(s,t,u(s,t)))dtds \\ &= \int_0^x \int_0^y \varphi\left(\frac{(x-s)^{\alpha_1-1}(y-t)^{\alpha_2-1}}{\Gamma(\alpha_1)\Gamma(\alpha_2)} f(s,t,u(s,t))\right) dtds \end{aligned}$$

exists for almost every $(x,y) \in J_a \times J_b$ and is an element from $L^1(J_a \times J_b)$, that is, for almost every $(x,y) \in J_a \times J_b$, $s \in (0,x)$, $t \in (0,y)$ the measurable function

$$\begin{aligned} &\varphi\left(\frac{(x-s)^{\alpha_1-1}(y-t)^{\alpha_2-1}}{\Gamma(\alpha_1)\Gamma(\alpha_2)} f(s,t,u(s,t))\right) \\ &= \frac{(x-s)^{\alpha_1-1}(y-t)^{\alpha_2-1}}{\Gamma(\alpha_1)\Gamma(\alpha_2)} \varphi(f(s,t,u(s,t))) \end{aligned}$$

is Lebesgue integrable, hence the function $(s,t) \to \frac{(x-s)^{\alpha_1-1}(y-t)^{\alpha_2-1}}{\Gamma(\alpha_1)\Gamma(\alpha_2)} f(s,t,u(s,t))$ is Pettis integrable on $J_a \times J_b$, and thus the operator T is well defined. Let $R > 0$, be such that

$$R \geq \frac{p^* a^{\alpha_1} b^{\alpha_2}}{\Gamma(\alpha_1+1)\Gamma(\alpha_2+1)(1-mG)},$$

and consider the set

$$\begin{aligned} Q &= \{u \in C(J,E) : \|u\|_\infty \leq R \\ &\quad \text{and } \|u(x_2,y_2) - u(x_1,y_1)\| \leq R \sum_{i=1}^{m} \|g_i(x_2,y_2) - g_i(x_1,y_1)\| \\ &\quad + \frac{p^*}{\Gamma(\alpha_1+1)\Gamma(\alpha_2+1)} [x_2^{\alpha_1} y_2^{\alpha_2} - x_1^{\alpha_1} y_1^{\alpha_2}]; \text{for } (x_1,y_1),(x_2,y_2) \in J_a \times J_b\}. \end{aligned}$$

Clearly, the subset Q is closed, convex and equicontinuous. The remainder of the proof will be given in three steps.

Step 1: T maps Q into itself.
To see this, take $u \in Q$, $(x,y) \in J_a \times J_b$ and assume that $Tu(x,y) \neq 0$. Then there exists $\varphi \in E^*$ with $\|\varphi\| = 1$ such that $\|Tu(x,y)\| = \varphi(Tu(x,y))$. Thus, we obtain:

$$\begin{aligned} & \|(Tu)(x,y)\| = \varphi(Tu(x,y)) \\ = & \ \varphi\Bigg(\sum_{i=1}^{m} g_i(x,y)u(x-\xi_i, y-\mu_i) \\ & + \frac{1}{\Gamma(\alpha_1)\Gamma(\alpha_2)} \int_0^x \int_0^y (x-s)^{\alpha_1-1}(y-t)^{\alpha_2-1} f(s,t,u(s,t))dtds\Bigg) \\ = & \ \varphi\left(\sum_{i=1}^{m} g_i(x,y)u(x-\xi_i, y-\mu_i)\right) \\ & + \varphi\left(\frac{1}{\Gamma(\alpha_1)\Gamma(\alpha_2)} \int_0^x \int_0^y (x-s)^{\alpha_1-1}(y-t)^{\alpha_2-1} f(s,t,u(s,t))dtds\right) \\ \leq & \ \sum_{i=1}^{m} \|g_i(x,y)\| \|u(x-\xi_i, y-\mu_i)\| \\ & + \frac{1}{\Gamma(\alpha_1)\Gamma(\alpha_2)} \int_0^x \int_0^y (x-s)^{\alpha_1-1}(y-t)^{\alpha_2-1} \|f(s,t,u(s,t))\| dtds \\ \leq & \ mGR + \frac{1}{\Gamma(\alpha_1)\Gamma(\alpha_2)} \int_0^x \int_0^y (x-s)^{\alpha_1-1}(y-t)^{\alpha_2-1} p^* dtds \\ \leq & \ mGR + \frac{p^* a^{\alpha_1} b^{\alpha_2}}{\Gamma(\alpha_1+1)\Gamma(\alpha_2+1)} \\ \leq & \ R. \end{aligned}$$

On the other hand, for $(x,y) \in \tilde{J}$, we have

$$\|(Tu)(x,y)\| = \varphi(Tu(x,y)) = \varphi(\Psi(x,y)) \leq R.$$

Next, suppose that $(x_1,y_1),(x_2,y_2) \in J_a \times J_b$ with $x_1 < x_2$ and $y_1 < y_2$, and let $u \in Q$, so $Tu(x_1,y_1) - Tu(x_2,y_2) \neq 0$. Then there exists $\varphi \in E^*$ such that $\|Tu(x_1,y_1) - Tu(x_2,y_2)\| = \varphi(Tu(x_1,y_1) - Tu(x_2,y_2))$ and $\|\varphi\| = 1$. Thus

$$
\begin{aligned}
&\|Tu(x_2,y_2) - Tu(x_1,y_1)\| \\
&= \varphi\Bigg(\sum_{i=1}^{m} g_i(x_2,y_2)u(x_2-\xi_i,y_2-\mu_i) \\
&\quad +\frac{1}{\Gamma(\alpha_1)\Gamma(\alpha_2)}\int_0^{x_2}\int_0^{y_2}(x_2-s)^{\alpha_1-1}(y_2-t)^{\alpha_2-1}f(s,t,u(s,t))dtds \\
&\quad -\sum_{i=1}^{m} g_i(x_1,y_1)u(x_1-\xi_i,y_1-\mu_i) \\
&\quad +\frac{1}{\Gamma(\alpha_1)\Gamma(\alpha_2)}\int_0^{x_1}\int_0^{y_1}(x_1-s)^{\alpha_1-1}(y_1-t)^{\alpha_2-1}f(s,t,u(s,t))dtds\Bigg) \\
&= \varphi\left(\sum_{i=1}^{m} g_i(x_2,y_2)u(x_2-\xi_i,y_2-\mu_i) - \sum_{i=1}^{m} g_i(x_1,y_1)u(x_1-\xi_i,y_1-\mu_i)\right) \\
&\quad +\varphi\Bigg(\frac{1}{\Gamma(\alpha_1)\Gamma(\alpha_2)}\int_{x_1}^{x_2}\int_{y_1}^{y_2}(x_2-s)^{\alpha_1-1}(y_2-t)^{\alpha_2-1}f(s,t,u(s,t))dtds \\
&\quad +\frac{1}{\Gamma(\alpha_1)\Gamma(\alpha_2)}\int_0^{x_1}\int_0^{y_1}[(x_2-s)^{\alpha_1-1}(y_2-t)^{\alpha_2-1}-(x_1-s)^{\alpha_1-1}(y_1-t)^{\alpha_2-1}] \\
&\quad \times f(s,t,u(s,t))dtds+\frac{1}{\Gamma(\alpha_1)\Gamma(\alpha_2)}\int_0^{x_1}\int_{y_1}^{y_2}(x_2-s)^{\alpha_1-1}(y_2-t)^{\alpha_2-1}f(s,t,u(s,t))dtds \\
&\quad +\frac{1}{\Gamma(\alpha_1)\Gamma(\alpha_2)}\int_{x_1}^{x_2}\int_0^{y_1}(x_2-s)^{\alpha_1-1}(y_2-t)^{\alpha_2-1}f(s,t,u(s,t))dtds\Bigg) \\
&\leq \sum_{i=1}^{m}\|g_i(x_2,y_2)u(x_2-\xi_i,y_2-\mu_i) - g_i(x_1,y_1)u(x_1-\xi_i,y_1-\mu_i)\| \\
&\quad +\frac{p^*}{\Gamma(\alpha_1)\Gamma(\alpha_2)}\int_0^{x_1}\int_0^{y_1}[(x_2-s)^{\alpha_1-1}(y_2-t)^{\alpha_2-1}-(x_1-s)^{\alpha_1-1}(y_1-t)^{\alpha_2-1}]dtds \\
&\quad +\frac{p^*}{\Gamma(\alpha_1)\Gamma(\alpha_2)}\int_{x_1}^{x_2}\int_{y_1}^{y_2}(x_2-s)^{\alpha_1-1}(y_2-t)^{\alpha_2-1}dtds \\
&\quad +\frac{p^*}{\Gamma(\alpha_1)\Gamma(\alpha_2)}\int_0^{x_1}\int_{y_1}^{y_2}(x_2-s)^{\alpha_1-1}(y_2-t)^{\alpha_2-1}dtds \\
&\quad +\frac{p^*}{\Gamma(\alpha_1)\Gamma(\alpha_2)}\int_{x_1}^{x_2}\int_0^{y_1}(x_2-s)^{\alpha_1-1}(y_2-t)^{\alpha_2-1}dtds \\
&\leq R\sum_{i=1}^{m}|g_i(x_2,y_2)-g_i(x_1,y_1)| + \frac{p^*}{\Gamma(\alpha_1+1)\Gamma(\alpha_2+1)}[x_2^{\alpha_1}y_2^{\alpha_2}-x_1^{\alpha_1}y_1^{\alpha_2}].
\end{aligned}
$$

Thus $T(Q) \subset Q$.

Step 2: T is weakly sequentially continuous.
Let (u_n) be a sequence in Q with $(u_n(x,y)) \to u(x,y)$ in (E,ω) for each $(x,y) \in J$. Obviously, $Tu_n \to Tu$ for any $(x,y) \in [-\xi,0]\times[-\mu,0]$. Fix $(x,y) \in J_a \times J_b$, we have

$$\sum_{i=1}^{m} g_i(x,y)u_n(x-\xi_i,y-\mu_i) \to \sum_{i=1}^{m} g_i(x,y)u(x-\xi_i,y-\mu_i).$$

and since $f(x,y,\cdot)$ is weakly sequentially continuous (Assumption (11.76.1))we have immediately $f(x,y,u_n(x,y))$ converging weakly uniformly to $f(x,y,u(x,y))$. Then, Lebesgue Dominated Convergence Theorem for the Pettis integral implies that $Tu_n(x,y)$ converging weakly uniformly to $Tu(x,y)$ in (E,ω). Since this holds, for each $(x,y)\in J_a\times J_b$ we have $Tu_n\to Tu$, i.e $T:Q\to Q$ is weakly sequentially continuous.

Step 3: The implication (1.6) holds.

Let V be a subset of Q such that $\overline{V}=\overline{conv}(T(V)\cup\{0\})$. Obviously $V(x,y)\subset\overline{conv}(T(V(x,y))\cup\{0\}),\forall(x,y)\in J$. Further, as V is bounded and equicontinuous , by Ambrosetti Lemma (cf.[118], Lemma 3) the function $(x,y)\to\upsilon(x,y)=\beta(V(x,y))$ is continuous on J. Since Ψ is continuous on $[-\xi,0]\times[-\mu,0]$, the set $\{\Psi(x,y),(x,y)\in[-\xi,0]\times[-\mu,0]\}\subset E$ is compact. By (11.76.3) and the properties of the measure β, for any $(x,y)\in J_a\times J_b$, we have

$$\begin{aligned}
&\upsilon(x,y)\leq\beta(T(V)(x,y)\cup\{0\})\\
\leq\ &\beta(T(V)(x,y))\\
\leq\ &\beta\left(\left\{\sum_{i=1}^{m}g_i(x,y)u(x-\xi_i,y-\mu_i)\right.\right.\\
&\left.\left.+\frac{1}{\Gamma(\alpha_1)\Gamma(\alpha_2)}\int_0^x\int_0^y(x-s)^{\alpha_1-1}(y-t)^{\alpha_2-1}f(s,t,u(s,t))dtds;u\in V\right\}\right)\\
\leq\ &\beta\left(\left\{\sum_{i=1}^{m}g_i(x,y)u(x-\xi_i,y-\mu_i);u\in V\right\}\right)\\
&+\beta\left(\left\{\frac{1}{\Gamma(\alpha_1)\Gamma(\alpha_2)}\int_0^x\int_0^y(x-s)^{\alpha_1-1}(y-t)^{\alpha_2-1}f(s,t,u(s,t))dtds;u\in V\right\}\right)\\
\leq\ &\sum_{i=1}^{m}\beta(\{g_i(x,y)u(x-\xi_i,y-\mu_i);u\in V\})\\
&+\frac{1}{\Gamma(\alpha_1)\Gamma(\alpha_2)}\beta\left(\left\{\int_0^x\int_0^y(x-s)^{\alpha_1-1}(y-t)^{\alpha_2-1}f(s,t,u(s,t))dtds;u\in V\right\}\right)\\
\leq\ &\sum_{i=1}^{m}g_i(x,y)\beta(V(x,y))\\
&+\frac{1}{\Gamma(\alpha_1)\Gamma(\alpha_2)}\int_0^x\int_0^y(x-s)^{\alpha_1-1}(y-t)^{\alpha_2-1}p(s,t)\beta(V(s,t))dtds\\
\leq\ &mG\|\upsilon\|_\infty+\frac{p^*a^{\alpha_1}b^{\alpha_2}}{\Gamma(\alpha_1+1)\Gamma(\alpha_2+1)}\|\upsilon\|_\infty.
\end{aligned}$$

In particular,

$$\|\upsilon\|_\infty\leq\|\upsilon\|_\infty\left(mG+\frac{p^*a^{\alpha_1}b^{\alpha_2}}{\Gamma(\alpha_1+1)\Gamma(\alpha_2+1)}\right).$$

By (11.82) it follows that $\|\upsilon\|_\infty=0$, that is $\upsilon(x,y)=\beta(V(x,y))=0$ for each $(x,y)\in J$ and then V is weakly relatively compact in $C(J,E)$. Applying now Theorem 1.64 we conclude that T has a fixed point which is a solution of problem (11.80)-(11.81). □

11.8. Notes and Remarks

The results of Chapter 11 are taken from [94, 95, 105, 106, 107]. Other results on the existence of weak solutions for differential and integral equations are reported in [61, 168, 169, 297, 298, 299, 332, 333, 335, 343].

Chapter 12

Weak Solutions for Nonlinear Fractional Differential Inclusions

12.1. Introduction

In this chapter, we consider the existence of weak solutions for some classes of fractional order nonlinear differential inclusions. The main tool is based of the measure of weak noncompactness together with appropriates fixed point theorems.

12.2. Weak Solutions for Boundary-Value Problems with Nonlinear Differential Inclusions

12.2.1. Introduction

This section is concerned with the existence of solutions of the boundary value problem with fractional order differential inclusions and nonlinear integral conditions of the form

$$^cD^{\alpha}x(t) \in F(t,x(t)), \quad \text{for a.e. } t \in J = [0,T],\ 1 < \alpha \leq 2, \tag{12.1}$$

$$x(0) - x'(0) = \int_0^T g(s,x(s))ds, \tag{12.2}$$

$$x(T) + x'(T) = \int_0^T h(s,x(s))ds, \tag{12.3}$$

where $^cD^{\alpha}$, $1 < \alpha \leq 2$ is the Caputo fractional derivative, $F : J \times E \to P(E)$ is a multivalued map, E is a Banach space with the norm $\|\cdot\|$, $P(E)$ is the family of all nonempty subsets of E, and $g,\ h : J \times E \to E$ are given functions satisfying some assumptions that will be specified later.

12.2.2. Existence of Solutions

Definition 12.1. *A function $x \in AC^1(J,E)$ is said to be a solution of (12.1)-(12.3), if there exists a function $\upsilon \in L^1(J,E)$ with $\upsilon(t) \in F(t,x(t))$ for a.e. $t \in J$, such that*

$$^cD^\alpha x(t) = \upsilon(t) \ a.e. \ t \in J, \ \ 1 < \alpha \leq 2,$$

and the function x satisfies the boundary conditions (12.2) and (12.3).

For any $x \in C(J,E)$, we define the set

$$S_{F,x} = \{\upsilon \in L^1(J,E) : \upsilon(t) \in F(t,x(t)) \ \text{ for a.e. } t \in J\}$$

This is known as the set of *selection functions*.
For the existence of solutions to the problem (12.1)-(12.3), we need the following auxiliary lemma.

Lemma 12.2. *Let $1 < \alpha \leq 2$ and let σ, σ_1, $\sigma_2 : J \to E$ be continuous. A function x is a solution of the fractional integral equation*

$$x(t) = P(t) + \int_0^T G(t,s)\sigma(s)ds \tag{12.4}$$

with

$$P(t) = \frac{(T+1-t)}{T+2}\int_0^T \sigma_1(s)ds + \frac{(t+1)}{T+2}\int_0^T \sigma_2(s)ds \tag{12.5}$$

and

$$G(t,s) = \begin{cases} \frac{(t-s)^{\alpha-1}}{\Gamma(\alpha)} - \frac{(1+t)(T-s)^{\alpha-1}}{(T+2)\Gamma(\alpha)} - \frac{(1+t)(T-s)^{\alpha-2}}{(T+2)\Gamma(\alpha-1)}, & 0 \leq s \leq t, \\ -\frac{(1+t)(T-s)^{\alpha-1}}{(T+2)\Gamma(\alpha)} - \frac{(1+t)(T-s)^{\alpha-2}}{(T+2)\Gamma(\alpha-1)}, & t \leq s < T, \end{cases} \tag{12.6}$$

if and only if x is a solution of the fractional boundary value problem

$$^cD^\alpha x(t) = \sigma(t), \quad t \in J,$$

$$x(0) - x'(0) = \int_0^T \sigma_1(s)ds,$$

$$x(T) + x'(T) = \int_0^T \sigma_2(s)ds.$$

Let

$$\tilde{G} = \sup\left\{\int_0^T |G(t,s)|ds, \, t \in J\right\}.$$

We are now in a position to state and prove our existence result for the problem (12.1)-(12.3). We first list the following hypotheses:

(12.3.1) $F : J \times E \to P_{cp,cl,cv}(E)$ has weakly sequentially closed graph.

(12.3.2) For each $t \in J$, $g(t,\cdot)$ and $h(t,\cdot)$ are weakly sequentially continuous.

(12.3.3) For each continuous $x : J \to E$, there exists a scalarly measurable function $\upsilon : J \to E$ with $\upsilon(t) \in F(t,x(t))$ a.e. on J and υ is Pettis integrable on J.

(12.3.4) For each $x \in C(J,E)$, $g(\cdot,x(\cdot))$ and $h(\cdot,x(\cdot))$ are Pettis integrable on J.

(12.3.5) There exist $p \in L^{\infty}(J,\mathbb{R}^+)$ and a continuous nondecreasing function $\psi : [0,\infty) \to [0,\infty)$ such that

$$\|F(t,x)\| = \sup\{|\upsilon| : \upsilon \in F(t,x)\} \le p(t)\psi(\|x\|).$$

(12.3.6) There exist $\phi_g \in L^1(J,\mathbb{R}^+)$ and a continuous nondecreasing function $\psi^* : [0,\infty) \to [0,\infty)$ such that

$$\|g(t,x)\| \le \phi_g(t)\psi^*(\|x\|).$$

(12.3.7) There exist $\phi_h \in L^1(J,\mathbb{R}^+)$ and a continuous nondecreasing function $\bar{\psi} : [0,\infty) \to [0,\infty)$ such that

$$\|h(t,x)\| \le \phi_h(t)\bar{\psi}(\|x\|).$$

(12.3.8) there exists a number $R > 0$ such that

$$\frac{R}{a\psi^*(R) + b\bar{\psi}(R) + c\tilde{G}\psi(R)} > 1, \tag{12.7}$$

where

$$a = \frac{T+1}{T+2}\int_0^T \phi_g(s)ds, \quad b = \frac{T+1}{T+2}\int_0^T \phi_h(s)ds, \quad \text{and} \quad c = \|p\|_{L^\infty}.$$

(12.3.9) For each bounded set $Q \subset E$ and each $t \in J$,

$$\beta(F(t,Q)) \le p(t)\beta(Q), \tag{12.8}$$

$$\beta(g(t,Q)) \le \phi_g(t)\beta(Q), \tag{12.9}$$

$$\beta(h(t,Q)) \le \phi_h(t)\beta(Q). \tag{12.10}$$

Theorem 12.3. *Let E be a Banach space. Assume that hypotheses (12.3.1)-(12.3.9) hold. If*

$$\frac{T+1}{T+2}\int_0^T [\phi_g(s) + \phi_h(s)]ds + \tilde{G}\|p\|_{L^\infty} < 1, \tag{12.11}$$

then the problem (12.1)-(12.3) has at least one solution.

Proof. We transform the problem (12.1)-(12.3) into fixed point problem by considering the multivalued operator $N : C(J,E) \to P_{cl,cv}(C(J,E))$ defined by

$$N(y) = \left\{ h \in C(J,E) : h(t) = P_x(t) + \int_0^T G(t,s)\upsilon(s)ds, \upsilon \in S_{F,x} \right\}, \tag{12.12}$$

where

$$P_x(t) = \frac{T+1-t}{T+2}\int_0^T g(s,x(s))ds + \frac{t+1}{T+2}\int_0^T h(s,x(s))ds$$

and the function $G(\cdot,\cdot)$ is given by (12.6). Clearly, from Lemma 12.2, the fixed points of N are solutions of the problem (12.1)-(12.3). We first show that (12.12) makes sense. To see this, let $x \in C(J,E)$; by (12.3.3) there exists a Pettis integrable function $\upsilon : J \to E$ such that $\upsilon(t) \in F(t,x(t))$ for a.e. $t \in J$. Since $G(t,\cdot) \in L^\infty(J)$, then $G(t,\cdot)\upsilon(\cdot)$ is Pettis integrable and thus N is well defined. Let $R \in \mathbb{R}_+^*$, and consider the set

$$\begin{aligned} Q = \{x \in C(J,E) : \|x\|_\infty \le R \text{ and } \|x(t_1) - x(t_2)\| \\ \le \frac{|t_1-t_2|}{T+2}\psi^*(R)\int_0^T \phi_g(s)ds \\ + \frac{|t_1-t_2|}{T+2}\bar{\psi}(R)\int_0^T \phi_h(s)ds \\ + \|p\|_{L^\infty}\psi(R)\int_0^T \|G(t_2,s)-G(t_1,s)\|ds \text{ for } t_1,t_2 \in J\}. \end{aligned}$$

Notice that Q is a closed, convex, bounded and equicontinuous subset of $C(J,E)$. We shall show that N satisfies the assumptions of Theorem 1.67.

Step 1: $N(x)$ is convex for each $x \in Q$. Indeed, if h_1 and h_2 belong to $N(x)$, then there exists Pettis integrable functions $\upsilon_1(t), \upsilon_2(t) \in F(t,x(t))$ such that, for all $t \in J$, we have

$$h_i(t) = P_x(t) + \int_0^T G(t,s)\upsilon_i(s)ds, \; i = 1,2.$$

Let $0 \le \lambda \le 1$; then, for each $t \in J$, we have

$$(\lambda h_1 + (1-\lambda)h_2)(t) = P_y(t) + \int_0^T G(t,s)[\lambda\upsilon_1(s) - (1-\lambda)\upsilon_2(s)]ds.$$

Since F has convex values, $(\lambda\upsilon_1 + (1-\lambda)\upsilon_2)(t) \in F(t,x(t))$, and we have $\lambda h_1 + (1-\lambda)h_2 \in N(x)$.

Step 2: N maps Q into Q. To see this, take $u \in NQ$. Then there exists $x \in Q$ with $u \in N(x)$ and there exists a Pettis integrable function $\upsilon : J \to E$ with $\upsilon(t) \in F(t,x(t))$ for a.e. $t \in J$. Without loss of generality, we assume $u(s) \neq 0$ for all $s \in J$. Then, there exists $\varphi_s \in E^*$ with $\|\varphi_s\| = 1$ and $\varphi_s(u(s)) = \|u(s)\|$. Hence, for each fixed $t \in J$, we have

$$\begin{aligned} \|u(t)\| &= \varphi_t(u(t)) \\ &= \varphi_t\left(P_x(t) + \int_0^T G(t,s)\upsilon(s)ds\right) \\ &\le \varphi_t(P_x(t)) + \varphi_t\left(\int_0^T G(t,s)\upsilon(s)ds\right) \\ &\le \|P_x(t)\| + \int_0^T \|G(t,s)\|\varphi_t(\upsilon(s))ds \\ &\le \frac{T+1}{T+2}\psi^*(\|x\|_\infty)\int_0^T \phi_g(s)ds + \frac{T+1}{T+2}\bar{\psi}(\|x\|_\infty)\int_0^T \phi_h(s)ds \\ &\quad + \tilde{G}\psi(\|x\|_\infty)\|p\|_{L^\infty}. \end{aligned}$$

Therefore, by (12.5.8),

$$\|u\|_\infty \leq \frac{T+1}{T+2}\psi^*(R)\int_0^T \phi_g(s)ds + \frac{T+1}{T+2}\bar{\psi}(R)\int_0^T \phi_h(s)ds + \tilde{G}\psi(R)\|p\|_{L^\infty} \leq R.$$

Next suppose $u \in NQ$ and $t_1,\ t_2 \in J$ with $t_1 < t_2$ so that $u(t_2) - u(t_1) \neq 0$. Then, there exist $\varphi \in E^*$ such that $\|u(t_2) - u(t_1)\| = \varphi(u(t_2) - u(t_1))$. Thus,

$$\begin{aligned}
\|u(t_2) - u(t_1)\| &= \varphi\left(P_x(t_2) - P_x(t_1) + \int_0^T (G(t_2,s) - G(t_1,s))\upsilon(s)ds\right) \\
&\leq \varphi(P_x(t_2) - P_x(t_1)) + \varphi\left(\int_0^T (G(t_2,s) - G(t_1,s))\upsilon(s)ds\right) \\
&\leq \|P_x(t_2) - P_x(t_1)\| + \int_0^T \|G(t_2,s) - G(t_1,s)\|\|\upsilon(s)\|ds \\
&\leq \frac{(t_2-t_1)}{T+2}\psi^*(R)\int_0^T \phi_g(s)ds + \frac{(t_2-t_1)}{T+2}\bar{\psi}(R)\int_0^T \phi_h(s)ds \\
&\quad + \psi(R)\|p\|_{L^\infty}\int_0^T \|G(t_2,s) - G(t_1,s)\|ds.
\end{aligned}$$

Therefore, $u \in Q$.

Step 3: N has a weakly sequentially closed graph. Let $(x_n, y_n)_1^\infty$ be a sequence in $Q \times Q$ with $x_n(t) \to x(t)$ in (E,ω) for each $t \in J$, $y_n(t) \to y(t)$ in (E,ω) for each $t \in J$, and $y_n \in N(x_n)$ for $n \in \{1,2,\dots\}$. We shall show that $y \in Nx$. By the relation $y_n \in N(x_n)$, we mean that there exists $\upsilon_n \in S_{F,x_n}$ such that

$$y_n(t) = P_{x_n}(t) + \int_0^T G(t,s)\upsilon_n(s)ds.$$

We must show that there exists $\upsilon \in S_{F,x}$ such that, for each $t \in J$,

$$y(t) = P_x(t) + \int_0^T G(t,s)\upsilon(s)ds.$$

Since $F(\cdot,\cdot)$ has compact values, there exists a subsequence υ_{n_m} such that

$$\upsilon_{n_m}(\cdot) \to \upsilon(\cdot) \text{ in } (E,\omega) \text{ as } m \to \infty$$

and

$$\upsilon_{n_m}(t) \in F(t, x_n(t)) \text{ a.e. } t \in J.$$

Since $F(t,\cdot)$ has a weakly sequentially closed graph, $\upsilon \in F(t,x)$. The Lebesgue Dominated Convergence Theorem for the Pettis integral then implies that for each $\varphi \in E^*$,

$$\varphi(y_n(t)) = \varphi\left(P_{x_n}(t) + \int_0^T G(t,s)\upsilon_n(s)ds\right) \to \varphi\left(P_x(t) + \int_0^T G(t,s)\upsilon(s)ds\right)$$

i.e., $y_n(t) \to Nx(t)$ in (E,ω). We can repeat this for each $t \in J$, so $y(t) \in Nx(t)$.

Step 4: The implication (1.6) holds. Now let V be a subset of Q such that $V = \overline{conv}(N(V) \cup \{0\})$. Clearly, $V(t) \subset \overline{conv}(N(V) \cup \{0\})$ for all $t \in J$. Also, $NV(t) \subset NQ(t)$, for each $t \in J$, and is bounded in $P(E)$. By (12.5.9) and the properties of the measure β, we have

$$\begin{aligned}
\beta(NV(t)) &= \beta\left\{P_x(t) + \int_0^T G(t,s)\upsilon(s)ds : \upsilon \in S_{F,x},\ x \in V,\ t \in J\right\} \\
&\leq \beta\{P_x(t) :\ x \in V,\ t \in J\} \\
&\quad + \beta\left\{\int_0^T G(t,s)\upsilon(s)ds : \upsilon \in S_{F,x},\ x \in V,\ t \in J\right\} \\
&\leq \beta\left\{\frac{T+1-t}{T+2}\int_0^T g(s,x(s))ds + \frac{t+1}{T+2}\int_0^T h(s,x(s))ds : x \in V\right\} \\
&\quad + \beta\left\{\int_0^T G(t,s)\upsilon(s)ds : \upsilon(t) \in F(t,x(t)),\ x \in V,\ t \in J\right\} \\
&\leq \int_0^T \frac{T+1-t}{T+2}\phi_g(s)\beta(V(s))ds + \int_0^T \frac{t+1}{T+2}\phi_h(s)\beta(V(s))ds \\
&\quad + \int_0^T \|G(t,s)\|p(s)\beta(V(s))ds \\
&\leq \frac{T+1}{T+2}\int_0^T \phi_g(s)\beta(V(s))ds + \frac{T+1}{T+2}\int_0^T \phi_h(s)\beta(V(s))ds \\
&\quad + \int_0^T \|G(t,s)\|p(s)\beta(V(s))ds.
\end{aligned}$$

for each $t \in J$. This means that

$$\|v\|_\infty \leq \|v\|_\infty \left[\frac{T+1}{T+2}\int_0^T (\phi_g(s) + \phi_h(s))ds + \tilde{G}\int_0^T p(s)ds\right]$$

i.e.,

$$\|v\|_\infty \left[1 - \frac{T+1}{T+2}\int_0^T (\phi_g(s) + \phi_h(s))ds + \tilde{G}\|p\|_\infty\right] \leq 0.$$

By (12.11) it follows that $\|v\|_\infty = 0$. Thus, V is weakly relatively compact. Applying Theorem 1.67, we conclude that N has a fixed point that is a solution of the problem (12.1)-(12.3). □

12.3. Weak Solutions for Hyperbolic Partial Differential Inclusions in Banach Spaces

12.3.1. Introduction

In this section, we consider the hyperbolic fractional differential inclusion of the form

$$({}^cD_0^\alpha u)(x,y) \in F(x,y,u), \quad (x,y) \in J_a \times J_b, \tag{12.13}$$

$$u(x,0) = \phi(x), \quad u(0,y) = \psi(y), \tag{12.14}$$

where ${}^cD_0^\alpha$ is the Caputo fractional derivative of order $\alpha = (\alpha_1,\alpha_2) \in (0,1] \times (0,1]$, $J_a = [0,a]$, $J_b = [0,b]$ for $a,b > 0$, $F : J_a \times J_b \times E \to P(E)$ is a multivalued map, ($P(E)$ is the family of all nonempty subsets of E), $\phi : J_a \to E$ and $\psi : J_b \to E$ are given absolutely continuous functions with $\phi(0) = \psi(0)$, and E is a real Banach space with norm $\|\cdot\|$.

12.3.2. Existence of Solutions

We first define what we mean by a solution of the initial value problem (12.13)-(12.14).

Definition 12.4. *A function $u \in AC(J_a \times J_b, E)$ is said to be a solution of (12.13)-(12.14), if there exists a function $\upsilon \in L^1(J_a \times J_b, E)$ with $\upsilon(x,y) \in F(x,y,u(x,y))$ and such that*

$$^cD^\alpha u(x,y) = \upsilon(x,y), \ a.e. \quad (x,y) \in J_a \times J_b, \ \ \alpha = (\alpha_1, \alpha_2) \in (0,1] \times (0,1],$$

and the function u satisfies conditions (9.2).

Set

$$\tilde{h}(x,y) = \phi(x) + \psi(y) - \phi(0) \tag{12.15}$$

for each $(x,y) \in J_a \times J_b$. We are now in the position to state and prove our existence result for the problem (12.13)-(12.14). We first list the following hypotheses:

(12.6.1) $F : J_a \times J_b \times E \to P_{cp,cl,cv}(E)$ has weakly sequentially closed graph.

(12.6.2) For each continuous $u : J_a \times J_b \to E$, there exists a measurable function $\upsilon : J_a \times J_b \to E$ with $\upsilon(x,y) \in F(x,y,u(x,y))$ a.e. on $J_a \times J_b$ and υ is Pettis integrable on $J_a \times J_b$.

(12.6.3) There exist $p \in L^\infty(J_a \times J_b, \mathbb{R}^+)$ with

$$(s,t) \to (x-s)^{\alpha_1 - 1}(y-t)^{\alpha_2 - 1} p(s,t) \in L^\infty(J_a \times J_b, \mathbb{R}^+),$$

and a continuous nondecreasing function $\bar{\psi} : [0,\infty) \to (0,\infty)$ such that:

$$\|F(x,y,u)\|_{\mathcal{P}} = \sup\{\|\upsilon\| : \upsilon \in F(x,y,u)\} \le p(x,y)\bar{\psi}(\|u\|).$$

(12.6.4) There exists a number $R > 0$ such that

$$\|\tilde{h}\|_\infty + \frac{p^* \bar{\psi}(R) a^{\alpha_1} b^{\alpha_2}}{\Gamma(\alpha_1 + 1)\Gamma(\alpha_2 + 1)} \le R, \tag{12.16}$$

where $p^* = \|p\|_\infty$.

(12.6.5) Let $r_0 > 0$ be arbitrary (but fixed). For any $\varepsilon > 0$ and for any subset $X \subset B_{r_0}$, there exists a closed subset $I_\varepsilon \subset J_a \times J_b$ such that $\mu(J_a \times J_b \backslash I_\varepsilon) < \varepsilon$ and

$$\beta(F(T \times X)) \le \sup_{(x,y) \in T} p(x,y) \beta(X),$$

for each closed subset T of I_ε, where μ denotes the Lebesgue measure in $\mathbb{R}^2$.

The following result reads as follows.

Theorem 12.5. *Assume that assumptions* (12.6.1) – (12.6.5) *hold. If*

$$\frac{p^* a^{\alpha_1} b^{\alpha_2}}{\Gamma(\alpha_1 + 1)\Gamma(\alpha_2 + 1)} < 1, \tag{12.17}$$

then the problem (12.13)-(12.14) has at least one solution.

Proof. To transform the problem (12.13)-(12.14) into a fixed point problem, we define a multivalued map $\Omega : C(J_a \times J_b, E) \to P_{cl}(C(J_a \times J_b, E))$ as

$$\begin{aligned}\Omega(u) &= \{h \in C(J_a \times J_b, E) : \\ & h(x,y) = \frac{1}{\Gamma(\alpha_1)\Gamma(\alpha_2)} \int_0^x \int_0^y (x-s)^{\alpha_1-1}(y-t)^{\alpha_2-1}\upsilon(s,t)dtds \\ & +\tilde{h}(x,y), \upsilon \in S_{F,u}\},\end{aligned}$$

where $\tilde{h}(\cdot,\cdot)$ is the function defined by (12.15). Now we prove that Ω satisfies all the assumptions of Theorem 1.65, and thus Ω has a fixed point which is a solution of the problem (12.13)-(12.14). First notice that, for all $u \in C(J_a \times J_b, E)$, there exists a Pettis integrable $\upsilon : J_a \times J_b \to E$ such that $\upsilon(x,y) \in F(x,y,u(x,y))$ for a.e. $(x,y) \in J_a \times J_b$ (Assumption (12.8.2)). Since

$$\|(x-s)^{\alpha_1-1}(y-t)^{\alpha_2-1}\upsilon(s,t)\| \le (x-s)^{\alpha_1-1}(y-t)^{\alpha_2-1}p(s,t)\bar{\psi}(\|u(s,t)\|),$$

then by (12.6.2) and Proposition 11.35, the function $(x-\cdot)^{\alpha_1-1}(y-\cdot)^{\alpha_2-1}\upsilon(\cdot,\cdot)$, for all $(x,y) \in J_a \times J_b$, is Pettis integrable and thus, the operator Ω is well defined. Let $R > 0$ and consider the set

$$\begin{aligned}Q &= \{u \in C(J_a \times J_b, E) : \|u\|_\infty \le R \\ & \text{and } \|u(x_1,y_1) - u(x_2,y_2)\| \le \|\phi(x_1) - \phi(x_2)\| + \|\psi(y_1) - \psi(y_2)\| \\ & + \frac{p^*\bar{\psi}(R)}{\Gamma(\alpha_1+1)\Gamma(\alpha_2+1)}[2y_2^{\alpha_2}(x_2-x_1)^{\alpha_1} + 2x_2^{\alpha_1}(y_2-y_1)^{\alpha_2} \\ & + x_2^{\alpha_1}y_2^{\alpha_2} - x_1^{\alpha_1}y_1^{\alpha_2} - 2(x_2-x_1)^{\alpha_1}(y_2-y_1)^{\alpha_2}]; \text{ for } (x_1,y_1),(x_2,y_2) \in J_a \times J_b\}.\end{aligned}$$

Clearly, the subset Q is closed, convex, bounded and equicontinuous. The remainder of the proof will be given in four steps.

Step 1: $\Omega(u)$ is convex for each $u \in Q$.
For that, let $h_1, h_2 \in \Omega(u)$. Then there exist $\upsilon_1, \upsilon_2 \in S_{F,u}$ such that, for each $(x,y) \in J_a \times J_b$, we have, for $i = 1,2$,

$$h_i(x,y) = \frac{1}{\Gamma(\alpha_1)\Gamma(\alpha_2)} \int_0^x \int_0^y (x-s)^{\alpha_1-1}(y-t)^{\alpha_2-1}\upsilon_i(s,t)dtds + \tilde{h}(x,y).$$

Let $0 \le \lambda \le 1$. Then, for each $(x,y) \in J_a \times J_b$, we have

$$\begin{aligned}[\lambda h_1 + (1-\lambda)h_2](x,y) &= \frac{1}{\Gamma(\alpha_1)\Gamma(\alpha_2)} \int_0^x \int_0^y (x-s)^{\alpha_1-1}(y-t)^{\alpha_2-1}[\lambda\upsilon_1(s,t) \\ &+ (1-\lambda)\upsilon_2(s,t)]dtds + \tilde{h}(x,y).\end{aligned}$$

Since $S_{F,u}$ is convex (because F has convex values), it follows that $\lambda h_1 + (1-\lambda)h_2 \in \Omega(u)$.

Step 2: Ω maps Q into Q.

To see this, take $h \in \Omega Q$. Then there exists $u \in Q$ with $h \in \Omega(u)$, and there exists a Pettis integrable $\upsilon : J_a \times J_b \to E$ with $\upsilon(x,y) \in F(x,y,u(x,y))$ for a.e. $(x,y) \in J_a \times J_b$. We can consider that $h(x,y) \neq 0$. From Hahn-Banach theorem, there exists $\varphi \in E^*$ with $\|\varphi\| = 1$ such that $\varphi(h(x,y)) = \|h(x,y)\|$. Thus by using (12.16), we obtain

$$\begin{aligned}
\|h(x,y)\| &= \varphi(h(x,y)) \\
&= \varphi\left(\tilde{h}(x,y) + \frac{1}{\Gamma(\alpha_1)\Gamma(\alpha_2)}\int_0^x\int_0^y (x-s)^{\alpha_1-1}(y-t)^{\alpha_2-1}\upsilon(s,t)dtds\right) \\
&= \varphi(\tilde{h}(x,y)) + \varphi\left(\frac{1}{\Gamma(\alpha_1)\Gamma(\alpha_2)}\int_0^x\int_0^y (x-s)^{\alpha_1-1}(y-t)^{\alpha_2-1}\upsilon(s,t)dtds\right) \\
&\leq \|\tilde{h}(x,y)\| + \frac{1}{\Gamma(\alpha_1)\Gamma(\alpha_2)}\int_0^x\int_0^y (x-s)^{\alpha_1-1}(y-t)^{\alpha_2-1}p(s,t)\bar{\psi}(R)dtds \\
&\leq \|\tilde{h}\|_\infty + \frac{p^*\bar{\psi}(R)}{\Gamma(\alpha_1)\Gamma(\alpha_2)}\int_0^x\int_0^y (x-s)^{\alpha_1-1}(y-t)^{\alpha_2-1}dtds \\
&\leq \|\tilde{h}\|_\infty + \frac{p^*\bar{\psi}(R)}{\Gamma(\alpha_1+1)\Gamma(\alpha_2+1)}a^{\alpha_1}b^{\alpha_2} \\
&\leq R.
\end{aligned}$$

Next suppose $(x_1,y_1),(x_2,y_2) \in J_a \times J_b$,with $x_1 < x_2$ and $y_1 < y_2$, and let $h \in \Omega(u)$, so $h(x_1,y_1) - h(x_2,y_2) \neq 0$. Then there exists $\varphi \in E^*$ such that $\|h(x_1,y_1) - h(x_2,y_2)\| = \varphi(h(x_1,y_1) - h(x_2,y_2))$ and $\|\varphi\| = 1$. Thus

$$\begin{aligned}
&\|h(x_2,y_2) - h(x_1,y_1)\| = \varphi(\phi(x_1) - \phi(x_2) + \psi(y_1) - \psi(y_2) \\
&+\frac{1}{\Gamma(\alpha_1)\Gamma(\alpha_2)}\int_{x_1}^{x_2}\int_{y_1}^{y_2}(x_2-s)^{\alpha_1-1}(y_2-t)^{\alpha_2-1}\upsilon(s,t)dtds \\
&+\frac{1}{\Gamma(\alpha_1)\Gamma(\alpha_2)}\int_0^{x_1}\int_0^{y_1}[(x_2-s)^{\alpha_1-1}(y_2-t)^{\alpha_2-1} - (x_1-s)^{\alpha_1-1}(y_1-t)^{\alpha_2-1}] \\
&\times\upsilon(s,t)dtds + \frac{1}{\Gamma(\alpha_1)\Gamma(\alpha_2)}\int_0^{x_1}\int_{y_1}^{y_2}(x_2-s)^{\alpha_1-1}(y_2-t)^{\alpha_2-1}\upsilon(s,t)dtds \\
&+\frac{1}{\Gamma(\alpha_1)\Gamma(\alpha_2)}\int_{x_1}^{x_2}\int_0^{y_2}(x_2-s)^{\alpha_1-1}(y_2-t)^{\alpha_2-1}\upsilon(s,t)dtds\Big) \\
\leq\ &\|\phi(x_1) - \phi(x_2) + \psi(y_1) - \psi(y_2)\| \\
&+\frac{p^*\bar{\psi}(R)}{\Gamma(\alpha_1)\Gamma(\alpha_2)}\int_{x_1}^{x_2}\int_{y_1}^{y_2}(x_2-s)^{\alpha_1-1}(y_2-t)^{\alpha_2-1}dtds \\
&+\frac{p^*\bar{\psi}(R)}{\Gamma(\alpha_1)\Gamma(\alpha_2)}\int_0^{x_1}\int_0^{y_1}[(x_1-s)^{\alpha_1-1}(y_1-t)^{\alpha_2-1} - (x_2-s)^{\alpha_1-1}(y_2-t)^{\alpha_2-1}]dtds \\
&+\frac{p^*\bar{\psi}(R)}{\Gamma(\alpha_1)\Gamma(\alpha_2)}\int_0^{x_1}\int_{y_1}^{y_2}(x_2-s)^{\alpha_1-1}(y_2-t)^{\alpha_2-1}dtdt \\
&+\frac{p^*\bar{\psi}(R)}{\Gamma(\alpha_1)\Gamma(\alpha_2)}\int_{x_1}^{x_2}\int_0^{y_1}(x_2-s)^{\alpha_1-1}(y_2-t)^{\alpha_2-1}dtds \\
\leq\ &\|\phi(x_1) - \phi(x_2)\| + \|\psi(y_1) - \psi(y_2)\| \\
&+\frac{p^*\bar{\psi}(R)}{\Gamma(\alpha_1+1)\Gamma(\alpha_2+1)}[2y_2^{\alpha_2}(x_2-x_1)^{\alpha_1} + 2x_2^{\alpha_1}(y_2-y_1)^{\alpha_2} \\
&+x_2^{\alpha_1}y_2^{\alpha_2} - x_1^{\alpha_1}y_1^{\alpha_2} - 2(x_2-x_1)^{\alpha_1}(y_2-y_1)^{\alpha_2}].
\end{aligned}$$

This implies that $h \in Q$. Hence $\Omega Q \subset Q$.

Step 3: Ω has weakly sequentially closed graph.

Let (u_n, w_n) be a sequence in $Q \times Q$, with $u_n(x,y) \to u(x,y)$ in (E,ω) for each $(x,y) \in J_a \times J_b$, $w_n(x,y) \to w(x,y)$ in (E,ω) for each $(x,y) \in J_a \times J_b$, and $w_n \in \Omega(u_n)$ for $n \in \{1,2,\ldots\}$. We show that $w \in \Omega(u)$. Since $w_n \in \Omega(u_n)$, there exists $\upsilon_n \in S_{F,u_n}$ such that

$$w_n(x,y) = \tilde{h}(x,y) + \frac{1}{\Gamma(\alpha_1)\Gamma(\alpha_2)} \int_0^x \int_0^y (x-s)^{\alpha_1-1}(y-t)^{\alpha_2-1}\upsilon_n(s,t)dtds.$$

We show that there exists $\upsilon \in S_{F,u}$ such that, for each $(x,y) \in J_a \times J_b$,

$$w(t) = \tilde{h}(x,y) + \frac{1}{\Gamma(\alpha_1)\Gamma(\alpha_2)} \int_0^x \int_0^y (x-s)^{\alpha_1-1}(y-t)^{\alpha_2-1}\upsilon(s,t)dtds.$$

Since $F(\cdot,\cdot,\cdot)$ has compact values, there exists a subsequence υ_{n_m} such that υ_{n_m} is Pettis integrable,

$$\upsilon_{n_m}(x,y) \in F(x,y,u_n(x,y) \text{ a.e. } (x,y) \in J_a \times J_b,$$
$$\upsilon_{n_m}(\cdot,\cdot) \to \upsilon(\cdot,\cdot) \text{ in } (E,\omega) \text{ as } m \to \infty.$$

As $F(x,y,\cdot)$ has weakly sequentially closed graph, $\upsilon(x,y) \in F(x,y,u(x,y))$. Then Lebesgue's Dominated Convergence Theorem for the Pettis integral implies

$$\varphi(w_n(x,y)) \to \varphi(\tilde{h}(x,y) + \frac{1}{\Gamma(\alpha_1)\Gamma(\alpha_2)} \int_0^x \int_0^y (x-s)^{\alpha_1-1}(y-t)^{\alpha_2-1}\upsilon(s,t)dtds).$$

i.e. $w_n(x,y) \to \Omega u(x,y)$ in (E,ω). Since this holds, for each $(x,y) \in J_a \times J_b$, we have $w \in \Omega(u)$.

Step 4: The implication (1.6) holds.

Let V be a subset of Q, such that $\overline{V} = \overline{conv}(\Omega(V) \cup \{0\})$. Obviously $V(x,y) \subset \overline{conv}(\Omega(V(x,y)) \cup \{0\})$ for each $(x,y) \in J_a \times J_b$. Further, as V is bounded and equicontinuous , the function $(x,y) \to v(x,y) = \beta(V(x,y))$ is continuous on $J_a \times J_b$. Since ϕ and ψ are continuous on $J_a \times J_b$, the set $\overline{\{\phi(x) + \psi(y) - \phi(0), (x,y) \in J_a \times J_b\}} \subset E$ is compact. By (12.8.5) and the properties of the measure β, for any $(x,y) \in J_a \times J_b$ we have

$$\begin{aligned}
\upsilon(x,y) &\leq \beta((\Omega V)(x,y) \cup \{0\}) \\
&\leq \beta((\Omega V)(x,y)) \\
&\leq \beta\{(\Omega u)(x,y) : u \in V\} \\
&\leq \beta\{\frac{1}{\Gamma(\alpha_1)\Gamma(\alpha_2)} \int_0^x \int_0^y (x-s)^{\alpha_1-1}(y-t)^{\alpha_2-1}\upsilon(s,t)dtds \\
&\quad \upsilon(x,y) \in F(x,y,u), u \in V\} \\
&\leq \beta\{\frac{1}{\Gamma(\alpha_1)\Gamma(\alpha_2)} \int_0^x \int_0^y (x-s)^{\alpha_1-1}(y-t)^{\alpha_2-1}F(s,t,V(s,t))dtds\} \\
&\leq \frac{1}{\Gamma(\alpha_1)\Gamma(\alpha_2)} \int_0^x \int_0^y (x-s)^{\alpha_1-1}(y-t)^{\alpha_2-1}p(s,t)\beta(V(s,t))dtds \\
&\leq \|\upsilon\|_\infty \frac{1}{\Gamma(\alpha_1)\Gamma(\alpha_2)} \int_0^x \int_0^y (x-s)^{\alpha_1-1}(y-t)^{\alpha_2-1}p(s,t)dtds \\
&\leq \|\upsilon\|_\infty \frac{p^* a^{\alpha_1} b^{\alpha_2}}{\Gamma(\alpha_1+1)\Gamma(\alpha_2+1)}.
\end{aligned}$$

In particular,

$$\|\upsilon\|_\infty \leq \|\upsilon\|_\infty \frac{p^* a^{\alpha_1} b^{\alpha_2}}{\Gamma(\alpha_1+1)\Gamma(\alpha_2+1)}.$$

By (12.17) it follows that $\|\upsilon\|_\infty = 0$, that is, $\upsilon(x,y) = \beta(V(x,y)) = 0$ for each $(x,y) \in J_a \times J_b$, and then V is weakly relatively compact in $C(J_a \times J_b, E)$. Applying now Theorem 1.64, we conclude that Ω has a fixed point which is a solution of the problem (12.13)-(12.14). □

12.4. Notes and Remarks

The results of Chapter 12 are taken from [97]. Other results can be found in [62].

Chapter 13

Ulam Stabilities for the Darboux Problem for Partial Fractional Differential Equations and Inclusions

13.1. Introduction

In this chapter, we shall present some Ulam's type stability concepts for the Darboux problem of partial differential equations and inclusions of Caputo's fractional order derivative.

13.2. Ulam Stabilities for the Darboux Problem for Partial Fractional Differential and Integro-Differential Equations

13.2.1. Introduction

In this section we discuss the Ulam stabilities for the following fractional partial differential equation

$$^{c}D^{r}_{\theta}u(x,y) = f(x,y,u(x,y)); \text{ if } (x,y) \in J := [0,a] \times [0,b], \qquad (13.1)$$

where $a,b > 0$, $\theta = (0,0)$, $^{c}D^{r}_{\theta}$ is the fractional Caputo derivative of order $r = (r_1,r_2) \in (0,1] \times (0,1]$, $f : J \times E \to E$ is a given function, E is a (real or complex) Banach space.

Next, we consider the following fractional partial integro-differential equation of the form

$$^{c}D^{r}_{\theta}u(x,y) = f(x,y,I^{r}_{\theta}u(x,y),u(x,y)); \text{ if } (x,y) \in J := [0,a] \times [0,b], \qquad (13.2)$$

where I^{r}_{θ} is the left-sided mixed Riemann-Liouville integral of order $r = (r_1,r_2) \in (0,1] \times (0,1]$, $f : J \times E \times E \to E$ is a given function.

Now, we consider the Ulam stability of fractional differential equations (13.1) and (13.2). Let ε be a positive real number and $\Phi : J \rightarrow [0,\infty)$ be a continuous function. We consider the following inequalities

$$\|^c D^r_\theta u(x,y) - f(x,y,u(x,y))\|_E \leq \varepsilon;\ \text{ if } (x,y) \in J. \tag{13.3}$$

$$\|^c D^r_\theta u(x,y) - f(x,y,u(x,y))\|_E \leq \Phi(x,y);\ \text{ if } (x,y) \in J. \tag{13.4}$$

$$\|^c D^r_\theta u(x,y) - f(x,y,u(x,y))\|_E \leq \varepsilon\Phi(x,y);\ \text{ if } (x,y) \in J. \tag{13.5}$$

$$\|^c D^r_\theta u(x,y) - f(x,y,I^r_\theta u(x,y),u(x,y))\|_E \leq \varepsilon;\ \text{ if } (x,y) \in J. \tag{13.6}$$

$$\|^c D^r_\theta u(x,y) - f(x,y,I^r_\theta u(x,y),u(x,y))\|_E \leq \Phi(x,y);\ \text{ if } (x,y) \in J. \tag{13.7}$$

$$\|^c D^r_\theta u(x,y) - f(x,y,I^r_\theta u(x,y),u(x,y))\|_E \leq \varepsilon\Phi(x,y);\ \text{ if } (x,y) \in J. \tag{13.8}$$

In the theory of functional differential equations and inclusions, there are some special kind of data dependence, (see [234, 280, 323, 328]). Following these results in mind, we shall present four types of Ulam stability for fractional differential equations (13.1) and (13.2).

Definition 13.1. *Problem (13.1) [or (13.2)] is Ulam-Hyers stable if there exists a real number $c_f > 0$ such that for each $\varepsilon > 0$ and for each solution $u \in C$ of the inequality (13.3) [or (13.6)], there exists a solution $v \in C$ of problem (13.1) [or (13.2)] with*

$$\|u(x,y) - v(x,y)\|_E \leq \varepsilon c_f;\ (x,y) \in J.$$

Definition 13.2. *Problem (13.1) [or (13.2)] is generalized Ulam-Hyers stable if there exists $\theta_f \in C([0,\infty),[0,\infty))$, $\theta_f(0) = 0$ such that for each $\varepsilon > 0$ and for each solution $u \in C$ of the inequality (13.3) [or (13.6)], there exists a solution $v \in C$ of problem (13.1) [or (13.2)] with*

$$\|u(x,y) - v(x,y)\|_E \leq \theta_f(\varepsilon);\ (x,y) \in J.$$

Definition 13.3. *Problem (13.1) [or (13.2)] is Ulam-Hyers-Rassias stable with respect to Φ if there exists a real number $c_{f,\Phi} > 0$ such that for each $\varepsilon > 0$ and for each solution $u \in C$ of the inequality (13.5) [or (13.8)], there exists a solution $v \in C$ of problem (13.1) [or (13.2)] with*

$$\|u(x,y) - v(x,y)\|_E \leq \varepsilon c_{f,\Phi}\Phi(x,y);\ (x,y) \in J.$$

Definition 13.4. *Problem (13.1) [or (13.2)] is generalized Ulam-Hyers-Rassias stable with respect to Φ if there exists a real number $c_{f,\Phi} > 0$ such that for each solution $u \in C$ of the inequality (13.4) [or (13.7)], there exists a solution $v \in C$ of problem (13.1) [or (13.2)] with*

$$\|u(x,y) - v(x,y)\|_E \leq c_{f,\Phi}\Phi(x,y);\ (x,y) \in J.$$

Remark 13.5. *It is clear that*

(i) Definition 13.1 $\Rightarrow$ Definition 13.2,

(ii) Definition 13.3 $\Rightarrow$ Definition 13.4,

(iii) Definition 13.3 for $\Phi(x,y) = 1 \Rightarrow$ Definition 13.1.

Remark 13.6. *A function $u \in C$ is a solution of the inequality (13.3) if and only if there exist a function $g \in C$ (which depend on u) such that*

(*i*) $\|g(x,y)\|_E \leq \varepsilon$,

(*ii*) ${}^cD_\theta^r u(x,y) = f(x,y,u(x,y)) + g(x,y)$; *if* $(x,y) \in J$.

One can have similar remarks for the inequalities (13.4)–(13.8). So, the Ulam stabilities of the fractional differential equations are some special types of data dependence of the solutions of fractional differential equations

13.2.2. Main Results

In this section, we present conditions for the Ulam stability of problem (13.1).

Lemma 13.7. *If $u \in C$ is a solution of the inequality (13.3) then u is a solution of the following integral inequality*

$$\|u(x,y) - \mu(x,y) - \frac{1}{\Gamma(r_1)\Gamma(r_2)} \int_0^x \int_0^y (x-s)^{r_1-1}(y-t)^{r_2-1} f(s,t,u(s,t))dtds\|_E$$
$$\leq \frac{\varepsilon a^{r_1} b^{r_2}}{\Gamma(1+r_1)\Gamma(1+r_2)}; \text{ if } (x,y) \in J. \tag{13.9}$$

Proof. By Remark 13.6, for $(x,y) \in J$, we have

$${}^cD_\theta^r u(x,y) = f(x,y,u(x,y)) + g(x,y).$$

Then, for $(x,y) \in J$, we get

$$u(x,y) = \mu(x,y) + \frac{1}{\Gamma(r_1)\Gamma(r_2)} \int_0^x \int_0^y (x-s)^{r_1-1}(y-t)^{r_2-1}(g(s,t) + f(s,t,u(s,t)))dtds.$$

Thus, for $(x,y) \in J$ we obtain

$$\left\| u(x,y) - \mu(x,y) - \frac{1}{\Gamma(r_1)\Gamma(r_2)} \int_0^x \int_0^y (x-s)^{r_1-1}(y-t)^{r_2-1} f(s,t,u(s,t))dtds \right\|_E$$
$$= \left\| \frac{1}{\Gamma(r_1)\Gamma(r_2)} \int_0^x \int_0^y (x-s)^{r_1-1}(y-t)^{r_2-1} g(s,t)dtds \right\|_E$$
$$\leq \frac{\varepsilon a^{r_1} b^{r_2}}{\Gamma(1+r_1)\Gamma(1+r_2)}.$$

Hence, we obtain (13.9). □

Remark 13.8. *We have similar results for the solutions of the inequalities (13.4)–(13.8).*

Consider the following Darboux problem of partial differential equations

$$\begin{cases} {}^cD^r_\theta u(x,y) = f(x,y,u(x,y)); \text{ if } (x,y) \in J, \\ u(x,0) = \varphi(x); \ x \in [0,a], \\ u(0,y) = \psi(y); \ y \in [0,b], \\ \varphi(0) = \psi(0), \end{cases} \tag{13.10}$$

where $\varphi : [0,a] \to E$, $\psi : [0,b] \to E$ are given absolutely continuous functions.

Theorem 13.9. *[2] Assume that the following hypotheses hold*

$(13.1.1)$ *The function $f : J \times E \to E$ is continuous,*

$(13.1.2)$ *There exists a constant $l_f > 0$ such that*

$$\|f(x,y,u) - f(x,y,\overline{u})\|_E \leq l_f \|u - \overline{u}\|_E, \textit{ for each } (x,y) \in J, \textit{and each } u, \overline{u} \in E.$$

If

$$\frac{l_f a^{r_1} b^{r_2}}{\Gamma(1+r_1)\Gamma(1+r_2)} < 1, \tag{13.11}$$

then (13.10) has a unique solution on J.

Theorem 13.10. *Assume that the assumptions* $(13.1.1), (13.1.2)$ *and the following hypothesis hold*

$(13.2.1)$ *$\Phi \in L^1(J, [0,\infty))$ and there exists $\lambda_\Phi > 0$ such that, for each $(x,y) \in J$ we have*

$$(I^r_\theta \Phi)(x,y) \leq \lambda_\Phi \Phi(x,y).$$

If the condition (13.11) holds, then equation (13.1) is generalized Ulam-Hyers-Rassias stable.

Proof. Let $u \in C$ be a solution of the inequality (13.4). By Theorem 13.9 there v is a unique solution of the problem (13.10). Then we have

$$v(x,y) = \mu(x,y) + \frac{1}{\Gamma(r_1)\Gamma(r_2)} \int_0^x \int_0^y (x-s)^{r_1-1}(y-t)^{r_2-1} f(s,t,v(s,t))dtds; \ (x,y) \in J.$$

By differential inequality (13.4), for each $(x,y) \in J$, we have

$$\begin{aligned} &\|u(x,y) - \mu(x,y) - \frac{1}{\Gamma(r_1)\Gamma(r_2)} \int_0^x \int_0^y (x-s)^{r_1-1}(y-t)^{r_2-1} f(s,t,u(s,t))dtds\|_E \\ &\leq \frac{1}{\Gamma(r_1)\Gamma(r_2)} \int_0^x \int_0^y (x-s)^{r_1-1}(y-t)^{r_2-1} \Phi(s,t)dtds. \end{aligned}$$

Thus, by $(13.2.1)$ for each $(x,y) \in J$, we get

$$\begin{aligned} &\|u(x,y) - \mu(x,y) - \frac{1}{\Gamma(r_1)\Gamma(r_2)} \int_0^x \int_0^y (x-s)^{r_1-1}(y-t)^{r_2-1} f(s,t,u(s,t))dtds\|_E \\ &\leq \lambda_\Phi \Phi(x,y). \end{aligned}$$

Hence for each $(x,y) \in J$, it follows that

$$\begin{aligned}
\|u(x,y)-v(x,y)\|_E &\leq \|u(x,y)-\mu(x,y) \\
&- \frac{1}{\Gamma(r_1)\Gamma(r_2)} \int_0^x \int_0^y (x-s)^{r_1-1}(y-t)^{r_2-1} f(s,t,u(s,t))dtds\|_E \\
&+ \frac{1}{\Gamma(r_1)\Gamma(r_2)} \int_0^x \int_0^y (x-s)^{r_1-1}(y-t)^{r_2-1} \|f(s,t,u(s,t))-f(s,t,v(s,t))\|_E dtds \\
&\leq \lambda_\Phi \Phi(x,y) + \frac{l_f}{\Gamma(r_1)\Gamma(r_2)} \int_0^x \int_0^y (x-s)^{r_1-1}(y-t)^{r_2-1} \|u(s,t)-v(s,t)\|_E dtds.
\end{aligned}$$

From Lemma 1.68, there exists a constant $\delta = \delta(r_1, r_2)$ such that

$$\begin{aligned}
\|u(x,y)-v(x,y)\|_E &\leq \lambda_\Phi \Phi(x,y) \\
&+ \frac{\delta l_f \lambda_\Phi}{\Gamma(r_1)\Gamma(r_2)} \int_0^x \int_0^y (x-s)^{r_1-1}(y-t)^{r_2-1} \Phi(s,t)dtds \\
&\leq (1+\delta l_f \lambda_\Phi)\lambda_\Phi \Phi(x,y) \\
&:= c_{f,\Phi}\Phi(x,y).
\end{aligned}$$

Finally, equation (13.1) is generalized Ulam-Hyers-Rassias stable. □

Now we present (without proof) the Ulam stability of the equation (13.2).

Theorem 13.11. *Assume that* (13.2.1) *and the following hypotheses hold*

(13.3.1) *The function* $f : J \times E \times E \to E$ *is continuous,*

(13.3.2) *There exist constants* $L_{1,f}, L_{2,f} > 0$ *such that*

$$\|f(x,x,u,v) - f(x,y,\overline{u},\overline{v})\|_E \leq L_{2,f}\|u-\overline{u}\|_E + L_{1,f}\|v-\overline{v}\|_E,$$

for each $(x,y) \in J$ *and* $u, v, \overline{u}, \overline{v} \in E$.

If

$$L_{3,f}(L_{1,f} + L_{2,f}L_{3,f}) < 1, \tag{13.12}$$

where

$$L_{3,f} := \frac{a^{r_1} b^{r_2}}{\Gamma(1+r_1)\Gamma(1+r_2)},$$

then equation (13.2) is generalized Ulam-Hyers-Rassias stable.

13.2.3. An Example

As an application of our results we consider the following partial hyperbolic functional differential equation of the form

$$({}^cD_\theta^r u)(x,y) = \frac{e^{-x-y-3}}{1+5|u(x,y)|}, \text{ if } (x,y) \in J = [0,1] \times [0,1], \tag{13.13}$$

where $(r_1, r_2) \in (0,1] \times (0,1]$. Set

$$f(x,y,u(x,y)) = \frac{1}{(1+5|u(x,y)|)e^{x+y+3}}, \ (x,y) \in [0,1] \times [0,1].$$

For each $u,\ \overline{u} \in \mathbb{R}$ and $(x,y) \in [0,1] \times [0,1]$ we have

$$|f(x,y,u) - f(x,y,\overline{u})| \leq \frac{5}{e^3}|u - \overline{u}|.$$

Hence conditions (13.1.1) and (13.1.2) are satisfied with $l_f = \frac{5}{e^3}$. We shall show that condition (13.11) holds with $a = b = 1$. Indeed

$$\frac{l_f a^{r_1} b^{r_2}}{\Gamma(1+r_1)\Gamma(1+r_2)} = \frac{5}{e^3 \Gamma(1+r_1)\Gamma(1+r_2)} < \frac{20}{e^3} < 1,$$

which is satisfied for each $(r_1, r_2) \in (0,1] \times (0,1]$. The hypothesis (13.2.1) is satisfied with $\Phi(x,y) = xy^2$ and $\lambda_\Phi = \frac{2}{\Gamma(2+r_1)\Gamma(3+r_2)}$. Indeed, for each $(x,y) \in [0,1] \times [0,1]$ we get

$$(I_\theta^r \Phi)(x,y) = \frac{\Gamma(2)\Gamma(3)}{\Gamma(2+r_1)\Gamma(3+r_2)} x^{1+r_1} y^{2+r_2} \leq \frac{2}{\Gamma(2+r_1)\Gamma(3+r_2)} xy^2 = \lambda_\Phi \Phi(x,y).$$

Consequently Theorem 13.10 implies that equation (13.13) is generalized Ulam-Hyers-Rassias stable.

13.3. Ulam-Hyers-Rassias Stabilities for the Darboux Problem for Partial Fractional Implicit Differential Equations

13.3.1. Introduction

This section studied the Ulam stabilities for the following fractional partial implicit differential equation

$$\overline{D}_\theta^r u(x,y) = f(x,y,u(x,y), \overline{D}_\theta^r u(x,y)); \text{ if } (x,y) \in J := [0,a] \times [0,b], \tag{13.14}$$

where $a, b > 0$, $\theta = (0,0)$, $\overline{D}_\theta^r$ is the mixed regularized derivative of order $r = (r_1, r_2) \in (0,1] \times (0,1]$, $f : J \times E \times E \to E$ is a given function. This work initiates the Ulam stabilities of the Darboux problem for hyperbolic implicit differential equations of fractional order. Denote $L^1(J)$ the space of Bochner-integrable functions $u : J \to E$ with the norm

$$\|u\|_{L^1} = \int_0^a \int_0^b \|u(x,y)\|_E dy dx.$$

13.3.2. Main Results

In this section, we present conditions for the Ulam stability of the equation (13.14). Let $\Phi : J \to [0,\infty)$ be a continuous function. We consider the following inequality

$$\|\overline{D}_\theta^r u(x,y) - f(x,y,u(x,y), \overline{D}_\theta^r u(x,y))\|_E \leq \Phi(x,y); \ \text{ if } (x,y) \in J. \tag{13.15}$$

Consider the following Darboux problem of partial differential equations

$$\begin{cases} \overline{D}_\theta^r u(x,y) = f(x,y,u(x,y),\overline{D}_\theta^r u(x,y)); \text{ if } (x,y)\in J, \\ u(x,0)=\varphi(x);\ x\in[0,a], \\ u(0,y)=\psi(y);\ y\in[0,b], \\ \varphi(0)=\psi(0), \end{cases} \tag{13.16}$$

where $\varphi:[0,a]\to E,\ \psi:[0,b]\to E$ are given absolutely continuous functions.

Theorem 13.12. *[50] Assume that the following hypotheses hold*

(13.4.1) *The function $f: J\times E\times E\to E$ is continuous,*

(13.4.2) *There exist constants $k_f>0$ and $0<l_f<1$ such that*

$$\|f(x,y,u,z)-f(x,y,v,w)\|_E \le k_f\|u-v\|_E + l_f\|z-w\|_E,\ u,v,w,z\in E,\ (x,y)\in J.$$

If

$$\frac{k_f a^{r_1} b^{r_2}}{(1-l_f)\Gamma(1+r_1)\Gamma(1+r_2)} < 1, \tag{13.17}$$

then there exists a unique solution for problem (13.16) on J.

Theorem 13.13. *Assume that the assumptions* (13.4.1), (13.4.2) *and the following hypothesis hold*

(13.5.1) *$\Phi\in L^1(J,[0,\infty))$ and there exists $\lambda_\Phi>0$ such that, for each $(x,y)\in J$ we have*

$$(I_\theta^r\Phi)(x,y)\le\lambda_\Phi\Phi(x,y).$$

If the condition (13.17) holds, then equation (13.14) is generalized Ulam-Hyers-Rassias stable.

Proof. Let $u\in C$ be a solution of the inequality (13.15). By Theorem 13.12 there v is a unique solution of the problem (13.16). Then for each $(x,y)\in J$, we have

$$v(x,y)=\mu(x,y)+\frac{1}{\Gamma(r_1)\Gamma(r_2)}\int_0^x\int_0^y (x-s)^{r_1-1}(y-t)^{r_2-1}g_v(s,t)dtds;$$

where $g_v\in C$ such that

$$g_v(x,y)=f(x,y,v(x,y),g_v(x,y)).$$

By differential inequality (13.15), for each $(x,y)\in J$, we have

$$\begin{aligned} &\|u(x,y)-\mu(x,y)-\frac{1}{\Gamma(r_1)\Gamma(r_2)}\int_0^x\int_0^y (x-s)^{r_1-1}(y-t)^{r_2-1}g(s,t)dtds\|_E \\ &\le \frac{1}{\Gamma(r_1)\Gamma(r_2)}\int_0^x\int_0^y (x-s)^{r_1-1}(y-t)^{r_2-1}\Phi(s,t)dtds, \end{aligned}$$

where $g\in C$ such that

$$g(x,y)=f(x,y,u(x,y),g(x,y)).$$

Thus, by $(13.5.1)$ for each $(x,y) \in J$, we get

$$\|u(x,y) - \mu(x,y) - \frac{1}{\Gamma(r_1)\Gamma(r_2)} \int_0^x \int_0^y (x-s)^{r_1-1}(y-t)^{r_2-1} \times g(s,t)dtds\|_E \leq \lambda_\Phi \Phi(x,y).$$

Hence for each $(x,y) \in J$, it follows that

$$\begin{aligned} \|u(x,y) - v(x,y)\|_E \leq & \|u(x,y) - \mu(x,y) \\ & - \frac{1}{\Gamma(r_1)\Gamma(r_2)} \int_0^x \int_0^y (x-s)^{r_1-1}(y-t)^{r_2-1} g(s,t)dtds\|_E \\ & + \frac{1}{\Gamma(r_1)\Gamma(r_2)} \int_0^x \int_0^y (x-s)^{r_1-1}(y-t)^{r_2-1} \\ & \times \|g(s,t) - g_v(s,t)\|_E dtds. \end{aligned}$$

By $(13.4.2)$, we get

$$\|g(x,y) - g_v(x,y)\|_E \leq k_f \|u(x,y) - v(x,y)\|_E + l_f \|g(x,y) - g_v(x,y)\|_E.$$

Then

$$\|g(x,y) - g_v(x,y)\|_E \leq \frac{k_f}{1-l_f} \|u(x,y) - v(x,y)\|_E.$$

Hence

$$\begin{aligned} \|u(x,y) - v(x,y)\|_E \leq & \lambda_\Phi \Phi(x,y) \\ & + \frac{k_f}{(1-l_f)\Gamma(r_1)\Gamma(r_2)} \int_0^x \int_0^y (x-s)^{r_1-1}(y-t)^{r_2-1} \\ & \times \|u(s,t) - v(s,t)\|_E dtds. \end{aligned}$$

From Lemma 1.68, there exists a constant $\delta = \delta(r_1, r_2)$ such that

$$\begin{aligned} \|u(x,y) - v(x,y)\|_E \leq & \lambda_\Phi \Phi(x,y) + \frac{\delta k_f \lambda_\Phi}{(1-l_f)\Gamma(r_1)\Gamma(r_2)} \\ & \times \int_0^x \int_0^y (x-s)^{r_1-1}(y-t)^{r_2-1} \Phi(s,t)dtds \\ & \leq \left(1 + \frac{\delta k_f \lambda_\Phi}{1-l_f}\right) \lambda_\Phi \Phi(x,y) \\ & := c_{f,\Phi} \Phi(x,y). \end{aligned}$$

Finally, equation (13.14) is generalized Ulam-Hyers-Rassias stable. □

13.3.3. An Example

Let

$$E = l^1 = \left\{ w = (w_1, w_2, \ldots, w_n, \ldots) : \sum_{n=1}^{\infty} |w_n| < \infty \right\},$$

be the Banach space with norm

$$\|w\|_E = \sum_{n=1}^{\infty} |w_n|.$$

Consider the following infinite system of partial hyperbolic fractional implicit differential equations of the form

$$\overline{D}_\theta^r u(x,y) = f(x,y,u(x,y),\overline{D}_\theta^r u(x,y)); \text{ if } (x,y) \in [0,1]\times[0,1], \tag{13.18}$$

where $(r_1,r_2) \in (0,1]\times(0,1]$,

$$u = (u_1,u_2,...,u_n,...),\ \overline{D}_\theta^r u = (\overline{D}_\theta^r u_1, \overline{D}_\theta^r u_2, ..., \overline{D}_\theta^r u_n, ...),\ f = (f_1,f_2,...,f_n,...),$$

and

$$f_n(x,y,u,\overline{D}_\theta^r u) = \frac{1}{10e^{x+y+3}(1+|u_n|+|\overline{D}_\theta^r u_n|)};\ (x,y) \in [0,1]\times[0,1] \textit{ and } n \in \mathbb{N}.$$

Clearly, the function f is continuous. For each $n \in \mathbb{N}$, $u,v,\overline{u},\overline{v} \in E$ and $(x,y) \in [0,1]\times[0,1]$ we have

$$|f_n(x,y,u(x,y),v(x,y)) - f_n(x,y,\overline{u}(x,y),\overline{v}(x,y))| \le \frac{1}{10e^3}(|u_n-\overline{u}_n| + |v_n-\overline{v}_n|).$$

Thus, For each $u,v,\overline{u},\overline{v} \in E$ and $(x,y) \in [0,1]\times[0,1]$ we get

$$\begin{aligned}
&\|f(x,y,u(x,y),v(x,y)) - f(x,y,\overline{u}(x,y),\overline{v}(x,y))\|_E \\
&= \sum_{n=1}^{\infty} |f_n(x,y,u(x,y),v(x,y)) - f_n(x,y,\overline{u}(x,y),\overline{v}(x,y))| \\
&\le \frac{1}{10e^3}\left(\sum_{n=1}^{\infty}|u_n-\overline{u}_n| + \sum_{n=1}^{\infty}|v_n-\overline{v}_n|\right) \\
&= \frac{1}{10e^3}(\|u-\overline{u}\|_E + \|v-\overline{v}\|_E).
\end{aligned}$$

Hence condition (13.4.2) is satisfied with $k_f = l_f = \frac{1}{10e^3}$. We shall show that condition (13.17) holds with $a = b = 1$. Indeed

$$\frac{k_f a^{r_1} b^{r_2}}{(1-l_f)\Gamma(1+r_1)\Gamma(1+r_2)} = \frac{1}{(10e^3-1)\Gamma(1+r_1)\Gamma(1+r_2)} < \frac{4}{10e^3-1} < 1,$$

which is satisfied for each $(r_1,r_2) \in (0,1]\times(0,1]$. The hypothesis (13.5.1) is satisfied with $\Phi(x,y) = xy^2$ and $\lambda_\Phi = \frac{2}{\Gamma(2+r_1)\Gamma(3+r_2)}$. Indeed, for each $(x,y) \in [0,1]\times[0,1]$ we get

$$(I_\theta^r \Phi)(x,y) = \frac{\Gamma(2)\Gamma(3)}{\Gamma(2+r_1)\Gamma(3+r_2)} x^{1+r_1} y^{2+r_2} \le \frac{2}{\Gamma(2+r_1)\Gamma(3+r_2)} xy^2 = \lambda_\Phi \Phi(x,y).$$

Consequently Theorem 13.13 implies that equation (13.18) is generalized Ulam-Hyers-Rassias stable.

13.4. Some Stability Concepts for Darboux Problem for Partial Fractional Differential Equations on Unbounded Domain

13.4.1. Introduction

In the present section, we investigate some Ulam's type stability concepts for the Darboux problem of partial fractional differential equations

$$^cD^r_\theta u(x,y) = f(x,y,u(x,y));\ \text{if } (x,y) \in J := [0,\infty) \times [0,\infty), \tag{13.19}$$

where $f : J \times E \to E$ is a given function. We use a nonlinear alternative of Leray-Schauder type for contraction maps in Fréchet spaces and a fractional version of Gronwall's inequality. We prove that the equation (13.19) is generalized Ulam-Hyers-Rassias stable.

For each $p \in \mathbb{N}^*$ define in the set C the semi-norms $\|x\|_p$ defined by

$$\|u\|_p = \sup_{(x,y)\in J_p} \|u(x,y)\|_E.$$

The set C is Fréchet space with the family of semi-norms $\{\|u\|_p\}_{p\in\mathbb{N}^*}$.

Let ε be a positive real number and $\Phi : J \to [0,\infty)$ be a continuous function. We consider the following inequalities

$$\|^cD^r_\theta u(x,y) - f(x,y,u(x,y))\|_E \le \varepsilon;\ \text{if } (x,y) \in J. \tag{13.20}$$

$$\|^cD^r_\theta u(x,y) - f(x,y,u(x,y))\|_E \le \Phi(x,y);\ \text{if } (x,y) \in J. \tag{13.21}$$

13.4.2. Main Results

For each $p \in \mathbb{N}^*$, consider the following Darboux problem of partial differential equations

$$\begin{cases} ^cD^r_\theta u(x,y) = f(x,y,u(x,y));\ \text{if } (x,y) \in J_p, \\ u(x,0) = \varphi(x);\ x \in [0,p], \\ u(0,y) = \psi(y);\ y \in [0,p], \\ \varphi(0) = \psi(0), \end{cases} \tag{13.22}$$

where $\varphi, \psi : [0,p] \to E$ are given absolutely continuous functions.

In the sequel, we need the following Lemma

Lemma 13.14. *[2] Let $0 < r_1, r_2 \le 1$, $\mu(x,y) = \varphi(x) + \psi(y) - \varphi(0)$. A function $u \in C_p$ is a solution of the fractional integral equation*

$$u(x,y) = \mu(x,y) + \frac{1}{\Gamma(r_1)\Gamma(r_2)} \int_0^x \int_0^y (x-s)^{r_1-1}(y-t)^{r_2-1} f(s,t,u(s,t))dtds, \tag{13.23}$$

if and only if u is a solution of the problem (13.22).

Lemma 13.15. *If $u \in C$ is a solution of the inequality (13.20) then u is a solution of the following integral inequality*

$$\|u(x,y)-\mu(x,y)-\frac{1}{\Gamma(r_1)\Gamma(r_2)}\int_0^x\int_0^y (x-s)^{r_1-1}(y-t)^{r_2-1}f(s,t,u(s,t))dtds\|_E$$
$$\leq \frac{\varepsilon\, p^{r_1+r_2}}{\Gamma(1+r_1)\Gamma(1+r_2)};\ if\ (x,y)\in J_p. \tag{13.24}$$

Consider now the following Darboux problem of partial differential equations

$$\begin{cases} {}^cD_\theta^r u(x,y) = f(x,y,u(x,y));\ \text{if } (x,y)\in J, \\ u(x,0)=\varphi(x);\ x\in[0,\infty), \\ u(0,y)=\psi(y);\ y\in[0,\infty), \\ \varphi(0)=\psi(0), \end{cases} \tag{13.25}$$

where $\varphi,\psi:[0,\infty)\to E$ are given absolutely continuous functions.

Now, we make use of the nonlinear alternative of Leray-Schauder type for contraction maps on Fréchet spaces Theorem 1.66, for proving the uniqueness of the solution of the system (13.25).

Theorem 13.16. *Assume that the following hypotheses hold*

(13.6.1) *The function $f: J\times E\times E\to E$ is continuous,*

(13.6.2) *For each $p\in \mathbb{N}^*$, there exists $l_{p,f}\in C_p$ such that for each $(x,y)\in J_p$*

$$\|f(x,y,u)-f(x,y,v)\|_E \leq l_{p,f}(x,y)\|u-v\|_E,\ for\ each\ u,\ v\in E.$$

If

$$\frac{l^*_{p,f}\,p^{r_1+r_2}}{\Gamma(1+r_1)\Gamma(1+r_2)} < 1, \tag{13.26}$$

where

$$l^*_{p,f} = \sup_{(x,y)\in J_p} l_{p,f}(x,y),$$

then the system (13.25) has a unique solution on J.

Proof. Transform the problem (13.25) into a fixed point problem. Consider the operator $N: C_p\to C_p$ defined by,

$$(Nu)(x,y) = \begin{cases} \phi(x,y), & (x,y)\in\tilde{J}, \\ \mu(x,y) & \\ +\frac{1}{\Gamma(r_1)\Gamma(r_2)}\int_0^x\int_0^y (x-s)^{r_1-1}(y-t)^{r_2-1}f(s,t,u(s,t))dtds, & (x,y)\in J. \end{cases}$$

Let u be a possible solution of the problem $u=\lambda N(u)$ for some $0<\lambda<1$. This implies that for each $(x,y)\in J_p$, we have

$$u(x,y)=\lambda\mu(x,y)+\frac{\lambda}{\Gamma(r_1)\Gamma(r_2)}\int_0^x\int_0^y (x-s)^{r_1-1}(y-t)^{r_2-1}f(s,t,u(s,t))dtds.$$

This implies by (13.6.2) that for each $(x,y) \in J_p$, we have

$$\begin{aligned}
\|u(x,y)\|_E &\leq \|\mu(x,y)\|_E + \frac{1}{\Gamma(r_1)\Gamma(r_2)}\int_0^x\int_0^y (x-s)^{r_1-1}(y-t)^{r_2-1}\|f(s,t,0)\|_E dtds \\
&+ \frac{1}{\Gamma(r_1)\Gamma(r_2)}\int_0^x\int_0^y (x-s)^{r_1-1}(y-t)^{r_2-1}\|f(s,t,u(s,t))-f(s,t,0)\|_E dtds \\
&\leq \|\mu(x,y)\|_E + \frac{f^* p^{r_1+r_2}}{\Gamma(1+r_1)\Gamma(1+r_2)} \\
&+ \frac{1}{\Gamma(r_1)\Gamma(r_2)}\int_0^x\int_0^y (x-s)^{r_1-1}(y-t)^{r_2-1} l_{p,f}(s,t)\|u(s,t)\|_E dtds \\
&\leq \|\mu\|_p + \frac{f^* p^{r_1+r_2}}{\Gamma(1+r_1)\Gamma(1+r_2)} \\
&+ \frac{l^*_{p,f}}{\Gamma(r_1)\Gamma(r_2)}\int_0^x\int_0^y (x-s)^{r_1-1}(y-t)^{r_2-1}(s,t)\|u(s,t)\|_E dtds,
\end{aligned}$$

where

$$f^* = \sup_{(x,y)\in J_p} \|f(x,y,0)\|_E.$$

Set

$$w = \|\mu\|_p + \frac{f^* p^{r_1+r_2}}{\Gamma(1+r_1)\Gamma(1+r_2)}.$$

From Lemma 1.68, there exists a constant $\delta = \delta(r_1,r_2)$ such that we have for $(x,y) \in J_p$,

$$\begin{aligned}
\|u(x,y)\|_E &\leq w\left(1 + \frac{\delta l^*_{p,f}}{\Gamma(r_1)\Gamma(r_2)}\int_0^x\int_0^y (x-s)^{r_1-1}(y-t)^{r_2-1} dtds\right) \\
&\leq w\left(1 + \frac{\delta l^*_{p,f} p^{r_1+r_2}}{\Gamma(1+r_1)\Gamma(1+r_2)}\right) := M_p.
\end{aligned}$$

Thus

$$\|u(x,y)\|_E \leq M_p.$$

Set

$$U = \{u \in C_p : \|u\|_p \leq 1 + M_p \text{ for all } p \in \mathbb{N}\}.$$

We shall show that $N : U \longrightarrow C_p$ is a contraction map. Indeed, consider $v,w \in U$. Then for each $(x,y) \in J_p$, we obtain

$$\|N(v) - N(w)\|_p \leq \frac{l^*_{p,f} p^{r_1+r_2}}{\Gamma(1+r_1)\Gamma(1+r_2)}\|v-w\|_p.$$

Hence by (13.26), $N : U \longrightarrow C_p$ is a contraction
By our choice of U, there is no $u \in \partial U^n$ such that $u = \lambda N(u)$, for $\lambda \in (0,1)$. As a consequence of Theorem 1.66, we deduce that N has a unique fixed point u in U which is a solution to problem (13.25). □

Theorem 13.17. *Assume that the assumptions* (13.6.1), (13.6.2) *and the following hypothesis hold*

(13.7.1) *For each* $p \in \mathbb{N}^*$, $\Phi \in L^1(J_p, [0,\infty))$ *and there exists* $\lambda_\Phi > 0$ *such that, for each* $(x,y) \in J_p$ *we have*

$$(I_\theta^r \Phi)(x,y) \leq \lambda_\Phi \Phi(x,y).$$

If the condition (13.26) holds, then equation (13.19) is generalized Ulam-Hyers-Rassias stable.

Proof. Let $u \in C$ be a solution of the inequality (13.21). By Theorem 13.16 there v is a unique solution of the problem (13.25). Then we have

$$v(x,y) = \mu(x,y) + \frac{1}{\Gamma(r_1)\Gamma(r_2)} \int_0^x \int_0^y (x-s)^{r_1-1}(y-t)^{r_2-1} f(s,t,v(s,t))dtds; \ (x,y) \in J.$$

By differential inequality (13.21), for each $(x,y) \in J$, we have

$$\begin{aligned} \|u(x,y) - \mu(x,y) - &\frac{1}{\Gamma(r_1)\Gamma(r_2)} \int_0^x \int_0^y (x-s)^{r_1-1}(y-t)^{r_2-1} f(s,t,u(s,t))dtds\|_E \\ &\leq \frac{1}{\Gamma(r_1)\Gamma(r_2)} \int_0^x \int_0^y (x-s)^{r_1-1}(y-t)^{r_2-1} \Phi(s,t)dtds. \end{aligned}$$

Thus, by (13.7.1) for each $(x,y) \in J_p$, we get

$$\begin{aligned} \|u(x,y) - \mu(x,y) - &\frac{1}{\Gamma(r_1)\Gamma(r_2)} \int_0^x \int_0^y (x-s)^{r_1-1}(y-t)^{r_2-1} f(s,t,u(s,t))dtds\|_E \\ &\leq \lambda_\Phi \Phi(x,y). \end{aligned}$$

Hence for each $(x,y) \in J_p$, it follows that

$$\begin{aligned} \|u(x,y) - v(x,y)\|_E \leq\ & \|u(x,y) - \mu(x,y) \\ &- \frac{1}{\Gamma(r_1)\Gamma(r_2)} \int_0^x \int_0^y (x-s)^{r_1-1}(y-t)^{r_2-1} \\ &\times f(s,t,u(s,t))dtds\|_E \\ &+ \frac{1}{\Gamma(r_1)\Gamma(r_2)} \int_0^x \int_0^y (x-s)^{r_1-1}(y-t)^{r_2-1} \\ &\times \|f(s,t,u(s,t)) - f(s,t,v(s,t))\|_E dtds \\ &\leq \lambda_\Phi \Phi(x,y) \\ &+ \frac{l_{p,f}^*}{\Gamma(r_1)\Gamma(r_2)} \int_0^x \int_0^y (x-s)^{r_1-1}(y-t)^{r_2-1} \\ &\times \|u(s,t) - v(s,t)\|_E dtds. \end{aligned}$$

From Lemma 1.68, there exists a constant $\gamma = \gamma(r_1, r_2)$ such that

$$\begin{aligned} \|u(x,y) - v(x,y)\|_E &\leq \lambda_\Phi \Phi(x,y) \\ &+ \frac{\gamma l_{p,f}^* \lambda_\Phi}{\Gamma(r_1)\Gamma(r_2)} \int_0^x \int_0^y (x-s)^{r_1-1}(y-t)^{r_2-1} \Phi(s,t)dtds \\ &\leq (1 + \gamma l_{p,f}^* \lambda_\Phi)\lambda_\Phi \Phi(x,y) \\ &:= c_{f,\Phi} \Phi(x,y). \end{aligned}$$

Consequently, equation (13.19) is generalized Ulam-Hyers-Rassias stable. □

13.4.3. An Example

Let

$$E = l^1 = \left\{ w = (w_1, w_2, ..., w_n, ...) : \sum_{n=1}^{\infty} |w_n| < \infty \right\},$$

be the Banach space with norm

$$\|w\|_E = \sum_{n=1}^{\infty} |w_n|.$$

Consider the following infinite system of partial hyperbolic fractional implicit differential equations of the form

$$^cD_\theta^r u(x,y) = f(x,y,u(x,y)); \text{ if } (x,y) \in J := [0,\infty) \times [0,\infty), \tag{13.27}$$

where $(r_1, r_2) \in (0,1] \times (0,1]$,

$$u = (u_1, u_2, ..., u_n, ...),\ ^cD_\theta^r u = (^cD_\theta^r u_1, ^cD_\theta^r u_2, ..., ^cD_\theta^r u_n, ...),\ f = (f_1, f_2, ..., f_n, ...),$$

and

$$f_n(x,y,u(x,y)) = \frac{c_p}{(1+5|u_n(x,y)|)e^{x+y+3}};\ (x,y) \in J,\ n \in \mathbb{N}^*,$$

where $c_p = \frac{\Gamma(1+r_1)\Gamma(1+r_2)}{p^{r_1+r_2}}$; $p \in \mathbb{N}^*$. For each $u,\ \overline{u} \in E,\ n \in \mathbb{N}^*$ and $(x,y) \in J_p$ we have

$$|f_n(x,y,u) - f_n(x,y,\overline{u})| \leq \frac{5c_p}{e^3}|u_n - \overline{u}_n|.$$

Thus, For each $u, \overline{u} \in E$ and $(x,y) \in J_p$, we get

$$\begin{aligned}
&\|f(x,y,u(x,y)) - f(x,y,\overline{u}(x,y))\|_E \\
&= \sum_{n=1}^{\infty} |f_n(x,y,u(x,y)) - f_n(x,y,\overline{u}(x,y))| \\
&\leq \frac{1}{10e^3} \sum_{n=1}^{\infty} |u_n - \overline{u}_n| \\
&= \frac{5c_p}{e^3} \|u - \overline{u}\|_E.
\end{aligned}$$

Hence conditions (13.6.1) and (13.6.2) are satisfied with $l_{p,f}^* = \frac{5c_p}{e^3}$. We shall show that condition (13.26) holds. Indeed

$$\frac{l_{p,f}^* p^{r_1+r_2}}{\Gamma(1+r_1)\Gamma(1+r_2)} = \frac{5}{e^3} < 1,$$

which is satisfied for each $(r_1, r_2) \in (0,1] \times (0,1]$. The hypothesis (13.7.1) is satisfied with $\Phi(x,y) = xy^2$ and $\lambda_\Phi = 8p^{r_1+r_2}$. Indeed, for each $(x,y) \in J_p$ we get

$$(I_\theta^r \Phi)(x,y) = \frac{\Gamma(2)\Gamma(3)}{\Gamma(2+r_1)\Gamma(3+r_2)} x^{1+r_1} y^{2+r_2} \leq 8p^{r_1+r_2} xy^2 = \lambda_\Phi \Phi(x,y).$$

Consequently Theorem 13.17 implies that equation (13.27) is generalized Ulam-Hyers-Rassias stable.

13.5. Ulam Stabilities for the Darboux Problem for Partial Fractional Differential Inclusions

13.5.1. Introduction

In this section, we investigate some Ulam's type stability concepts for the following fractional partial differential inclusion

$$^{c}D_{\theta}^{r}u(x,y) \in F(x,y,u(x,y)); \text{ if } (x,y) \in J := [0,a] \times [0,b], \tag{13.28}$$

where $F : J \times E \to \mathcal{P}(E)$ is a set-valued function with nonempty values in a (real or complex) separable Banach space E, and $\mathcal{P}(E)$ is the family of all nonempty subsets of E. We initiate the Ulam stabilities of the Darboux problem for hyperbolic fractional differential inclusions in Banach spaces when the right hand side is nonconvex valued and we prove that the inclusion (13.28) is generalized Ulam-Hyers-Rassias stable.

Consider the following Darboux problem of partial differential equations

$$\begin{cases} ^{c}D_{\theta}^{r}u(x,y) = f(x,y,u(x,y)); \text{ if } (x,y) \in J, \\ u(x,0) = \varphi(x); \ x \in [0,a], \\ u(0,y) = \psi(y); \ y \in [0,b], \\ \varphi(0) = \psi(0), \end{cases} \tag{13.29}$$

where $\varphi : [0,a] \to E$, $\psi : [0,b] \to E$ are given absolutely continuous functions.

In the sequel, we need the following Lemma given by the authors [2].

Lemma 13.18. *[2] Let $0 < r_1, r_2 \leq 1$, $\mu(x,y) = \varphi(x) + \psi(y) - \varphi(0)$. A function $u \in C$ is a solution of the fractional integral equation*

$$u(x,y) = \mu(x,y) + \frac{1}{\Gamma(r_1)\Gamma(r_2)} \int_0^x \int_0^y (x-s)^{r_1-1}(y-t)^{r_2-1} f(s,t,u(s,t))dtds, \tag{13.30}$$

if and only if u is a solution of the problem (13.29).

Now, we consider the Ulam stability of fractional differential inclusion (13.28). Let ε be a positive real number and $\Phi : J \to [0,\infty)$ be a continuous function. We consider the following inequalities

$$H_d(^{c}D_{\theta}^{r}u(x,y), F(x,y,u(x,y))) \leq \varepsilon; \text{ if } (x,y) \in J. \tag{13.31}$$

$$H_d(^{c}D_{\theta}^{r}u(x,y), F(x,y,u(x,y))) \leq \Phi(x,y); \text{ if } (x,y) \in J. \tag{13.32}$$

$$H_d(^{c}D_{\theta}^{r}u(x,y), F(x,y,u(x,y))) \leq \varepsilon\Phi(x,y); \text{ if } (x,y) \in J. \tag{13.33}$$

Definition 13.19. *Problem (13.28) is Ulam-Hyers stable if there exists a real number $c_F > 0$ such that for each $\varepsilon > 0$ and for each solution $u \in C$ of the inequality (13.31), there exists a solution $v \in C$ of problem (13.28) with*

$$\|u(x,y) - v(x,y)\|_E \leq \varepsilon c_F; \ (x,y) \in J.$$

Definition 13.20. *Problem (13.28) is generalized Ulam-Hyers stable if there exists* $\theta_F \in C([0,\infty),$
$[0,\infty))$, $\theta_F(0) = 0$ *such that for each* $\varepsilon > 0$ *and for each solution* $u \in C$ *of the inequality (13.31), there exists a solution* $v \in C$ *of problem (13.28) with*

$$\|u(x,y) - v(x,y)\|_E \leq \theta_F(\varepsilon);\ (x,y) \in J.$$

Definition 13.21. *Problem (13.28) is Ulam-Hyers-Rassias stable with respect to* Φ *if there exists a real number* $c_{F,\Phi} > 0$ *such that for each* $\varepsilon > 0$ *and for each solution* $u \in C$ *of the inequality (13.33), there exists a solution* $v \in C$ *of problem (13.28) with*

$$\|u(x,y) - v(x,y)\|_E \leq \varepsilon c_{F,\Phi}\Phi(x,y);\ (x,y) \in J.$$

Definition 13.22. *Problem (13.28) is generalized Ulam-Hyers-Rassias stable with respect to* Φ *if there exists a real number* $c_{f,\Phi} > 0$ *such that for each solution* $u \in C$ *of the inequality (13.32), there exists a solution* $v \in C$ *of problem (13.28) with*

$$\|u(x,y) - v(x,y)\|_E \leq c_{F,\Phi}\Phi(x,y);\ (x,y) \in J.$$

Remark 13.23. *It is clear that*

(i) Definition 13.19 $\Rightarrow$ *Definition 13.20,*

(ii) Definition 13.21 $\Rightarrow$ *Definition 13.22,*

(iii) Definition 13.21 for $\Phi(x,y) = 1$ $\Rightarrow$ *Definition 13.19.*

Remark 13.24. *A function* $u \in C$ *is a solution of the inequality (13.31) if and only if there exist a function* $g \in C$ *(which depend on* u*) such that*

(i) $\|g(x,y)\|_E \leq \varepsilon$,

(ii) ${}^cD_\theta^r u(x,y) - g(x,y) \in F(x,y,u(x,y));$ *if* $(x,y) \in J$.

One can have similar remarks for the inequalities (13.32) and (13.33). So, the Ulam stabilities of the fractional differential equations are some special types of data dependence of the solutions of fractional differential equations.

13.5.2. Main Results

Now, we present conditions for the Ulam stability of problem (13.28).

Lemma 13.25. *If* $u \in C$ *is a solution of the inequality (13.31) then there exists* $f \in S_{F,u}$ *such that* u *is a solution of the following integral inequality*

$$\begin{aligned}\|u(x,y) - \mu(x,y) - \frac{1}{\Gamma(r_1)\Gamma(r_2)}\int_0^x\int_0^y (x-s)^{r_1-1}(y-t)^{r_2-1}f(s,t)dtds\|_E \\ \leq \frac{\varepsilon a^{r_1}b^{r_2}}{\Gamma(1+r_1)\Gamma(1+r_2)};\ if\ (x,y) \in J. \qquad (13.34)\end{aligned}$$

Proof. By Remark 13.24, for $(x,y) \in J$, we have

$$^{c}D_{\theta}^{r}u(x,y) - g(x,y) \in F(x,y,u(x,y)).$$

Then, there exists $f \in S_{F,u}$ such that for $(x,y) \in J$, we get

$$u(x,y) = \mu(x,y) + \frac{1}{\Gamma(r_1)\Gamma(r_2)} \int_0^x \int_0^y (x-s)^{r_1-1}(y-t)^{r_2-1}(g(s,t)+f(s,t))dtds.$$

Thus, for $(x,y) \in J$ we obtain

$$\begin{aligned} &\left\| u(x,y) - \mu(x,y) - \frac{1}{\Gamma(r_1)\Gamma(r_2)} \int_0^x \int_0^y (x-s)^{r_1-1}(y-t)^{r_2-1} f(s,t,u(s,t))dtds \right\|_E \\ &= \left\| \frac{1}{\Gamma(r_1)\Gamma(r_2)} \int_0^x \int_0^y (x-s)^{r_1-1}(y-t)^{r_2-1} g(s,t)dtds \right\|_E \\ &\leq \frac{\varepsilon a^{r_1} b^{r_2}}{\Gamma(1+r_1)\Gamma(1+r_2)}. \end{aligned}$$

Hence, we obtain (13.34). □

Remark 13.26. *We have similar results for the solutions of the inequalities (13.32) and (13.33).*

Consider the following Darboux problem of partial differential inclusions

$$\begin{cases} ^{c}D_{\theta}^{r}u(x,y) \in F(x,y,u(x,y)); \text{ if } (x,y) \in J, \\ u(x,0) = \varphi(x); \ x \in [0,a], \\ u(0,y) = \psi(y); \ y \in [0,b], \\ \varphi(0) = \psi(0), \end{cases} \tag{13.35}$$

where $\varphi : [0,a] \to E$, $\psi : [0,b] \to E$ are given absolutely continuous functions.

In the sequel, we present a result for the problem (13.35) with a nonconvex valued right hand side. Our considerations are based on the fixed point theorem for contraction multivalued maps given by Covitz and Nadler [142].

Theorem 13.27. *Assume that the following hypotheses hold*

(13.8.1) *there exists $l \in L^{\infty}(J,[0,\infty))$ such that*

$$H_d(F(x,y,u),F(x,y,\overline{u})) \leq l(x,y)\|u-\overline{u}\|_E \text{ for every } u,\overline{u} \in E,$$

and

$$d(0,F(x,y,0)) \leq l(x,y), \text{ a.e. } (x,y) \in J,$$

(13.8.2) *$F : J \times E \longrightarrow \mathcal{P}_{cp}(E)$ has the property that $F(.,.,u) : J \to \mathcal{P}_{cp}(E)$ is measurable for each $u \in E$.*

If

$$\frac{l^* a^{r_1} b^{r_2}}{\Gamma(r_1+1)\Gamma(r_2+1)} < 1, \tag{13.36}$$

where $l^* = \|l\|_{L^\infty}$, *then problem (13.29) has at least one solution on* J.

Remark 13.28. *For each* $u \in C$, *the set* $S_{F,u}$ *is nonempty since by* $(13.8.2)$, F *has a measurable selection (see [134], Theorem III.6).*

Proof. Transform the problem (13.35) into a fixed point problem. Consider the multivalued operator $N : C \to \mathcal{P}(C)$ defined by

$$N(u) = \{h \in C : h(x,y) = \mu(x,y) + I_\theta^r f(x,y);\ (x,y) \in J\},$$

where $f \in S_{F,u}$.

Remark 13.29. *Clearly, from Lemma 13.18, the fixed points of N are solutions to (13.35).*

We shall show that N satisfies the assumptions of Lemma 1.63. The proof will be given in two steps.

Step 1. $N(u) \in \mathcal{P}_{cl}(C)$ *for each* $u \in C$.

Indeed,let $(u_n)_{n\geq 0} \in N(u)$ such that $u_n \longrightarrow \tilde{u}$ in C. Then, $\tilde{u} \in C$ and there exists $f_n(.,.) \in S_{F,u}$ such that, for each $(x,y) \in J$,

$$u_n(x,y) = \mu(x,y) + \frac{1}{\Gamma(r_1)\Gamma(r_2)} \int_0^x \int_0^y (x-s)^{r_1-1}(y-t)^{r_2-1} f_n(s,t)dtds.$$

Using the fact that F has compact values and from $(13.8.1)$, we may pass to a subsequence if necessary to get that $f_n(.,.)$ converges to f in $L^1(J,E)$, and hence $f \in S_{F,u}$. Then, for each $(x,y) \in J$,

$$u_n(x,y) \longrightarrow \tilde{u}(x,y) = \mu(x,y) + \frac{1}{\Gamma(r_1)\Gamma(r_2)} \int_0^x \int_0^y (x-s)^{r_1-1}(y-t)^{r_2-1} f(s,t)dtds.$$

So, $\tilde{u} \in N(u)$.

Step 2. *There exists* $\gamma < 1$ *such that*

$$H_d(N(u), N(\overline{u})) \leq \gamma \|u - \overline{u}\|_E \text{ for each } u, \overline{u} \in C.$$

Let $u, \overline{u} \in C$ and $h \in N(u)$. Then, there exists $f(x,y) \in F(x,y,u(x,y))$ such that for each $(x,y) \in J$

$$h(x,y) = \mu(x,y) + \frac{1}{\Gamma(r_1)\Gamma(r_2)} \int_0^x \int_0^y (x-s)^{r_1-1}(y-t)^{r_2-1} f(s,t)dtds.$$

From $(13.8.1)$ it follows that

$$H_d(F(x,y,u(x,y)), F(x,y,\overline{u}(x,y))) \leq l(x,y)\|u(x,y) - \overline{u}(x,y)\|_E.$$

Hence, there exists $w \in F(x,y,\overline{u}(x,y))$ such that

$$\|f(x,y) - w\|_E \leq l(x,y)\|u(x,y) - \overline{u}(x,y)\|_E;\ (x,y) \in J.$$

Consider $U : J \to \mathcal{P}(E)$ given by

$$U(x,y) = \{w \in E : \|f(x,y) - w\|_E \leq l(x,y)\|u(x,y) - \overline{u}(x,y)\|_E\}.$$

Since the multivalued operator $u(x,y) = U(x,y) \cap F(x,y,\overline{u}(x,y))$ is measurable (see Proposition III.4 in [134]), there exists a function $\overline{f}(x,y)$ which is a measurable selection for $\overline{u}$. So, $\overline{f}(x,y) \in F(x,y,\overline{u}(x,y))$, and for each $(x,y) \in J$,

$$\|f(x,y) - \overline{f}(x,y)\|_E \leq l(x,y)\|u(x,y) - \overline{u}(x,y)\|_E.$$

Let us define for each $(x,y) \in J$

$$\overline{h}(x,y) = \mu(x,y) + \frac{1}{\Gamma(r_1)\Gamma(r_2)}\int_0^x\int_0^y (x-s)^{r_1-1}(y-t)^{r_2-1}\overline{f}(s,t)dtds.$$

Then for each $(x,y) \in J$, we have

$$\begin{aligned}
\|h(x,y) - \overline{h}(x,y)\|_E &\leq \frac{1}{\Gamma(r_1)\Gamma(r_2)}\int_0^x\int_0^y (x-s)^{r_1-1}(y-t)^{r_2-1} \\
&\times \|f(s,t) - \overline{f}(s,t)\|_E dtds \\
&\leq \frac{1}{\Gamma(r_1)\Gamma(r_2)}\int_0^x\int_0^y (x-s)^{r_1-1}(y-t)^{r_2-1}l(s,t) \\
&\times \|u(s,t) - \overline{u}(s,t)\|_E dtds \\
&\leq \frac{1}{\Gamma(r_1)\Gamma(r_2)}\int_0^x\int_0^y (x-s)^{r_1-1}(y-t)^{r_2-1} \\
&\times l(s,t)\|u(s,t) - \overline{u}(s,t)\|_E dtds \\
&\leq \frac{\|u_{(.,.)}) - \overline{u}_{(.,.)})\|_E}{\Gamma(r_1)\Gamma(r_2)}\int_0^x\int_0^y (x-s)^{r_1-1}(y-t)^{r_2-1}l(s,t)dtds.
\end{aligned}$$

Thus, for each $(x,y) \in J$

$$\|h - \overline{h}\|_\infty \leq \frac{l^* a^{r_1} b^{r_2}}{\Gamma(1+r_1)\Gamma(1+r_2)}\|u - \overline{u}\|_\infty.$$

By an analogous relation, obtained by interchanging the roles of u and $\overline{u}$, it follows that

$$H_d(N(u), N(\overline{u})) \leq \frac{l^* a^{r_1} b^{r_2}}{\Gamma(r_1+1)\Gamma(r_2+1)}\|u - \overline{u}\|_\infty.$$

So by (13.36), N is a contraction and thus, by Lemma 1.63, N has a fixed point u which is solution to problem (13.35). □

Theorem 13.30. *Assume that the assumptions* (13.8.1), (13.8.2) *and the following hypothesis hold*

(13.9.1) $\Phi \in L^1(J, [0,\infty))$ *and there exists* $\lambda_\Phi > 0$ *such that, for each* $(x,y) \in J$ *we have*

$$(I_\theta^r \Phi)(x,y) \leq \lambda_\Phi \Phi(x,y).$$

If the condition (13.36) holds, then (13.28) is generalized Ulam-Hyers-Rassias stable.

Proof. Let $u \in C$ be a solution of the inequality (13.32). By Theorem 13.27 there v is a solution of the problem (13.35). Then, for each $(x,y) \in J$, we have

$$v(x,y) = \mu(x,y) + \frac{1}{\Gamma(r_1)\Gamma(r_2)} \int_0^x \int_0^y (x-s)^{r_1-1}(y-t)^{r_2-1} f_v(s,t)dtds,$$

where $f_v \in S_{F,v}$. By differential inequality (13.32), for each $(x,y) \in J$, we have

$$\begin{aligned}\|u(x,y) - \mu(x,y) - &\frac{1}{\Gamma(r_1)\Gamma(r_2)} \int_0^x \int_0^y (x-s)^{r_1-1}(y-t)^{r_2-1} f_u(s,t)dtds\|_E \\ &\leq \frac{1}{\Gamma(r_1)\Gamma(r_2)} \int_0^x \int_0^y (x-s)^{r_1-1}(y-t)^{r_2-1} \Phi(s,t)dtds,\end{aligned}$$

where $f_u \in S_{F,u}$. Thus, by (13.9.1) for each $(x,y) \in J$, we get

$$\begin{aligned}\|u(x,y) - \mu(x,y) - &\frac{1}{\Gamma(r_1)\Gamma(r_2)} \int_0^x \int_0^y (x-s)^{r_1-1}(y-t)^{r_2-1} f_u(s,t)dtds\|_E \\ &\leq \lambda_\Phi \Phi(x,y).\end{aligned}$$

Hence for each $(x,y) \in J$, it follows that

$$\begin{aligned}\|u(x,y) - v(x,y)\|_E &\leq \|u(x,y) - \mu(x,y) \\ &- \frac{1}{\Gamma(r_1)\Gamma(r_2)} \int_0^x \int_0^y (x-s)^{r_1-1}(y-t)^{r_2-1} f_u(s,t)dtds\|_E \\ &+ \frac{1}{\Gamma(r_1)\Gamma(r_2)} \int_0^x \int_0^y (x-s)^{r_1-1}(y-t)^{r_2-1} \\ &\times \|f_u(s,t) - f_v(s,t)\|_E dtds \\ &\leq \lambda_\Phi \Phi(x,y) + \frac{l_f}{\Gamma(r_1)\Gamma(r_2)} \int_0^x \int_0^y (x-s)^{r_1-1}(y-t)^{r_2-1} \\ &\times \|u(s,t) - v(s,t)\|_E dtds.\end{aligned}$$

From Lemma 1.68, there exists a constant $\delta = \delta(r_1, r_2)$ such that

$$\begin{aligned}\|u(x,y) - v(x,y)\|_E &\leq \lambda_\Phi \Phi(x,y) \\ &+ \frac{\delta l_f \lambda_\Phi}{\Gamma(r_1)\Gamma(r_2)} \int_0^x \int_0^y (x-s)^{r_1-1}(y-t)^{r_2-1} \Phi(s,t)dtds \\ &\leq (1 + \delta l_f \lambda_\Phi)\lambda_\Phi \Phi(x,y) \\ &:= c_{F,\Phi} \Phi(x,y).\end{aligned}$$

Finally, the inclusion (13.28) is generalized Ulam-Hyers-Rassias stable. □

13.5.3. An Example

Let

$$E = l^1 = \left\{ w = (w_1, w_2, \ldots, w_n, \ldots) : \sum_{n=1}^{\infty} |w_n| < \infty \right\},$$

be the Banach space with norm

$$\|w\|_E = \sum_{n=1}^{\infty} |w_n|.$$

Consider the following infinite system of partial hyperbolic fractional differential inclusions of the form

$$^cD^r_\theta u(x,y) \in F(x,y,u(x,y)); \text{ a.e. } (x,y) \in J = [0,1]\times[0,1], \tag{13.37}$$

where $(r_1,r_2) \in (0,1]\times(0,1]$,

$$u = (u_1,u_2,\ldots,u_n,\ldots), \quad ^cD^r_\theta u = (^cD^r_\theta u_1, ^cD^r_\theta u_2,\ldots, ^cD^r_\theta u_n,\ldots),$$

$$F(x,y,u(x,y)) = \{v \in E : \|f_1(x,y,u(x,y))\|_E \le \|v\|_E \le \|f_2(x,y,u(x,y))\|_E\};$$

$(x,y) \in [0,1]\times[0,1]$, and $f_1, f_2 : [0,1]\times[0,1]\times E \to E$,

$$f_k = (f_{k,1}, f_{k,2},\ldots,f_{k,n},\ldots); \; k \in \{1,2\}, \; n \in \mathbb{N},$$

$$f_{1,n}(x,y,u_n(x,y)) = \frac{xy^2u_n}{e^{10+x+y}(1+\|u_n\|_E)}; \; n \in \mathbb{N},$$

and

$$f_{2,n}(x,y,u_n(x,y)) = \frac{xy^2u_n}{e^{10+x+y}}; \; n \in \mathbb{N}.$$

We assume that F is compact valued. For each $(x,y) \in J$ and all $z_1, z_2 \in E$, we have

$$\|f_2(x,y,z_2) - f_1(x,y,z_1)\|_E \le xy^2e^{-10-x-y}\|z_2 - z_1\|_E,$$

then the hypotheses (13.8.1) and (13.8.2) are satisfied with $l(x,y) = \frac{xy^2}{e^{10+x+y}}$. We shall show that condition (13.36) holds with $a = b = 1$. Indeed $l^* = \frac{1}{e^{10}}$, and

$$\frac{l^*a^{r_1}b^{r_2}}{\Gamma(1+r_1)\Gamma(1+r_2)} = \frac{1}{e^{10}\Gamma(1+r_1)\Gamma(1+r_2)} < \frac{4}{e^{10}} < 1,$$

which is satisfied for each $(r_1,r_2) \in (0,1]\times(0,1]$. The hypothesis (13.9.1) is satisfied with $\Phi(x,y) = xy^2$ and $\lambda_\Phi = 8$. Indeed, for each $(x,y) \in [0,1]\times[0,1]$ we get

$$(I^r_\theta\Phi)(x,y) = \frac{\Gamma(2)\Gamma(3)}{\Gamma(2+r_1)\Gamma(3+r_2)}x^{1+r_1}y^{2+r_2} \le 8xy^2 = \lambda_\Phi\Phi(x,y).$$

Consequently Theorem 13.30 implies that the inclusion (13.37) is generalized Ulam-Hyers-Rassias stable.

13.6. Ulam Stabilities for the Darboux Problem for Impulsive Partial Fractional Differential Equations

13.6.1. Introduction

In the present section we investigate some Ulam's type stability concepts for the Darboux problem of partial differential equations of Caputo's fractional order derivative. Our results are based upon the Banach contraction principle and a fractional version of Gronwall's inequality. In [34], Abbas *et al.* studied the existence and the uniqueness of solutions of the following Darboux problem of partial impulsive differential equations

$$\begin{cases} {}^cD^r_{\theta_k}u(x,y) = f(x,y,u(x,y)); \text{ if } (x,y) \in J_k, \; k = 0,\dots,m, \\ u(x_k^+,y) = u(x_k^-,y) + I_k(u(x_k^-,y)); \text{ if } y \in [0,b], \; k = 1,\dots,m, \\ u(x,0) = \varphi(x); \; x \in [0,a], \\ u(0,y) = \psi(y); \; y \in [0,b], \\ \varphi(0) = \psi(0), \end{cases} \tag{13.38}$$

where $J_0 = [0,x_1] \times [0,b]$, $J_k := (x_k, x_{k+1}] \times [0,b]$; $k = 1,\dots,m$, $a,b > 0$, $\theta_k = (x_k,0)$; $k = 0,\dots,m$, ${}^cD^r_{\theta_k}$ is the fractional Caputo derivative of order $r = (r_1,r_2) \in (0,1] \times (0,1]$, $0 = x_0 < x_1 < \cdots < x_m < x_{m+1} = a$, $f : J \times \mathbb{R}^n \times \mathbb{R}^n \to \mathbb{R}^n$, $J = [0,a] \times [0,b]$, $I_k : \mathbb{R}^n \to \mathbb{R}^n$; $k = 1,\dots,m$ are given continuous functions, $\varphi : [0,a] \to \mathbb{R}^n$ and $\psi : [0,b] \to \mathbb{R}^n$ are given absolutely continuous functions. Here $u(x_k^+,y)$ and $u(x_k^-,y)$ denote the right and left limits of $u(x,y)$ at $x = x_k$, respectively. Recently, in [364], Wang *et al.* introduced some new concepts about Ulam stability of impulsive fractional differential equations

Motivated by the above papers, in this article, we discuss Ulam stability for impulsive fractional partial differential equations

$$\begin{cases} {}^cD^r_{\theta_k}u(x,y) = f(x,y,u(x,y)); \text{ if } (x,y) \in J_k, \; k = 0,\dots,m, \\ u(x_k^+,y) = u(x_k^-,y) + I_k(u(x_k^-,y)); \text{ if } y \in [0,b], \; k = 1,\dots,m, \end{cases} \tag{13.39}$$

where $f : J \times E \to E$, $I_k : E \to E$; $k = 1,\dots,m$ are given continuous functions.
Consider the space

$$\begin{aligned} PC = \{ u : J \to E : \; & u \in C(J_k); \; k = 0,1,\dots,m, \text{ and there} \\ & \text{exist } u(x_k^-,y) \text{ and } u(x_k^+,y); \; k = 1,\dots,m, \\ & \text{with } u(x_k^-,y) = u(x_k,y) \text{ for each } y \in [0,b] \}. \end{aligned}$$

This set is a Banach space with the norm

$$\|u\|_{PC} = \sup_{(x,y)\in J} \|u(x,y)\|_E.$$

Let $a_1 \in [0,a]$, $z^+ = (a_1,0) \in J$, $J_z = (a_1,a] \times [0,b]$, $r_1,r_2 > 0$ and $r = (r_1,r_2)$. For $u \in L^1(J_z)$, the expression

$$(I^r_{z^+}u)(x,y) = \frac{1}{\Gamma(r_1)\Gamma(r_2)} \int_{a_1^+}^{x} \int_0^y (x-s)^{r_1-1}(y-t)^{r_2-1}u(s,t)\,dt\,ds,$$

is called the left-sided mixed Riemann-Liouville integral of order r of u.

Definition 13.31. *[360]. For $u \in L^1(J_z)$ where $D^2_{xy}u$ is Lebesque integrable on $[x_k,x_{k+1}] \times [0,b]$, $k = 0,\dots,m$, the Caputo fractional order derivative of order r of u is defined by the expression*

$$({}^cD^r_{z^+}f)(x,y) = (I^{1-r}_{z^+}D^2_{xy}f)(x,y).$$

Now, we consider the Ulam stability of impulsive fractional differential equations (13.39). Let ε be a positive real number and $\Phi : J \rightarrow [0,\infty)$ be a continuous function. We consider the following inequalities

$$\begin{cases} \|{}^cD^r_{\theta_k}u(x,y) - f(x,y,u(x,y))\|_E \leq \varepsilon; \text{ if } (x,y) \in J_k, \ k = 0,\dots,m, \\ \|u(x_k^+,y) - u(x_k^-,y) - I_k(u(x_k^-,y))\|_E \leq \varepsilon; \text{ if } y \in [0,b], \ k = 1,\dots,m. \end{cases} \tag{13.40}$$

$$\begin{cases} \|{}^cD^r_{\theta_k}u(x,y) - f(x,y,u(x,y))\|_E \leq \Phi(x,y); \text{ if } (x,y) \in J_k, \ k = 0,\dots,m, \\ \|u(x_k^+,y) - u(x_k^-,y) - I_k(u(x_k^-,y))\|_E \leq \Phi(x,y); \text{ if } y \in [0,b], \ k = 1,\dots,m. \end{cases} \tag{13.41}$$

$$\begin{cases} \|{}^cD^r_{\theta_k}u(x,y) - f(x,y,u(x,y))\|_E \leq \varepsilon\Phi(x,y); \text{ if } (x,y) \in J_k, \ k = 0,\dots,m, \\ \|u(x_k^+,y) - u(x_k^-,y) - I_k(u(x_k^-,y))\|_E \leq \varepsilon\Phi(x,y); \text{ if } y \in [0,b], \ k = 1,\dots,m. \end{cases} \tag{13.42}$$

Definition 13.32. *[364] Problem (13.39) is Ulam-Hyers stable if there exists a real number $c_{f,m} > 0$ such that for each $\varepsilon > 0$ and for each solution $u \in PC$ of the inequality (13.40), there exists a solution $v \in PC$ Problem (13.39) with*

$$\|u(x,y) - v(x,y)\|_E \leq \varepsilon c_{f,m}; \ (x,y) \in J.$$

Definition 13.33. *[364] Problem (13.39) is generalized Ulam-Hyers stable if there exists $\theta_{f,m} \in PC([0,\infty),[0,\infty))$, $\theta_{f,m}(0) = 0$ such that for each $\varepsilon > 0$ and for each solution $u \in PC$ of the inequality (13.40) there exists a solution $v \in PC$ Problem (13.39) with*

$$\|u(x,y) - v(x,y)\|_E \leq c_{f,m}(\varepsilon); \ (x,y) \in J.$$

Definition 13.34. *[364] Problem (13.39) is Ulam-Hyers-Rassias stable with respect to Φ if there exists a real number $c_{f,m,\Phi} > 0$ such that for each $\varepsilon > 0$ and for each solution $u \in PC$ of the inequality (13.42) there exists a solution $v \in PC$ Problem (13.39) with*

$$\|u(x,y) - v(x,y)\|_E \leq \varepsilon c_{f,m,\Phi}\Phi(x,y); \ (x,y) \in J.$$

Definition 13.35. *[364] Problem (13.39) is generalized Ulam-Hyers-Rassias stable with respect to Φ if there exists a real number $c_{f,m,\Phi} > 0$ such that for each solution $u \in PC$ of the inequality (13.41) there exists a solution $v \in PC$ Problem (13.39) with*

$$\|u(x,y) - v(x,y)\|_E \leq c_{f,m,\Phi}\Phi(x,y); \ (x,y) \in J.$$

Remark 13.36. *It is clear that*

(i) Definition 13.32 $\Rightarrow$ Definition 13.33,

(ii) Definition 13.34 $\Rightarrow$ Definition 13.35,

(iii) Definition 13.34 for $\Phi(x,y) = 1$ $\Rightarrow$ Definition 13.32.

Remark 13.37. *A function $u \in PC$ is a solution of the inequality (13.40) if and only if there exist a function $g \in PC$ and a sequence g_k; $k = 1,\dots,m$ (which depend on u) such that*

(i) $\|g(x,y)\|_E \leq \varepsilon$ and $\|g_k\|_\infty \leq \varepsilon$; $k = 1,\dots,m$,

(ii) ${}^cD^r_{\theta_k}u(x,y) = f(x,y,u(x,y)) + g(x,y);$ *if* $(x,y) \in J_k,\ k = 0,\ldots,m,$

(iii) $u(x_k^+,y) = u(x_k^-,y) + I_k(u(x_k^-,y)) + g_k;$ *if* $y \in [0,b],\ k = 1,\ldots,m.$

One can have similar remarks for the inequalities (13.41) and (13.42). So, the Ulam stabilities of the impulsive fractional differential equations are some special types of data dependence of the solutions of impulsive fractional differential equations.

13.6.2. Main Results

Now, we present conditions for the Ulam stability of problem (13.39).

We need the following auxiliary Lemma whose proof in when E is of a finite dimension was given in [34].

Lemma 13.38. *[34] Let* $0 < r_1, r_2 \leq 1$, $\mu(x,y) = \phi(x) + \psi(y) - \phi(0)$ *and let* $h : J \to E$ *be continuous. A function* $u \in PC(J)$ *is a solution of the fractional integral equation*

$$u(x,y) = \begin{cases} \mu(x,y) + \frac{1}{\Gamma(r_1)\Gamma(r_2)} \int_0^x \int_0^y (x-s)^{r_1-1}(y-t)^{r_2-1} f(s,t,u(s,t))dtds; \\ \text{if } (x,y) \in [0,x_1] \times [0,b], \\ \mu(x,y) + \sum_{i=1}^k (I_i(u(x_i^-,y)) - I_i(u(x_i^-,0))) \\ + \frac{1}{\Gamma(r_1)\Gamma(r_2)} \sum_{i=1}^k \int_{x_{i-1}}^{x_i} \int_0^y (x_i-s)^{r_1-1}(y-t)^{r_2-1} f(s,t,u(s,t))dtds \\ + \frac{1}{\Gamma(r_1)\Gamma(r_2)} \int_{x_k}^x \int_0^y (x-s)^{r_1-1}(y-t)^{r_2-1} f(s,t,u(s,t))dtds; \\ \text{if } (x,y) \in (x_k,x_{k+1}] \times [0,b],\ k = 1,\ldots,m, \end{cases} \tag{13.43}$$

if and only if u is a solution of the problem (13.38).

Lemma 13.39. *If* $u \in PC$ *is a solution of the inequality (13.40) then u is a solution of the following integral inequality*

$$\begin{cases} \|u(x,y) - \mu(x,y) - \frac{1}{\Gamma(r_1)\Gamma(r_2)} \int_0^x \int_0^y (x-s)^{r_1-1}(y-t)^{r_2-1} f(s,t,u(s,t))dtds\|_E \\ \leq \frac{\varepsilon a^{r_1} b^{r_2}}{\Gamma(1+r_1)\Gamma(1+r_2)};\ \text{if } (x,y) \in [0,x_1] \times [0,b], \\ \|u(x,y) - \mu(x,y) - \sum_{i=1}^k (I_i(u(x_i^-,y)) - I_i(u(x_i^-,0))) \\ - \frac{1}{\Gamma(r_1)\Gamma(r_2)} \sum_{i=1}^k \int_{x_{i-1}}^{x_i} \int_0^y (x_i-s)^{r_1-1}(y-t)^{r_2-1} f(s,t,u(s,t))dtds \\ - \frac{1}{\Gamma(r_1)\Gamma(r_2)} \int_{x_k}^x \int_0^y (x-s)^{r_1-1}(y-t)^{r_2-1} f(s,t,u(s,t))dtds\|_E \\ \leq \frac{2\varepsilon a^{r_1} b^{r_2}}{\Gamma(1+r_1)\Gamma(1+r_2)};\ \text{if } (x,y) \in (x_k,x_{k+1}] \times [0,b],\ k = 1,\ldots,m. \end{cases} \tag{13.44}$$

Proof. by Remark 13.37 we have that

$$\begin{cases} {}^cD^r_{\theta_k}u(x,y) = f(x,y,u(x,y)) + g(x,y);\ \text{if } (x,y) \in J_k,\ k = 0,\ldots,m, \\ u(x_k^+,y) = u(x_k^-,y) + I_k(u(x_k^-,y)) + g_k;\ \text{if } y \in [0,b],\ k = 1,\ldots,m. \end{cases}$$

Then

$$u(x,y)=\begin{cases}\mu(x,y)+\frac{1}{\Gamma(r_1)\Gamma(r_2)}\int_0^x\int_0^y(x-s)^{r_1-1}(y-t)^{r_2-1}(g(s)+f(s,t,u(s,t)))dtds;\\ \text{if } (x,y)\in[0,x_1]\times[0,b],\\ \mu(x,y)+\sum_{i=1}^k(I_i(u(x_i^-,y))-I_i(u(x_i^-,0)))\\ +\frac{1}{\Gamma(r_1)\Gamma(r_2)}\sum_{i=1}^k\int_{x_{i-1}}^{x_i}\int_0^y(x_i-s)^{r_1-1}(y-t)^{r_2-1}(g(s)+f(s,t,u(s,t)))dtds\\ +\frac{1}{\Gamma(r_1)\Gamma(r_2)}\int_{x_k}^x\int_0^y(x-s)^{r_1-1}(y-t)^{r_2-1}(g(s)+f(s,t,u(s,t)))dtds;\\ \text{if } (x,y)\in(x_k,x_{k+1}]\times[0,b],\ k=1,\ldots,m.\end{cases}$$

Thus, it follows that

$$\begin{cases}\|u(x,y)-\mu(x,y)-\frac{1}{\Gamma(r_1)\Gamma(r_2)}\int_0^x\int_0^y(x-s)^{r_1-1}(y-t)^{r_2-1}f(s,t,u(s,t))dtds\|_E\\ =\|\frac{1}{\Gamma(r_1)\Gamma(r_2)}\int_0^x\int_0^y(x-s)^{r_1-1}(y-t)^{r_2-1}g(s,t)dtds\|_E;\\ \text{if } (x,y)\in[0,x_1]\times[0,b],\\ \|u(x,y)-\mu(x,y)-\sum_{i=1}^k(I_i(u(x_i^-,y))-I_i(u(x_i^-,0)))\\ +\frac{1}{\Gamma(r_1)\Gamma(r_2)}\sum_{i=1}^k\int_{x_{i-1}}^{x_i}\int_0^y(x_i-s)^{r_1-1}(y-t)^{r_2-1}f(s,t,u(s,t))dtds\\ -\frac{1}{\Gamma(r_1)\Gamma(r_2)}\int_{x_k}^x\int_0^y(x-s)^{r_1-1}(y-t)^{r_2-1}f(s,t,u(s,t))dtds\|_E\\ =\|\frac{1}{\Gamma(r_1)\Gamma(r_2)}\sum_{i=1}^k\int_{x_{i-1}}^{x_i}\int_0^y(x_i-s)^{r_1-1}(y-t)^{r_2-1}g(s,t)dtds\\ +\frac{1}{\Gamma(r_1)\Gamma(r_2)}\int_{x_k}^x\int_0^y(x-s)^{r_1-1}(y-t)^{r_2-1}g(s,t)dtds\|_E;\\ \text{if } (x,y)\in(x_k,x_{k+1}]\times[0,b],\ k=1,\ldots,m.\end{cases}$$

Hence, we obtain (13.44). □

Remark 13.40. *We have similar results for the solutions of the inequalities (13.41) and (13.42).*

In the sequel, we need the following Theorem.

Theorem 13.41. *[34] Assume that the following hypotheses hold*

(13.10.1) *There exists a constant $l_f>0$ such that*

$$\|f(x,y,u)-f(x,y,\overline{u})\|_E\le l_f\|u-\overline{u}\|_E,\ \textit{for each } (x,y)\in J, \textit{and each } u,\overline{u}\in E,$$

(13.10.2) *There exists a constant $l^*>0$ such that*

$$\|I_k(u)-I_k(\overline{u})\|_E\le l^*\|u-\overline{u}\|_E,\ \textit{for each } u,\overline{u}\in E,\ k=1,\ldots,m.$$

If

$$2ml^*+\frac{2l_fa^{r_1}b^{r_2}}{\Gamma(1+r_1)\Gamma(1+r_2)}<1, \tag{13.45}$$

then (13.38) has a unique solution on J.

Theorem 13.42. *Assume that assumptions* (13.10.1), (13.10.2) *and the following hypotheses hold*

(13.11.1) *there exists* $\lambda_\Phi > 0$ *such that*

$$\frac{1}{\Gamma(r_1)\Gamma(r_2)}\int_0^x\int_0^y (x-s)^{r_1-1}(y-t)^{r_2-1}\|u(s,t)-v(s,t)\|_E dtds \leq \Phi(x,y).$$

(13.11.2) $\|I_k(u)\|_E \leq \Phi(x,y)$; *for each* $u \in E$, $k = 1,\ldots,m$.

If the condition (13.45) is satisfied, then problem (13.39) is generalized Ulam-Hyers-Rassias stable.

Proof. Let $u \in PC$ be a solution of the inequality (13.41). By Theorem 13.41 there v is a unique solution of the problem (13.38). Then we have

$$v(x,y) = \begin{cases} \mu(x,y) + \frac{1}{\Gamma(r_1)\Gamma(r_2)}\int_0^x\int_0^y (x-s)^{r_1-1}(y-t)^{r_2-1}f(s,t,v(s,t))dtds; \\ \text{if } (x,y) \in [0,x_1]\times[0,b], \\ \mu(x,y) + \sum_{i=1}^k (I_i(v(x_i^-,y)) - I_i(v(x_i^-,0))) \\ +\frac{1}{\Gamma(r_1)\Gamma(r_2)}\sum_{i=1}^k \int_{x_{i-1}}^{x_i}\int_0^y (x_i-s)^{r_1-1}(y-t)^{r_2-1}f(s,t,v(s,t))dtds \\ +\frac{1}{\Gamma(r_1)\Gamma(r_2)}\int_{x_k}^x\int_0^y (x-s)^{r_1-1}(y-t)^{r_2-1}f(s,t,v(s,t))dtds; \\ \text{if } (x,y) \in (x_k,x_{k+1}]\times[0,b],\ k = 1,\ldots,m. \end{cases}$$

By differential inequality (13.41), for each $(x,y) \in J$, we have

$$\begin{cases} \|u(x,y) - \mu(x,y) - \frac{1}{\Gamma(r_1)\Gamma(r_2)}\int_0^x\int_0^y (x-s)^{r_1-1}(y-t)^{r_2-1}f(s,t,u(s,t))dtds\|_E \\ \leq \frac{1}{\Gamma(r_1)\Gamma(r_2)}\int_0^x\int_0^y (x-s)^{r_1-1}(y-t)^{r_2-1}\Phi(s,t)dtds \text{ if } (x,y) \in [0,x_1]\times[0,b], \\ \textit{and} \\ \|u(x,y) - \mu(x,y) - \sum_{i=1}^k (I_i(u(x_i^-,y)) - I_i(u(x_i^-,0))) \\ -\frac{1}{\Gamma(r_1)\Gamma(r_2)}\sum_{i=1}^k \int_{x_{i-1}}^{x_i}\int_0^y (x_i-s)^{r_1-1}(y-t)^{r_2-1}f(s,t,u(s,t))dtds \\ -\frac{1}{\Gamma(r_1)\Gamma(r_2)}\int_{x_k}^x\int_0^y (x-s)^{r_1-1}(y-t)^{r_2-1}f(s,t,u(s,t))dtds\|_E \\ \leq \frac{1}{\Gamma(r_1)\Gamma(r_2)}\sum_{i=1}^k \int_{x_{i-1}}^{x_i}\int_0^y (x_i-s)^{r_1-1}(y-t)^{r_2-1}\Phi(s,t)dtds \\ +\frac{1}{\Gamma(r_1)\Gamma(r_2)}\int_{x_k}^x\int_0^y (x-s)^{r_1-1}(y-t)^{r_2-1}\Phi(s,t)dtds; \\ \text{if } (x,y) \in (x_k,x_{k+1}]\times[0,b];\ k = 1,\ldots,m. \end{cases}$$

Thus, by (13.11.1) for each $(x,y) \in J$, we get

$$\begin{cases} \|u(x,y) - \mu(x,y) - \frac{1}{\Gamma(r_1)\Gamma(r_2)}\int_0^x\int_0^y (x-s)^{r_1-1}(y-t)^{r_2-1}f(s,t,u(s,t))dtds\|_E \\ \leq \lambda_\Phi\Phi(x,y); \text{ if } (x,y) \in [0,x_1]\times[0,b], \\ \textit{and} \\ \|u(x,y) - \mu(x,y) - \sum_{i=1}^k (I_i(u(x_i^-,y)) - I_i(u(x_i^-,0))) \\ -\frac{1}{\Gamma(r_1)\Gamma(r_2)}\sum_{i=1}^k \int_{x_{i-1}}^{x_i}\int_0^y (x_i-s)^{r_1-1}(y-t)^{r_2-1}f(s,t,u(s,t))dtds \\ -\frac{1}{\Gamma(r_1)\Gamma(r_2)}\int_{x_k}^x\int_0^y (x-s)^{r_1-1}(y-t)^{r_2-1}f(s,t,u(s,t))dtds\|_E \\ \leq (1+k)\lambda_\Phi\Phi(x,y); \text{ if } (x,y) \in (x_k,x_{k+1}]\times[0,b];\ k = 1,\ldots,m. \end{cases}$$

Hence for each $(x,y) \in [0,x_1] \times [0,b]$, it follows that

$$\begin{aligned}
\|u(x,y)-v(x,y)\|_E \leq & \|u(x,y)-\mu(x,y) \\
& -\frac{1}{\Gamma(r_1)\Gamma(r_2)}\int_0^x\int_0^y (x-s)^{r_1-1}(y-t)^{r_2-1} \\
& \times f(s,t,u(s,t))dtds\|_E \\
& +\frac{1}{\Gamma(r_1)\Gamma(r_2)}\int_0^x\int_0^y (x-s)^{r_1-1}(y-t)^{r_2-1} \\
& \times \|f(s,t,u(s,t))-f(s,t,v(s,t))\|_E dtds \\
& \leq \lambda_\Phi \Phi(x,y)+\frac{l_f}{\Gamma(r_1)\Gamma(r_2)}\int_0^x\int_0^y (x-s)^{r_1-1}(y-t)^{r_2-1} \\
& \times \|u(s,t)-v(s,t)\|_E dtds,
\end{aligned}$$

and for each $(x,y) \in (x_k,x_{k+1}] \times [0,b]$, $k=1,\dots,m$, we obtain

$$\begin{aligned}
\|u(x,y)-v(x,y)\|_E \leq & \|u(x,y)-\mu(x,y)-\sum_{i=1}^{k}(I_i(u(x_i^-,y))-I_i(u(x_i^-,0))) \\
& -\frac{1}{\Gamma(r_1)\Gamma(r_2)}\sum_{i=1}^{k}\int_{x_{i-1}}^{x_i}\int_0^y (x_i-s)^{r_1-1}(y-t)^{r_2-1} \\
& \times f(s,t,u(s,t))dtds \\
& -\frac{1}{\Gamma(r_1)\Gamma(r_2)}\int_{x_k}^x\int_0^y (x-s)^{r_1-1}(y-t)^{r_2-1} \\
& \times f(s,t,u(s,t))dtds\|_E \\
& +\sum_{i=1}^{k}\|I_i(u(x_i^-,y))-I_i(v(x_i^-,y))\|_E \\
& +\sum_{i=1}^{k}\|I_i(u(x_i^-,0))-I_i(v(x_i^-,0))\|_E \\
& +\frac{1}{\Gamma(r_1)\Gamma(r_2)}\sum_{i=1}^{k}\int_{x_{i-1}}^{x_i}\int_0^y (x_i-s)^{r_1-1}(y-t)^{r_2-1} \\
& \times \|f(s,t,u(s,t))-f(s,t,v(s,t))\|_E dtds \\
& -\frac{1}{\Gamma(r_1)\Gamma(r_2)}\int_{x_k}^x\int_0^y (x-s)^{r_1-1}(y-t)^{r_2-1} \\
& \times \|f(s,t,u(s,t))-f(s,t,v(s,t))\|_E dtds \\
& \leq (1+k)\lambda_\Phi\Phi(x,y)+4k\Phi(x,y) \\
& +\frac{l_f(1+k)}{\Gamma(r_1)\Gamma(r_2)}\int_0^x\int_0^y (x-s)^{r_1-1}(y-t)^{r_2-1} \\
& \times \|u(s,t)-v(s,t)\|_E dtds.
\end{aligned}$$

Hence, for each $(x,y) \in (x_k,x_{k+1}] \times [0,b]$, $k=0,\dots,m$, we get

$$\|u(x,y)-v(x,y)\|_E \leq (1+k)\lambda_\Phi\Phi(x,y)+4k\Phi(x,y)$$

$$+\frac{l_f(1+k)}{\Gamma(r_1)\Gamma(r_2)}\int_0^x\int_0^y(x-s)^{r_1-1}(y-t)^{r_2-1}$$
$$\times\|u(s,t)-v(s,t)\|_E dtds.$$

From Lemma 1.68, there exists a constant $\delta=\delta(r_1,r_2)$ such that

$$\begin{aligned}\|u(x,y)-v(x,y)\|_E &\leq (4k+(1+k)\lambda_\Phi)\Phi(x,y)+\frac{\delta l_f(1+k)(4k+(1+k)\lambda_\Phi)}{\Gamma(r_1)\Gamma(r_2)}\\ &\times\int_0^x\int_0^y(x-s)^{r_1-1}(y-t)^{r_2-1}\Phi(s,t)dtds\\ &\leq [(4m+(1+m)\lambda_\Phi)\Phi(x,y)\\ &+\delta l_f\lambda_\Phi(1+m)(4m+(1+m)\lambda_\Phi)]\Phi(x,y)\\ &:= c_{f,m,\Phi}\Phi(x,y).\end{aligned}$$

Finally, problem (13.39) is generalized Ulam-Hyers-Rassias stable. □

13.6.3. An Example

Let

$$E=l^1=\left\{w=(w_1,w_2,\dots,w_n,\dots):\sum_{n=1}^{\infty}|w_n|<\infty\right\},$$

be the Banach space with norm

$$\|w\|_E=\sum_{n=1}^{\infty}|w_n|.$$

Consider the following infinite system of partial hyperbolic fractional impulsive differential equations of the form

$$\begin{cases}{}^cD^r_{\theta_k}u(x,y)=f(x,y,u(x,y)); & \text{if } (x,y)\in J_k,\ k=0,\dots,m,\\ u(x_k^+,y)=u(x_k^-,y)+I_k(u(x_k^-,y)); & \text{if } y\in[0,1],\ k=1,\dots,m,\end{cases}\tag{13.46}$$

where $J=[0,1]\times[0,1]$, $(r_1,r_2)\in(0,1]\times(0,1]$, $u=(u_1,u_2,\dots,u_n,\dots)$,

$$f=(f_1,f_2,\dots,f_n,\dots),\ {}^cD^r_{\theta_k}u=({}^cD^r_{\theta_k}u_1,{}^cD^r_{\theta_k}u_2,\dots,{}^cD^r_{\theta_k}u_n,\dots);\ k=1,\dots,m,$$

$$f_n(x,y,u)=\frac{1}{(10e^{x+y+4})(1+\|u_n\|_E)};\ (x,y)\in[0,1]\times[0,1],\ n\in\mathbb{N}.$$

and

$$I_k(u(x_k^-,y))=\frac{xy^2}{(30e^{x+y+4})(1+\|u(x_k^-,y)\|_E)},\ y\in[0,1],\ k=1,\dots,m.$$

Clearly, the function f is continuous. For each $n\in\mathbb{N}$, $u,\overline{u}\in E$ and $(x,y)\in[0,1]\times[0,1]$ we have

$$\|f(x,y,u(x,y))-f(x,y,\overline{u}(x,y))\|_E\leq\frac{1}{10e^4}\|u-\overline{u}\|_E.$$

and

$$\|I_k(u)-I_k(\overline{u})\|_E\leq\frac{1}{30e^4}\|u-\overline{u}\|_E.$$

Then, the hypotheses (13.10.1) and (13.10.2) are satisfied with $l_f = \frac{1}{10e^4}$ and $l^* = \dfrac{1}{30e^4}$. We shall show that condition (13.45) holds with $a = b = 1$. Indeed, if we assume, for instance, that the number of impulses $m = 3$, then we have

$$2ml^* + \frac{2l_f a^{r_1} b^{r_2}}{\Gamma(1+r_1)\Gamma(1+r_2)} = \frac{1}{e^4} + \frac{1}{5e^4\Gamma(1+r_1)\Gamma(1+r_2)} < \frac{9}{5e^4} < 1,$$

which is satisfied for each $(r_1, r_2) \in (0,1] \times (0,1]$. The hypothesis (13.11.1) is satisfied with $\Phi(x,y) = xy^2$ and $\lambda_\Phi = 8$. Indeed, for each $(x,y) \in [0,1] \times [0,1]$ we get

$$(I_\theta^r \Phi)(x,y) = \frac{\Gamma(2)\Gamma(3)}{\Gamma(2+r_1)\Gamma(3+r_2)} x^{1+r_1} y^{2+r_2} < 8xy^2 = \lambda_\Phi \Phi(x,y).$$

Finally, for each $(x,y) \in [0,1] \times [0,1]$ and $u \in E$, we get

$$\|I_k(u)\|_E \leq xy^2 = \Phi(x,y).$$

Hence, the hypothesis (13.11.2) is satisfied. Consequently Theorem 13.42 implies that the problem (13.46) is generalized Ulam-Hyers-Rassias stable.

13.7. Notes and Remarks

The results of Chapter 13 are taken from Abbas *et al.* [29, 30, 31, 46]. Other results may be found in [234, 235, 328, 364, 365, 366, 367, 368].

Appendix A

Interpretations and Motivations for Fractional Derivative

A.1. Some Physical Interpretations

A.1.1. Physical Interpretation of the Stieltjes Integral

Imagine a car equipped with two devices for measurements: the speedometer recording the velocity $v(s)$ and the clock which should show the time s. The clock, however, shows the time incorrectly; let us suppose that the relationship between the wrong time s, which is shown by the clock and which the driver considers as the correct time, on one hand, and the true time T, on the other, is described by the function $T = g(s)$. This means that where the driver "measures" the time interval ds, the real time interval is given by $dT = dg(s)$. The driver A, who do not know about wrong operation of the clock, will compute the passed distance as the classical integral:

$$S_A(t) = \int_0^t v(s)ds. \tag{A.1}$$

However, the observer 0 knowing about the wrong clock and having the function $g(s)$, which restores the correct values of time from the driver's wrong time s, will compute the really passed distance as

$$S_0(t) = \int_0^t v(s)dg(s). \tag{A.2}$$

This example shows that the Stieltjes integral (A.2) can be interpreted as the real distance passed by a moving object, for which we have recorded correct values of speed and incorrect values of time; the relationship between the wrongly recorded time s and the correct time T is given by a known function $T = g(s)$.

A.1.2. Physical Interpretation of Fractional Integration: Shadows of the Past

Now let us consider the left-sided Riemann-Liouville fractional integral

$$S_0(t) = \int_0^t v(s)dg_t(s) = I^{\alpha}v(t); \tag{A.3}$$

where

$$g_t(s) = \frac{1}{\Gamma(\alpha)}(t^\alpha - (t-s)^\alpha). \tag{A.4}$$

The fractional integral $S_0(t)$ of the function $v(s)$ can be interpreted as the real distance passed by a moving object, for which we have recorded the local values of its speed $v(s)$ (individual speed) and the local values of its time s (individual time); the relationship between the locally recorded time s (which is considered as owing equably) and the cosmic time (which flows non-equably) is given by a known function $g_t(s)$. The function $g_t(s)$ describes the inhomogeneous time scale, which depends not only on s , but also on the parameter t representing the last measured value of the individual time of the moving object. When t changes, the entire preceding cosmic time interval changes as well. This is in agreement with the current views in physics. Indeed, B. N. Ivanov [[233], p. 33] mentioned that time intervals depend on gravitational fields. Similarly, S. Hawking [[211], p. 32-33] wrote that:

": . time should appear to run slower near a massive body like the earth." [. . .] ": . there is no unique absolute time, but instead each individual has his own personal measure of time that depends on where he is and how he is moving."

When a moving body changes its position in space-time, the gravitational fields in the entire space-time also changes due to this movement. As a consequence, the cosmic time interval, which corresponds to the history of the movement of the moving object, changes. This affects the calculation (using formula (A.3)) of the real distance $S_0(t)$ passed by such a moving object.

In other words, the left-sided Riemann-Liouville fractional integral of the individual speed $v(s)$ of a moving object, for which the relationship between its individual time s and the cosmic time T at each individual time instance t is given by the known function $T = g_t(s)$, represents the real distance $S_0(t)$ passed by that object.

A.2. Physical Interpretation of the Riemann-Liouville FD

On the other hand, we can use the properties of fractional differentiation and integration [337, 314] and express $v(t)$ from the equation (A.3) as a left-sided Riemann Liouville fractional derivative of

$$S_0(t)v(t) = \int_0^t S_0(s)ds \tag{A.5}$$

where D^α denotes the Riemann-Liouville fractional derivative [337, 314], which is for $0 < \alpha < 1$ defined by D^α

$$D^\alpha f(t) = \frac{1}{\Gamma(1-\alpha)}\frac{d}{dt}\int_0^t \frac{f(s)}{(t-s)^\alpha}ds. \tag{A.6}$$

This shows that the left-sided Riemann-Liouville fractional derivative of the real distance $S_0(t)$ passed by a moving object, for which the relationship between its individual time s and the cosmic time T at each individual time instance t is given by the known function $T = g_t(s)$ described by equation (A.4), is equal to the individual speed $v(s)$ of that object. On the other hand, we can differentiate the relationship (A.3) with respect to the cosmic time variable t, which gives the relationship between $v_0(t) = S_0'(t)$ the velocity of the movement from the viewpoint of the independent observer 0 and the individual velocity $v(t)$:

$$v_0(t) = \frac{d}{dt} I^{\alpha} v(t) = D^{1-\alpha} v(t) \tag{A.7}$$

Therefore, the $(1-\alpha)$-th-order Riemann-Liouville derivative of the individual velocity $v(t)$ is equal to the velocity $v_0(t)$ from the viewpoint of the independent observer, if the individual time s and the cosmic time T are related by the function $T = g_t(s)$ described by equation (A.4). For $\alpha = 1$, when there is no dynamic deformation of the time scale, both velocities coincide:

$$v_0(t) = v(t).$$

A.2.1. Physical Interpretation of the Caputo Fractional Derivative

Applying fractional integration of order $\beta = 1-\alpha$ to both parts of the relationship (A.7) gives:

$$v(t) = I^{1-\alpha} v_0(t) = I^{1-\alpha} S_0'(t) =^c D^{\alpha} S_0(t), \tag{A.8}$$

where ${}^cD^{\alpha}$ denotes the Caputo fractional derivative [56, 73], which is for $0 < \alpha < 1$ defined by ${}^cD^{\alpha}$

$${}^cD^{\alpha} f(t) = \frac{1}{\Gamma(1-\alpha)} \frac{d}{dt} \int_0^t \frac{f'(s)}{(t-s)^{\alpha}} ds. \tag{A.9}$$

The relationship (A.8) is similar to (A.7). Therefore, the Caputo fractional derivative has the same physical interpretation as the Riemann-Liouville fractional derivative. This coincidence becomes more obvious, if we recall [314] that if $f(0) = 0$, then the Riemann-Liouville derivative and the Caputo derivative of order α $(0 < \alpha < 1)$, coincide:

$${}^cD^{\alpha} f(t) =^c D^{\alpha} f(t).$$

A.3. Some Motivations

Brain tissue modeling is described in [263] as a fractional derivation operator and it would find applications in neurosurgery. A viscoelastic model of the hamstring muscle group is introduced in [203] where the stretching force in time domain involves Mittag-Leffler functions. In [311] the viscoelastic properties of the human dentin is studied via a fractional mathematical model. Some constants describing those mechanical properties are determined from a highly nonlinear system involving Mittag-Leffler functions. In [310] the authors formulate a new model for the shrinkage strain developed during photo polymeration in dental composites using a diffusion equation of fractional order. In [139] the authors study an infinitely long capacitor-resistor transmission line are recasted in terms of fractional calculus. There the characteristic of the transmission lines are expressed in terms of Mittag-Leffler functions and it is improved in [317]. A dynamical system of fractional order modeling the HIV infection of $CD4^{+}T$-cells is proposed in [167] and a fractional logistic equation in [182]. Other models on fractional diffusion are proposed in [198, 199, 200, 273, 289].

A.3.1. Spring-Pot Model

We shall start with a spring-pot alone, which is a linear viscoelastic element whose behaviour is intermediate between that of an elastic element (spring) and a viscous element (dashport). The term "spring-pot" was introduced by Koeller [249], although the concept of an element with intermediate properties had been introduced some time earlier. The constitutive equation of a spring-pot is

$$\sigma(t) = K_0 D^{\alpha}\varepsilon(t) \ \text{ or } \ \varepsilon(t) = \frac{1}{K_0} D^{-\alpha}\sigma(t). \tag{A.10}$$

where σ is stress, ε is strain, and K is the model constant. The spring-pot is the viscoelastic version of Westerund's "simple model" (Westerlund [369]). If $\alpha = 0$ the element is linear elastic (Hookean spring) whereas if $\alpha = 1$ it is purely viscous (Newtonian dashpot). Insight can be gained from the response of a spring-pot in few simple cases, using the general relationship (Podlubny [314]; Samko *et al* [337])

$$D^{\alpha}(at^p) = a\frac{\Gamma(1+p)}{\Gamma(1+p-\alpha)}t^{p-\alpha}. \tag{A.11}$$

Creep or general finite load. In the case of creep, a stress step σ_0 is applied at initial time $t = 0$. The strain response is hence $\varepsilon(t) = \dfrac{\sigma_0}{\Gamma(1+\alpha)}t^{\alpha}$. The initial value of the strain, only an elastic (retarded) response. However, the first ordinary derivative of strain is unbounded, so that a finite though undefined strain can be described in an arbitrarily small time interval.

The change of $\varepsilon(t)$ is described by the fractional differential equation

$$D^{\alpha}\varepsilon(t) = \frac{\sigma_0}{K}. \tag{A.12}$$

In accordance with the theory of fractional differential equation in terms of Riemann-Liouville derivatives, an initial condition involving $D^{1-\alpha}\varepsilon(t)$ is required. This condition can be found by taking the first-order integral of the constitutive equation as

$$[D^{\alpha-1}\varepsilon(t)]_{t\to 0} = [D^{-1}\frac{\sigma_0}{K}]_{t\to 0}. \tag{A.13}$$

In the case under consideration stress is finite at all times, hence $[D^{-1}\sigma_0]_{t\to 0} = 0$, which leads to zero initial condition for $D^{\alpha-1}\varepsilon(t)$, namely

$$[D^{\alpha-1}\varepsilon(t)]_{t\to 0} = 0. \tag{A.14}$$

the same consideration apply to a general finite load $\sigma(t)$. In the latter case the equation to be solved is

$$D^{\alpha}\varepsilon(t) = \frac{\sigma(t)}{K}. \tag{A.15}$$

and the initial condition to be attached to this equation is the zero initial condition (A.14).

Stress relaxation or general deformation The stress response to a strain step ε_0 is $\sigma(t) = \dfrac{\varepsilon_0 K}{\Gamma(1-\alpha)t^{-\alpha}}$. The initial stress is unbounded reflecting the fact that a spring-pot (just like a dashpot) cannot respond immediately to a bounded stress: it has an infinite initial modulus or a vanishing initial compliance. However, relaxation to a finite though undefined stress occurs in an arbitrarily small time interval. The change of $\sigma(t)$ is described by the fractional differential equation

$$D^{-\alpha}\sigma(t) = K\varepsilon_0. \tag{A.16}$$

From the known value of ε_0 we can obtain the initial value (as t approaches zero) of $D^{-\alpha}\sigma(t)$). Clearly, if $D^{-\alpha}\sigma(t)$ is to be finite although it is defined over a vanishingly small time interval, $\sigma(t)$ must be unbounded. On the contrary, the initial value of $D^{-\alpha}\sigma(t)$ is well defined and finite, and that of $D^{-\alpha}\sigma(t)$ is zero. Thus, contrary to the idea expressed by some authors (e.g. Glöckle and Nonnenmacher [197]), initial value problems expressed in terms of fractional integrals are not better posed than those expressed in terms of fractional derivatives. If strain increases linearly with time, stress increases as t1)a. Stress is bounded, but the initial values of its integer order derivatives are unbounded. The known strain rate allows us to define the initial value of $D^{-\alpha}\sigma(t)$. In fact, in this case, zero initial conditions are found both for $D^{-\alpha}\sigma(t)$ and $D^{-\alpha}\sigma(t)$. For any general finite strain $\varepsilon(t)$, following the same reasoning, again zero initial conditions are found. In all three examples given here, initial conditions expressed in terms of fractional derivatives or integrals arise naturally when taking measurable quantities into account.

A.3.2. Impulse Response

The impulse response is seldom used in viscoelasticity except as a mathematical convenience, because it is even more problematic to apply a homogeneous impulse of stress or strain on a sample than it is to apply a step. However, we shall investigate the impulse response following the same reasoning as for the step response above. Consider an impulse of stress defined as $B\delta(t)$ applied to the spring-pot at time $t = 0$. After that, the stress remains zero. The strain response is $\varepsilon(t) = \dfrac{B}{K\Gamma(\alpha)}t^{\alpha-1}$. The initial stress singularity gives rise to a lower order strain singularity, since a spring-pot cannot deform immediately. The strain $\varepsilon(t)$ for $t > 0$ is the solution to the fractional differential equation

$$D^{\alpha}\varepsilon(t) = 0. \tag{A.17}$$

In accordance with the theory of fractional differential equations with Riemann-Liouville derivatives, an initial condition involving $[D^{\alpha-1}\varepsilon(t)]_{t\to 0}$ is required. This can be found through integration of the constitutive equation, as

$$[D^{\alpha-1}\varepsilon(t)]_{t\to 0} = \left[\frac{D^{-1}\sigma(t)}{K}\right]_{t\to 0} = \frac{B}{K}. \tag{A.18}$$

which gives the following initial condition to equation (A.17):

$$[D^{\alpha-1}\varepsilon(t)] = \frac{B}{K}. \tag{A.19}$$

In this problem Riemann-Liouville derivatives B is the initial impulse of stress $\sigma(t)$, $\varepsilon(t)$ is the strain after application of this impulse, and the known impulse of stress yields a nonzero initial condition (A.19) involving a fractional derivative of strain. This fractional derivative is nonzero, well defined, and bounded. Note that both strain and its integer-order derivatives are unbounded, and its first-order integral is zero, so that a meaningful initial condition expressing the loading conditions cannot be obtained using integral-order derivatives.

The physically unrealistic stress response to a prescribed strain impulse will not be considered here. In fact, the analytical solution has a strong $t^{-(1+\alpha)}$ divergence, reflecting the fact that a strain impulse cannot be applied to a spring-pot.

A.4. Some General Interpretation of Riemann-Liouville Derivative

Now, after introducing the above simple example, let us formulate present the general approach to interpretation of initial conditions involving the RiemannLiouville fractional derivatives due to Heymans and Podlubny [223]. In a general case, when we consider some fractional differential equation for, say, $U(t)$, we have to consider also some function $V(t)$, for which some dual relation exists between $U(t)$ and $V(t)$. For example, in viscoelasticity we have to consider the pair of stress $r(t)$ and strain $e(t)$; in electrical circuits the pair of current $i(t)$ and voltage $v(t)$; in heat conduction the pair of the temperature difference $T(t)$ and the heat flux $q(t)$, etc. Functions $U(t)$ and $V(t)$ are normally related by some basic physical law for the particular field of science. In each scientific field there are such pairs of functions like the aforementioned, which are as inseparable as Siamese twins: the left-hand side of the initial condition involves one of them, whereas the evaluation of the right-hand side is related to the other. This concept is not restricted to the spring-pot treated above, but is further applied in the subsequent sections to more elaborate models of viscoelastic behavior. Indeed, a spring-pot is a particularly crude model, which has several unrealistic and un physical characteristics. As pointed out above, it has a vanishing initial compliance or an infinite initial modulus. Viscoelastic solids, on the contrary, have a well-defined instantaneous modulus. (Note that an unbounded initial modulus is no more of a problem when describing a viscoelastic fluid than it is when describing Newtonian viscosity: if a step strain is applied to a dashpot, the initial stress is also unbounded). At long times there is no limit to an elastic strain of a spring-pot: creep continues indefinitely. Also, stress relaxes to vanishingly small values. The increase of stress in constant strain-rate conditions means that if attempting to describe a viscoelastic fluid, steady-state flow is never attained. When describing a viscoelastic solid, again we find an unbounded modulus at t=0. In spite of these limitations, the spring-pot can give an approximate description of polymer viscoelasticity in the intermediate time range. Several slightly more elaborate models, which alleviate some oversimplifications of a single spring-pot, will be investigated below.

A.4.1. The Fractional Order Voigt Model

The fractional Voigt model (a spring and a spring-pot in parallel) is nowadays generally understood as a long term approximation to the fractional Zener model and it might seem irrelevant to express concern over initial value problems for the Voigt model. However, as the purpose of this note is mainly to examine how initial conditions endowed with physical meaning may be expressed in systems whose constitutive equations contain fractional derivatives, we shall continue to examine the fractional Voigt model. The constitutive equation of this model is

$$\sigma(t) = E\varepsilon(t) + KD^{\alpha}\varepsilon(t) \tag{A.20}$$

The Voigt element or associations thereof are considered in viscoelasticity modeling to be appropriate to obtain the strain response to a prescribed stress program, so we shall investigate only such cases here. Assume a stress impulse $B\delta(t)$ is applied to a Voigt element at time t=0. Then the fractional order equation we need to solve for $\varepsilon(t)$ $(t > 0)$ is

$$E\varepsilon(t) + KD^{\alpha}\varepsilon(t) = 0. \tag{A.21}$$

In agreement with the theory of fractional differential equations in terms of RiemannLiouville derivatives, we need an initial condition, which will involve the value of

$D^{\alpha-1}\varepsilon(t)$ for $t \to 0$. This condition can be obtained by integration of the constitutive equation as

$$[ED^{-1}\varepsilon(t) + KD^{\alpha-1}\varepsilon(t) = D^{-1}\sigma(t)]_{t\to 0}. \tag{A.22}$$

The limit of the right-hand side is the magnitude B of the stress impulse. On physical grounds, the spring-pot cannot deform instantaneously under a finite stress, and, as is the case for a spring-pot alone, any singularity of $\varepsilon(t)$ must be weaker than that of the stress impulse, thus

$$[D^{\alpha-1}\varepsilon(t)]_{t\to 0} = 0.$$

This can also be found from examination of the behavior of the left-hand side of the relationship (A.22): if $[D^{\alpha-1}\varepsilon(t)]_{t\to 0} = 0$ is nonzero, then $[D^{\alpha-1}\varepsilon(t)]_{t\to 0} = 0$ is unbounded and equation A.22 cannot be fulfilled. Hence the initial condition finally takes on the form of $[KD^{\alpha-1}\varepsilon(t)]_{t\to 0} = B$. This condition expresses the initial value of the fractional derivative of strain, $D^{\alpha-1}\varepsilon(t)$, in terms of the stress impulse. We see that the initial condition is obtained by expressing a fractional derivative of the unknown strain in terms of a measurable, physically meaningful value of its inseparable Siamese twinthe stress. The obtained initial condition is, in fact, identical to the initial condition of the spring-pot alone. This reflects the known fact that the spring in the Voigt model only affects long-term behavior. Now let us consider creep, i.e. the response to a stress step σ_0 applied at $t = 0$. The equation to be solved for the strain $\varepsilon(t)$ is

$$E\varepsilon(t) + KD^{\alpha}\varepsilon(t) = \sigma_0. \tag{A.23}$$

and the initial condition for equation A.23 can be found from

$$[ED^{-1}\varepsilon(t) + KD^{\alpha-1}\varepsilon(t) = D^{-1}\sigma(t)]_{t\to 0}. \tag{A.24}$$

where the limit of the right-hand side is zero. A bounded stress can produce only a bounded strain, so the limit of the first-order ordinary integral of strain in the left-hand side is also zero. Thus, the initial condition has the form:

$$[D^{\alpha-1}\varepsilon(t)]_{t\to 0} = 0. \tag{A.25}$$

Once more, knowledge of a measurable quantity (σ_0) leads to an initial condition expressed in terms of a fractional order derivative of the unknown $(\varepsilon(t))$, its inseparable Siamese twin.

The case of a general finite stress program is similar to that of creep. The equation to be solved is now

$$E\varepsilon(t)+KD^{\alpha}\varepsilon(t)=\sigma(t). \tag{A.26}$$

and the initial condition is identical to the condition (A.26) obtained in creep.

A.4.2. The Fractional Order Maxwell Model

To keep to a simple model while describing realistic behavior for a viscoelastic solid, a spring expressing instantaneous elasticity must be associated in series with the spring-pot. This eliminates the unbounded initial stress in description of relaxation. The Maxwell element or associations thereof are considered in viscoelasticity modeling to be appropriate to obtain the stress response to a prescribed strain program, so we shall investigate only such cases here. The constitutive equation of the Maxwell model is

$$\varepsilon(t)=\frac{1}{E}\sigma(t)+\frac{1}{K}D^{-\alpha}\sigma(t)$$

or

$$\frac{1}{E}D^{\alpha}\sigma(t)+\frac{1}{K}\sigma(t)=D^{\alpha}\varepsilon(t) \tag{A.27}$$

In stress relaxation, a step strain e0 is applied at $t=0$. Then the equation to be solved is

$$\frac{1}{E}D^{\alpha}\sigma(t)+\frac{1}{K}\sigma(t)=\frac{\varepsilon_0 t^{-\alpha}}{\Gamma(1-\alpha)}. \tag{A.28}$$

An initial condition is required, involving the value of $[D^{1-\alpha}\sigma(t)]_{t\to 0}$. Integrating the constitutive equation A.27 and considering the limit as t approaches zero, we have

$$[\frac{1}{E}D^{\alpha}\sigma(t)+\frac{1}{K}\sigma(t)=D^{\alpha}\varepsilon(t)]_{t\to 0}. \tag{A.29}$$

Since strain remains bounded during loading, and $\alpha<1$, the right-hand side inside brackets is bounded and vanishes when $t\to 0$. Since the left-hand side is a linear combination of positive functions with positive coefficients, it can only vanish if each term vanishes. This means that stress remains bounded during loading, and hence that we obtain the following initial condition:

$$[D^{\alpha-1}\varepsilon(t)]_{t\to 0}. \tag{A.30}$$

Here again we observe that the initial condition on the unknown stress arises naturally from its Siamese twin, the known strain. Now we shall consider the strain impulse response. A strain impulse of magnitude $A\delta(t)$ is applied at time $t=0$. Thereafter, the equation to be solved is

$$\frac{1}{E}\sigma(t)+\frac{1}{K}D^{-\alpha}\sigma(t)=0. \tag{A.31}$$

The required initial condition is obtained as above by integrating the constitutive equation (A.27) and considering the limit as t approaches zero:

$$\left[\frac{1}{E}D^{-1}\sigma(t)+\frac{1}{K}D^{-\alpha-1}\sigma(t)=D^{-1}\varepsilon(t)\right]_{t\to 0}. \tag{A.32}$$

The limit of the right-hand side of equation (A.32) is A. Hence the limit of the left-hand side must also be bounded. This means that the limit of the first-order integral of stress must be bounded, and the $\alpha+1$ integral must vanish, and the initial condition is finally

$$\left[\frac{1}{E}D^{-1}\sigma(t)\right]_{t\to 0}=A. \tag{A.33}$$

The singularity in the stress response to a strain impulse is now of the same order as that of the strain impulse itself: adding a spring in series with the springpot has weakened the singularity. The strain response to a stress impulse is identical to that of a spring-pot alone since the impulse response of the spring vanishes.

A.4.3. The Fractional Order Zener Model

Among the fractional order models of viscoelasticity considered in this article, the most general is the Zener model. Its constitutive equation is

$$\sigma(t)+\nu_0 D^{\alpha}\sigma(t)=\lambda\varepsilon(t)+\mu_0 D^{\alpha}\varepsilon(t), \tag{A.34}$$

where $\lambda=E_\infty$ long-term modulus $\mu=\dfrac{K(E_0-E_\infty)}{E_0}$, $\nu=\dfrac{\mu}{E_0}$ and E_0 is the instantaneous modulus. Let us first investigate the response to a stress impulse $B\delta(t)$ applied to the Zener element at time $t=0$. Then the fractional differential equation we need to solve for $\varepsilon(t)$ $(t>0)$ is

$$\lambda\varepsilon(t)+\mu_0 D^{\alpha}\varepsilon(t)=0. \tag{A.35}$$

In accordance with the theory of fractional differential equations, we need an initial condition involving the initial value of $D^{\alpha-1}\varepsilon(t)$. Integration of the constitutive equation gives

$$D^{-1}\sigma(t)+\nu_0 D^{\alpha-1}\sigma(t)=\lambda_0 D^{-1}\varepsilon(t)+\mu_0 D^{\alpha-1}\varepsilon(t). \tag{A.36}$$

The initial condition can be found by considering equation (A.36) as $t\to 0$:

$$[D^{-1}\sigma(t)+\nu_0 D^{\alpha-1}\sigma(t)=\lambda_0 D^{-1}\varepsilon(t)+\mu_0 D^{\alpha-1}\varepsilon(t)]_{t\to 0}. \tag{A.37}$$

Using considerations similar to those in case of the Voigt model under stress impulse, we obtain the initial condition in the form

$$[\mu_0 D^{\alpha-1}\varepsilon(t)]_{t\to 0}=B. \tag{A.38}$$

As in case of the Voigt model, this condition gives the initial value of the fractional derivative of unknown strain, $D^{\alpha-1}\varepsilon(t)$ in terms of its inseparable twinthe stress. The right- and left-hand sides of equations (A.35) and (A.36) are formally identical, hence following the same reasoning as above the response to a strain impulse $A\delta(t)$ applied to the Zener element at time $t=0$ will be the solution to the equation

$$\sigma(t)+\nu D^{\alpha}\sigma(t)=0. \tag{A.39}$$

with the initial condition

$$[\nu D^{\alpha-1}\varepsilon(t)]=A. \tag{A.40}$$

This formal equivalence between response to a stress or strain impulse reflects the well-known fact that the Zener model is the simplest model capable of describing response to a stress or strain program equally well. In case of creep, i.e. a step-stress $\sigma(t) = \sigma_0$ for $\sigma > 0$, we have the equation for $\varepsilon(t)$:

$$\lambda\varepsilon(t) + \mu_0 D^{\alpha}\varepsilon(t) = \sigma_0 + \nu\sigma_0 \frac{t^{-\alpha}}{\Gamma(1-\alpha)}. \tag{A.41}$$

The initial condition can also be found by considering equation (A.36) as $t \to 0$. Following a similar reasoning to that given above for the Maxwell model in stress relaxation, we find a zero initial condition to accompany equation (A.41):

$$[D^{\alpha-1}\varepsilon(t)]_{t\to 0} = 0. \tag{A.42}$$

This initial condition in terms of fractional derivative of $\varepsilon(t)$ appeared from consideration of its inseparable twin $\sigma(t)$. Similarly, for stress relaxation, $\varepsilon(t) = \varepsilon_0$, we obtain the equation for $\sigma(t)$:

$$\sigma(t) + \mu_0 D^{\alpha}\sigma(t) = \lambda\varepsilon_0 + \nu\varepsilon_0 \frac{t^{-\alpha}}{\Gamma(1-\alpha)}. \tag{A.43}$$

The initial condition for the unknown stress $\sigma(t)$,

$$[D^{\alpha-1}\sigma(t)]_{t\to 0} = 0, \tag{A.44}$$

appears naturally from consideration of the initial value of strain. Let us now consider the case of general load $\sigma(t) = \sigma_*(t)$. The equation to be solved for $\varepsilon(t)$ is

$$\lambda\varepsilon(t) + \mu_0 D^{\alpha}\varepsilon(t) = \sigma_* + \nu\sigma_* \frac{t^{-\alpha}}{\Gamma(1-\alpha)}. \tag{A.45}$$

The corresponding initial condition can be obtained using the following procedure. Consider some small $t = a$. Starting at $t = 0$, stress $\sigma(t)$ must be recorded until $t = a$, and based on the recorded values the left-hand side of the relationship (A.36) must be evaluated. The obtained quantity provides an approximation of the initial value for the expression in the right-hand side of equation (A.36). In some cases it is possible to find the limit of such approximation as $a \to 0$. For example, for a physically realisable continuous load $\sigma_*(t)$ we obtain a zero initial condition in the form:

$$[D^{\alpha-1}\varepsilon(t)]_{t\to 0} = 0. \tag{A.46}$$

It is worth mentioning that this procedure amounts, in fact, to the same as measuring the initial value of, for example, the first derivative in the case of classical differential equations of integer order. From the examples given above, it can be seen that for any physically realistic model, zero initial conditions will be found for a continuous loading program or even in the case of a step discontinuity. Nonzero conditions will only be found in the case of an impulse. Conditions expressed in terms of RiemannLiouville fractional derivatives. To summarize, expressing initial conditions in terms of fractional derivatives of a function $U(t)$

is not a problem, because it does not require a direct experimental evaluation of these fractional derivatives. Instead, one should consider its inseparable twin $V(t)$ related to $U(t)$ via a basic physical law, and measure (or consider) its initial values. It is worth noting that the only case where nonzero initial conditions appeared in our considerations, is the case of impulse response. In other cases (including the Zener model under physically realisable load program), the initial conditions are zero, and in such cases the use of the RiemannLiouville derivatives, the Grü nwald- Letnikov derivatives, and the Caputo derivatives is equivalent.

A.5. Diffusion Phenomena

Diffusion phenomena, heat transfer, and fluid motion are all represented by a partial differential equation of the parabolic type:

$$\frac{\partial}{\partial t}f(x,t) - k^2\frac{\partial^2}{\partial^2 x}f(x,t) = 0 \tag{A.47}$$

and the operator factorization form is the following:

$$\left(\frac{\partial^{0.5}}{\partial^{0.5} + k\dfrac{\partial}{\partial x}}\right)\left(\frac{\partial^{0.5}}{\partial^{0.5} - k\dfrac{\partial}{\partial x}}\right) f(x,t) = 0. \tag{A.48}$$

It can be shown that only the first operator needs to be considered:

$$\frac{\partial^{0.5}}{\partial^{0.5} f(x,t) + k\dfrac{\partial}{\partial x}} f(x,t) = 0 \tag{A.49}$$

Since most of transport phenomena are also governed by flux-force equation like $J(x,t) = \frac{\partial}{\partial x}f(x,t)$, substituting this relation in (A.49) we obtain:

$$\frac{\partial^{0.5}}{\partial t^{0.5}}f(x,t) = -kJ(x,t). \tag{A.50}$$

A.6. Notes and Remarks

The models of Appendix A are taken from [73, 223, 315]. Other results and motivations of fractional differential equation can be found in [147, 148, 300, 314, 318, 337, 340, 344].

References

[1] S. Abbas, D. Baleanu and M. Benchohra, Global attractivity for fractional order delay partial integro-differential equations, *Adv. Difference Equ.* 2012, 19 pages doi:10.1186/1687-1847-2012-62.

[2] S. Abbas and M. Benchohra, Darboux problem for perturbed partial differential equations of fractional order with finite delay, *Nonlinear Anal. Hybrid Syst.* 3 (2009), 597-604.

[3] S. Abbas and M. Benchohra, Partial hyperbolic differential equations with finite delay involving the Caputo fractional derivative, *Commun. Math. Anal.* 7 (2009), 62-72.

[4] S. Abbas and M. Benchohra, Upper and lower solutions method for impulsive partial hyperbolic differential equations with fractional order, *Nonlinear Anal. Hybrid Syst.* 4 (2010), 406-413.

[5] S. Abbas and M. Benchohra, The method of upper and lower solutions for partial hyperbolic fractional order differential inclusions with impulses, *Discuss. Math. Differ. Incl. Control Optim.* 30 (1) (2010), 141-161.

[6] S. Abbas and M. Benchohra, Impulsive partial hyperbolic differential inclusions of fractional order, *Demonstratio Math.* XLIII (4) (2010), 775-797.

[7] S. Abbas and M. Benchohra, Darboux problem for partial functional differential equations with infinite delay and Caputo's fractional derivative, *Adv. Dyn. Syst. Appl.* 5 (1) (2010), 1-19.

[8] S. Abbas and M. Benchohra, Impulsive partial hyperbolic functional differential equations of fractional order with state-dependent delay, *Frac. Calc. Appl. Anal.* 13 (3) (2010), 225-244.

[9] S. Abbas and M. Benchohra, Upper and lower solutions method for the darboux problem for fractional order partial differential inclusions, *Int. J. Modern Math.* 5 (3) (2010), 327-338.

[10] S. Abbas and M. Benchohra, Upper and lower solutions method for partial hyperbolic functional differential equations with Caputo's fractional derivative, *Libertas Math.* 31 (2011), 103-110.

[11] S. Abbas and M. Benchohra, Darboux problem for implicit impulsive partial hyperbolic differential equations, *Electron. J. Differential Equations* 2011 (2011), 15 pp.

[12] S. Abbas and M. Benchohra, On the set of solutions of fractional order Riemann-Liouville integral inclusions, *Demonstratio Math.* XLVI (2) (2013), 271-281.

[13] S. Abbas and M. Benchohra, On the set of solutions for the Darboux problem for fractional order partial hyperbolic functional differential inclusions, *Fixed Point Theory* 14 (2) (2013), 253-262.

[14] S. Abbas and M. Benchohra, Fractional order Riemann-Liouville integral equations with multiple time delay, *Appl. Math. E-Notes* 12 (2012), 79-87.

[15] S. Abbas and M. Benchohra, Fractional order Riemann-Liouville integral inclusions with two independent variables and multiple delay, *Opuscula Math.* 33 (2) (2013), 209-222.

[16] S. Abbas and M. Benchohra, On the existence and local asymptotic stability of solutions of fractional order integral equations, *Comment. Math.* 52 (1) (2012), 91-100.

[17] S. Abbas and M. Benchohra, Fractional Order partial hyperbolic differential equations involving Caputo's derivative, *Stud. Univ. Babeş-Bolyai Math.* 54 (2012), 469-479.

[18] S. Abbas and M. Benchohra, Impulsive partial functional integro-differential equations of fractional order, *Commun. Appl. Anal.* 16 (2) (2012), 249-260.

[19] S. Abbas and M. Benchohra, Global stability results for nonlinear partial fractional order Riemann-Liouville Volterra-Stieltjes functional integral equations, *Math. Sci. Res. J.* 16 (2012), 82–92.

[20] S. Abbas and M. Benchohra, Nonlinear quadratic Volterra Riemann-Liouville integral equations of fractional order, *Nonlinear Anal. Forum* 17 (2012), 107-115.

[21] S. Abbas and M. Benchohra, A global uniqueness result for fractional order implicit differential equations, *Comment. Math. Univ. Carolinae* 53 (2012), 605-614.

[22] S. Abbas and M. Benchohra, Uniqueness results for Fredholm type fractional order Riemann-Liouville integral equations, *Fasciculi Math.* 51 (2013), 5-19.

[23] S. Abbas and M. Benchohra, Existence and stability for partial functional integro-differential equations of fractional order with multiple time delay, *Nonlinear Studies* 20 (2) (2013), 149-162.

[24] S. Abbas and M. Benchohra, Global asymptotic stability for nonlinear multi-delay differential equations of fractional order, *Proc. A. Ramadze Math. Inst.* 161 (2013), 1-13.

[25] S. Abbas and M. Benchohra, Fractional order integral equations of two independent variables, *Appl. Math. Comput.* 227 (2014), 755-761.

[26] S. Abbas and M. Benchohra, Qualitative theory for fractional order Riemann-Liouville integral equations in two independent variables, *Tamsui Oxford J. Inf. Math. Sci.* 29 (2) (2013), .

[27] S. Abbas and M. Benchohra, Nonlinear fractional order Riemann-Liouville Volterra-Stieltjes partial integral equations on unbounded domains, *Commun. Math. Anal.* 14 (1) (2013), 104-117.

[28] S. Abbas and M. Benchohra, Existence and stability of nonlinear fractional order Riemann-Liouville Volterra-Stieltjes multi-delay lntegral equations, *J. Integral Equ. Appl.*, 25 (2) (2013), 143-158.

[29] S. Abbas and M. Benchohra, Ulam-Hyers stability for the Darboux problem for partial fractional differential and Integro-differential equations via Picard operators, *Results. Math.* 65 (1-2) (2014), 67-79.

[30] S. Abbas and M. Benchohra, On the generalized Ulam-Hyers-Rassias stability for Darboux problem for partial fractional implicit differential equations, *Appl. Math. E-Notes* 14 (2014), 20-28.

[31] S. Abbas and M. Benchohra, Some stability concepts for Darboux problem for partial fractional differential equations on unbounded domain, *Fixed Point Theory.* (to appear).

[32] S. Abbas, M. Benchohra and A. Cabada, Partial neutral functional integro-differential equations of fractional order with delay, *Bound. Value Prob.* Vol. 2012 (2012), 128, 15 pp.

[33] S. Abbas , M. Benchohra and T. Diagana, Existence and attractivity results for some fractional order partial integro-differential equations with delay, *Afr. Diaspora J. Math.* 15 (2) (2013), 87-100.

[34] S. Abbas, M. Benchohra and L. Gorniewicz, Existence theory for impulsive partial hyperbolic functional differential equations involving the Caputo fractional derivative, *Sci. Math. Jpn..* online e- 2010, 271-282.

[35] S. Abbas, M. Benchohra and J. R. Graef, Integro-differential equations of fractional order, *Differ. Equ. Dyn. Syst.* 20 (2) (2012), 139-148.

[36] S. Abbas, M. Benchohra and J. Henderson, Global asymptotic stability of solutions of nonlinear quadratic Volterra integral equations of fractional order, *Comm. Appl. Nonlinear Anal.* 19 (2012), 79-89.

[37] S. Abbas, M. Benchohra and J. Henderson, Asymptotic behavior of solutions of nonlinear fractional order Riemann-Liouville Volterra-Stieltjes Quadratic integral equations, *Int. Electron. J. Pure Appl. Math.* 4 (3) (2012), 195-209.

[38] S. Abbas, M. Benchohra and J. Henderson, Asumptotic attractive nonlinear fractional order Riemann-Liouville integral equations in Banach algebras, *Nonlinear Stud.* 20 (2013), 1-10.

[39] S. Abbas, M. Benchohra and J. Henderson, Ulam stability for partial fractional integral inclusions via Picard operators, *J. Frac. Calc. Appl.*, 5 (2)(2014), 133-144.

[40] S. Abbas, M. Benchohra and G.M. N'Guérékata, *Topics in Fractional Differential Equations*, Developments in Mathematics, 27, Springer, New York, 2012.

[41] S. Abbas, M. Benchohra and G.M. N'Guérékata, Asymptotic stability in nonlinear delay differential equations of fractional order, *J. Nonlinear Evolution Equ. Appl.* 2012, 85-96.

[42] S. Abbas, M. Benchohra, G.M. N'Guérékata and B.A. Slimani, Darboux problem for fractional order discontinuous hyperbolic partial differential equations in Banach algebras, *Complex Var. Elliptic Equ.* 57 (2012), 337-350.

[43] S. Abbas, M. Benchohra and J.J. Nieto, Global uniqueness results for fractional order partial hyperbolic functional differential equations, *Adv. Difference Equ.* 2011, Art. ID 379876, 25 pp.

[44] S. Abbas, M. Benchohra and J.J. Nieto, Global attractivity of solutions for nonlinear fractional order Riemann-Liouville Volterra-Stieltjes partial integral equations, *Electron. J. Qual. Theory Differ. Equ.* 81 (2012), 1-15.

[45] S. Abbas, M. Benchohra and J.J. Nieto, Functional implicit hyperbolic fractional order differential equations with delay, *Afr. Diaspora J. Math.* 15 (1) (2013), 74-96.

[46] S. Abbas, M. Benchohra and J.J. Nieto, Ulam stabilities for impulsive partial fractional differential equations, *Acta Univ. Palacki. Olomuc.* 53 (1) (2014), 5-17.

[47] S. Abbas, M. Benchohra, M. Rivero and J. J. Trujillo, Existence and stability results for nonlinear fractional order Riemann-Liouville Volterra-Stieltjes quadratic integral equations, submitted.

[48] S. Abbas, M. Benchohra and S. Sivasundaram, Ulam stability for partial fractional differential inclusions with multiple delay and impulses via Picard operators, *J. Nonlinear Stud.* 20 (4) (2013), 623-641.

[49] S. Abbas, M. Benchohra and J.J. Trujillo, Fractional order impulsive hyperbolic implicit differential equations with state-dependent delay, in *Proceedings of the Sixth International Conference on Analytic Methods of Analysis and Differential Equations* (AMADE-2011; Minsk, Belarus) [Dedicated to the Memory of Anatoly A. Kilbas] (Segei V. Rogosin, Editor), Publ. Center of the Belarusian State University, Minsk, Belarus, 2012, pp. 207-228 [ISBN 978-985-476-993-6].

[50] S. Abbas, M. Benchohra and A. N. Vityuk, On fractional order derivatives and Darboux problem for implicit differential equations, *Frac. Calc. Appl. Anal.* 15 (2) (2012), 168-182.

[51] S. Abbas, M. Benchohra and Y. Zhou, Darboux problem for fractional order neutral functional partial hyperbolic differential equations, *Int. J. Dyn. Syst. Differ. Equ.* 2 (3&4) (2009), 301-312.

[52] S. Abbas, M. Benchohra and Y. Zhou, Fractional order partial functional differential inclusions with infinite delay, *Proc. A. Razmadze Math. Inst.* 154 (2010), 1-19.

[53] S. Abbas, M. Benchohra and Y. Zhou, Fractional order partial hyperbolic functional differential equations with state-dependent delay, *Int. J. Dyn. Syst. Differ. Equ.* 3 (4) (2011), 459-490.

[54] N.H. Abel, Solutions de quelques problèmes à l'aide d'intégrales définies (1823). *Œuvres complètes de Niels Henrik Abel*, 1, Grondahl, Christiania, 1881, 11-18.

[55] R. P. Agarwal, B. Andradec, G. Siracusa, On fractional integro-differential equations with state-dependent delay, *Comput. Math. Appl.* 63 (3) (2011), 1142-1149.

[56] R.P Agarwal, M. Benchohra and S. Hamani, Boundary value problems for fractional differential inclusions, *Adv. Stud. Contemp. Math.*, 16 (2) (2008), 181-196.

[57] R.P Agarwal, M. Benchohra and S. Hamani, A survey on existence result for boundary value problems of nonlinear fractional differential equations and inclusions, *Acta. Appl. Math.* 109 (2010), 973-1033.

[58] R. P. Agarwal, M. Benchohra and B.A. Slimani, Existence results for differential equations with fractional order and impulses, *Mem. Differential Equations Math. Phys.* 44 (2008), 1-21.

[59] R. P. Agarwal, M. Meehan, and D. O'Regan, Fixed Point Theory and Applications, *Cambridge Tracts in Mathematics* 141, Cambridge University Press, 2001.

[60] R.P. Agarwal, D. O'Regan, S. Stanek, Positive solutions for Dirichlet problem of singular nonlinear fractional differential equations, *J. Math. Anal. Appl.* 371 (2010), 57-68.

[61] R. P. Agarwal, D. O'Regan, M. A. Taoudi. Fixed point theorems for convex-power condensing operators relative to the weak topology and applications to Volterra integral equations. *J. Integral Equations Appl.* 24 (2012), 167-181.

[62] R. P. Agarwal, D. O'Regan, M. A. Taoudi. Fixed point theorems for condensing multivalued mappings under weak topology features. *Fixed Point Theory* 12 (2011), 247–254.

[63] R.P Agarwal, Y. Zhou and Y. He, Existence of fractional neutral functional differential equations, *Comput. Math. Appl.* 59 (3) (2010), 1095–1100.

[64] R.P. Agarwal, Y. Zhou, J. Wang, X. Luo. Fractional functional differential equations with causal operators in Banach spaces. *Math. Comput. Modelling* 54 (2011), no. 5-6, 1440–1452.

[65] O.P. Agrawal, O. Defterli, D. Baleanu. Fractional optimal control problems with several state and control variables. *J. Vib. Control* 16 (2010), no. 13, 1967–1976.

[66] B. Ahmad, J.J. Nieto. Existence of solutions for impulsive anti-periodic boundary value problems of fractional order. *Taiwanese J. Math.* 15 (2011), no. 3, 981–993.

[67] B. Ahmad, S. Sivasundaram, Existence results for nonlinear impulsive hybrid boundary value problems involving fractional differential equations. *Nonlinear Anal. Hybrid Syst.* 3 (2009), 251-258.

[68] E. Ait Dads, M. Benchohra and S. Hamani, Impulsive fractional differential inclusions involving the Caputo fractional derivative, *Fract. Calc. Appl. Anal.* 12 (1) (2009), 15-38.

[69] R. Almeida, D.F.M. Torres, Fractional variational calculus for nondifferentiable functions. *Comput. Math. Appl.* 61 (2011), 3097-3104.

[70] R. Almeida, D.F.M. Torres, Necessary and sufficient conditions for the fractional calculus of variations with Caputo derivatives. *Commun. Nonlinear Sci. Numer. Simul.* 16 (2011), 1490-1500.

[71] G. A. Anastassiou, *Advances on Fractional Inequalities*, Springer, New York, 2011.

[72] D. Araya and C. Lizama, Almost automorphic mild solutions to fractional differential equations, *Nonlinear Anal.* 69 (2008), 3692-3705.

[73] P. Arena, R. Caponetto, L. Fortuna, D. Porto, *Nonlinear Noninteger Order Circuits and Systems. An Introduction*. World Scientific Series on Nonlinear Science. Series A: Monographs and Treatises, 38. World Scientific Publishing Co., Inc., River Edge, NJ, 2000. viii+203 pp.

[74] O. Arino, S. Gautier, and J.P. Penot, A Fixed-point theorem for sequentially continuous mappings with application to ordinary differential equations, *Funkc. Ekvac.* 27 (1984), 273-279.

[75] W. Arendt, C. Batty, M., Hieber, and F. Neubrander, Vector-Valued Laplace Transforms and Cauchy Problems,*Monographs in Mathematics*, 96 Birkhauser, Basel, 2001.

[76] S. Arshad, V. Lupulescu, On the fractional differential equations with uncertainty, *Nonlinear Anal.* 74 (2011), 3685-3693.

[77] J. P. Aubin, *Impulse differential inclusions and hybrid systems: a viability approach, Lecture Notes*, Universit Paris-Dauphine, 2002.

[78] J. P. Aubin and A. Cellina, *Differential Inclusions*, Springer-Verlag, Berlin-Heidelberg, New York, 1984.

[79] J.P. Aubin and H. Frankowska, *Set-Valued Analysis*, Birkhauser, Boston, 1990.

[80] K. Balachandran, S. Kiruthika, J.J. Trujillo, Existence results for fractional impulsive integrodifferetial equations in Banach spaces, *Commun. Nonlinear Sci. Numer. Simul.* 16 (2011), 1970-1977.

[81] K. Balachandran, J.J. Trujillo, The nonlocal Cauchy problem for nonlinear fractional integrodifferential equations in Banach spaces, *Nonlinear Anal.* 72 (2010), 4587-4593.

[82] D. Baleanu, K. Diethelm, E. Scalas, J.J. Trujillo, *Fractional Calculus Models and Numerical Methods*, World Scientific Publishing, New York, 2012.

[83] D. Baleanu, S.I. Vacaru. Fractional curve flows and solitonic hierarchies in gravity and geometric mechanics. *J. Math. Phys.* 52 (2011), no. 5, 053514, 15 pp.

[84] J. Banaś, Existence results for Volterra-Stieltjes quadratic integral equations on an unbounded interval, *Math. Scand.* 98(2006), 143-160.

[85] J. Banaś and B.C. Dhage, Global asymptotic stability of solutions of a functional integral equation, *Nonlinear Anal.* 69 (7) (2008), 1945-1952.

[86] J. Banaś and K. Goebel, *Measures of Noncompactness in Banach Spaces*, Marcel Dekker, New York, 1980.

[87] J. Banas and K. Sadarangani, Solvability of Volterra-Stieltjes operator-integral equations and their applications, *Comput. Math. Appl.* 41 (2001), 1535-1544.

[88] J. Banaś and T. Zając, A new approach to the theory of functional integral equations of fractional order, *J. Math. Anal. Appl.* 375 (2011), 375-387.

[89] V. Barbu, *Nonlinear Semigroups and Differential Equations in Banach Spaces*, Noordhoff International Publishing, Leiden, 1976.

[90] E. Bazhlekova, *Fractional Evolution Equations in Banach Spaces*, University Press Facilities, Eindhoven University of Technology, 2001.

[91] A. Belarbi, M. Benchohra and A. Ouahab, Uniqueness results for fractional functional differential equations with infinite delay in Fréchet spaces, *Appl. Anal.* 85 (2006), 1459-1470.

[92] M. Benchohra, K. Ezzinbi and S. Litimein, Existence and controllability results for fractional order integro-differential inclusions in Fréchet spaces, *Proc. A. Ramadze Math. Inst.* 162 (2013), 1-23.

[93] M. Benchohra, J. R. Graef and S. Hamani, Existence results for boundary value problems of nonlinear fractional differential equations with integral conditions, *Appl. Anal.* 87 (7) (2008), 851-863.

[94] M. Benchohra, J.R. Graef, F-Z. Mostefai. Weak solutions for nonlinear fractional differential equations on reflexive Banach spaces. *Electron. J. Qual. Theory Differ. Equ.* 2010, No. 54, 10 pp.

[95] M. Benchohra, J. Graef and F-Z. Mostefai, Weak solutions for boundary-value problems with nonlinear fractional differential inclusions, *Nonlinear Dyn. Syst. Theory* 11 (3) (2011), 227-237.

[96] M. Benchohra, S. Hamani and S.K. Ntouyas, boundary value problems for differential equations with fractional order, *Surv. Math. Appl.* 3 (2008), 1-12.

[97] M. Benchohra, J. Henderson and F-Z. Mostefai, Weak solutions for hyperbolic partial fractional differential inclusions in Banach spaces, *Comput. Math. Appl.* 64 (2012), 3101-3107.

[98] M. Benchohra, J. Henderson and S. K. Ntouyas, *Impulsive Differential Equations and Inclusions*, Hindawi Publishing Corporation, Vol 2, New York, 2006.

[99] M. Benchohra, J. Henderson, S.K. Ntouyas and A. Ouahab, Existence results for functional differential equations of fractional order, *J. Math. Anal. Appl.* 338 (2008), 1340-1350.

[100] M. Benchohra and S. Litimein, A global uniqueness result for abstract integral equations of Volterra type in Banach spaces, *Comm. Appl. Nonlinear Anal.* 18 (1) (2011), 37-44.

[101] M. Benchohra and S. Litimein, Fractional integro-differential equations with state-dependent delay on an unbounded domain, *Afr. Diaspora J. Math.* 12 (2) (2011), 13-25.

[102] M. Benchohra and S. Litimein, A global existence and uniqueness result for functional integro-differential equations in Banach spaces, *An. Stiint,. Univ. Al. I. Cuza Iasi. Mat. (N.S.)* LVIII (2012), 55-66.

[103] M. Benchohra, S. Litimein and G. N'Gueŕékata, On fractional integro-differential inclusions with state-dependent delay in Banach spaces, *Appl. Anal.* 92 (2) (2013), 335-350.

[104] M. Benchohra, S. Litimein, J. J. Trujillo and M.P. Velasco, Abstract fractional integro-differential equations with state-dependent delay, *Intern. J. Evol. Equa.* 6(2) (2012), 25-38.

[105] M. Benchohra and F-Z. Mostefai, Weak solutions for nonlinear differential equations with integral boundary conditions in Banach spaces, *Opuscula Math.* 32 (1) (2012), 31-40.

[106] Benchohra and F-Z. Mostefai, Measure of weak noncompactness and fractional order Pettis integral equations with multiple time delay in Banach spaces, *An. Stiint,. Univ. Al. I. Cuza Iasi. Mat. (N.S.)* (to appear).

[107] Benchohra, F-Z. Mostefai, J.J. Nieto and A. Ouahab, Weak solutions to boundary value problem of fractional order on unbounded domains in Banach spaces, *Panamer. Math J.* 23 (2013), 75-94.

[108] M. Benchohra, J.J. Nieto, D. Seba. Measure of noncompactness and hyperbolic partial fractional differential equations in Banach spaces. *Panamer. Math. J.* 20 (2010), no. 3, 27–37.

[109] M. Benchohra and S.K. Ntouyas, An existence theorem for an hyperbolic differential inclusion in Banach spaces. *Discuss. Math. Differ. Incl. Control Optim.* 22 (2002), 5-16.

[110] M. Benchohra and S.K. Ntouyas, On an hyperbolic functional differential inclusion in Banach spaces. *Fasc. Math.* 33 (2002), 27-35.

[111] M. Benchohra and S.K. Ntouyas, An existence result for hyperbolic functional differential inclusions. *Comment. Math. Prace Mat.* 42 (2002), 1-16.

[112] M. Benchohra and B. A. Slimani, Existence and uniqueness of solutions to impulsive fractional differential equations, *Electron. J. Differential Equations*, Vol. 2009 (2009), No. 10, 11 pp.

[113] A. Bica, V. A. Caus and S. Muresan, Application of a trapezoid inequality to neutral Fredholm integro-differential equations in Banach spaces, *J. Inequal. Pure Appl. Math.* 7, 5 (2006), Art. 173.

[114] H.F. Bohnenblust and S. Karlin , On a theorem of ville. Contribution to the theory of games.155-160, *Annals of Mathematics Studies,* no. 24. Priceton University Press, Princeton. N. G. 1950.

[115] M.F. Bota-Boriceanu, A. Petrusel, Ulam-Hyers stability for operatorial equations and inclusions, *Analele Univ. I. Cuza Iasi* 57 (2011), 65-74.

[116] N. Bourbaki, *Integration*, Hermann, Paris, 1968, 2eme edition.

[117] A. Bressan and G. Colombo, Extensions and selections of maps with decomposable values, *Studia Math.* 90 (1988), 69-86.

[118] D. Bugajewski, S. Szufla, Kneser's theorem for weak solutions of the Darboux problem in a Banach space, *Nonlinear Anal.* 20 (2)(1993), 169-173.

[119] T.A. Burton, Fractional differential equations and Lyapunov functionals. *Nonlinear Anal.* 74 (2011), 5648-5662.

[120] T. A. Burton, A fixed point theorem of Krasnoselskii, *Appl. Math. Lett.* 11 (1998) 85-88.

[121] T. A. Burton. *Volterra Integral and Differential Equations*, Academic Press, New York, 1983.

[122] T. A. Burton and T. Furumochi, Krasnosel'skii's fixed point theorem and stability, *Nonlinear Anal.* 49 (2002), 445-454.

[123] T.A. Burton and C. Kirk, A fixed point theorem of Krasnoselskii-Schaefer type, *Math. Nachr.* 189 (1998), 23-31.

[124] L.W. Busbridge, *The Mathematics of Radiative Transfer*, Cambridge Univ. Press, Cambridge, 1960.

[125] L. Byszewski, Existence and uniqueness of solutions of nonlocal problems for hyperbolic equation $u_{xt} = F(x,t,u,u_x)$, *J. Appl. Math. Stochastic Anal.* 3 (1990), 163-168.

[126] L. Byszewski, Theorem about existence and uniqueness of continuous solutions of nonlocal problem for nonlinear hyperbolic equation, *Appl. Anal.*, 40 (1991), 173-180.

[127] L. Byszewski, Existence and uniqueness of mild and classical solutions of semilinear functional differential evolution nonlocal Cauchy problem, *Selected Problems in Mathematics, Cracow Univ. of Tech. Monographs, Anniversary Issue* 6 (1995), 25-33.

[128] L. Byszewski and V. Lakshmikantam, Monotone iterative technique for non-local hyperbolic differential problem, *J. Math. Phys. Sci* 26 (1992), 345-359.

[129] L. Byszewski and S. N. Papageorgiou, An application of a noncompactness technique to an investigation of the existence of solutions to a nonlocal multivalued Darboux problem, *J. Appl. Math. Stoch. Anal.* 12 (1999), 179-180.

[130] M. Caputo, Linear models of dissipation whose Q is almost frequency independent II. *Geophys. J. Roy. Astron. Soc.* 13, 529-539 (1967), reprinted in *Fract. Calc. Appl. Anal.* 11 (2008), 4-14.

[131] M. Caputo, Linear models of dissipation whose $\mathbb{Q}$ is almost frequency independent-II, *Geophys. J. R. Astr. Soc.* 13 (1967), 529-539.

[132] M. Caputo, *Elasticità e Dissipazione*, Zanichelli, Bologna, 1969.

[133] K.M. Case, P.F. Zweifel, *Linear Transport Theory*, Addison-Wesley, Reading, MA, 1967.

[134] C. Castaing and M. Valadier, *Convex Analysis and Measurable Multifunctions*, Lecture Notes in Mathematics 580, Springer-Verlag, Berlin-Heidelberg-New York, 1977.

[135] L.P. Castro, A. Ramos, Hyers-Ulam-Rassias stability for a class of Volterra integral equations, *Banach J. Math. Anal.* 3(2009), 36-43.

[136] S. Chandrasekhar, *Radiative Transfer*, Oxford University Press, London, 1950.

[137] Y.-K. Chang and J.J. Nieto, Some new existence results for fractional differential inclusions with boundary conditions, *Math. Comput. Model.* 49 (2009), 605-609.

[138] F. Chen, J.J. Nieto and Y. Zhou, Global attractivity for nonlinear fractional differential equations, *Nonlinear Anal. Real World Appl.* 13 (2012) 287-298.

[139] T. Clarke, B. N. N. Achar, and J. W. Hanneken, Mittag-Leffler functions and transmission lines, *J. Molecular Liquids* 114 (1-3)(2004), 159-163.

[140] C. Corduneanu, *Integral Equations and Stability of Feedback Systems*, Acedemic Press, New York, 1973.

[141] C. Corduneanu, *Integral Equations and Applications*, Cambridge University Press, 1991.

[142] H. Covitz and S. B. Nadler Jr., Multivalued contraction mappings in generalized metric spaces, *Israel J. Math.* 8 (1970), 5-11.

[143] T. Czlapinski, On the Darboux problem for partial differential-functional equations with infinite delay at derivatives. *Nonlinear Anal.* 44 (2001), 389–398.

[144] T. Czlapinski, Existence of solutions of the Darboux problem for partial differential-functional equations with infinite delay in a Banach space. *Comment. Math. Prace Mat.* 35 (1995), 111-122.

[145] M.F. Danca, K. Diethelm, Kai. Fractional-order attractors synthesis via parameter switchings. *Commun. Nonlinear Sci. Numer. Simul.* 15 (2010), no. 12, 3745–3753.

[146] M.A. Darwish, J. Henderson, D. O'Regan, Existence and asymptotic stability of solutions of a perturbed fractional functional-integral equation with linear modification of the argument. *Bull. Korean Math. Soc.* 48 (2011), 539-553.

[147] S. Das, *Functional Fractional Calculus for System Identification and Controls*, Springer-Verlag Berlin Heidelberg 2008.

[148] S. Das, *Functional Fractional Calculus*, Springer-Verlag Berlin Heidelberg 2011.

[149] M. Dawidowski and I. Kubiaczyk, An existence theorem for the generalized hyperbolic equation $z''_{xy} \in F(x,y,z)$ in Banach space, *Ann. Soc. Math. Pol. Ser. I, Comment. Math.*, 30 (1) (1990), 41-49.

[150] A. Debbouche. Fractional evolution integro-differential systems with nonlocal conditions. *Adv. Dyn. Syst. Appl.* 5 (2010), no. 1, 49–60.

[151] A. Debbouche and D. Baleanu,Controllability of fractional evolution nonlocal impulsive quasilinear delay integro-differential systems, *Comput. Math. Appl.* 62 (2011), 1442-1450.

[152] F. S. De Blasi, G. Pianigiani and V. Staicu: On the solution sets of some nonconvex hyperbolic differential inclusions, *Czechoslovak Math. J.* 45 (1995), 107-116.

[153] K. Deimling, *Multivalued Differential Equations*, Walter De Gruyter, Berlin-New York, 1992.

[154] K. Deimling, *Nonlinear Functional Analysis*, Springer-Verlag, Berlin, 1985.

[155] D. Delbosco and L. Rodino, Existence and uniqueness for a nonlinear fractional differential equation, *J. Math. Anal. Appl.* 204 (1996), 609-625.

[156] Z. Denton, A.S. Vatsala. Monotone iterative technique for finite systems of nonlinear Riemann-Liouville fractional differential equations. *Opuscula Math.* 31 (2011), no. 3, 327–339.

[157] B. C. Dhage, A nonlinear alternative in Banach algebras with applications to functional differential equations, *Nonlinear Funct. Anal. Appl.* 8 (2004), 563-575.

[158] B. C. Dhage, Some algebraic fixed point theorems for multi-valued mappings with applications, *Diss. Math. Differential Inclusions, Control & Optim.* 26 (2006), 5-55.

[159] B. C. Dhage, Nonlinear functional boundary value problems in Banach algebras involving Carathéodories, *Kyungpook Mathematical Journal*, 46 (4) (2006), 527-541.

[160] B. C. Dhage, Existence theorems for hyperbolic differential inclusions in Banach algebras, *J. Math. Anal. Appl.* 335 (2007), 225-242.

[161] B. C. Dhage, Attractivity and positivity results for nonlinear functional integral equations via measure of noncompactness, *Diff. Equ. Appl.* 2 (3)(2010), 299-318.

[162] T. Diagana, G.M. Mophou and G.M. N'Guérékata, On the existence of mild solutions to some semilinear fractional integro-differential equations. *Electron. J. Qual. Theory Differ. Equ.* 2010, No. 58, 17 pp.

[163] J. Diestel and J. J. Uhl Jr., Vector measures, in: *Math. Surveys*, 15, *Amer. Math. Soc., Providence*, R.I. 1977.

[164] K. Diethelm, The Analysis of Fractional Differential Equations, *Lecture Notes in Mathematics*, Springer, Berlin, 2010.

[165] K. Diethelm and N. J. Ford, Analysis of fractional differential equations, *J. Math. Anal. Appl.* 265 (2002), 229-248.

[166] K. Diethelm and A.D. Freed, On the solution of nonlinear fractional order differential equations used in the modeling of viscoplasticity, in *"Scientifice Computing in Chemical Engineering II-Computational Fluid Dynamics, Reaction Engineering and Molecular Properties"* (F. Keil, W. Mackens, H. Voss, and J. Werther, Eds), pp 217-224, Springer-Verlag, Heidelberg, 1999.

[167] Y. Ding and H. He, A fractional-order differential equation model of HIV infection of CD4+ T-cells, *Math. Comput. Modelling* 50 (3-4) (2009), 386-392.

[168] S. Djebali, D. O'Regan and Z. Sahnoun, Fixed point theorems for some operator equations and inclusions in Banach spaces relative to the weak topology, *Commun. Appl. Anal.* 16 (1) (2012), 1-10.

[169] S. Djebali, D. O'Regan and Z. Sahnoun, On the solvability of some operator equations and inclusions in Banach spaces with the weak topology, *Commun. Appl. Anal.* 15 (1) (2011), 125-140.

[170] X. Dong, J. Wang, Y. Zhou, Yong. On nonlocal problems for fractional differential equations in Banach spaces. *Opuscula Math.* 31 (2011), no. 3, 341–357.

[171] S. Dugowson, L'élaboration par Riemann d'une définition de la dérivation d'ordre non entier, *revue d'histoire des Mathématiques* 3 (1997), 49-97.

[172] G. A. Edgar, Geometry and the Pettis-integral, *Indiana Univ. Math. J.* 26 (1977), 663-677.

[173] G. A. Edgar, Geometry and the Pettis-integral II, *Indiana Univ. Math. J.* 28 (1979), 559-579.

[174] M. M. El-Borai and A. Debbouche, On some fractional integro-differential equations with analytic semigroups, *Interna. J. Contemporary Math. Sci.* 4 (2009), 1361-1371.

[175] M. M. El-Borai, On some fractional evolution equations with nonlocal conditions. *Int. J. Pure Appl. Math.* 24 (2005), 405–413.

[176] M. M. El-Borai, The fundamental solutions for fractional evolution equations of parabolic type. *J. Appl. Math. Stoch. Anal.* 2004 (3), 197–211.

[177] M.M. El-Borai, K. El-Said El-Nadi, E.G. El-Akabawy On some fractional evolution equations. *Comput. Math. Appl.* 59 (2010), no. 3, 1352–1355.

[178] M.M. El-Borai, K. El-Nadi, H.A. Fouad. On some fractional stochastic delay differential equations. *Comput. Math. Appl.* 59 (2010), no. 3, 1165–1170.

[179] A. M. A. El-Sayed, Fractional order evolution equations, *J. Fract. Calc.* 7 (1995), 89-100.

[180] A. M. A. El-Sayed, Fractional order diffusion-wave equations, *Intern. J. Theo. Physics* 35 (1996), 311-322.

[181] A. M. A. El-Sayed, Nonlinear functional differential equations of arbitrary orders, *Nonlinear Anal.* 33 (1998), 181-186.

[182] A. M. A. El-Sayed, A. E. M. El-Mesiry, H. A. A. El-Saka, On the fractional-order logistic equation, *Appl. Math. Lett.* 20 (7) (2007), 817-823.

[183] K. J. Engel and R. Nagel, One-Parameter Semigroups for Linear Evolution Equations, *Graduate Texts in Mathematics*, 194, Springer-Verlag, New York, 2000.

[184] H.O. Fattorini, Second Order Linear Differential Equations in Banach Spaces, North-Holland, *Mathematical Studies*, 108, North-Holland, Amsterdam, 1985.

[185] J.B.J. Fourier, *Théorie Analytique de la Chaleur*, Didot, Paris (1822), 499-508.

[186] M. Frigon, Théorèmes d'existence de solutions d'inclusions différentielles, Topological Methods in Differential Equations and Inclusions (edited by A. Granas and M. Frigon), *NATO ASI Series C*, Vol. 472, Kluwer Academic Publishers, Dordrecht, 1995, 51-87.

[187] M. Frigon and A. Granas, Théorèmes d'existence pour des inclusions différentielles sans convexité, *C. R. Acad. Sci. Paris, Ser. I* 310 (1990), 819-822.

[188] M. Frigon and D. O'Regan, Impulsive differential equations with variable times, *Nonlinear Anal.* 26 (1996), 1913-1922.

[189] M. Frigon and D. O'Regan, First order impulsive initial and periodic problems with variable moments, *J. Math. Anal. Appl.* 233 (1999), 730-739.

[190] M. Frigon and D. O'Regan, Second order Sturm-Liouville BVP's with impulses at variable moments, *Dynam. Contin. Discrete Impuls. Systems* 8 (2), (2001), 149-159.

[191] K. M. Furati and N.-eddine Tatar, Behavior of solutions for a weighted Cauchy-type fractional differential problem, *J. Frac. Calc.* 28 (2005), 23-42.

[192] K. M. Furati and N.-eddine Tatar, Power type estimates for a nonlinear fractional differential equation, *Nonlinear Anal.* 62 (2005), 1025-1036.

[193] L. Gaul, P. Klein and S. Kempfle, Damping description involving fractional operators, *Mech. Systems Signal Processing* 5 (1991), 81-88.

[194] R. F. Geitz, Pettis integration, *Proc Amer. Math. Soc* 82 (1981), 81-86.

[195] R. F. Geitz, Geometry and the Pettis integration, *Trans. Amer. Math. Soc.* 269 (1982), 535-548.

[196] W. G. Glockle and T. F. Nonnenmacher, A fractional calculus approach of self-similar protein dynamics, *Biophys. J.* 68 (1995), 46-53.

[197] W. G. Glóckle and T. F. Nonnenmacher, Fractional integral operators and Fox functions in the theory of viscoelasticity. *Macromolecules*, 24 (1991) 64266434.

[198] R. Gorenflo and F. Mainardi, Some recent advances in theory and simulation of fractional diffusion processes, *J. Comput. Appl. Math.* 229 (2) (2009), 400-415.

[199] R. Gorenflo and F. Mainardi, Continuous time random walk, Mittag-Leffler waiting time and fractional diffusion: mathematical aspects, Chap. 4 In R. Klages, G. Radons and I.M. Sokolov (Editors): *Anomalous Transport: Foundations and Applications*, Wiley-VCH, Weinheim, Germany, 2008, pp. 93-127.

[200] R. Gorenflo, F. Mainardi and A. Vivoli, Continuous time random walk and parametric subordination in fractional diffusion, *Chaos, Solitons Fractals* 34 (2007), 87-103.

[201] R. Gorenflo and S. Vessella, Abel Integral Equations, *Lecture Notes in Math.*, 1461, Springer, Berlin, 1991.

[202] L. Gorniewicz, *Topological Fixed Point Theory of Multivalued Mappings, Mathematics and its Applications*, 495, Kluwer Academic Publishers, Dordrecht, 1999.

[203] N. M. Grahovac, Modelling of the hamstring muscle group by use of fractional derivatives *Comput. Math. Appl.*, 59 (2010), 1695-1700.

[204] A. Granas and J. Dugundji, *Fixed Point Theory*, Springer-Verlag, New York, 2003.

[205] A.K. Grunwald, Dérivationen und deren Anwendung, *Zeitschrift für Mathematik und Phisik*, 12 (1867), 441-480.

[206] D.J. Guo, V. Lakshmikantham, X. Liu, *Nonlinear Integral Equations in Abstract Spaces*, Kluwer Academic Publishers, Dordrecht, 1996.

[207] J. Hale and J. Kato, Phase space for retarded equationswith infinite delay, *Funkcial. Ekvac.* 21, (1978),11-41.

[208] J. K. Hale and S. Verduyn Lunel, Introduction to Functional -Differential Equations, *Applied Mathematical Sciences*, 99, Springer-Verlag, New York, 1993.

[209] F. Hartung, Differentiability of solutions with respect to parameters in neutral differential equations with state-dependent delays, *J. Math. Anal. Appl.* 324 (1) (2006), 504-524.

[210] F. Hartung, Linearized stability in periodic functional differential equations with state-dependent delays, *J. Comput. Appl. Math.* 174 (2) (2005), 201-211

[211] S.W. Hawking, *A Brief History of Time*, Bantam Press, London (1988).

[212] D. Henry, *Geometric Theory of Semilinear Parabolic Partial Differential Equations*, Springer-Verlag, Berlin/New York, 1989.

[213] S. Heikkila and V. Lakshmikantham, *Monotone Iterative Technique for Nonlinear Discontinuous Differential Equations*, Marcel Dekker Inc., New York, 1994.

[214] D. Henry, *Geometric Theory of Semilinear Parabolic Partial Differential Equations*, Springer-Verlag, Berlin/New York, 1989.

[215] J. Henderson and A. Ouahab, Fractional functional differential inclusions with finite delay, *Nonlinear Anal.* 70 (2009) 2091-2105.

[216] J. Henderson, A. Ouahab, Impulsive differential inclusions with fractional order, *Comput. Math. Appl.* 59 (2010), 1191-1226.

[217] J. Henderson, A. Ouahab, A Filippov's theorem, some existence results and the compactness of solution sets of impulsive fractional order differential inclusions. *Mediterr. J. Math.* 9 (2012), 453–485.

[218] J. Henderson and C. Tisdell, Topological transversality and boundary value problems on time scales, *J. Math. Anal. Appl.* 289 (2004), 110-125.

[219] E. Hernández, A. Prokopczyk and L. Ladeira, A note on partial functional differential equations with state-dependent delay. *Nonlinear Anal. Real World Applications* 7 (2006), 510-519.

[220] E. Hernandez M., R. Sakthivel, S. Tanaka Aki, Existence results for impulsive evolution differential equations with state-dependent delay, *Electron. J. Differential Equations* 2008 (2008) No. 28 pp. 1-11.

[221] M.A.E. Herzallah, D. Baleanu, Fractional-order variational calculus with generalized boundary conditions. *Adv. Difference Equ.* 2011, Article ID 357580, 9 p.

[222] M.A.E. Herzallah, A.M.A. El-Sayed, D. Baleanu. Perturbation for fractional-order evolution equation. *Nonlinear Dynam.* 62 (2010), no. 3, 593–600.

[223] N. Heymans and I. Podlubny, Physical interpretation of initial conditions for fractional differential equations with Riemann-Liouville fractional derivatives. *Rheologica Acta* 45 (5) (2006), 765-772.

[224] R. Hilfer, *Applications of Fractional Calculus in Physics*, World Scientific, Singapore, 2000.

[225] E. Hille and R. S. Phillips, Functional analysis and semi-groups, in: *Amer. Math. Soc. Colloq.*, Publ. 31, *Amer. Math. Soc.*, Providence, RI, 1957.

[226] Y. Hino, S. Murakami and T. Naito, Functional Differential Equations with Infinite Delay, in: *Lecture Notes in Mathematics*, 1473, Springer-Verlag, Berlin, 1991.

[227] S. Hu, M. Khavanin and W. Zhuang, Integral equations arising in the kinetic theory of gases, *Appl. Anal.* 34 (1989), 261-266.

[228] Sh. Hu and N. Papageorgiou, *Handbook of Multivalued Analysis, Theory I*, Kluwer, Dordrecht, 1997.

[229] D.H. Hyers, On the stability of the linear functional equation, *Proc. Nat. Acad. Sci.* 27 (1941), 222-224.

[230] D.H. Hyers, G. Isac, Th.M. Rassias, *Stability of Functional Equations in Several Variables*, Birkhuser, 1998.

[231] R.W. Ibrahim, Existence and uniqueness of holomorphic solutions for fractional Cauchy problem. *J. Math. Anal. Appl.* 380 (2011), 232-240.

[232] R.W. Ibrahim and H. A. Jalab, Existence of the solution of fractiona integral inclusion with time delay, *Misk. Math. Notes* 11 (2) (2010), 139-150.

[233] B.N. Ivanov, *Printsipy Sovremennoi Fziki (Principles of Modern Physics)*. Nauka, Moscow (1973) (In Russian).

[234] S.-M. Jung, *Hyers-Ulam-Rassias Stability of Functional Equations in Mathematical Analysis*, Hadronic Press, Palm Harbor, 2001.

[235] S.-M. Jung, *Hyers-Ulam-Rassias Stability of Functional Equations in Nonlinear Analysis*, Springer, New York, 2011.

[236] S.-M. Jung, A fixed point approach to the stability of a Volterra integral equation, *Fixed Point Theory Appl.* 2007 (2007), Article ID 57064, 9 pages.

[237] T. Kaczorek, *Selected Problems of Fractional Systems Theory*, Springer-Verlag, 2011.

[238] A. Kadem, D. Baleanu. Homotopy perturbation method for the coupled fractional Lotka-Volterra equations. *Romanian J. Phys.* 56 (2011), no. 3-4, 332–338.

[239] Z. Kamont, *Hyperbolic Functional Differential Inequalities and Applications*. Kluwer Academic Publishers, Dordrecht, 1999.

[240] Z. Kamont, and K. Kropielnicka, Differential difference inequalities related to hyperbolic functional differential systems and applications. *Math. Inequal. Appl.* 8 (4) (2005), 655-674.

[241] E.R. Kaufmann and E. Mboumi, Positive solutions of a boundary value problem for a nonlinear fractional differential equation, *Electron. J. Qual. Theory Differ. Equ.* 2007, No. 3, 11 pp.

[242] A. A. Kilbas, B. Bonilla and J. Trujillo, Nonlinear differential equations of fractional order in a space of integrable functions, *Dokl. Ross. Akad. Nauk,* 374 (4) (2000), 445-449.

[243] A. A. Kilbas and S. A. Marzan, Nonlinear differential equations with the Caputo fractional derivative in the space of continuously differentiable functions, *Differential Equations* 41 (2005), 84-89.

[244] A. A. Kilbas, Hari M. Srivastava, and Juan J. Trujillo, Theory and Applications of Fractional Differential Equations. *North-Holland Mathematics Studies*, 204. Elsevier Science B.V., Amsterdam, 2006.

[245] M. Kirane, M. Medved, N. Tatar. Semilinear Volterra integrodifferential problems with fractional derivatives in the nonlinearities. *Abstr. Appl. Anal.* 2011, Art. ID 510314, 11 pp.

[246] V.S. Kiryakova, Y.F. Luchko. The multi-index Mittag-Leffler functions and their applications for solving fractional order problems in applied analysis. *Application of mathematics in technical and natural sciences*, 597–613, *AIP Conf. Proc.*, 1301, *Amer. Inst. Phys.*, Melville, NY, 2010.

[247] M. Kisielewicz, *Differential Inclusions and Optimal Control*, Kluwer, Dordrecht, The Netherlands, 1991.

[248] J. Klafter, S. C. Lim, R. Metzler, *Fractional Dynamics in Physics: Recent Advances,* World Scientific, Singapore 2011.

[249] R.C. Koeller, Application of fractional calculus to the theory of viscoelasticity, *J. Appl. Mechanics* 51, (1984), 299307.

[250] S. Labidi, N. Tatar. Blow-up of solutions for a nonlinear beam equation with fractional feedback. *Nonlinear Anal.* 74 (2011), no. 4, 1402–1409.

[251] S.F. Lacroix, Traité du Calcul Différentiel et du Calcul Intégral, *Courcier, Paris*, t.3 (1819), 409-410.

[252] G.S. Ladde, V. Lakshmikanthan and A. S. Vatsala, *Monotone Iterative Techniques for Nonliner Differential Equations*, Pitman Advanced Publishing Program, London, 1985.

[253] V. Lakshmikantham, Theory of fractional differential equations, *Nonlinear Anal.* 60 (2008), 3337- 3343.

[254] V. Lakshmikantham, S. Leela and J. Vasundhara, *Theory of Fractional Dynamic Systems*, Cambridge Academic Publishers, Cambridge, 2009.

[255] V. Lakshmikantham and S. G. Pandit, The method of upper, lower solutions and hyperbolic partial differential equations, *J. Math. Anal. Appl.* 105 (1985), 466-477.

[256] V. Lakshmikantham, N.S. Papageorgiou, and J. Vasundhara, The method of upper and lower solutions and monotone technique for impulsive differential equations with variable moments, *Appl. Anal.* 15 (1993), 41-58.

[257] V. Lakshmikantham and A. S. Vatsala, Basic theory of fractional differential equations, *Nonlinear Anal.* 69 (2008), 2677-2682.

[258] V. Lakshmikantham, L. Wen and B. Zhang, Theory of Differential Equations with Unbounded Delay, *Mathematics and its Applications*, Kluwer Academic Publishers, Dordrecht, 1994.

[259] A. Lasota and Z. Opial, An application of the Kakutani-Ky Fan theorem in the theory of ordinary differential equations, *Bull. Acad. Pol. Sci. Ser. Sci. Math. Astronom. Phys.* 13 (1965), 781–786.

[260] G.W. Leibniz, Letter from Hanover, Germany, Deptember 30, 1695 to G.A. L'Hospital, *JLeibnizen Mathematische Schriften* 2, 301-302. Olms Verlag, Hildesheim, Germany, (1962). First published in 1849.

[261] F. Li and G.M. N'Guérékata, An existence result for neutral delay integrodifferential equations with fractional order and nonlocal conditions *Abst. Appl. Anal.* (2011), Article ID 952782, 20 pages.

[262] Y. Li, Y. Chen, I. Podlubny, Igor. Stability of fractional-order nonlinear dynamic systems: Lyapunov direct method and generalized Mittag-Leffler stability. *Comput. Math. Appl.* 59 (2010), no. 5, 1810–1821.

[263] V. Libertiaux and F. Pascon, Differential versus integral formulation of fractional hyperviscoelastic constitutive laws for brain tissue modeling *J. Comput. Appl. Math.*, 234 (2010), 2029-2035.

[264] T.C. Lim, On fixed point stability for set-valued contractive mappings with applications to generalized differential equations. *J. Math. Anal. Appl.* 110 (1985), 436-441.

[265] J. Liouville, Mémoire sur le calcul des différentielles à indices quelconques, *J. l'Ecole Roy. Polytéchn.* 13 (1832), 529-539.

[266] C. Lizama, Regularized solutions for abstract Volterra equations, *J. Math. Anal. Appl.* 243 (2000), 278-292.

[267] C. Lizama and J. Sánchez, On perturbation of k-regularized resolvent families, *Taiwan. J. Math.* 7, (2003), 217-227.

[268] Y. Luchko, Initial-boundary-value problems for the generalized multi-term time-fractional diffusion equation. *J. Math. Anal. Appl.* 374 (2011), 538–548.

[269] R. Magin, *Fractional Calculus in Bioengineering*. Begell House Publishers, Redding, 2006.

[270] R. Magin, M.D. Ortigueira, I. Podlubny, J.J. Trujillo, On the fractional signals and systems. *Signal Process* 91 (2011), 350–371.

[271] F. Mainardi, Fractional calculus: Some basic problems in continuum and statistical mechanics, in "*Fractals and Fractional Calculus in Continuum Mechanics* " (A. Carpinteri and F. Mainardi, Eds), pp. 291-348, Springer-Verlag, Wien, 1997.

[272] F. Mainardi, *Fractional Calculus and Waves in Linear Viscoelasticity: An Introduction to Mathematical Models,* Imperial College Press, London 2010.

[273] F. Mainardi, G. Pagnini and R. Gorenflo, Some aspects of fractional diffusion equations of single and distributed order, *Appl. Math. Comput.* 187 (2007), 295-305.

[274] S. Marano and V. Staicu, On the set of solutions to a class of nonconvex nonclosed differential inclusions, *Acta Math. Hungar.* 76 (1997), 287-301.

[275] F. Metzler, W. Schick, H. G. Kilian and T. F. Nonnenmacher, Relaxation in filled polymers: A fractional calculus approach, *J. Chem. Phys.* 103 (1995), 7180-7186.

[276] K. S. Miller and B. Ross, *An Introduction to the Fractional Calculus and Differential Equations*, John Wiley & Sons, New York, 1993.

[277] A.R. Mitchell, Ch. Smith, An existence theorem for weak solutions of differential equations in Banach spaces in: V. Lakshmikantham (Ed.), *Nonlinear Equations in Abstract Spaces*, 1978, pp. 387-404.

[278] G.M. Mittag-Leffler, Sur la nouvelle function E_α, *C. R. Acad. Sci. Paris*, 137 (1903), 554-558.

[279] G.M. Mittag-Leffler, Sopra la funzione $E_\alpha(x)$, *Rend. Accad. Lincei, ser. 5*, 13 (1904), 3-5.

[280] T. Miura, S.-M. Jung and S.-E. Takahasi, Hyers-Ulam-Rassias stability of the Banach space valued linear differential equation $y' = \lambda y$, *J. Korean Math. Soc.,* 41 (2004), 995-1005.

[281] K. Moaddy, S. Momani, I. Hashim. The non-standard finite difference scheme for linear fractional PDEs in fluid mechanics. *Comput. Math. Appl.* 61 (2011), no. 4, 1209–1216.

[282] G.M. Mophou, Optimal control of fractional diffusion equation. *Comput. Math. Appl.* 61 (2011), 68–78.

[283] M. Mophou, O. Nakoulima and G. M. N'Guérékata, Existence results for some fractional differential equations with nonlocal conditions, *Nonlinear Stud.* 17 (2010), 15-22.

[284] G.M. Mophou and G.M. N'Guérékata, Existence of the mild solution for some fractional differential equations with nonlocal conditions. *Semigroup Forum* 79 (2009), 315–322.

[285] G. M. Mophou, G.M. N'Guérékata, On some classes of almost automorphic functions and applications to fractional differential equations, *Comput. Math. Appl.* 59 (2010), 1310-1317.

[286] G.M. Mophou and G.M. N'Guérékata, On integral solutions of some nonlocal fractional differential equations with nondense domain. *Nonlinear Anal.* 71 (2009), 4668–4675.

[287] G.M. Mophou, G.N. N'Guérékata, Controllability of semilinear neutral fractional functional evolution equations with infinite delay. *Nonlinear Stud.* 18 (2011), 195-209.

[288] G. M. Mophou, G.M. N'Guérékata and V. Valmorin. Pseudo almost automorphic solutions of a neutral functional fractional differential equations. *Intern. J. Evol. Equ.* 4 (2009), 129-139.

[289] A. Mura and F. Mainardi, A class of self-similar stochastic processes with stationary increments to model anomalous diffusion in physics, *Integral Transforms Special Funct.* 20 (3-4)(2009), 185-198.

[290] V. Mureşan, *Functional-integral equations*, Ed. Mediamira, Cluj-Napoca, 2003.

[291] V. Mureşan, Some results on the solutions of a functional-integral equation *Stud. Univ. Babes-Bolyai Math.* 56 (2011), 157-164.

[292] S. Muslih, O.P. Agrawal. Riesz fractional derivatives and fractional dimensional space. *Internat. J. Theoret. Phys.* 49 (2010), no. 2, 270–275.

[293] S. Muslih, O. P. Agrawal, D. Baleanu. A fractional Schrödinger equation and its solution. *Internat. J. Theoret. Phys.* 49 (2010), no. 8, 1746–1752.

[294] I. P. Natanson, *Theory of Functions of a Real Variable*, Ungar, New York, 1960.

[295] J.J. Nieto, Maximum principles for fractional differential equations derived from Mittag-Leffler functions, *Appl. Math. Lett.*, 23 (2010), 1248-1251.

[296] K.B. Oldham and J. Spanier, *The Fractional Calculus*, Academic Press, New York, London, 1974.

[297] D. O'Regan, Fixed point theory for weakly sequentially continuous mapping, *Math. Comput. Model.* 27 (5) (1998), 1-14.

[298] D. O'Regan, Weak solutions of ordinary differential equations in Banach spaces, *Appl. Math. Lett.* 12 (1999), 101–105.

[299] D. O'Regan, M.-A. Taoudi. Fixed point theorems for the sum of two weakly sequentially continuous mappings. *Nonlinear Anal.* 73 (2010), 283–289.

[300] M. D. Ortigueira, *Fractional Calculus for Scientists and Engineers*, Springer-Verlag, Dordrecht, 2011.

[301] A. Ouahab. Fractional semilinear differential inclusions. *Comput. Math. Appl.* 64 (2012), 3235–3252.

[302] B. G. Pachpatte, On Volterra-Fredholm integral equation in two variables, *Demonstratio Math.* XL (4) (2007), 839-852.

[303] B. G. Pachpatte, On Fredholm type integrodifferential equation, *Tamkang J. Math.* 39 (2008), 85-94.

[304] B.G. Pachpatte, On Fredholm type integral equation in two variables *Diff. Equ. Appl.* 1 (2009), 27-39.

[305] B. G. Pachpatte, On a general partial integral equation of Barbashin type,*Tamsui Oxford J. Math. Sci.* 27 (2011), 99-115.

[306] S. G. Pandit, Monotone methods for systems of nonlinear hyperbolic problems in two independent variables, *Nonlinear Anal.*, 30 (1997), 235-272.

[307] A. Pazy, *Semigroups of Linear Operators and Applications to Partial Differential Equations*, Springer-Verlag, New York, 1983.

[308] T.P. Petru, and M.-F. Bota, Ulam-Hyers stabillity of operational inclusions in complete gauge spaces, *Fixed Point Theory* 13 (2012), 641-650.

[309] T.P. Petru, A. Petrusel. J.-C. Yao, Ulam-Hyers stability for operatorial equations and inclusions via nonself operators, *Taiwanese J. Math.* 15 (2011), 2169-2193.

[310] L. M. Petrovic, and T. M. Atanackovic, A model for shrinkage strain in photo polymerization of dental composites, *Dental Materials* 24 (4) (2008), 556-560.

[311] L. M. Petrovic, D. T. Spasic, and T. M. Atanackovic, On a mathematical model of a human root dentin, *Dental Materials* 21 (2) (2005), 125-128.

[312] B. J. Pettis, On integration in vector spaces, *Trans. Amer. Math. Soc.* 44 (1938), 277-304.

[313] R.S. Phillips, A decomposition of additive set functions, *Bull. Amer. Math. Soc.* 46 (1940), 274-277.

[314] I. Podlubny, Fractional Differential Equations, *Mathematics in science and engineering* 198, Academic Press, San Diego, 1999.

[315] I. Podlubny, Geometric and physical interpretation of fractional integration and fractional differentiation, *Fract. Calculus Appl. Anal.* 5 (2002), 367-386.

[316] I. Podlubny, I. Petraš, B. M. Vinagre, P. O'Leary and L. Dorčak, Analogue realizations of fractional-order controllers. Fractional order calculus and its applications, *Nonlinear Dynam.* 29 (2002), 281-296.

[317] Y. Z. Povstenko, Fractional radial diffusion in a cylinder. *J. Molecular Liquids* 137 (1-3) (2008), 46-50.

[318] J. Prüss, *Evolutionary Integral Equations and Applications* Monographs Math. 87, Bikhäuser Verlag, 1993.

[319] C. Qian, Global attractivity in nonlinear delay differential equations, *J. Math. Anal. Appl.* 197 (1996), 529-547.

[320] C. Qian, Global attractivity in a delay differential equation with application in a commodity model, *Appl. Math. Lett.* 24 (2011), 116-121.

[321] C. Qian, Global attractivity in a nonlinear delay differential equation with applications. *Nonlinear Anal.* 71 (2009), 1893-1900.

[322] M.D. Quinn and N. Carmichael, An approach to nonlinear control problem using fixed point methods, degre theory, pseudo-inverse,*Numer. Funct. Anal. Optim.*, 7 (1984-85), 197-219.

[323] Th.M. Rassias, On the stability of linear mappings in Banach spaces, *Proc. Amer. Math. Soc.* 72 (1978), 297-300.

[324] J.D. Ramirez, A.S. Vatsala. Monotone method for nonlinear Caputo fractional boundary value problems. *Dynam. Systems Appl.* 20 (2011), no. 1, 73–88.

[325] A. Razminia, V.J. Majd, D. Baleanu. Chaotic incommensurate fractional order Rössler system: active control and synchronization. *Adv. Difference Equ.* 2011, 2011:15, 12 pp.

[326] M. Rivero, J.J. Trujillo, L. Vazquez, M.P. Velasco. Fractional dynamics of populations. *Appl. Math. Comput.* 218, (2011), 1089-1095.

[327] W. Rudin, *Real and Complex Analysis*, McGraw-Hill, New York, 1970.

[328] I. A. Rus, Ulam stability of ordinary differential equations, *Studia Univ. Babes-Bolyai, Math.* LIV (4)(2009), 125-133.

[329] I. A. Rus, Remarks on Ulam stability of the operatorial equations, *Fixed Point Theory* 10 (2009), 305-320.

[330] J. Sabatier, O. Agrawal, J. Machado (Eds.), *Advances in Fractional Calculus. Theoretical Developments and Applications in Physics and Engineering*. Springer, Dordrecht, 2007.

[331] J. Sabatier, M. Merveillaut, R. Malti, A. Oustaloup. How to impose physically coherent initial conditions to a fractional system? *Commun. Nonlinear Sci. Numer. Simul.* 15 (2010), no. 5, 1318–1326.

[332] H.A.H. Salem, On the fractional order m-point boundary value problem in reflexive Banach spaces and weak topologies, *Comput. Math. Appl.* 224 (2009), 565-572.

[333] H.A.H. Salem. On the fractional calculus in abstract spaces and their applications to the Dirichlet-type problem of fractional order. *Comput. Math. Appl.* 59 (2010), no. 3, 1278–1293.

[334] H.A.H. Salem Global monotonic solutions of multi term fractional differential equations. *Appl. Math. Comput.* 217 (2011), no. 14, 6597–6603.

[335] H. A. H. Salem, A. M. A. El-Sayed and O. L. Moustafa, A note on the fractional calculus in Banach spaces, *Studia Sci. Math. Hungar.* 42 (2) (2005), 115-130.

[336] H. A. H. Salem and M. Väth, An abstract Gronwall lemma and applications to global existence results for functional differential and integrals of fractional order, *J. Integral Equations Appl.* 16 (2004), 411-439.

[337] S. G. Samko, A. A. Kilbas and O. I. Marichev, *Fractional Integrals and Derivatives. Theory and Applications*, Gordon and Breach, Yverdon, 1993.

[338] N. Samko, S. Samko, B. Vakulov. Fractional integrals and hypersingular integrals in variable order Hlder spaces on homogeneous spaces. *J. Funct. Spaces Appl.* 8 (2010), no. 3, 215–244.

[339] N. P. Semenchuk, On one class of differential equations of noninteger order, *Differents. Uravn.*, 10 (1982), 1831-1833.

[340] H. Sheng, Y. Chen, T. Qiu, *Fractional Processes and Fractional-order Signal Processing; Techniques and Applications*. Springer-Verlag, 2011.

[341] Jong Son Shin, An existence theorem of functional differential equations with infinite delay in a Banach space, *Funkcial. Ekvac.* 30 (1987), 19-29.

[342] X. Su and S. Zhang, Unbounded solutions to a boundary value problem of fractional order on the half-line, *Comput. Math. Appl.* 61 (2011), 1079-1087.

[343] M.-A. Taoudi. Krasnoselskii type fixed point theorems under weak topology features. *Nonlinear Anal.* 72 (2010), 478–482.

[344] V. E. Tarasov, *Fractional dynamics: Application of Fractional Calculus to Dynamics of Particles, Fields and Media*, Springer, Heidelberg; Higher Education Press, Beijing, 2010.

[345] V. E. Tarasov. Fractional dynamics of relativistic particle. Internat. *J. Theoret. Phys.* 49 (2010), no. 2, 293–303.

[346] V. E. Tarasov. M. Edelman. Fractional dissipative standard map. *Chaos* 20 (2010), no. 2, 023127, 7 pp.

[347] J.A. Tenreiro Machado. Time-delay and fractional derivatives. *Adv. Difference Equ.* 2011, Art. ID 934094, 12 pp.

[348] J.A. Tenreiro Machado. Entropy analysis of integer and fractional dynamical systems. *Nonlinear Dynam.* 62 (2010), no. 1-2, 371–378.

[349] J.A. Tenreiro Machado Time-delay and fractional derivatives. *Adv. Difference Equ.* 2011, Art. ID 934094, 12 pp.

[350] J.A. Tenreiro Machado, Time-delay and fractional derivatives. *Adv. Difference Equ.* 2011, Article ID 934094, 12 p.

[351] J.A. Tenreiro Machado, V. Kiryakova, F. Mainardi. A poster about the old history of fractional calculus. *Fract. Calc. Appl. Anal.* 13 (2010), no. 4, 447–454.

[352] J.A. Tenreiro Machado, V. Kiryakova, F. Mainardi. Recent history of fractional calculus. *Commun. Nonlinear Sci. Numer. Simul.* 16 (2011), no. 3, 1140–1153.

[353] J.C. Trigeassou, N. Maamri, J. Sabatier, A.A. Oustaloup, Lyapunov approach to the stability of fractional differential equations. *Signal Process* 91 (2011), 437–445.

[354] V. V. Uchaikin, *Fractional Derivatives for Physicists and Engineers.* Volume I. Background and theory. Nonlinear Physical Science. Higher Education Press, Beijing; Springer, Heidelberg, 2013.

[355] V. V. Uchaikin, *Fractional Derivatives for Physicists and Engineers.* Volume II. Applications. Nonlinear Physical Science. Higher Education Press, Beijing; Springer, Heidelberg, 2013.

[356] S.M. Ulam, *A Collection of Mathematical Problems*, Interscience Publishers, New York, 1968.

[357] L. Vazquez. From Newton's equation to fractional diffusion and wave equations. *Adv. Difference Equ.* 2011, Art. ID 169421, 13 pp.

[358] A. N. Vityuk, On solutions of hyperbolic differential inclusions with a nonconvex right-hand side. (Russian) *Ukran. Mat. Zh.* 47 (1995), no. 4, 531-534; translation in *Ukrainian Math. J.* 47 (1995), no. 4, 617-621 (1996).

[359] A. N. Vityuk, Existence of Solutions of partial differential inclusions of fractional order, *Izv. Vyssh. Uchebn. , Ser. Mat.* ,8 (1997), 13-19.

[360] A. N. Vityuk and A. V. Golushkov, Existence of solutions of systems of partial differential equations of fractional order, *Nonlinear Oscil.* 7 (3) (2004), 318-325.

[361] A. N. Vityuk and A. V. Golushkov, The Darboux problem for a differential equation containing a fractional derivative, *Nonlinear Oscil.* 8 (2005), 450-462.

[362] A. N. Vityuk and A. V. Mykhailenko, On one class of differential quations of fractional order, *Nonlinear Oscil.* 11 (3) (2008), 307-319.

[363] A. N. Vityuk and A. V. Mykhailenko, The Darboux problem for an implicit fractional-order differential equation, *J. Math. Sci.* 175 (4) (2011), 391-401.

[364] J. Wang, Y. Zhou and M. Fečkan, Nonlinear impulsive problems for fractional differential equations and Ulam stability, *Comput. Math. Appl.* 64 (2012), 3389-3405.

[365] J. Wang, Y. Zhou, W. Wei. A class of fractional delay nonlinear integrodifferential controlled systems in Banach spaces, *Commun. Nonlinear Sci. Numer. Simul.* 16 (2011), no. 10, 4049–4059.

[366] J. Wang, L. Lv, Y. Zhou, Ulam stability and data dependence for fractional differential equations with Caputo derivative, *E. J. Qual. Theory Diff. Equ.* (63) (2011) 1-10.

[367] J. Wang, L. Lv, Y. Zhou, New concepts and results in stability of fractional differential equations, *Commun. Nonlinear Sci. Numer. Simul.* 17 (2012), 2530-2538.

[368] W. Wei, X. Li, X. Li, New stability results for fractional integral equation, *Comput. Math. Appl.* 64 (2012), 3468-3476.

[369] S. Westerlund, *Dead Matter has Memory! Causal Consulting*, Kalmar, Sweden (2002).

[370] T.-J. Xiao, J. Liang, and J. van Casteren, Time dependent Desch-Schappacher type perturbations of Volterra integral equations, *Integral Equations Operator Theory* 44 (4) (2002), 494-506.

[371] K. Yosida, *Functional Analysis*, 6th edn. Springer-Verlag, Berlin, 1980.

[372] C. Yu and G. Gao, Existence of fractional differential equations, *J. Math. Anal. Appl.* 310 (2005), 26-29.

[373] P.P. Zabrejko, A.I. Koshelev, M.A. Krasnosel'skii, S.G. Mikhlin, L.S. Rakovschik and V.J. Stecsenko, *Integral Equations*, Noordhoff, Leyden, 1976.

[374] G. Zaslavsky, *Hamiltonian Chaos and Fractional Dynamics*. Oxford University Press, New York, 2005.

[375] S. Zhang, Positive solutions for boundary-value problems of nonlinear fractional diffrential equations, *Electron. J. Differential Equations* 2006, No. 36, pp. 1-12.

[376] S. Zhang, Existence of positive solutions of a singular partial differential equation, *Math. Bohemica* 133 (1) (2008), 29-40.

[377] Y. Zhou, Existence and uniqueness of fractional functional differential equations with unbounded delay. *Int. J. Dyn. Syst. Differ. Equ.* 1 (4) (2008), 239–244.

[378] Y. Zhou, *Basic Theory of Fractional Differential Equations*, World Scientific, Singapore, 2014.

[379] Y. Zhou, F. Jiao and J. Li, Existence and uniqueness for p-type fractional neutral differential equations. *Nonlinear Anal.* 71 (2009), 2724–2733.

[380] Y. Zhou, F. Jiao and J. Li, Existence and uniqueness for fractional neutral differential equations with infinite delay. *Nonlinear Anal.* 71 (2009), 3249–3256.

Index